Lecture Notes in Computational Science and Engineering

90

Editors:

For further volumes:
http://www.springer.com/series/3527

Clemens Pechstein

Finite and Boundary Element Tearing and Interconnecting Solvers for Multiscale Problems

Springer

Clemens Pechstein
Institute of Computational Mathematics
Johannes Kepler University
Linz
Austria

ISSN 1439-7358
ISBN 978-3-642-23587-0 ISBN 978-3-642-23588-7 (eBook)
DOI 10.1007/978-3-642-23588-7
Springer Heidelberg New York Dordrecht London

Library of Congress Control Number: 2012953096

Math. Subj. Class. (2010): 65F08, 65N22, 65N30, 65N38, 65N55, 65Y05

Printed on acid-free paper

Springer is part of Springer Science+Business Media (www.springer.com)

Für Astrid und Johanna

Preface

The purpose of this book is to give a detailed and self-contained presentation of tearing and interconnecting methods for finite and boundary element discretizations of elliptic partial differential equations. This includes a description of the corresponding algorithms as well as a rigorous convergence theory. In particular, two issues are addressed which have not been treated in any previous monograph. Firstly, we consider also the case of boundary element tearing and interconnecting (BETI), the coupling of finite and boundary elements within the tearing and interconnecting framework, and the application to exterior problems. Secondly, we consider the case of highly varying (multiscale) coefficients not resolved by the subdomain partitioning. In this sense, this book serves as a good complement to the existing monographs and surveys by Farhat and Roux [FR94], Toselli and Widlund [TW05], Korneev and Langer [KL04], and Mathew [Mat08], which already treat the finite element tearing and interconnecting (FETI) method, dual-primal FETI (FETI-DP), and balancing domain decomposition by constraints (BDDC).

As model problem throughout, we use the well-known scalar elliptic problem $-\text{div}(\alpha\,\nabla u) = f$, discretized by continuous piecewise linear finite elements or boundary elements. Extensions to other PDEs and other discretizations will be discussed briefly.

Tearing and interconnecting methods are special non-overlapping domain decomposition methods. They can be perceived as

(i) Coupling methods between possibly different discretizations, and/or
(ii) Iterative solvers for large-scale equations well-suited for parallelization.

The main idea of tearing and interconnecting methods is to subdivide the computational domain into non-overlapping subdomains. The equation is posed locally on each subdomains, and the coupling of these local problems is performed by Lagrangian multipliers. This procedure corresponds to aspect (i). Using different strategies, the original variables can be eliminated, and the remaining equation involves only the Lagrangian multipliers. To solve this "dual" problem, one usually employs an iterative method with a certain preconditioner. In many cases, the resulting parallel method is quasi-optimal, i.e., the total computational effort grows only polylogarithmically in the number of unknowns per subdomain.

We will not elaborate the general subject of domain decomposition methods (but refer to [Mat08, SBG96, QV99, TW05]). However, to ensure a self-contained presentation, we have collected the basics of Sobolev spaces, projections, finite and boundary elements, Schur complements, etc. into Chap. 1. The remainder of the book is organized as follows.

Chapter 2 discusses in detail the formulation of (one-level) FETI/BETI methods. Besides the classical formulation, we also consider the total FETI method and its boundary element counterpart, all-floating BETI, as well as the closely related balancing Neumann-Neumann methods. Furthermore, we provide a rigorous convergence analysis for the model problem under certain geometric assumptions and on the assumption that the coefficient α is constant in each subdomain.

The scope of Chap. 3 is to relax the assumptions on α and allow for more rapid or noisy variation, not necessarily resolved by the subdomains. Note that preconditioning such *multiscale problems* is in itself quite an old discipline. However, many earlier works assume that the coefficient is resolved by the subdomains or the coarse mesh. The branch of research that considers coefficients being not resolved by the subdomains/coarse mesh is rather young and growing (see the references in Sect. 3.1.1). In this context, the book offers a detailed view to an active and up-to-date area of research. The main tool in this chapter, a weighted Poincaré inequality (WPI), will be discussed in detail.

Chapter 4 is devoted to exterior problems. Here, one of the subdomains is unbounded and treated by boundary elements. We discuss how and to which extent the theory of Chap. 2 can be generalized. Due to the fact that the unbounded subdomain can have many neighboring subdomains, there are new difficulties compared to the bounded case, which have to be treated suitably.

In Chap. 5, we formulate the dual-primal FETI/BETI methods and the closely related BDDC methods. As in Chap. 2, we treat both theoretical and algorithmic aspects. We also discuss extensions to the unbounded case from Chap. 4, and to some extent to the multiscale problems from Chap. 3.

Some comments on the bibliography. I have tried to provide the basic and original references in the field, as up-to-date as possible. However, there are probably some more (beyond the 300) that I have missed, and I apologize for this.

Thanks are due to several people. My dear parents Barbara and Gerd gave me the opportunity of high-level education at school and university. Ulrich Langer taught me numerical analysis, supervised my doctoral thesis, and he encouraged me to write this monograph. My collaborator and friend Robert Scheichl invested his time and energy for our joint research, motivated and pushed me. I have profited tremendously from numerous scientific discussions with Clark Dohrmann, Axel Klawonn, Günther Of, Oliver Rheinbach, Marcus Sarkis, Olaf Steinbach, and Olof Widlund. Helpful remarks on the manuscript were provided by Rob Scheichl, Clark Dohrmann, Clemens Hofreither, and Bedřich Sousedík. At Springer-Verlag, Martin Peters and Thanh-Ha Le Thi took care about the organization around this monograph; thanks also for their patience. My colleagues at the Institute of

Computational Mathematics and the associated group of the Radon Institute for Computational and Applied Mathematics (RICAM) at Linz spent time to discuss now and then or to cheer me up, especially Walter Zulehner. It should also be mentioned that the Austrian Science Fund (FWF) supported a lot of my research.

Finally, there is my wonderful wife Astrid, without whom this monograph would have never been finished. Not only did she support me with love, faith, and understanding, but she provided enough pressure (to complete the book) and read several times through the manuscript! Moreover, she gave birth to the sunshine of our life, our lovely daughter Johanna.

Linz, Austria Clemens Pechstein

Contents

Chapter 1
Preliminaries

This chapter contains some standard results that we need in subsequent chapters. The material is arranged such that an unexperienced reader can read through it linearly. The experienced reader my bypass the chapter in the beginning and return to it when necessary.

Section 1.1 introduces the Lax-Milgram theorem, Galerkin's method, projections, and pseudo inverses. In Sect. 1.2 we discuss Sobolev spaces, the potential equation in classical and weak form, the finite element method (FEM), solvers for linear systems, and finally the variational setting of the exterior Laplace problem. Section 1.3 is a survey on boundary integral equations for the Laplace problem and Sect. 1.3.7 a brief introduction to the boundary element method (BEM).

A few words on notation. The dual of a Banach space V is denoted by V^*, and the duality product on $V^* \times V$ by $\langle \cdot, \cdot \rangle$. Recall that the dual norm is given by

$$\|\psi\|_{V^*} := \sup_{v \in V} \frac{|\langle \psi, v \rangle|}{\|v\|_V}.$$

For the sake of simplicity, we silently exclude $v = 0$ in the supremum above and in all the following suprema and infima of this form.

As usual, $\top$ indicates the transpose of a vector or matrix, but we will also use it for the adjoint operator (for $A : V \to W^*$, $A^\top : W \to V^*$ is given by $\langle A^\top w, v \rangle = \langle A\, v, w \rangle$). Throughout the book, a *domain* is understood as an open and connected subset of $\mathbb{R}^d$, but not necessarily bounded. A prominent example is the exterior $\Omega^{\text{ext}} := \mathbb{R}^d \setminus \overline{\Omega}$ of a bounded domain Ω. For a list of symbols see also p. 303ff.

C. Pechstein, *Finite and Boundary Element Tearing and Interconnecting Solvers for Multiscale Problems*, Lecture Notes in Computational Science and Engineering 90,
DOI 10.1007/978-3-642-23588-7_1, © Springer-Verlag Berlin Heidelberg 2013

1.1 Abstract Tools

1.1.1 The Lax-Milgram Theorem

The following fundamental theorem discusses the solvability of variational problems with coercive bilinear forms.

Theorem 1.1 (Lax-Milgram). *Let V be a Hilbert space and $V_0 \subset V$ a closed subspace. Let the bilinear form $a(\cdot,\cdot) : V \times V \to \mathbb{R}$ be* continuous, *i.e., there exists $\bar{c}_a > 0$ such that*

$$a(v, w) \le \bar{c}_a \, \|v\|_V \, \|w\|_V \qquad \forall v, w \in V,$$

and coercive *on V_0, i.e., there exists $\underline{c}_a > 0$ such that*

$$a(v, v) \ge \underline{c}_a \, \|v\|_V^2 \qquad \forall v \in V_0.$$

Then, for any $g \in V$ and $F \in V_0^*$, the variational problem

$$\text{find } u \in g + V_0 : \quad a(u, v) = \langle F, v \rangle \qquad \forall v \in V_0 \tag{1.1}$$

has a unique solution $u \in g + V_0$ which fulfills the a priori estimate

$$\|u\|_V \le \frac{1}{\underline{c}_a} \|F\|_{V_0^*} + \left(1 + \frac{\bar{c}_a}{\underline{c}_a}\right) \|g\|_V \, .$$

Additionally, if $g = 0$ we have the lower bound $\frac{1}{\bar{c}_a} \|F\|_{V_0^*} \le \|u\|_V$.

Proof. See, e.g., [BS02, Sect. 2.7].

Definition 1.2. We call a linear operator $A : V \to V^*$ *positive definite* if the associated bilinear form $a(v, w) := \langle A\, v, w \rangle$ is coercive on V.

Theorem 1.1 (for $V_0 = V$ and $g = 0$) implies that the inverse A^{-1} of a linear, continuous, and positive definite operator A exists and is again linear, continuous, and positive definite. Be aware that in the leading literature, a coercive *operator* is not necessarily positive definite.

Notation. We will abbreviate "self-adjoint and positive definite" simply by SPD. Note that the operator A in Definition 1.2 is self-adjoint ($A = A^\top$) if and only if the associated bilinear form $a(\cdot,\cdot)$ is symmetric.

Lemma 1.3. *Let the assumptions of Theorem 1.1 hold and assume additionally that $a(\cdot,\cdot)$ is symmetric and $F \in V^*$. Then the solution u of* (1.1) *is the unique minimizer of the* Ritz energy functional*:*

$$u = \operatorname*{argmin}_{v \in g+V_0} J(v), \qquad \textit{where } J(v) = \tfrac{1}{2}a(v,v) - \langle F, v\rangle.$$

Proof. See, e.g., [BS02, Sect. 2.5].

The statement of Theorem 1.1 can be generalized to nonlinear strongly monotone and Lipschitz continuous operators; for a proof see [Zei90].

Theorem 1.4. *Let V be a Hilbert space, $V_0 \subset V$ a closed subspace and $g \in V$. Let the operator $A : V \to V_0^*$ be* Lipschitz continuous *on $g + V_0$, i.e., there exists $\bar{c}_A > 0$ with*

$$\|A(v) - A(w)\|_{V_0^*} \;\le\; \bar{c}_A \,\|v - w\|_V \qquad \forall v, w \in g + V_0\,,$$

and strongly monotone *on $g + V_0$, i.e., there exists $\underline{c}_A > 0$ with*

$$\langle A(v) - A(w), v - w\rangle \;\ge\; \underline{c}_A \,\|v - w\|_V^2 \qquad \forall v, w \in g + V_0\,.$$

Then, for any $F \in V_0^*$, the (nonlinear) operator equation

$$\textit{find } u \in g + V_0 : \quad A(u) \;=\; F \tag{1.2}$$

has a unique solution and $\frac{1}{\bar{c}_A}\|F - A(g)\|_{V_0^*} \le \|u - g\|_V \le \frac{1}{\underline{c}_A}\|F - A(g)\|_{V_0^*}$.

1.1.2 Galerkin's Method

Let the assumptions of Theorem 1.1 be fulfilled. We choose a finite-dimensional subspace $V_0^h \subset V_0$ and consider the projected problem:

$$\text{find } u_h \in g + V_0^h : \quad a(u_h, v_h) \;=\; \langle F, v_h\rangle \qquad \forall v_h \in V_0^h\,. \tag{1.3}$$

Due to Theorem 1.1 the solution u_h is unique. The following a-priori estimate (see, e.g., [BS02, Sect. 2.8]) states that the discretization error is controlled in terms of the approximation error of the space V_0^h.

Lemma 1.5 (Céa). *Let the assumptions of Theorem 1.1 hold, let $u \in g + V_0$ and $u_h \in g + V_0^h$ be the solutions of* (1.1) *and* (1.3)*, respectively. Then*

$$\|u - u_h\|_V \;\le\; \frac{\bar{c}_a}{\underline{c}_a} \inf_{v_h \in V_0^h} \|u - g - v_h\|_V\,.$$

The following lemma relates the energies of the two solutions (see e.g. [BS02, Sect. 2.5]).

Lemma 1.6. *Let the assumptions of Theorem 1.1 hold and assume additionally that $a(\cdot,\cdot)$ is symmetric and $F \in V^*$. Let u and u_h be the solutions of* (1.1) *and* (1.3), *respectively. Then, for $J(v) = \frac{1}{2}a(v,v) - \langle F, v\rangle$,*

$$u = \operatorname*{argmin}_{v\in g+V_0} J(v), \qquad J(u) \le J(u_h),$$

$$u_h = \operatorname*{argmin}_{v_h\in g+V_0^h} J(v_h), \qquad a(u_h,u_h) \le a(u,u).$$

1.1.3 Polars, Projections, and Pseudo Inverses

In this subsection, we give definitions and properties of the polar, the factor space, certain projections and pseudo inverses, mostly on finite-dimensional Hilbert spaces.

1.1.3.1 The Polar

Definition 1.7. Let V be a Hilbert space. For a subset (or subspace) $Z \subset V$, the *polar* Z° is given by

$$Z^\circ := \{f \in V^* : \langle f, z\rangle = 0 \quad \forall z \in Z\}.$$

Note that the "universe" V should be clear from the context.

Lemma 1.8. *Let V, W be Hilbert spaces and let $B : V \to W^*$ be a bounded linear operator. Then*

$$\operatorname{range}(B)^\circ = \ker(B^\top).$$

If $\operatorname{range}(B^\top)$ *is closed then*

$$\operatorname{range}(B^\top) = \ker(B)^\circ.$$

Proof. The first statement follows from the definitions of the polar and the adjoint. The second statement is known as the closed range theorem, see, e.g., [BF91].

1.1.3.2 The Factor Space and Its Dual

Definition 1.9. Let V be a finite-dimensional Hilbert space and $Z \subset V$ a subspace. The factor space $V_{/Z}$ is the space of equivalence classes of the relation

$$v_1 \sim v_2 \quad\Longleftrightarrow\quad v_2 - v_1 \in Z.$$

For a Hilbert space Y, a linear operator $F : V \to Y$ is well-defined on the factor space $V_{/Z}$ if $F(Z) = \{0\}$.

Corollary 1.10. *For $Z \subset V$ as in Definition 1.9, we have*

$$(V_{/Z})^* = Z^\circ.$$

In particular, for finite-dimensional Hilbert spaces V, W and for any linear operator $A : V \to W^*$,

$$\begin{aligned} (V_{/\ker(A)})^* &= \mathrm{range}(A^\top), \\ W^*_{/\mathrm{range}(A)} &= \ker(A^\top)^*. \end{aligned}$$

1.1.3.3 Projections

Let U and Z be finite-dimensional Hilbert spaces with $\dim(Z) < \dim(U)$, and let $G : Z \to U^*$ be an injective linear operator. Due to our assumptions, $\ker(G) = \{0\}$ but $\ker(G^\top) \neq \{0\}$. Let $Q : U^* \to U$ be self-adjoint and assume that Q is positive definite on $\mathrm{range}(G)$. We are interested in operators $P : U \to U$ and their adjoints $P^\top : U^* \to U^*$ of the form

$$\begin{aligned} P &= I - Q\,G\,(G^\top Q\,G)^{-1} G^\top, \\ P^\top &= I - G\,(G^\top Q\,G)^{-1} G^\top Q. \end{aligned} \tag{1.4}$$

Note that the term $(G^\top Q\,G)^{-1}$ is well-defined because G is injective and Q is SPD on $\mathrm{range}(G)$. By construction we have

$$G^\top P = 0, \qquad P\,Q\,G = 0, \qquad P\,Q = Q\,P^\top. \tag{1.5}$$

The operator P is a *projection* ($P^2 = P$). Also $P^\top$, $I - P$, and $I - P^\top$ are projections, and we have the direct space decompositions

$$U = \mathrm{range}(P) \oplus \mathrm{range}(I - P), \qquad U^* = \mathrm{range}(P^\top) \oplus \mathrm{range}(I - P^\top). \tag{1.6}$$

From the identities in (1.5) we obtain further that

$$\begin{aligned} \mathrm{range}(P) &= \ker(G^\top), & \mathrm{range}(I - P) &= \mathrm{range}(Q\,G), \\ \mathrm{range}(P^\top) &= \ker(G^\top Q), & \mathrm{range}(I - P^\top) &= \mathrm{range}(G). \end{aligned} \tag{1.7}$$

Lemma 1.11. *Let P be defined according to* (1.4)*. Then the space* $\mathrm{range}(P^\top)$ *is a realization of* $\mathrm{range}(P)^*$*. The same holds if P is replaced by $I - P$. For simplicity, we write*

$$\operatorname{range}(P)^* = \operatorname{range}(P^\top), \qquad \operatorname{range}(I-P)^* = \operatorname{range}(I-P^\top).$$

Proof. The dual of $\operatorname{range}(P) = \ker(G^\top)$ contains all the linear functionals from U^*, but two functionals $\varphi_1, \varphi_2 \in U^*$ are regarded equal in $\operatorname{range}(P)^*$ if

$$\langle \varphi_2 - \varphi_1, u\rangle = 0 \qquad \forall u \in \ker(G^\top),$$

i.e., $\varphi_2 - \varphi_1 \in \ker(G^\top)^\circ = \operatorname{range}(G)$. Hence, $\operatorname{range}(P)^* = U^*_{/\operatorname{range}(G)}$ is the space of the corresponding equivalence classes. Since $\operatorname{range}(G) = \operatorname{range}(I - P^\top)$ and because of (1.6), every such equivalence class can be represented by a unique element in $\operatorname{range}(P^\top)$. The proof for second case is analogous. □

Remark 1.12. Lemma 1.11 is extremely helpful when working in a subspace $W = \operatorname{range}(P)$ of $\mathbb{R}^n$ and its dual. In an implementation, instead of finding a new basis for W and one for W^*, we can still represent elements in W, W^* by vectors in $\mathbb{R}^n$ and realize the duality pairing by the Euclidean inner product.

If Q is SPD on the whole of U^*, it defines inner products $(v, w)_Q := \langle v, Q\, w\rangle$ and $(v, w)_{Q^{-1}} := \langle Q^{-1} v, w\rangle$ on U^* and U, respectively. Then, $Q^{-1}P = P^\top Q^{-1}$ and the decompositions in (1.6) are orthogonal in $(\cdot,\cdot)_{Q^{-1}}$ and $(\cdot,\cdot)_Q$, respectively. This implies that the projections have norms less or equal than one, i.e.,

$$\begin{aligned} \|P\, v\|_{Q^{-1}} &\le \|v\|_{Q^{-1}} & \|(I-P)v\|_{Q^{-1}} &\le \|v\|_{Q^{-1}} & \forall v &\in U,\\ \|P^\top w\|_Q &\le \|w\|_Q & \|(I-P^\top)w\|_Q &\le \|w\|_Q & \forall w &\in U^*. \end{aligned}$$

1.1.3.4 Pseudo Inverses

For the general definition of the Moore-Penrose pseudo inverse see, e.g., [EHN96]. The following (different) definition is sufficient for the purpose of this book.

Definition 1.13. For a finite-dimensional Hilbert space V, let the linear operator $A : V \to V^*$ be self-adjoint and positive semi-definite, i.e., $\langle A\, v, v\rangle \ge 0$ for all $v \in$V. A self-adjoint and positive semi-definite operator $A^\dagger : V^* \to V$ is called a *pseudo inverse* of A if

$$A\, A^\dagger f = f \qquad \forall f \in \operatorname{range}(A).$$

If A is positive definite, the only pseudo inverse is of course A^{-1}.

Lemma 1.14. *Let V be a finite-dimensional Hilbert space, $A : V \to V^*$ self-adjoint and positive semi-definite, and let $A^\dagger : V^* \to V$ be a pseudo inverse of A in the sense of Definition 1.13. Furthermore, let $V^\perp \subset V$ a subspace be such that the sum $V = V^\perp \oplus \ker(A)$ is direct. Then*

$$\langle f, A^\dagger f\rangle = \sup_{v\in V^\perp} \frac{\langle f, v\rangle^2}{\langle A v, v\rangle} \qquad \forall f \in \mathrm{range}(A).$$

Proof. For $W := A^\dagger(\mathrm{range}(A))$, the space range$(A)$ is isomorphic to W^*, and we simply write $W^* = \mathrm{range}(A)$. We define $B : W \to W^*$ as $B := A_{|W}$. Then B is positive definite and $B^{-1} = A^\dagger{}_{|W^*}$. With the Cauchy-Schwarz inequality, for all $f \in W^* = \mathrm{range}(A)$:

$$\langle f, A^\dagger f\rangle = \langle f, B^{-1} f\rangle = \sup_{g\in W^*} \frac{\langle f, B^{-1}g\rangle^2}{\langle g, B^{-1}g\rangle} = \sup_{v\in W} \frac{\langle f, v\rangle^2}{\langle B v, v\rangle} = \sup_{v\in W} \frac{\langle f, v\rangle^2}{\langle A v, v\rangle}.$$

Since for each $v \in W$ there is a unique element $v_0 \in \ker(A)$ such that $v - v_0 \in V^\perp$ and since $f \in \mathrm{range}(A)$ implies that $\langle f, v\rangle = 0$ for all $v \in \ker(A)$, we can replace W in the rightmost supremum by $V^\perp$, which concludes the proof. □

Remark 1.15. Formally, we may choose $V^\perp = V_{/\ker(A)}$, see Sect. 1.1.3.2.

Let the assumptions of Definition 1.13 hold. We give a recipe for a general construction of $A^\dagger$ (for a an alternative approach see, e.g., [FR92, Appendix A]). Fix a set of functionals $r_k \in V^*$, $k = 1, \dots, \dim(\ker(A))$ such that

$$\forall v \in \ker(A) : \Big(\forall k = 1, \dots, \dim(\ker(A)) : \langle r_k, v\rangle = 0\Big) \quad \Longrightarrow \quad v = 0.$$

Let $\widetilde{A} : V \to V^*$ be given by

$$\langle \widetilde{A} v, w\rangle = \langle A v, w\rangle + \sum_{k=1}^{\dim(\ker(A))} \beta_k \langle r_k, v\rangle \langle r_k, w\rangle \qquad \text{for } v, w \in V,$$

with arbitrary coefficients $\beta_k > 0$ ($\widetilde{A}$ is a rank-dim(ker(A)) correction of A). One shows without problems that $\widetilde{A}$ is positive definite and that $\widetilde{A}^{-1}$ is a pseudo inverse of A. Furthermore, if $f \in \mathrm{range}(A)$ then $r_k(\widetilde{A}^{-1} f) = 0$ for all $k = 1, \dots, \dim(\ker(A))$.

1.1.4 Saddle Point Problems

Let V, W be finite-dimensional Hilbert spaces and

$$A : V \to V^*, \qquad B : V \to W^*$$

bounded linear operators where A is self-adjoint and positive semi-definite. We are concerned with the following problem. For given $f \in V^*$ and $g \in W^*$, find $(u, p) \in V \times W$ such that

$$\begin{bmatrix} A & B^\top \\ B & 0 \end{bmatrix} \begin{bmatrix} u \\ p \end{bmatrix} = \begin{bmatrix} f \\ g \end{bmatrix}. \tag{1.8}$$

Lemma 1.16. *(i) If* $\ker(B^\top) = \{0\}$*, then Problem* (1.8) *is uniquely solvable if and only if* $\ker(A) \cap \ker(B) = \{0\}$.
(ii) If $\ker(A) \cap \ker(B) = \{0\}$*, then Problem* (1.8) *is solvable and the solution* (u, p) *is unique up to adding elements from* $\ker(B^\top)$ *to* p.

Proof. See [BF91].

Problem (1.8) is equivalent to the *saddle point problem* find $(u, p) \in V \times W$ such that

$$\mathscr{L}(u, q) \leq \mathscr{L}(u, p) \leq \mathscr{L}(v, p) \qquad \forall (v, q) \in V \times W,$$

where $\mathscr{L}(v, q) := \frac{1}{2}\langle A\, v, v\rangle + \langle B\, v, q\rangle - \langle f, v\rangle - \langle g, q\rangle$. Usually, one calls (1.8) a saddle point problem too. Furthermore, if $\ker(A) \cap \ker(B) = \{0\}$, then the solution u of (1.8) solves the constrained minimization problem

$$u = \operatorname*{argmin}_{v \in V : B\, v = g} \tfrac{1}{2}\langle A\, v, v\rangle - \langle f, v\rangle. \tag{1.9}$$

Vice versa, if u solves (1.9), then there exists $p \in W$ such that (u, p) solves (1.8).

1.2 The Potential Equation: Variational Framework and Finite Elements

1.2.1 Sobolev Spaces

1.2.1.1 Lebesgue Spaces and Lipschitz Domains

For a Lebesgue measurable domain $\Omega \subset \mathbb{R}^d$, let $L^p(\Omega)$, $1 \leq p \leq \infty$ denote the usual Lebesgue spaces with norm

$$\|v\|_{L^p(\Omega)} := \left(\int_\Omega |v|^p \,\mathrm{d}x \right)^{1/p} \tag{1.10}$$

for $1 \leq p < \infty$ and $\|v\|_{L^\infty(\Omega)} := \operatorname*{ess.sup}_{x \in \Omega} |v(x)|$. Furthermore, $(v, w)_{L^2(\Omega)} := \int_\Omega v\, w \,\mathrm{d}x$ defines an inner product and $L^2(\Omega)$ is a Hilbert space.

In the following, we specialize to more regular domains. A domain Ω is a *Lipschitz domain* if its boundary $\partial\Omega$ can be represented by a finite family of Lipschitz continuous functions, see, e.g., [Eva98, McL00]. Figure 1.1 shows two famous examples of domains that are not Lipschitz. However, both are the union

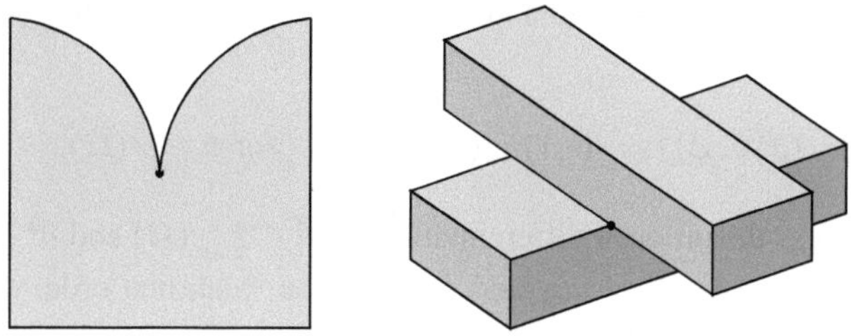

Fig. 1.1 Two domains that fail to be Lipschitz

of two Lipschitz domains. If all functions in the family have Lipschitz continuous derivatives up to order k, we speak of a $\mathscr{C}^{k,1}$ domain.

Every Lipschitz domain Ω has a well-defined surface measure on $\partial\Omega$ (hence, $L^p(\partial\Omega)$ is well-defined) and there exists a unique outward unit normal vector $n \in L^\infty(\partial\Omega)^d$, cf. [McL00, p. 96f].

1.2.1.2 Schwartz Distributions and Weak Derivatives

Let $\mathscr{D}(\Omega)$ denote the space $\mathscr{C}_0^\infty(\Omega)$ (i.e., $\mathscr{C}^\infty(\Omega)$ with compact support in Ω) equipped with a special topology: for a sequence $(\phi_n)_{n\in\mathbb{N}}$ in $\mathscr{D}(\Omega)$, we write

$$\phi_n \overset{\mathscr{D}(\Omega)}{\to} 0$$

if for every multi-index α, the sequence $\partial^\alpha \phi_n$ converges uniformly to zero on every compact subset of Ω. A linear functional $\psi : \mathscr{D}(\Omega) \to \mathbb{R}$ is said to be a *Schwartz distribution* if it is continuous in the above topology, i.e.,

$$\phi_n \overset{\mathscr{D}(\Omega)}{\to} 0 \qquad \Longrightarrow \qquad \langle \psi, \phi_n \rangle \to 0,$$

where $\langle \psi, \phi_n \rangle := \psi(\phi_n)$. The set of all Schwartz distributions is denoted by $\mathscr{D}^*(\Omega)$.

Some Schwartz distributions have a representation as a *function*. Let $L^1_{\text{loc}}(\Omega)$ be the functions that are in $L^1(K)$ for every compact subset K of Ω. We say that a distribution $\psi \in \mathscr{D}^*(\Omega)$ is *regular* if there exists a function $\widetilde{\psi} \in L^1_{\text{loc}}(\Omega)$:

$$\langle \psi, \phi \rangle = \int_\Omega \widetilde{\psi}\, \phi \,\mathrm{d}x \qquad \forall \phi \in \mathscr{D}(\Omega).$$

In that sense, every locally integrable function can be identified with a regular distribution and vice versa (but not all distributions are regular). If ψ above is regular, we simply write $\psi \in L^1_{\text{loc}}(\Omega)$.

For a distribution $\psi \in \mathscr{D}^*(\Omega)$ and a multi-index α, the *distributional derivative* $\partial^\alpha \psi \in \mathscr{D}^*(\Omega)$ is given by

$$\langle \partial^\alpha \psi, \phi \rangle := (-1)^{|\alpha|} \langle \psi, \partial^\alpha \phi \rangle \qquad \text{for } \phi \in \mathscr{D}(\Omega),$$

where $|\alpha|$ denotes the order of differentiation. If $f \in L^1_{\text{loc}}(\Omega)$ and $\partial^\alpha f \in L^1_{\text{loc}}(\Omega)$, then $\partial^\alpha f$ is called *weak derivative* of f. Thus, the weak first order derivative $\frac{\partial f}{\partial x_i}$ satisfies the integration by parts formula

$$\int_\Omega \frac{\partial f}{\partial x_i} \phi \, \mathrm{d}x = -\int_\Omega f \frac{\partial \phi}{\partial x_i} \, \mathrm{d}x \qquad \forall \phi \in \mathscr{C}_0^\infty(\Omega).$$

1.2.1.3 Sobolev Spaces on a Domain

There are several definitions of Sobolev spaces, which are all equivalent for Lipschitz domains. The following one is the probably most convenient one.

Definition 1.17. For a Lipschitz domain Ω and an integer $k \geq 0$, the Sobolev space $H^k(\Omega)$ is given by

$$H^k(\Omega) := \left\{ v \in L^2(\Omega) : \partial^\alpha v \in L^2(\Omega) \quad \forall \text{ multi-indices } \alpha, \ |\alpha| \leq k \right\}.$$

Equipped with the norm

$$\|v\|_{H^k(\Omega)} := \Big(\sum_{|\alpha| \leq k} \|\partial^\alpha v\|^2_{L^2(\Omega)} \Big)^{1/2}$$

it is a Hilbert space. We will frequently use the seminorm

$$|v|_{H^k(\Omega)} := \Big(\sum_{|\alpha| = k} \|\partial^\alpha v\|^2_{L^2(\Omega)} \Big)^{1/2}.$$

The Sobolev spaces have the property that

$$\overline{\mathscr{C}^\infty(\overline{\Omega})}^{\|\cdot\|_{H^k(\Omega)}} = H^k(\Omega) \qquad \forall k \geq 0$$

($\overline{W}^{\|\cdot\|_V}$ denotes the closure of $W \subset V$ with respect to $\|\cdot\|_V$). Furthermore, we set

$$H_0^k(\Omega) := \overline{\mathscr{C}_0^\infty(\Omega)}^{\|\cdot\|_{H^k(\Omega)}}.$$

For $k = 0$, $H_0^k(\Omega) = H^k(\Omega) = L^2(\Omega)$ and $\mathscr{C}_0^\infty(\Omega)$ is dense in $L^2(\Omega)$. For an integer $k > 0$, $H_0^k(\Omega)$ is a genuine subspace of $H^k(\Omega)$: the extension of a function from $H_0^k(\Omega)$ by zero is in $H^k(\mathbb{R}^d)$, which is not in general true for a function from $H^k(\Omega)$. We will see later on that functions from $H_0^k(\Omega)$ vanish on the boundary $\partial\Omega$.

The dual spaces $H^k(\Omega)^*$ and $H_0^k(\Omega)^*$ are defined in the usual way. Note that any functional $\psi \in H^k(\Omega)^*$ or $H_0^k(\Omega)^*$ can be restricted to $\mathcal{D}^*(\Omega)$ and thus defines a distribution. In some cases, this distribution might be identifiable with a function $f \in L^2(\Omega)$ such that $\langle \psi, v\rangle = (f, v)_{L^2(\Omega)}$ for all $v \in \mathcal{C}_0^\infty(\Omega)$ and by density also for all $v \in L^2(\Omega)$. Hence,

$$L^2(\Omega) \subset H^k(\Omega)^* \subset H_0^k(\Omega)^* \subset \mathcal{D}^*(\Omega).$$

In order to mark the dependence on Ω, we use $\langle \cdot, \cdot\rangle_\Omega$ for the duality product.

Note that the above definitions include the case where Ω is unbounded or even $\Omega = \mathbb{R}^d$. The following theorem states that Sobolev functions can be extended from a domain to the entire space. For a proof see [Ste70].

Theorem 1.18 (Stein). *Let Ω be a Lipschitz domain. Then there exists a linear extension operator $E : L^2(\Omega) \to L^2(\mathbb{R}^d)$ such that for all integers $k \geq 0$, the operator E maps $H^k(\Omega)$ to $H^k(\mathbb{R}^d)$ and there is a constant C_E (depending on k) with*

$$(Ev)_{|\Omega} = v, \qquad \|Ev\|_{H^k(\mathbb{R}^d)} \leq C_E \|v\|_{H^k(\Omega)} \qquad \forall v \in H^k(\Omega).$$

For extension operators on less regular domains (ε-δ domains or *uniform* domains) see [Jon81, Rog06].

In some cases we will also need Sobolev spaces with real indices.

Definition 1.19. For $s = \lfloor s \rfloor + \sigma$ and $\sigma \in (0, 1)$, we define the norm

$$\|v\|_{H^s(\Omega)} := \left(\|v\|^2_{H^{\lfloor s \rfloor}(\Omega)} + \sum_{|\alpha| = \lfloor s \rfloor} \int_\Omega \int_\Omega \frac{|\partial^\alpha v(x) - \partial^\alpha v(y)|^2}{|x-y|^{d+2\sigma}} \,\mathrm{d}x \,\mathrm{d}y \right)^{1/2}$$

and the corresponding Hilbert spaces

$$H^s(\Omega) := \{v \in L^2(\Omega) : \|v\|_{H^s(\Omega)} < \infty\}, \qquad H_0^s(\Omega) := \overline{\mathcal{C}_0^\infty(\Omega)}^{\|\cdot\|_{H^s(\Omega)}}.$$

1.2.1.4 Sobolev Spaces on Manifolds

We say that a measurable hypersurface $\widetilde{\Gamma}$ of dimension $d-1$ is *Lipschitz* if there exists a Lipschitz domain $\Omega \subset \mathbb{R}^d$ with $\widetilde{\Gamma} \subset \partial\Omega$.

Definition 1.20. For a Lipschitz hypersurface $\widetilde{\Gamma}$ of dimension $d-1$ and a real number $s \in (0, 1)$, the Sobolev-Slobodeckiĭ norm is given by

$$\|v\|_{H^s(\tilde{\Gamma})} := \left(\|v\|^2_{L^2(\tilde{\Gamma})} + |v|^2_{H^s(\tilde{\Gamma})}\right)^{1/2},$$

with the seminorm

$$|v|_{H^s(\tilde{\Gamma})} := \left(\int_{\tilde{\Gamma}}\int_{\tilde{\Gamma}} \frac{|v(x)-v(y)|^2}{|x-y|^{d-1+2s}}\,\mathrm{d}s_x\,\mathrm{d}s_y\right)^{1/2}.$$

The space

$$H^s(\widetilde{\Gamma}) := \left\{v \in L^2(\widetilde{\Gamma}) : \|v\|_{H^s(\tilde{\Gamma})} < \infty\right\}$$

equipped with that norm is then a Hilbert space. For $s = 1$, the space $H^1(\widetilde{\Gamma})$ must be defined using the tangential derivative. For $s > 1$, we need more smoothness of the hypersurface $\widetilde{\Gamma}$ to properly define $H^s(\widetilde{\Gamma})$, see e.g. [SS11, Definition 2.4.1].

Mostly, we will need $H^{1/2}(\partial\Omega)$ which turns out to be the trace space of $H^1(\Omega)$, see Sect. 1.2.1.5. The next lemma shows that $H^s(\widetilde{\Gamma})$ functions are extendable to $\partial\Omega$.

Lemma 1.21. *Let $\widetilde{\Gamma} \subset \partial\Omega$ be as in Definition 1.20. If $\widetilde{\Gamma}$ is open, then*

$$H^s(\widetilde{\Gamma}) = \left\{v \in L^2(\widetilde{\Gamma}) : \exists\,\tilde{v} \in H^s(\partial\Omega),\ \tilde{v}_{|\tilde{\Gamma}} = v\right\}.$$

Moreover, the norm

$$\inf\left\{\|\tilde{v}\|_{H^s(\partial\Omega)} : \tilde{v} \in H^s(\partial\Omega),\ \tilde{v}_{|\tilde{\Gamma}} = u\right\}$$

is equivalent to $\|u\|_{H^s(\tilde{\Gamma})}$.

Let $H^s_0(\widetilde{\Gamma}) := \overline{\mathscr{C}^\infty_0(\widetilde{\Gamma})}^{\|\cdot\|_{H^s(\tilde{\Gamma})}}$. For $s \leqq 1/2$, $H^s_0(\widetilde{\Gamma}) = H^s(\widetilde{\Gamma})$. However, the extension by zero of a function from $H^s(\widetilde{\Gamma})$ to $\partial\Omega$ does only belong to $H^s(\partial\Omega)$ if $s \neq 1/2$. This is why for $s = 1/2$ we need an extra space.

Definition 1.22. Let $\widetilde{\Gamma}$ be an open Lipschitz hypersurface. We define the norm

$$\|u\|_{H^{1/2}_{00}(\tilde{\Gamma})} := \left(\|u\|^2_{H^{1/2}(\tilde{\Gamma})} + \int_{\tilde{\Gamma}} \frac{u(x)^2}{\mathrm{dist}(x,\partial\widetilde{\Gamma})}\,\mathrm{d}s_x\right)^{1/2}.$$

The space

$$H^{1/2}_{00}(\widetilde{\Gamma}) := \left\{v \in L^2(\widetilde{\Gamma}) : \|v\|_{H^{1/2}_{00}(\tilde{\Gamma})} < \infty\right\}$$

equipped with that norm is a Hilbert space.

For $\widetilde{\Gamma} \subset \partial\Omega$, one can show that the extension of $u \in H^{1/2}_{00}(\widetilde{\Gamma})$ by zero to $\partial\Omega$ is still in $H^{1/2}(\partial\Omega)$. Note, however, that $H^{1/2}_{00}(\partial\Omega) = H^{1/2}(\partial\Omega)$. Finally, we define the dual spaces

$$H^{-s}(\widetilde{\Gamma}) := \left(H^s(\widetilde{\Gamma})\right)^*, \qquad H^{-1/2}_{00}(\widetilde{\Gamma}) := \left(H^{1/2}_{00}(\widetilde{\Gamma})\right)^*,$$

and denote the corresponding duality products by $\langle\cdot,\cdot\rangle_{\widetilde{\Gamma}}$. Note that

$$L^2(\widetilde{\Gamma}) \subset H^{-1/2}(\widetilde{\Gamma}) \subset H_{00}^{-1/2}(\widetilde{\Gamma})$$

and that for any functional $g \in H^{-1/2}(\partial\Omega)$, its restriction to $H_{00}^{-1/2}(\widetilde{\Gamma})$ is well defined (by restricting the test functions to $H_{00}^{1/2}(\widetilde{\Gamma})$). For more comprehensive references see, e.g., [LM72, GR86, Gri85, McL00].[1]

1.2.1.5 The Trace Operator

The following theorem explains the meaning of traces of H^1-functions and discusses their extendability.

Theorem 1.23 (Trace theorem). *Let Ω be a Lipschitz domain. Then the* Dirichlet trace operator $\gamma_0 : \mathscr{C}^\infty(\overline{\Omega}) \to \mathscr{C}^\infty(\partial\Omega)$ *defined by*

$$\gamma_0 u := u_{|\partial\Omega},$$

has a unique extension $\gamma_0 : H^1(\Omega) \to H^{1/2}(\partial\Omega)$ that is linear and continuous, i.e., there exists a constant $C_T > 0$ with

$$\|\gamma_0 u\|_{H^{1/2}(\partial\Omega)} \leq C_T \, \|u\|_{H^1(\Omega)}^2 \qquad \forall u \in H^1(\Omega).$$

Furthermore, γ_0 has a continuous right inverse, i.e., there exists a linear extension operator $\mathscr{E} : H^{1/2}(\partial\Omega) \to H^1(\Omega)$ and a constant $C_{IT} > 0$ such that

$$\gamma_0(\mathscr{E}u) = u, \qquad \|\mathscr{E}u\|_{H^1(\Omega)}^2 \leq C_{IT} \, \|u\|_{H^{1/2}(\partial\Omega)} \qquad \forall u \in H^{1/2}(\partial\Omega).$$

Analogous inequalities hold if we replace $\partial\Omega$ by $\widetilde{\Gamma} \subset \partial\Omega$, or if we replace the norms by the respective seminorms. Moreover, the trace operator is continuous as a mapping

$$\gamma_0 : H^s(\Omega) \to H^{s-1/2}(\partial\Omega)$$

for all real numbers $1/2 < s < 3/2$. If Ω is a $\mathscr{C}^{k-1,1}$ domain, the trace operator is continuous as a a mapping $\gamma_0 : H^s(\Omega) \to H^{s-1/2}(\partial\Omega)$ for $1/2 < s \leq k$.

Proof. For a proof of $1/2 \leq s \leq 1$ see e.g. [McL00, Theorem 3.37]. The statement for the seminorms follows from Lemma 1.35 in the next subsection. For a proof of $1 < s < 3/2$ see [Cos88].

[1] Note that in the literature of boundary integral equations (e.g., [McL00, Ste03b, Ste08]), the space $H_{00}^{1/2}(\widetilde{\Gamma})$ is often denoted by $\widetilde{H}^{1/2}(\widetilde{\Gamma})$ and the two dual spaces by $H^{-1/2}(\widetilde{\Gamma}) := (\widetilde{H}^{1/2}(\widetilde{\Gamma}))^*$ and $\widetilde{H}^{-1/2}(\widetilde{\Gamma}) := (H^{1/2}(\widetilde{\Gamma}))^*$.

Notation. In the sequel, we will often write $u_{|\partial\Omega}$ as a short hand for $\gamma_0 u$.

One can show that

$$\inf\left\{\|\tilde{u}\|_{H^1(\Omega)} : \tilde{u} \in H^1(\Omega),\, \tilde{u}_{|\partial\Omega} = u\right\} \tag{1.11}$$

is an equivalent norm to $\|u\|_{H^{1/2}(\partial\Omega)}$, and that

$$H_0^1(\Omega) \;=\; \overline{\mathscr{C}_0^\infty(\Omega)}^{\|\cdot\|_{H^1(\Omega)}} \;=\; \left\{u \in H^1(\Omega) : u_{|\partial\Omega} = 0\right\}, \tag{1.12}$$

see, e.g., [McL00, Theorem 3.40]. In particular, functions from $H_0^1(\Omega)$ can be extended by zero to $H^1(\mathbb{R}^d)$.

1.2.1.6 Poincaré's Inequality and the Bramble-Hilbert Lemma

Poincaré's inequality and the so-called Bramble-Hilbert lemma are powerful tools for the analysis of variational problems, finite element approximations, and domain decomposition methods. In this subsection, we derive these tools from the following result going back to Sobolev [Sob91].

Theorem 1.24. *Let Ω be a bounded Lipschitz domain and let $\psi : H^1(\Omega) \to \mathbb{R}$ be a bounded linear functional such that*

$$\forall c \in \mathbb{R} :\; \psi(c) = 0 \implies c = 0.$$

Then the norm $(|u|^2_{H^1(\Omega)} + |\psi(u)|^2)^{1/2}$ is equivalent to $\|u\|_{H^1(\Omega)}$. Furthermore there exists a constant $C_\psi > 0$ (depending on Ω and on ψ) such that

$$\|u\|_{L^2(\Omega)} \;\le\; C_\psi\, |u|_{H^1(\Omega)} \qquad \forall u \in H^1(\Omega)\,,\; \psi(u) = 0.$$

Proof. The proof is performed by contradiction using the compact imbedding of $H^1(\Omega)$ in $L^2(\Omega)$; see [Mor66, Neč̌67], [TW05, Theorem A.12], [BS02, Sect. 5.3], or [AF03, Chap. 4].

With the following scaling argument, the dependence on the diameter of the domain can be made explicit; this will be important later on when applying the inequality on subdomains.

Corollary 1.25. *Let the assumptions of Theorem 1.24 hold. If ψ reproduces constants, i.e., $\psi(c) = c$ for all $c \in \mathbb{R}$, then there exists a constant C_ψ independent of* $\operatorname{diam}(\Omega)$ *such that*

$$\left.\begin{aligned} \|u - \psi(u)\|_{L^2(\Omega)} &\le C_\psi \operatorname{diam}(\Omega)\, |u|_{H^1(\Omega)} \\ \|u\|_{L^2(\Omega)} &\le C_\psi \operatorname{diam}(\Omega)\, |u|_{H^1(\Omega)} + |\Omega|^{1/2}\, |\psi(u)| \end{aligned}\right\} \quad \forall u \in H^1(\Omega).$$

Proof. We define $\widehat{\Omega} := \{\frac{x}{\operatorname{diam}(\Omega)} : x \in \Omega\}$ and $F : \hat{x} \mapsto \operatorname{diam}(\Omega)\,\hat{x}$ such that $F(\widehat{\Omega}) = \Omega$. With $\hat{u} := u \circ F$ and $\widehat{\psi}(\hat{u}) := \psi(\hat{u} \circ F^{-1})$ we can apply Theorem 1.24 on $\widehat{\Omega}$:

$$\|u - \psi(u)\|^2_{L^2(\Omega)} = \operatorname{diam}(\Omega)^{d/2}\, \|\hat{u} - \widehat{\psi}(\hat{u})\|_{L^2(\hat{\Omega})} \le C_{\hat{\psi}} \operatorname{diam}(\Omega)^{d/2}\, |\hat{u}|_{H^1(\hat{\Omega})}\,.$$

Finally, $|u|_{H^1(\Omega)} = \operatorname{diam}(\Omega)^{d/2-1}\, |\hat{u}|_{H^1(\hat{\Omega})}$. The second estimate follows from the first one by the triangle inequality. □

Definition 1.26. For a bounded domain Ω and $u \in L^1(\Omega)$, we define the *average*

$$\bar{u}^{\Omega} := \frac{1}{|\Omega|} \int_{\Omega} u \,\mathrm{d}x.$$

Note that by Cauchy's inequality,

$$|\bar{u}^{\Omega}| \le |\Omega|^{-1/2}\, \|u\|_{L^2(\Omega)} \qquad \forall u \in L^2(\Omega). \tag{1.13}$$

Lemma 1.27 (Poincaré's inequality). *Let Ω be a bounded Lipschitz domain. Then there exists a constant $C_P(\Omega) > 0$ independent of* $\operatorname{diam}(\Omega)$ *such that*

$$\inf_{c\in\mathbb{R}} \|u-c\|_{L^2(\Omega)} = \|u-\bar{u}^{\Omega}\|_{L^2(\Omega)} \le C_P(\Omega) \operatorname{diam}(\Omega)\, |u|_{H^1(\Omega)} \qquad \forall u \in H^1(\Omega).$$

The smallest possible constant $C_P(\Omega)$ is called Poincaré constant *of Ω.*

Proof. The quadratic functional $c \mapsto \|c - u\|^2_{L^2(\Omega)}$ is convex. Hence, it attains its minimum at its only stationary point, where $\int_{\Omega}(c - u)\,\mathrm{d}x = 0$, i.e., $c = \bar{u}^{\Omega}$. The remaining statement follows from Corollary 1.25 by $\psi(u) := \bar{u}^{\Omega}$ and $C_P(\Omega) := C_{\psi}$. □

Remark 1.28. For Poincaré's inequality, it is essential that Ω is *connected*. The inequality does not hold for sets with more than one component.

Remark 1.29. The Poincaré constant $C_P(\Omega)$ is related to an eigenvalue problem: if $\lambda_1 < \lambda_2 < \ldots$ are the eigenvalues of

$$-\Delta u = \lambda\, u \quad \text{in } \Omega, \qquad \frac{\partial u}{\partial n} = 0 \quad \text{on } \partial\Omega,$$

then $\lambda_1 = 0$ and $\lambda_2 = C_P(\Omega)^{-2} \operatorname{diam}(\Omega)^{-2}$.

Remark 1.30. From the indirect proof of Theorem 1.24, we have no information on how the constant $C_P(\Omega)$ depends on the *shape* of Ω. For a certain class of simple domains, direct proofs with explicit constants exist; see, [Beb03, PW60] for convex domains, and [BS02, VV11] for domains star-shaped with respect to a ball. Due to a result found independently by Maz'ya [Maz60, Maz85] and Federer and

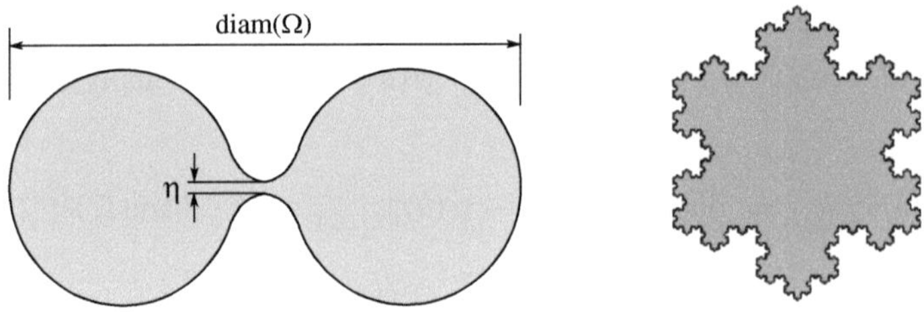

Fig. 1.2 *Left:* Two-dimensional dumbbell domain whose isoperimetric constant scales with $\mathrm{diam}(\Omega)/\eta$. *Right:* Von Koch snowflake domain

Fleming [FF60], one can get Poincaré's inequality from the so-called *isoperimetric inequality*. The isoperimetric constant $C_{IP}(\Omega)$ is the smallest constant such that for all measurable sets $A, B \subset \Omega$,

$$\min\left(\mathrm{meas}_d(A), \mathrm{meas}_d(B)\right)^{(d-1)/d} \le C_{IP}\,|\partial A \cap \partial B|.$$

The isoperimetric inequality is equivalent to the Sobolev inequality

$$\inf_{c\in\mathbb{R}} \left(\int_\Omega |u-c|^{d/(d-1)}\,\mathrm{d}x\right)^{(d-1)/d} \le C_{IP} \int_\Omega |\nabla u|\,\mathrm{d}x \qquad \forall u \in C^1(\overline{\Omega}).$$

For $d = 2$, the above Soblev inequality combined with Cauchy's inequality implies Poincaré's inequality. For $d = 3$, more technical arguments are needed including a Hölder inequality (for a proof by H. H. Kim see [Pec12, Lemma 3.4]). Summarizing,

$$C_P(\Omega) \le \left(\tfrac{4}{3}\right)^{d-2} \frac{\mathrm{meas}_d(\Omega)^{1/d}}{\mathrm{diam}(\Omega)}\, C_{IP}(\Omega) \qquad \text{for } d = 2, 3.$$

Obviously, the isoperimetric constant gets small whenever two large parts of the domain are only linked by a narrow channel, like the dumbbell domain in Fig. 1.2 (left). The isoperimetric constant is bounded for the large class of so-called John domains (see [Boj88, BK95, Haj01]) which can have quite irregular, not even measurable boundaries like the snowflake domain in Fig. 1.2 (right).

Lemma 1.31 (Friedrichs' inequality). *Let Ω be a bounded Lipschitz domain and let $\Gamma_D \subset \partial\Omega$ have positive surface measure. Then there exists a constant $C_F(\Omega, \Gamma_D) > 0$ independent of* $\mathrm{diam}(\Omega)$ *such that*

$$\|u\|_{L^2(\Omega)} \le C_F(\Omega, \Gamma_D)\,\mathrm{diam}(\Omega)\,|u|_{H^1(\Omega)} \qquad \forall u \in H^1(\Omega),\ u_{|\Gamma_D} = 0.$$

Proof. Follows from Corollary 1.25 by setting $\psi(u) = |\Gamma_D|^{-1} \int_{\Gamma_D} u\,\mathrm{d}s$. □

Corollary 1.32. *Let Ω be a bounded Lipschitz domain. Then the following (squared) norms are equivalent,*

$$|u|^2_{H^1(\Omega)} + \frac{1}{\operatorname{diam}(\Omega)^2} \|u\|^2_{L^2(\Omega)}$$

$$|u|^2_{H^1(\Omega)} + \frac{1}{\operatorname{diam}(\Omega)} \|u\|^2_{L^2(\partial\Omega)},$$

$$|u|^2_{H^1(\Omega)} + \frac{1}{|\partial\Omega| \operatorname{diam}(\Omega)} \Big(\int_{\partial\Omega} u \, ds \Big)^2,$$

and the equivalence constants are independent of $\operatorname{diam}(\Omega)$.

Proof. Follows from Corollary 1.25 by choosing the functional ψ appropriately. □

For a bounded linear functional $\psi \in H^1(\Omega)^*$, we define the (dimensionless) norm

$$|||\psi|||_{H^1(\Omega)^*} := \sup_{v \in H^1(\Omega)} \frac{|\Omega|^{1/2} |\psi(v)|}{\Big(\|v\|^2_{L^2(\Omega)} + \operatorname{diam}(\Omega)^2 |v|^2_{H^1(\Omega)} \Big)^{1/2}}. \tag{1.14}$$

The following two lemmas are closely related to a result by Bramble and Hilbert [BH70]. The first lemma sometimes helps to get a bound for the constant C_ψ from Corollary 1.25 in terms of the rather well-studied Poincaré constant $C_P(\Omega)$. The second lemma allows to estimate a linear functional in terms of a seminorm.

Lemma 1.33. *Let Ω be a Lipschitz domain and suppose that the linear functional $\psi \in H^1(\Omega)^*$ reproduces constants. Then for all $u \in H^1(\Omega)$,*

$$\|u - \psi(u)\|_{L^2(\Omega)} \le \Big(C_P(\Omega) + \sqrt{1 + C_P(\Omega)^2}\, |||\psi|||_{H^1(\Omega)^*} \Big) \operatorname{diam}(\Omega)\, |u|_{H^1(\Omega)}.$$

Proof. Using the triangle inequality, the fact that ψ reproduces constants, and the definition of the norm (1.14), we derive

$$\|u - \psi(u)\|_{L^2(\Omega)} \le \|u - \bar{u}^\Omega\|_{L^2(\Omega)} + |\Omega|^{1/2} \underbrace{|\bar{u}^\Omega - \psi(u)|}_{= |\psi(u - \bar{u}^\Omega)|}$$

$$\le \|u - \bar{u}^\Omega\|_{L^2(\Omega)} + |||\psi|||_{H^1(\Omega)^*} \Big(\|u - \bar{u}^\Omega\|^2_{L^2(\Omega)} + \operatorname{diam}(\Omega)^2 |u|^2_{H^1(\Omega)} \Big)^{1/2}.$$

The proof is concluded by Poincaré's inequality. □

Lemma 1.34 (Bramble-Hilbert). *Let Ω be a bounded Lipschitz domain and suppose that the linear functional $\psi \in H^1(\Omega)^*$ fulfills*

$$\psi(c) = 0 \qquad \forall c \in \mathbb{R}.$$

Then

$$|\psi(u)| \le \sqrt{1 + C_P(\Omega)^2}\, \|\psi\|_{H^1(\Omega)^*} \frac{\operatorname{diam}(\Omega)}{|\Omega|^{1/2}} |u|_{H^1(\Omega)} \qquad \forall u \in H^1(\Omega).$$

Proof. Since $\psi(\bar{u}^{\Omega}) = 0$, the definition of the norm (1.14) yields

$$|\psi(u)| = |\psi(u - \bar{u}^{\Omega})| \le \frac{\|\psi\|_{H^1(\Omega)^*}}{|\Omega|^{1/2}} \left(\|u - \bar{u}^{\Omega}\|^2_{L^2(\Omega)} + \operatorname{diam}(\Omega)^2 |u|^2_{H^1(\Omega)} \right)^{1/2}.$$

The proof is concluded by Poincaré's inequality. □

For the original result by Bramble and Hilbert, where the domains have to satisfy a strong cone condition, see [BH70], for a proof on domains which are star-shaped with respect to a ball see [BS02, Lemma 4.3.8]. These references even treat the more general case of $H^k(\Omega)$. For a comprehensive survey on the Bramble-Hilbert lemma containing novel explicit estimates see [DW06].

The following two lemmas provide variants of Poincaré's and Friedrichs' inequality in the space $H^{1/2}(\partial\Omega)$.

Lemma 1.35 (Poincaré type inequality in $H^{1/2}$). *Let Ω be a bounded Lipschitz domain. Then there exists a constant $C_P(\partial\Omega) > 0$ independent of* $\operatorname{diam}(\Omega)$ *such that*

$$\inf_{c \in \mathbb{R}} \|u - c\|_{L^2(\partial\Omega)} \le C_P(\partial\Omega) \operatorname{diam}(\Omega)^{1/2} |u|_{H^{1/2}(\partial\Omega)} \qquad \forall u \in H^{1/2}(\partial\Omega).$$

The infimum is attained at $c = \bar{u}^{\partial\Omega}$.

Lemma 1.36 (Friedrichs type inequality in $H^{1/2}$). *Let Ω be a bounded Lipschitz domain and let $\Gamma_D \subset \partial\Omega$ have positive surface measure. Then there exists a constant $C_F(\partial\Omega, \Gamma_D) > 0$ independent of* $\operatorname{diam}(\Omega)$ *such that*

$$\|u\|_{L^2(\partial\Omega)} \le C_F(\partial\Omega, \Gamma_D) \operatorname{diam}(\Omega)^{1/2} |u|_{H^{1/2}(\partial\Omega)} \qquad \forall u \in H^{1/2}(\partial\Omega),\ u_{|\Gamma_D} = 0.$$

Proofs for the two lemmas can be found with the analogous strategy as in Theorem 1.24 using that the embedding $H^{1/2}(\partial\Omega) \subset L^2(\partial\Omega)$ is compact; see, e.g., [McL00, Theorem 3.27].

1.2.2 The Potential Equation

1.2.2.1 Classical Formulation

Let Ω be a bounded domain with smooth boundary $\partial\Omega = \Gamma_D \cup \Gamma_N$ and $\Gamma_D \cap \Gamma_N = \emptyset$. For smooth functions $\mathscr{A} : \Omega \to \mathbb{R}^{d\times d}_{\text{sym}}$ and $f : \Omega \to \mathbb{R}$, the classical formulation of the potential equation reads

$$-\mathrm{div}(\mathscr{A}\nabla u) = f \qquad \text{in } \Omega, \tag{1.15}$$

subject to the Dirichlet and Neumann boundary conditions

$$\begin{aligned} u &= g_D \text{ on } \Gamma_D\,, \\ \mathscr{A}\nabla u \cdot n &= g_N \text{ on } \Gamma_N\,, \end{aligned} \tag{1.16}$$

where g_D, g_N are given, smooth functions. Note that here, Eq. (1.15) holds in the strong sense on the whole domain Ω. Often, one considers interface problems with non-smooth data. Let Ω be composed from two subdomains Ω_1, Ω_2 and let $\mathscr{A}$ be smooth on these subdomains, but discontinuous across the interface $\Gamma_{12} := \partial\Omega_1 \cap \partial\Omega_2$. The interface problem is to find $u : \Omega \to \mathbb{R}$ such that (1.15) holds in (the interior of) Ω_1 and Ω_2 together with boundary conditions (1.16) as well as the interface conditions

$$\begin{aligned} u_1 &= u_2 \text{ on } \Gamma_{12}\,, \\ \mathscr{A}_1\nabla u_1 \cdot n_1 + \mathscr{A}_2\nabla u_2 \cdot n_2 &= 0 \text{ on } \Gamma_{12}\,, \end{aligned} \tag{1.17}$$

where $\mathscr{A}_i$, u_i are the respective limits from Ω_i to Γ_{12} and n_i denotes the outward unit vector on $\partial\Omega_i$. The case of many subdomains is analogous.

1.2.2.2 Weak Formulation

In the following, we discuss the weak formulation, where neither the data nor the solution need to be smooth; we assume that

$$f \in L^2(\Omega), \qquad \mathscr{A} = \mathscr{A}^\top \in L^\infty(\Omega)^{d\times d}.$$

When looking for solutions $u \in H^1(\Omega)$ and replacing the classical derivatives by distributional ones, we obtain

$$-\mathrm{div}(\mathscr{A}\nabla u) = f \qquad \text{in } \mathscr{D}^*(\Omega). \tag{1.18}$$

By the definition of $H_0^1(\Omega)$ this is equivalent to

$$\int_\Omega \mathscr{A}\nabla u \cdot \nabla v \,\mathrm{d}x = \int_\Omega f\, v \,\mathrm{d}x \qquad \forall v \in H_0^1(\Omega). \tag{1.19}$$

Since all the test functions vanish on the boundary, we need to include the boundary conditions explicitly (whereas interface conditions like (1.17) above are already built in). The Dirichlet boundary condition reads

$$u_{|\Gamma_D} = g_D,$$

for a given function $g_D \in H^{1/2}(\Gamma_D)$ (cf. Theorem 1.23).

Before we can incorporate the Neumann boundary condition, we need a sound definition of the Neumann trace operator. The next theorem gives such a definition. Note that the theorem is applicable to any solution $u \in H^1(\Omega)$ of (1.18): the distribution $-\operatorname{div}(\mathscr{A}\nabla u)$ is identifiable with $f \in L^2(\Omega)$, and so $\operatorname{div}(\mathscr{A}\nabla u) \in L^2(\Omega)$.

Theorem 1.37. *For each $u \in H^1(\Omega)$ with $\operatorname{div}(\mathscr{A}\nabla u) \in L^2(\Omega)$, there exists a unique functional $\gamma_1^{\mathscr{A}} u \in H^{-1/2}(\partial\Omega)$ such that*

$$\int_\Omega \mathscr{A}\nabla u \cdot \nabla v \, \mathrm{d}x = -\int_\Omega \operatorname{div}(\mathscr{A}\nabla u)\, v \, \mathrm{d}x + \langle \gamma_1^{\mathscr{A}} u, \gamma_0 v\rangle_{\partial\Omega} \qquad \forall v \in H^1(\Omega). \tag{1.20}$$

For smooth functions u, we have $\gamma_1^{\mathscr{A}} u = \mathscr{A}\nabla u \cdot n$, and (1.20) is Green's classical identity. Furthermore, there is a constant C such that

$$\|\gamma_1^{\mathscr{A}} u\|_{H^{-1/2}(\partial\Omega)} \le C\left(\|u\|_{H^1(\Omega)} + \|\operatorname{div}(\mathscr{A}\nabla u)\|_{L^2(\Omega)}\right).$$

This implies that the Neumann trace operator

$$\gamma_1^{\mathscr{A}} : \{v \in H^1(\Omega) : \operatorname{div}(\mathscr{A}\nabla u) \in L^2(\Omega)\} \to H^{-1/2}(\partial\Omega)$$

is linear and continuous. For $\mathscr{A} = I$, we simply write γ_1.

Proof. See, e.g., [McL00, Lemma 4.3] (or also [GR86]). The functional is defined by

$$\langle \gamma_1^{\mathscr{A}} u, w\rangle_{\partial\Omega} = \int_\Omega \mathscr{A}\nabla u \cdot \nabla(\mathscr{E}w)\, \mathrm{d}x + \int_\Omega \operatorname{div}(\mathscr{A}\nabla u)\, \mathscr{E}w \, \mathrm{d}x$$

for $w \in H^{1/2}(\partial\Omega)$, where $\mathscr{E}$ is the extension operator from Theorem 1.23. This definition is independent of the particular choice of $\mathscr{E}$. □

Remark 1.38. The assumption $f \in L^2(\Omega)$ can be weakened to $f \in H^1(\Omega)^*$, but this introduces some complications, see Sect. 1.2.2.3 and [McL00, Lemma 4.3].

Recall that the restriction of $\gamma_1^{\mathscr{A}} u \in H^{-1/2}(\partial\Omega)$ to $H_{00}^{-1/2}(\Gamma_N)$ is well-defined. Combining (1.19) and (1.20), we obtain the complete variational formulation. Given $f \in L^2(\Omega)$, $g_D \in H^{1/2}(\Gamma_D)$, and $g_N \in H_{00}^{-1/2}(\Gamma_N)$, find $u \in H^1(\Omega)$, $u_{|\Gamma_D} = g_D$:

$$\underbrace{\int_\Omega \mathscr{A}\nabla w \cdot \nabla v \, \mathrm{d}x}_{=:\, a(u,v)} = \underbrace{\int_\Omega f\, v \, \mathrm{d}x + \langle g_N, v\rangle_{\Gamma_N}}_{=:\, \langle F, v\rangle_\Omega} \qquad \forall v \in H^1(\Omega),\ v_{|\Gamma_D} = 0. \tag{1.21}$$

Lemma 1.39. *Assume that the Dirichlet boundary Γ_D has positive surface measure and that there exists a uniform constant $\alpha_{\min} > 0$ such that*

$$(\mathscr{A}(x)\xi)\cdot\xi \;\geq\; \alpha_{\min}\,|\xi|^2 \qquad \forall\xi\in\mathbb{R}^d, \quad \forall x\in\Omega \text{ a.e.} \tag{1.22}$$

Then there exists a unique solution to the variational problem (1.21).

Proof. Due to Theorem 1.23, we can seek $u \in \mathscr{E}g_D + V_D$, where $V_D := \{v \in H^1(\Omega) : v_{|\Gamma_D} = 0\}$ is equipped with the norm $\|\cdot\|_{H^1(\Omega)}$. The variational formulation now fits into the framework of Theorem 1.1. The continuity of the bilinear form and the linear functional follow from Hölder's inequality and Theorem 1.23, coercivity is shown using the assumption on $\mathscr{A}$ and Friedrichs' inequality. We get

$$\underline{c}_a = \frac{\alpha_{\min}}{1 + C_F(\Omega,\,\Gamma_D)^2}, \qquad \bar{c}_a = \|\mathscr{A}\|_{L^\infty(\Omega)},$$

and $\|F\|_{V_D^*} \leq \|f\|_{L^2(\Omega)} + C_T\,\|g_N\|_{H_{00}^{-1/2}(\Gamma_N)}$. □

The pure Neumann problem ($\Gamma_D = \emptyset$) is treated in Sect. 1.2.5.

1.2.2.3 A Generalized Conormal Derivative

Occasionally, we will need to allow more general source functions than $f \in L^2(\Omega)$. The following theorem generalizes Theorem 1.37.

Theorem 1.40. *For $f \in H^1(\Omega)^*$ and $u \in H^1(\Omega)$ with $-\operatorname{div}(\mathscr{A}\nabla u) = f$ in $\mathscr{D}^*(\Omega)$, there exists a unique generalized conormal derivative $\gamma_1^{\mathscr{A}}(u,f) \in H^{-1/2}(\partial\Omega)$ such that*

$$a(u,v) \;=\; \langle f, v\rangle_\Omega + \langle\gamma_1^{\mathscr{A}}(u,f),\,\gamma_0 v\rangle_{\partial\Omega} \qquad \forall v\in H^1(\Omega) \tag{1.23}$$

and

$$\|\gamma_1^{\mathscr{A}}(u,f)\|_{H^{-1/2}(\partial\Omega)} \;\leq\; C\left(\|u\|_{H^1(\Omega)} + \|f\|_{H^1(\Omega)^*}\right).$$

Proof. See [McL00, Lemma 4.3]. The functional is defined by $\langle\gamma_1^{\mathscr{A}}(u,f), w\rangle := \int_\Omega \mathscr{A}\nabla u\cdot\nabla(\mathscr{E}w)\,\mathrm{d}x - \langle f, \mathscr{E}w\rangle_\Omega$ for $w\in H^{-1/2}(\partial\Omega)$.

We warn the reader explicitly that the definition of $\gamma_1^{\mathscr{A}}(u,f)$ above does not work for any $u\in H^1(\Omega)$ and $f\in H^1(\Omega)^*$ in general, but requires the relating equation $-\operatorname{div}(\mathscr{A}\nabla u) = f$ in $\mathscr{D}^*(\Omega)$.

1.2.2.4 Operators Associated to the Potential Equation

Definition 1.41. (i) For $v\in H^{1/2}(\partial\Omega)$, we define the *PDE-harmonic extension* $\mathscr{H}^a v\in H^1(\Omega)$ as the unique solution of

$$(\mathscr{H}^a v)_{|\partial\Omega} = v, \qquad a(\mathscr{H}^a v, w) \;=\; 0 \qquad \forall w\in H_0^1(\Omega).$$

Hence, $-\operatorname{div}(\mathscr{A}\nabla\mathscr{H}^a v) = 0$ in $\mathscr{D}^*(\Omega)$ and

$$\mathscr{H}^a v = \underset{\substack{\tilde{v}\in H^1(\Omega)\\ \tilde{v}_{|\partial\Omega}=v}}{\operatorname{argmin}}\, a(\tilde{v}, \tilde{v}),$$

where the latter follows from Lemma 1.3. Furthermore, $\mathscr{H}^a : H^{1/2}(\partial\Omega) \to H^1(\Omega)$ is a linear and continuous operator. For $\mathscr{A} = I$ we simply write $\mathscr{H}$ and call it the *harmonic extension operator*. Note that $|\mathscr{H}v|_{H^1(\Omega)}$ is an equivalent seminorm to $|v|_{H^{1/2}(\partial\Omega)}$. A function $v \in H^1(\Omega)$ is called *(PDE-) harmonic* in Ω, if $v = \mathscr{H}^a \gamma_0 v$.

(ii) The linear and continuous map $S := \gamma_1^{\mathscr{A}} \mathscr{H}^a : H^{1/2}(\Omega) \to H^{-1/2}(\partial\Omega)$ is called *Steklov-Poincaré operator*. From

$$\langle S\, v, w\rangle_{\partial\Omega} = a(\mathscr{H}^a v, \mathscr{H}^a w) \qquad \forall v, w \in H^{1/2}(\partial\Omega),$$

we see that S is self-adjoint. Furthermore,

$$|v|_S := \sqrt{\langle S\, v, v\rangle_{\partial\Omega}} = \min_{\substack{\tilde{v}\in H^1(\Omega)\\ \tilde{v}_{|\partial\Omega}=v}} \sqrt{a(\tilde{v}, \tilde{v})}.$$

is a seminorm on $H^{1/2}(\partial\Omega)$. For $\mathscr{A} = I$, $|\cdot|_S$ is equivalent to $|\cdot|_{H^{1/2}(\partial\Omega)}$ with equivalence constants depending only on the shape of Ω.

(iii) We define the *Newton potential*

$$N : H^1(\Omega)^* \to H^{-1/2}(\partial\Omega) : f \mapsto -\gamma_1^{\mathscr{A}}(u_f, f),$$

where $u_f \in H_0^1(\Omega)$ is the unique solution fulfilling

$$a(u_f, v_0) = \langle f, v_0\rangle_\Omega \qquad \forall v \in H_0^1(\Omega).$$

Theorems 1.1 and 1.40 imply that N is linear and continuous.

Note that both operators S and N depend on the coefficient $\mathscr{A}$.

Lemma 1.42. *(i) Any function $u \in H^1(\Omega)$ admits the $a(\cdot,\cdot)$-orthogonal decomposition*

$$u = \mathscr{H}^a \gamma_0 u + u_0,$$

where $u_0 \in H_0^1(\Omega)$. For two functions u, v decomposed like this,

$$a(u, v) = \langle S\, \gamma_0 u, \gamma_0 v\rangle_{\partial\Omega} + a(u_0, v_0).$$

(ii) For $f \in H^1(\Omega)^*$ and $u \in H^1(\Omega)$ with $-\operatorname{div}(\mathscr{A}\nabla u) = f$ in $\mathscr{D}^*(\Omega)$,

$$a(u, v) - \langle f, v\rangle_\Omega = \langle S\, \gamma_0 u, \gamma_0 v\rangle_{\partial\Omega} - \langle Nf, \gamma_0 v\rangle_{\partial\Omega} \qquad \forall v \in H^1(\Omega),$$

and so

$$\gamma_1^{\mathscr{A}}(u, f) = S\gamma_0 u - Nf. \tag{1.24}$$

If $f \in L^2(\Omega)$, then $\gamma_1^{\mathscr{A}} u = S\gamma_0 u - Nf$.

(iii) *If $f \in H^1(\Omega)^*$ is defined by $\langle f, v\rangle_\Omega = \langle g, \gamma_0 v\rangle_{\partial\Omega}$ for some $g \in H^{-1/2}(\partial\Omega)$, then*

$$Nf = g.$$

Proof. The properties follow from Definition 1.41 and Theorem 1.40.

Note that identity (1.24) relates the Cauchy data $(\gamma_0 u, \gamma_1^{\mathscr{A}}(u, f))$. Once having $\gamma_0 u$, the solution in the interior is given as the unique solution of the Dirichlet problem with $\gamma_0 u$ as data.

1.2.3 The Finite Element Method

This subsection deals with the basics of the finite element method (FEM). For more details we refer the reader, e.g., to [Bra01, BS02, Cia87]. Additionally we review the Scott-Zhang quasi-interpolant, high order FEM, and boundary concentrated FEM.

1.2.3.1 Triangulations

From here on, we assume that $\Omega \subset \mathbb{R}^2$ (or $\mathbb{R}^3$) is a Lipschitz polygon (polyhedron), i.e., the boundary is the union of finitely many straight line segments or planar faces. We consider a family $\{\mathscr{T}^h(\Omega)\}_{h\in\Theta}$ of *regular triangulations* of the domain Ω, which means that

(i) $\mathscr{T}^h(\Omega)$ is a finite set of closed, non-degenerate triangles/tetrahedra (the *elements*) such that $\overline{\Omega} = \bigcup_{\tau\in\mathscr{T}^h(\Omega)} \tau$,

(ii) The intersection between two different elements in $\mathscr{T}^h(\Omega)$ is either empty or equal to a vertex, an edge, or a face common to both elements.

The *mesh parameter* h equals the maximal diameter of the elements in $\mathscr{T}^h(\Omega)$. The set of nodes (element vertices) on $\overline{\Omega}$ is denoted by Ω^h, the set of nodes on $\partial\Omega$ by $\partial\Omega^h$. Note that other element types such as quadrilaterals, pyramids, prisms, or hexahedra are in use, but we do not consider them for the sake of simplicity.

Definition 1.43. (i) A family $\{\mathscr{T}^h(\Omega)\}_{h\in\Theta}$ of regular triangulations is called *shape regular* if there exists a uniform constant $c > 0$ (independent of h) such that

$$\rho_\tau \geq c\, h_\tau \qquad \forall \tau \in \mathscr{T}^h(\Omega) \quad \forall h \in \Theta,$$

where h_τ denotes the diameter of the element τ, and ρ_τ the radius of the largest ball inscribed in τ.

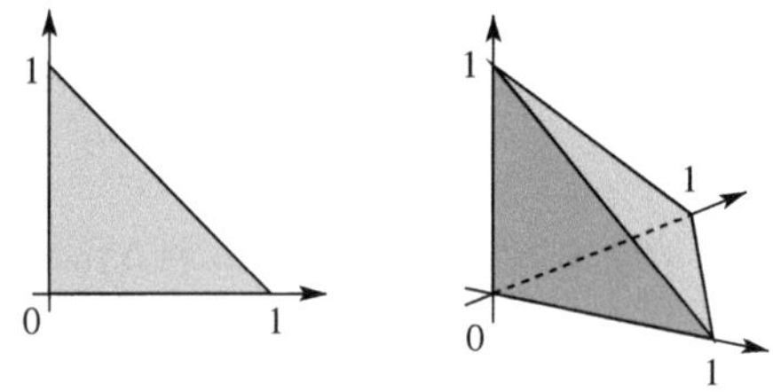

Fig. 1.3 Reference elements. *Left:* reference triangle. *Right:* reference tetrahedron

(ii) A family $\{\mathscr{T}^h(\Omega)\}_{h\in\Theta}$ of regular triangulations is called *quasi-uniform* if it is shape regular and if there exists a uniform constant $c > 0$ such that

$$h_\tau \geq c\,h \qquad \forall\tau \in \mathscr{T}^h(\Omega) \quad \forall h \in \Theta.$$

In the following we will formally consider one *fixed* triangulation $\mathscr{T}^h(\Omega)$ at a time, but have a whole *family* in mind.

Each triangle/tetrahedron in $\mathscr{T}^h(\Omega)$ can be transformed to a *reference element* (see Fig. 1.3) by an affine linear transformation. This fact is used frequently in the analysis and also in implementations.

1.2.3.2 Lowest Order H^1-Conforming Finite Elements

For a given triangulation $\mathscr{T}^h(\Omega)$, we define the finite element (FE) space

$$V^h(\Omega) := \left\{v \in \mathscr{C}(\overline{\Omega}) : v_{|\tau} \in \mathscr{P}_1 \quad \forall\tau \in \mathscr{T}^h(\Omega)\right\}, \tag{1.25}$$

where $\mathscr{P}_1$ is the space of affine linear polynomials. It can be shown that $V^h(\Omega) \subset H^1(\Omega)$. For each node $x^h \in \Omega^h$, the *nodal basis function* $\varphi_{x^h} \in V^h(\Omega)$ is given by

$$\varphi_{x^h}(y^h) = \begin{cases} 1 & \text{for } y^h = x^h, \\ 0 & \text{else,} \end{cases} \tag{1.26}$$

where $y^h \in \Omega^h$ (this defines φ_{x^h} uniquely). Then, $\{\varphi_{x^h}\}_{x^h\in\Omega^h}$ is a basis for V^h.

The linear functionals $\psi_{x^h} : \mathscr{C}(\Omega) \to \mathbb{R} : \psi_{x^h}(v) = v(x^h)$ (that can also be defined locally on each element) are called *nodal variables* or *degrees of freedom* (DOFs), cf. [BS02, Cia87].

1.2.3.3 Application to the Potential Equation

We consider the weak formulation (1.21) of the potential equation with mixed boundary conditions and assume that condition (1.22) holds. For simplicity, we assume that the Dirichlet boundary $\Gamma_D = \overline{\Gamma}_D$ is the union of edges/faces of the elements of $\mathscr{T}^h$. Also, we assume that the Dirichlet data g_D is the trace of a finite element function $g_h \in V^h(\Omega)$. Then the Galerkin projection to the above FE space reads

$$\text{find } u_h \in g_h + V_D^h(\Omega): \quad a(u_h, v_h) = \langle F, v_h \rangle_\Omega \qquad \forall v_h \in V_D^h(\Omega), \tag{1.27}$$

where $V_D^h(\Omega) := \{v_h \in V^h(\Omega) : v_{h|\Gamma_D} = 0\}$.

Let $x_1, \ldots, x_{n_F}$ be the *free* nodes on $\Omega^h \setminus \Gamma_D$. Then $\{\varphi_{x_i}\}_{i=1}^{n_F}$ is a basis of $V_D^h(\Omega)$, i.e., each function $v_h \in V_D^h(\Omega)$ is associated to a vector $\mathbf{v} = (v_1, \ldots, v_{n_F})^\top \in \mathbb{R}^{n_F}$ (and vice versa) such that

$$v_h(x) = \sum_{i=1}^{n_F} v_i \, \varphi_{x_i}(x).$$

This map is known as *Ritz isomorphism.* Note that $v_i = v_h(x_i)$. To express this relation for general vectors or functions, we write in short

$$\mathbf{v} \leftrightarrow v_h \,.$$

The *stiffness matrix* **K** and the *load vector* **f** are given by their entries

$$K_{ij} = a(\varphi_{x_j}, \varphi_{x_i}), \qquad f_i = \langle F, \varphi_{x_i} \rangle_\Omega - a(g_h, \varphi_{x_i}) \qquad \text{for } i,\, j = 1, \ldots, n_F \,. \tag{1.28}$$

The linear system

$$\text{find } \mathbf{u} \in \mathbb{R}^{n_F}: \quad \mathbf{K}\mathbf{u} = \mathbf{f} \tag{1.29}$$

is then equivalent to (1.27) in the sense that $u_h = g_h + \sum_{i=1}^{n_F} u_i \, \varphi_{x_i}$. Usually, one chooses g_h to vanish on the free nodes (this makes it unique). Each entry u_i of the coefficient vector **u** is associated to a node x_i and the corresponding degree of freedom (DOF). Note that in practice, the integrals appearing in the stiffness matrix and load vector are evaluated by numerical quadrature, see [Cia87, Sect. 4.1].

Note that properties of the stiffness matrix **K** will be treated in Sect. 1.2.3.5, solution methods for (1.29) in Sect. 1.2.4.

1.2.3.4 The FE Discretization Error

For the problem treated in the previous section with $\Gamma_D \neq \emptyset$, Lemma 1.5 (Céa) implies

$$\|u - u_h\|_{H^1(\Omega)} \le \frac{\|\mathscr{A}\|_{L^\infty(\Omega)}}{\alpha_{\min}} \left(1 + C_F(\Omega, \Gamma_D)^2\right) \inf_{v_h \in V_D^h(\Omega)} \|u - g_h - v_h\|_{H^1(\Omega)}.$$

The approximation error is treated in the next lemma.

Lemma 1.44. *Let $\mathscr{T}^h(\Omega)$ be a shape regular triangulation, and assume that $u \in H^m(\Omega)$ for some $m \in [1, 2]$. Then there exists a positive constant C depending only on the shape regularity constant and on m such that*

$$\inf_{v_h \in V^h(\Omega)} \|u - v_h\|_{H^k(\Omega)} \leq C\, h^{m-k}\, |u|_{H^m(\Omega)} \qquad \forall k \in [0,\, 1],$$

where $V^h(\Omega)$ is defined in (1.25)*. The inequality also holds for real Sobolev indices $m \in [1,\, 2]$ and $k \in [0,\, 1]$.*

Proof. For a proof see [BS02, Sect. 4.4]. It uses a nodal interpolator, the affine linear map to the reference element, a particular version of the Bramble-Hilbert lemma, and space interpolation.

For $k = 1$, the lemma states that if the solution u has regularity $H^{1+\delta}(\Omega)$, where $\delta \in (0, 1]$, we get a convergence proportional to h^δ. For $u \in H^1(\Omega)$, one can still show by a density argument that

$$\|u - u_h\|_{H^1(\Omega)} \to 0 \qquad \text{as } h \to 0.$$

Discretization estimates in the L^2-norm can be shown using the Aubin-Nitsche trick (see, e.g., [BS02, Sect. 5.8]) under further regularity assumptions.

For Lipschitz polytopes, the regularity of solutions depends in particular on the interior angles between the edges/faces of Ω and on the boundary conditions. A short overview can be found in [TW05, Sect. A.8]; for a comprehensive treatment we refer the reader to [Dau88, Gri92].

1.2.3.5 Properties of the Stiffness Matrix

Due to the fact that the basis functions have local support and due to the structure of the bilinear form $a(\cdot, \cdot)$, the stiffness matrix is *sparse*, i.e., there are only a few non-zero entries per row. From Ritz' isomorphism $\mathbf{u} \leftrightarrow u_h$, $\mathbf{v} \leftrightarrow v_h$ we can conclude that

$$(\mathbf{K}\,\mathbf{u},\, \mathbf{v})_{\ell^2} = a(u_h, v_h) \qquad \forall u_h, v_h \in V_D^h(\Omega).$$

Consequently, $\mathbf{K}$ is symmetric and positive definite. In can be shown that for a quasi-uniform triangulation $\mathscr{T}^h(\Omega)$, the spectral condition number $\kappa(\mathbf{K})$ of the stiffness matrix fulfills

$$\kappa(\mathbf{K}) = \mathscr{O}(h^{-2}) \qquad \text{as } h \to 0, \tag{1.30}$$

i.e., it grows when decreasing the mesh size, or equivalently when increasing the number of DOFs. For the sharpness of this bound see, e.g., [BS02]. In addition, $\kappa(\mathbf{K})$ usually grows linearly with the ratio $\frac{\|\mathscr{A}\|_{L^\infty(\Omega)}}{\alpha_{\min}}$.

1.2.3.6 The Scott-Zhang Quasi-Interpolation Operator

The nodal interpolation operator $I^h : C(\overline{\Omega}) \to V^h(\Omega)$, given by the relation

$$(I^h v)(x^h) = v(x^h) \qquad \text{for all nodes } x^h \in \overline{\Omega},$$

is not bounded in $H^1(\Omega)$. The quasi-interpolant by Clément [Clé75] defines the point value at node x^h as the mean value over the *node patch*

$$\omega_{x^h} := \bigcup_{\tau \in \mathcal{T}^h(\Omega):\, x^h \in \tau} \tau.$$

This way, the interpolator is continuous from $H^1(\Omega)$ to $V^h(\Omega)$ and encounters an L^2 approximation property. This result was further refined by Scott and Zhang [SZ90] (see also [BS02]): by averaging over faces rather than patches and using a dual basis, piecewise linear boundary values can be preserved (property (ii) in the lemma below) while still having stability and approximation (property (iii) below).

Lemma 1.45. *For a shape regular triangulation $\mathcal{T}^h(\Omega)$, there exists a quasi-interpolation operator $\Pi^h : H^1(\Omega) \to V^h(\Omega)$ with the following properties.*

(i) We have the projection property $\Pi^h v_h = v_h$ for all $v_h \in V^h(\Omega)$.
(ii) For $u \in H^1(\Omega)$, the values $(\Pi^h u)_{|\partial\Omega}$ only depend on $u_{|\partial\Omega}$, and

$$(\Pi^h u)_{|\partial\Omega} = u_{|\partial\Omega} \qquad \forall u \in H^1(\Omega),\ u_{|\partial\Omega} \in V^h(\partial\Omega),$$

where $V^h(\partial\Omega) := \{v_{h|\partial\Omega} : v_h \in V^h(\Omega)\}$.
(iii) There exists a uniform constant $C_{SZ} > 0$, depending only on the shape regularity constant of $\mathcal{T}^h(\Omega)$ such that for all $0 \le \ell \le k \le 2$,

$$|u - \Pi^h u|_{H^\ell(\tau)} \le C_{SZ}\, h^{k-\ell}\, |u|_{H^k(\omega_\tau)} \qquad \forall \tau \in \mathcal{T}^h(\Omega),\ \forall u \in H^k(\Omega).$$

where ω_τ is the union of elements (in Ω) touching τ. The last estimate holds true even for real Sobolev indices k and ℓ. In particular, we have the stability estimates

$$\left.\begin{aligned} |\Pi^h u|_{H^1(\Omega)} &\le C_{SZ}\, |u|_{H^1(\Omega)} \\ \|\Pi^h u\|_{L^2(\Omega)} &\le C_{SZ}\, \|u\|_{L^2(\Omega)} \end{aligned}\right\} \quad \forall u \in H^1(\Omega).$$

Remark 1.46. We see that the result of Lemma 1.44 can also be achieved using the Scott-Zhang quasi-interpolant.

1.2.3.7 High Order FEM

The high order FEM or *hp*-FEM was introduced in [BG86, BS87], see also the monographs [Dem07, KS99, Mel02, Sch98b].

Let $\mathcal{T}^h(\Omega)$ be a shape regular triangulation of Ω. For each element $\tau \in \mathcal{T}^h(\Omega)$ we choose a polynomial degree $p_\tau \in \mathbb{N}$. The *polynomial degree vector* is denoted by

$$\mathbf{p} := (p_\tau)_{\tau \in \mathcal{T}^h(\Omega)} .$$

The corresponding H^1-conforming high order finite element space is defined by

$$V^{hp}(\Omega) := \{v \in \mathcal{C}(\overline{\Omega}) : (v \circ F_\tau)_{|\hat{\tau}} \in \mathcal{P}_{p_\tau} \quad \forall \mathcal{T}^h(\Omega)\},$$

where $\mathcal{P}_k$ are the polynomials (in d variables) of total degree at most k, $\widehat{\tau}$ denotes the reference element, and $F_\tau : \widehat{\tau} \to \tau$ is the (bijective) affine linear map.[2]

Obviously, the continuity constraint in $V^{hp}(\Omega)$ implies that on a face f (edge if $d = 2$) between two elements τ_1, τ_2 we have to limit the polynomial degree to

$$p_f := \min(p_{\tau_1}, p_{\tau_2}),$$

i.e., $(v \circ F_{\tau_1})_{|F_{\tau_1}^{-1}(f)} \in \mathcal{P}_{p_f}$ and $(v \circ F_{\tau_2})_{|F_{\tau_2}^{-1}(f)} \in \mathcal{P}_{p_f}$. Analogously, the polynomial degree on a three-dimensional edge is the minimum p_e of the degrees on all the adjacent elements.

A hierarchical construction of a basis for $V^{hp}(\Omega)$ using vertex-, edge-, face-, and cell-based basis functions can, e.g., be found in [Dem07, Zag06].

If the solution is sufficiently regular, high-order methods achieve a much better approximation than low order FEM, in fact in case of high regularity, one can expect exponential convergence, cf. [Sch98b]. Hence, a good strategy to save degrees of freedom would be to use large elements with high order polynomial degree where the solution is (known to be) smooth and small elements with low order polynomial degree on the remainder. Several strategies for adaptive *hp*-refinement are around, see, e.g., [AS97, DORH89, ODRW89, ROD89, EM07a, HS05]. An a priori choice is made by the boundary concentrated FEM.

1.2.3.8 Boundary Concentrated FEM

The boundary concentrated FEM was introduced in [KM03] and further developed in [Eib06, EM06, EM07b]. The essential idea is based on the fact that analyticity of the coefficients and the right hand side guarantee that the solution is smooth in

---

[2]Often, the map F_τ itself is a polynomial of degree $\leq p_\tau$, which allows physical elements with curved boundaries.

the interior of Ω. Since the solution may have low regularity at the boundary, a safe choice is to use a mesh that is refined towards the boundary.

Definition 1.47 (geometric mesh). A shape regular triangulation $\mathscr{T}(\Omega)$ is called a *geometric mesh* with *boundary mesh size* $h > 0$ if there exist constants $c_1, c_2 > 0$ such that for all $\tau \in \mathscr{T}(\Omega)$,

(i) If $\quad \tau \cap \partial\Omega \neq \emptyset \quad$ then $\quad c_1\, h \le h_\tau \le h,$

(ii) If $\quad \tau \cap \partial\Omega = \emptyset \quad$ then $\quad c_1 \inf_{x\in\tau} \operatorname{dist}(x, \partial\Omega) \le h_\tau \le c_2 \sup_{x\in\tau} \operatorname{dist}(x, \partial\Omega).$

Property (i) from above implies that the corresponding boundary mesh $\mathscr{T}(\partial\Omega)$ is quasi-uniform with mesh parameter h.

Definition 1.48 (linear degree vector). Let $\mathscr{T}(\Omega)$ be a geometric mesh with boundary mesh size h and let $\mathbf{p} = (p_\tau)_{\tau\in\mathscr{T}(\Omega)}$ a polynomial degree vector. We call $\mathbf{p}$ a *linear degree vector* with *slope* $\alpha > 0$ if

$$1 + c_1\, \alpha \, \log(\tfrac{h_\tau}{h}) \;\le\; p_\tau \;\le\; 1 + c_2\, \alpha \, \log(\tfrac{h_\tau}{h}),$$

where c_1 and c_2 are the constants from Definition 1.47.

In practical applications, one usually sets $\alpha = 1$. In particular, the polynomial degree equals one in the elements touching the boundary.

Due to [KM03, Proposition 2.7], there exists a constant $C > 0$ depending on the shape regularity constant of $\mathscr{T}(\Omega)$ and on the constants in Definitions 1.47 and 1.48 such that

$$\begin{aligned} \sum_{\tau\in\mathscr{T}(\Omega)} 1 \;&\le\; C\,(H/h)^{d-1}, \\ \dim(V^{hp}(\Omega)) \;&\le\; C\,(H/h)^{d-1}, \\ \max_{\tau\in\mathscr{T}(\Omega)} p_\tau \;&\le\; C\,(1 + \log(H/h)), \end{aligned}$$

where $H = \operatorname{diam}(\Omega)$. In particular, there are essentially as many elements in $\mathscr{T}(\Omega)$ and DOFs in $V^{hp}(\Omega)$, as there are elements on the boundary $\partial\Omega$, and the maximal polynomial degree grows only logarithmically with the number of elements on the boundary. The number of DOFs is therefore comparable to that of a boundary element method (see Sect. 1.3.7).

In case that the coefficients and the right hand side of the PDE are sufficiently smooth (analytic), the discretization error of the boundary concentrated FEM is of the same order of magnitude as the discretization error of a low-order FEM on a quasi-uniform triangulation $\mathscr{T}^h(\Omega)$. Due to [BEL08, Theorem 3.13], the Schur complement of the hp-stiffness matrix is spectrally equivalent to the Schur complement of the stiffness matrix of the quasi-uniform FEM.

1.2.4 *Solvers for Linear Systems*

In order to solve a linear system of the form

$$\mathbf{K}\,\mathbf{u} = \mathbf{f},$$

with a sparse stiffness matrix $\mathbf{K} \in \mathbb{R}^{n \times n}$, we can use two paradigms.

Direct solvers first compute an *LU* or Cholesky factorization of $\mathbf{K} = \mathbf{L}\,\mathbf{U}$. The problem of solving $\mathbf{K}\,\mathbf{u} = \mathbf{f}$ is then reduced to solving the two triangular systems $\mathbf{L}\,\mathbf{v} = \mathbf{f}$ and $\mathbf{U}\,\mathbf{u} = \mathbf{v}$.

Iterative solvers rely on the fact that the *application* of $\mathbf{A}$ can be performed in optimal complexity, they produce a sequence of vectors that converges to the solution $\mathbf{u}$. The most well-known iterative solvers are *Richardson's method*, the method of steepest descent, and the *conjugate gradient method*, as well as their preconditioned versions. For symmetric indefinite or non-symmetric problems, we mention MinRes, BICGStab, and GMRes. Advanced preconditioners for elliptic problems are domain decomposition preconditioners (including overlapping Schwarz and Schur complement preconditioners) [KL04, Mat08, SBG96, QV99, TW05] as well as multilevel and multigrid preconditioners [BZ00, Hac03, KM07, Vas08].

In the following, we briefly treat direct solvers as well as the preconditioned conjugate gradient (PCG) method (which is well-suited for SPD problems).

1.2.4.1 Direct Solvers

As well-known, standard Gaussian elimination computes the *LU* factorization of a matrix of dimension n in $\frac{2}{3}n^3 + \mathcal{O}(n^2)$ arithmetic operations. The Cholesky factorization (for SPD matrices) has the same asymptotic complexity. Typically, the (lowest order) FE stiffness matrix $\mathbf{K}$ has a *band structure*. If the domain Ω is the unit square/cube, discretized by a structured mesh with a suitable numbering of the nodes/DOFs, the band width is $\mathcal{O}(h^{-d-1}) = \mathcal{O}(n_h^{(d-1)/d})$, where n_h is the total number of DOFs. Exploiting this band structure, the complexity of the Gauss algorithm can be reduced, cf. Table 1.1. The resulting complexity in two and three dimensions is, however, still not satisfactory.

Even for a sparse matrix the *fill-in* phenomenon arises: the factors L, U are dense. Nevertheless, there exist pivoting techniques minimizing the fill-in, e.g., *nested dissection*; cf. [GL81]. The resulting complexity is displayed in the last row of Table 1.1. Note that, as a special case, the finite difference discretization of the two-dimensional Laplacian on a square leads to the further reduced complexity $\mathcal{O}(n_h \log n_h)$. Note also that the table shows the computational complexity for *factorizing* the matrix. The complexity of pure *solving* is smaller in many cases.

Table 1.1 Computational complexity of direct solvers for one-, two-, and three-dimensional model problems (Source: [Zum03, p. 23])

Gauss	$d = 1$	$d = 2$	$d = 3$
Dense matrix	$\mathcal{O}(n_h^3)$	$\mathcal{O}(n_h^3)$	$\mathcal{O}(n_h^3)$
Using band structure	$\mathcal{O}(n_h)$	$\mathcal{O}(n_h^2)$	$\mathcal{O}(n_h^{7/3})$
Using sparsity	$\mathcal{O}(n_h)$	$\mathcal{O}(n_h^{3/2})$	$\mathcal{O}(n_h^2)$

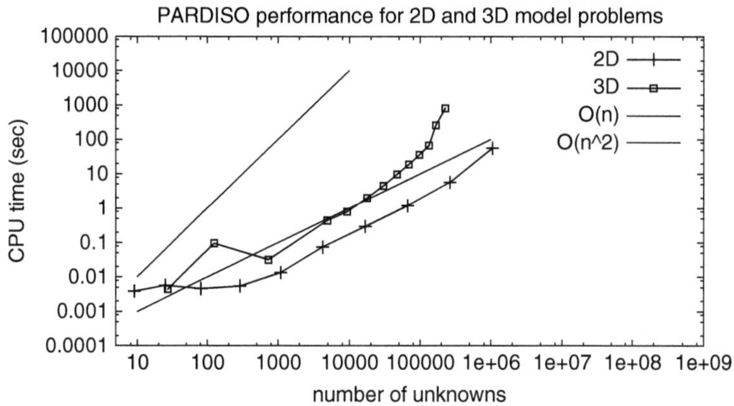

Fig. 1.4 CPU times of the factorization and solve of a finite element system by PARDISO 4.0 for a two- and three-dimensional model problem, computed on a laptop (double-logarithmic scale)

For implementations of such sparse direct solvers we mention (in alphabetical order)

MUMPS (multifrontal massively parallel sparse direct solver):
http://graal.ens-lyon.fr/MUMPS/
PARDISO (parallel sparse direct solver):
http://www.pardiso-project.org/
SUPERLU:
http://crd.lbl.gov/~xiaoye/SuperLU/
SPOOLES (sparse object oriented linear equations solver):
http://www.netlib.org/linalg/spooles/
UMFPACK (UMF = unsymmetric multifrontal):
http://www.cise.ufl.edu/research/sparse/umfpack/

In practice, these solvers work rather well (with almost linear preasymptotic complexity) until the number of unknowns goes beyond a certain barrier. Figure 1.4 shows the CPU times of PARDISO for a two- and three-dimensional model problem. One can see that for 100,000 unknowns in two dimensions as well as 10,000 unknowns in three dimensions, the algorithm has quite a good runtime complexity. In practice, memory is the most crucial bottleneck: in the study shown in Fig. 1.4, the factorization of problems larger than the displayed ones could not be stored in memory anymore.

Algorithm 1: PCG method

$\mathbf{x}_0$ given
$\mathbf{r}_0 = \mathbf{b} - \mathbf{A}\,\mathbf{x}_0$
$k = 0$
repeat
  $\mathbf{s}_k = \mathbf{C}^{-1}\mathbf{r}_k$
  $\mathbf{p}_k = \mathbf{s}_k + \beta_{k-1}\,\mathbf{p}_{k-1}$ where $\beta_{-1} = 0,\ \beta_{k-1} = \frac{(\mathbf{r}_k, \mathbf{s}_k)_{\ell^2}}{(\mathbf{r}_{k-1}, \mathbf{s}_{k-1})_{\ell^2}}$ for $k > 0$
  $\mathbf{x}_{k+1} = \mathbf{x}_k + \alpha_k\,\mathbf{p}_k$ where $\alpha_k = \frac{(\mathbf{r}_k, \mathbf{s}_k)_{\ell^2}}{(\mathbf{A}\,\mathbf{p}_k, \mathbf{p}_k)_{\ell^2}}$
  $\mathbf{r}_{k+1} = \mathbf{r}_k - \alpha_k\,\mathbf{A}\,\mathbf{p}_k$ $(= \mathbf{b} - \mathbf{A}\,\mathbf{x}_{k+1})$
  $k = k + 1$
until *stopping criterion fulfilled*

1.2.4.2 Preconditioned Conjugate Gradients

In this paragraph, we consider the general linear system

$$\mathbf{A}\,\mathbf{x} = \mathbf{b}$$

with a symmetric positive definite (SPD) matrix $\mathbf{A} \in \mathbb{R}^{n\times n}$, a given vector $\mathbf{b} \in \mathbb{R}^n$ and the unknown solution $\mathbf{x} \in \mathbb{R}^n$. We call an SPD matrix $\mathbf{C}^{-1}$ *preconditioner* for the matrix $\mathbf{A}$ if

- Applying $\mathbf{C}^{-1}$ is cheaper than applying $\mathbf{A}^{-1}$,
- $\kappa(\mathbf{C}^{-1}\,\mathbf{A}) \ll \kappa(\mathbf{A})$,

where $\kappa(\cdot)$ denotes the spectral condition number. Note that the matrix $\mathbf{C}^{-1}\mathbf{A}$ is self-adjoint and positive definite with respect to the inner product $(\mathbf{C}\,\cdot,\cdot)_{\ell^2}$ and has positive eigenvalues.

The *preconditioned conjugate gradient (PCG) method* is displayed in Algorithm 1. It is usually the method of choice (for SPD problems) due to its accelerated convergence in comparison to Richardson's method and the method of steepest descent.

Lemma 1.49. *Let* $\mathbf{A}$ *and* $\mathbf{C}^{-1} \in \mathbb{R}^{n\times n}$ *be SPD and let* $\mathbf{x}_0 \in \mathbb{R}^n$ *be given. For the exact solution* $\mathbf{x} = \mathbf{A}^{-1}\mathbf{b}$ *and the sequence* $\mathbf{x}_k$ *created by the PCG method in Algorithm 1,*

$$\|\mathbf{x} - \mathbf{x}_k\|_{\mathbf{A}} \leq 2\left(\frac{\sqrt{\kappa(\mathbf{C}^{-1}\mathbf{A})} - 1}{\sqrt{\kappa(\mathbf{C}^{-1}\mathbf{A})} + 1}\right)^k \|\mathbf{x} - \mathbf{x}_0\|_{\mathbf{A}} \qquad \forall k \geq 1,$$

where $\|v\|_{\mathbf{A}} := \sqrt{(\mathbf{A}\,v, v)_{\ell^2}}$ *is the* energy norm. *For a reduction of the error in the energy norm by a factor of* ε*, at most* $\mathcal{O}(|\log(\varepsilon/2)|\sqrt{\kappa(\mathbf{C}^{-1}\mathbf{A})})$ *iterations are required.*

Proof. See, e.g., [Saa03, SK04]. The main reason for the accelerated convergence is the fact that the search directions are orthogonal with respect to $(\mathbf{A}\cdot,\cdot)_{\ell^2}$, and the error in the energy norm is minimized over the so-called *Krylov space*.

Corollary 1.50. *Let V be a finite-dimensional Hilbert space and consider the operator equation*

$$\text{find } x \in V: \quad A\,x = b,$$

where $A : V \to V^*$ is an SPD operator and $b \in V^*$ is given. In Algorithm 1, replace the matrix $\mathbf{A}$ by A and the preconditioner $\mathbf{C}^{-1}$ by the inverse C^{-1} of an SPD operator $C : V \to V^*$. Replace the vectors $\mathbf{x}_k$, $\mathbf{p}_k$, and $\mathbf{s}_k$ by corresponding elements from V, the residual vectors $\mathbf{r}_k$ by corresponding elements from V^*, and the Euclidean inner product $(\cdot,\cdot)_{\ell^2}$ by the duality product $\langle\cdot,\cdot\rangle_{V^*\times V}$. Then the statement of Lemma 1.49 holds true with the energy norm defined by

$$\|v\|_A := \langle A\,v, v\rangle^{1/2},$$

and the condition number defined by

$$\kappa(C^{-1}A) = \sup_{v\in V}\frac{\langle A\,v, v\rangle}{\langle C\,v, v\rangle} \Big/ \inf_{v\in V}\frac{\langle A\,v, v\rangle}{\langle C\,v, v\rangle}.$$

Proof. The statement follows immediately by choosing a basis for V which encounters an isomorphism between V and $\mathbb{R}^{\dim(V)}$. □

We see that the trivial choice $\mathbf{C} = \mathbf{I}$ in the original algorithm actually corresponds to choosing C^{-1} as the Riesz isomorphism between the dual of $\mathbb{R}^n$ and $\mathbb{R}^n$.

Remark 1.51 (Lanczos' sequence). Let $(\alpha_k)_{k=0}^{n-1}$ and $(\beta_k)_{k=0}^{n-2}$ the coefficients from Algorithm 1 stopped after n steps. Furthermore let us define the tridiagonal matrix

$$\mathbf{L}_n := \begin{bmatrix} \delta_0 & \gamma_0 & & 0 \\ \gamma_0 & \delta_1 & \ddots & \\ & \ddots & \ddots & \gamma_{n-2} \\ 0 & & \gamma_{n-2} & \delta_{n-1} \end{bmatrix} := \begin{bmatrix} \frac{1}{\alpha_0} & -\frac{\sqrt{\beta_0}}{\alpha_0} & & 0 \\ -\frac{\sqrt{\beta_0}}{\alpha_0} & \frac{1}{\alpha_1}+\frac{\beta_0}{\alpha_0} & \ddots & \\ & \ddots & \ddots & -\frac{\sqrt{\beta_{n-2}}}{\alpha_{n-2}} \\ 0 & & -\frac{\sqrt{\beta_{n-2}}}{\alpha_{n-2}} & \frac{1}{\alpha_{n-1}}+\frac{\beta_{n-2}}{\alpha_{n-2}} \end{bmatrix}.$$

It can be shown that if the PCG has "converged", i.e., if the residual $\mathbf{r}_n$ is small, then the extremal eigenvalues $\lambda_{\min}(\mathbf{L}_n)$ and $\lambda_{\max}(\mathbf{L}_n)$ are good approximations of $\lambda_{\min}(\mathbf{C}^{-1}\mathbf{A})$ and $\lambda_{\max}(\mathbf{C}^{-1}\mathbf{A})$. Therefore, $\lambda_{\max}(\mathbf{L}_n)/\lambda_{\min}(\mathbf{L}_n)$ is a good approximation for $\kappa(\mathbf{C}^{-1}\mathbf{A})$. This fact has been widely used in many numerical studies. For further references see, e.g., [PS72, Van82, GV96]. Note that the two extremal eigenvalues can be efficiently computed using the characteristic polynomial $p_n(\lambda) = \det(\mathbf{L}_n - \lambda\,\mathbf{I})$, given by the recursion formula

$$\begin{aligned} p_0(\lambda) &= 1, \\ p_1(\lambda) &= \delta_0 - \lambda, \\ p_k(\lambda) &= (\delta_{k-1} - \lambda)\, p_{k-1}(\lambda) - \gamma_{k-1}^2\, p_{k-2}(\lambda) \qquad \text{for } k \geq 2. \end{aligned}$$

The Newton method for solving $p_n(\lambda) = 0$ started at 0 delivers $\lambda_{\min}(\mathbf{L}_n)$, and started at the row-sum norm $\|\mathbf{L}_n\|_\infty$, it delivers $\lambda_{\max}(\mathbf{L}_n)$.

1.2.5 The Pure Neumann Problem

1.2.5.1 Continuous Setting

If $\Gamma_D = \emptyset$, i.e., $\Gamma_N = \partial\Omega$, the variational problem

$$\text{find } u \in H^1(\Omega): \quad a(u, v) = \langle F, v \rangle \qquad \forall v \in H^1(\Omega) \tag{1.31}$$

is not uniquely solvable. Testing the variational equation with the constant function 1_Ω we obtain the *compatibility condition*

$$\langle F, 1 \rangle_\Omega = 0, \tag{1.32}$$

i.e., for the assumptions in Sect. 1.2.2.2,

$$\int_\Omega f \,\mathrm{d}x + \langle g_N, 1 \rangle_{\partial\Omega} = 0.$$

If the data (f, g_N) fulfills this condition, the solution is u unique up to an additive constant. A particular solution can be found by fixing some *gauge*, e.g.,

$$\int_\Omega u \,\mathrm{d}x = 0. \tag{1.33}$$

Using the regularized (coercive) bilinear form

$$\tilde{a}(u, v) := a(u, v) + \beta \int_\Omega u \,\mathrm{d}x \int_\Omega v \,\mathrm{d}x,$$

for some regularization parameter $\beta > 0$, the Galerkin equation

$$\text{find } \tilde{u} \in H^1(\Omega): \quad \tilde{a}(\tilde{u}, v) = \langle F, v \rangle \qquad \forall v \in H^1(\Omega)$$

is always solvable. If the data (f, g_N) fulfills the compatibility condition, its solution $\tilde{u}$ solves (1.31) and satisfies the gauge condition (1.33).

1.2.5.2 Discrete Setting

The finite element system corresponding to (1.31) reads

$$\mathbf{K}\,\mathbf{u} = \mathbf{f}.$$

The kernel of the stiffness matrix $\mathbf{K}$ has dimension one and is spanned by the vector $\mathbf{1}$ of ones. We assume that the compatibility condition $(\mathbf{f},\, \mathbf{1})_{\ell^2} = 0$ holds. In order to get a pseudo inverse of $\mathbf{K}$ as in Sect. 1.1.3.4, we need to choose a single vector $\mathbf{r}$ corresponding to a functional $r \in V_h(\Omega)^*$ and use the regularized matrix

$$\widetilde{\mathbf{K}} := \mathbf{K} + \beta\,\mathbf{r}\,\mathbf{r}^\top.$$

In order to preserve the sparsity, it is recommendable to use a functional with small support ($\mathbf{r}$ should have only a few entries). Let y^h be an arbitrary node of the mesh, then the functional

$$\langle r, v\rangle := v(y^h) \qquad \text{for } v \in V^h(\Omega) \tag{1.34}$$

corresponds to a vector $\mathbf{r}$ with just one entry, and so $\widetilde{\mathbf{K}}$ is created from $\mathbf{K}$ by simply adding β to the diagonal entry corresponding to y^h.

The matrix $\widetilde{\mathbf{K}}^{-1}$ is a pseudo inverse of $\mathbf{K}$. Since $\widetilde{\mathbf{K}}$ is sparse and has a condition number comparable to that of a stiffness matrix for the Dirichlet problem, we can use the same kind of solvers. The following lemma discusses the conditioning of $\widetilde{\mathbf{K}}$.

Lemma 1.52. *Let $\mathscr{T}^h(\Omega)$ be quasi-uniform, let the matrix $\widetilde{\mathbf{K}}$ be constructed as above using* (1.34)*, and assume that*

$$\alpha_{\min}\, h^{d-2} \;\lesssim\; \beta \;\lesssim\; \|\mathscr{A}\|_{L^\infty(\Omega)}\, h^{d-2}.$$

Then

$$\kappa(\widetilde{\mathbf{K}}) \;\le\; \begin{cases} C\,\dfrac{\|\mathscr{A}\|_{L^\infty(\Omega)}}{\alpha_{\min}}\, h^{-2}(1+\log(h^{-1})) & \text{if } d = 2,\\[2ex] C\,\dfrac{\|\mathscr{A}\|_{L^\infty(\Omega)}}{\alpha_{\min}}\, h^{-3} & \text{if } d = 3.\end{cases}$$

Proof. The proof uses the discrete Poincaré type inequality from Lemma 2.70, p. 121.

The above lemma justifies to choose β as the diagonal entry of $\mathbf{K}$ corresponding to y^h, which makes the implementation easy. The increase of the condition number compared to a Dirichlet problem will not dramatically affect the numerical stability of the direct solver.

An alternative to factorizing $\widetilde{\mathbf{K}}$ is solving the (sparse) saddle point problem

$$\begin{bmatrix} \mathbf{K} & \mathbf{r}\\ \mathbf{r}^\top & 0\end{bmatrix}\begin{bmatrix}\mathbf{u}\\ \boldsymbol{\lambda}\end{bmatrix} = \begin{bmatrix}\mathbf{f}\\ 0\end{bmatrix}.$$

Direct solvers like PARDISO or UMFPACK are able to treat such problems efficiently.

1.2.6 The Schur Complement System

In this subsection, we will eliminate degrees of freedom from the stiffness matrix that are in the interior of the domain, thus leading to a Schur complement matrix. As we will see there is a relation between this matrix and a discrete harmonic extension.

1.2.6.1 Algebraic Definition

Let $\mathbf{K}$ denote the stiffness matrix corresponding to the pure Neumann problem (1.31). We reorder and partition $\mathbf{K}$ into DOFs corresponding to nodes on the boundary $\partial\Omega$ (subscript B) and the remaining interior DOFs (subscript I):

$$\mathbf{K} = \begin{bmatrix} \mathbf{K}_{BB} & \mathbf{K}_{BI} \\ \mathbf{K}_{IB} & \mathbf{K}_{II} \end{bmatrix}.$$

Furthermore, we denote by $\mathbf{f}$ be the load vector contribution corresponding to the functional $\int_\Omega f \cdot \mathrm{d}x$ and partition it into the blocks $\mathbf{f}_B$ and $\mathbf{f}_I$. Finally, let $\mathbf{t}$ be the "B"-block of the load vector contribution corresponding to the Neumann data $g_N \in H^{-1/2}(\partial\Omega)$. If we partition the solution vector $\mathbf{u}$ as well, then the discrete system takes the form

$$\begin{bmatrix} \mathbf{K}_{BB} & \mathbf{K}_{BI} \\ \mathbf{K}_{IB} & \mathbf{K}_{II} \end{bmatrix} \begin{bmatrix} \mathbf{u}_B \\ \mathbf{u}_I \end{bmatrix} = \begin{bmatrix} \mathbf{f}_B + \mathbf{t} \\ \mathbf{f}_I \end{bmatrix}. \tag{1.35}$$

It can be seen as a discrete analogue of the generalized Green identity (1.23) from Theorem 1.40. Suppose, we have a solution $\mathbf{u}_I$ to the pure Dirichlet problem with given Dirichlet data $\mathbf{u}_B$, then the discrete conormal derivative is given by

$$\mathbf{t} = \mathbf{K}_{BB}\,\mathbf{u}_B + \mathbf{K}_{BI}\,\mathbf{u}_I - \mathbf{f}_B\,.$$

Substituting the second equation into this formula we obtain

$$\mathbf{t} = \mathbf{S}\,\mathbf{u}_B - \mathbf{N}\mathbf{f} \tag{1.36}$$

with the *Schur complement matrix* $\mathbf{S}$ and the matrix $\mathbf{N}$ given by

$$\mathbf{S} := \mathbf{K}_{BB} - \mathbf{K}_{BI}\,(\mathbf{K}_{II})^{-1}\,\mathbf{K}_{IB}, \tag{1.37}$$

$$\mathbf{N} := \left[\mathbf{I}_B \middle| -\mathbf{K}_{BI}\,(\mathbf{K}_{II})^{-1}\right]. \tag{1.38}$$

We see that $\mathbf{S}$ is a discrete analogue to the Steklov-Poincaré operator and $\mathbf{N}$ to the Newton potential. The system

$$\mathbf{S}\,\mathbf{u}_B \;=\; (\mathbf{f}_B + \mathbf{t}) - \mathbf{K}_{BI}(\mathbf{K}_{II})^{-1}\mathbf{f}_I \tag{1.39}$$

which is equivalent to (1.36) is also called *Schur complement system.*

1.2.6.2 Relation to the PDE-Harmonic Extension

Apparently, the "B" unknowns parametrize the space $V^h(\partial\Omega)$. Hence, there is a one-to-one correspondence (a Ritz isomorphism) between functions $v_h \in V^h(\partial\Omega)$ and vectors $\mathbf{v}$ in the "B" block. Again we write $\mathbf{v} \leftrightarrow v_h$ as a short hand.

Definition 1.53. We define $S_{\mathrm{FEM}} : V^h(\partial\Omega) \to V^h(\partial\Omega)^*$ by

$$\langle S_{\mathrm{FEM}} v_h, w_h\rangle \;=\; (\mathbf{S}\,\mathbf{v}, \mathbf{w})_{\ell^2} \qquad \forall \mathbf{v} \leftrightarrow v_h \in V^h(\partial\Omega),\ \mathbf{w} \leftrightarrow w_h \in V^h(\partial\Omega).$$

Lemma 1.54. *We have the minimizing property*

$$\langle S_{FEM} v_h, v_h\rangle \;=\; (\mathbf{S}\,\mathbf{v}, \mathbf{v})_{\ell^2} \;=\; \min_{\substack{\tilde v_h \in V^h(\Omega)\\ \tilde v_{h|\partial\Omega} = v_h}} a(\tilde v_h, \tilde v_h) \qquad \forall \mathbf{v} \leftrightarrow v_h \in V^h(\partial\Omega). \tag{1.40}$$

Proof. Follows from the definition of $\mathbf{S}$ and from Lemma 1.3.

Definition 1.55. The *discrete PDE-harmonic extension* of $v_h \in V^h(\partial\Omega)$ is given by

$$\mathscr{H}^{a,h} v_h := \operatorname*{argmin}_{\substack{\tilde v_h \in V^h(\Omega)\\ \tilde v_{h|\partial\Omega} = v_h}} a(\tilde v_h, \tilde v_h).$$

If $\mathscr{A} = I$, we simply write $\mathscr{H}^h$. It can be shown that

$$(\mathbf{S}\,\mathbf{v}, \mathbf{w})_{\ell^2} \;=\; a(\mathscr{H}^{a,h} v_h, \mathscr{H}^{a,h} w_h) \qquad \forall \mathbf{v} \leftrightarrow v_h \in V^h(\partial\Omega),\ \mathbf{w} \leftrightarrow w_h \in V^h(\partial\Omega).$$

Lemma 1.56 (Existence of a bounded discrete extension). *Let $\mathscr{A} = I$ and let $\mathscr{T}^h(\Omega)$ be shape regular. Then there exists a bounded linear extension operator $\mathscr{E}^h : V^h(\partial\Omega) \to V^h(\Omega)$ such that*

$$(\mathscr{E}^h v_h)_{|\partial\Omega} = v_h, \qquad |\mathscr{E}^h v_h|^2_{H^1(\Omega)} \le C_{SZ}^2 \,\langle S\, v_h, v_h\rangle \qquad \forall v_h \in V^h(\partial\Omega).$$

where C_{SZ} is the constant from Lemma 1.45 and S is the Steklov-Poincaré operator from Definition 1.41.

Proof. The operator $\mathscr{E}^h := \Pi^h \mathscr{H}$ fulfills the stated properties. Here, Π^h is the Scott-Zhang operator and $\mathscr{H}$ is the harmonic extension. □

Corollary 1.57. *Let $\mathscr{A} = I$ and let $\mathscr{T}^h(\Omega)$ be shape regular. Then, there exist positive constants c, C that depend only on the shape of Ω and on C_{SZ} such that*

$$\left.\begin{array}{rl} \langle S\, v_h, v_h \rangle_{\partial\Omega} \le \langle S_{FEM} v_h, v_h \rangle_{\partial\Omega} & \le\; C_{SZ}^2 \, \langle S\, v_h, v_h \rangle_{\partial\Omega} \\ c\, |v_h|^2_{H^{1/2}(\partial\Omega)} \le \langle S_{FEM}\, v_h, v_h \rangle_{\partial\Omega} & \le\; C\, |v_h|^2_{H^{1/2}(\partial\Omega)} \end{array}\right\} \quad \forall v_h \in V^h(\partial\Omega).$$

Proof. The minimization properties of S and S_{FEM} yield

$$\langle S\, v_h, v_h \rangle_{\partial\Omega} = \min_{\substack{\tilde{v} \in H^1(\Omega) \\ \tilde{v}_{|\partial\Omega} = v_h}} |\tilde{v}|^2_{H^1(\Omega)} \le \min_{\substack{\tilde{v}_h \in V^h(\Omega) \\ \tilde{v}_{h|\partial\Omega} = v_h}} |\tilde{v}_h|^2_{H^1(\Omega)} \le |\mathscr{E}^h v_h|^2_{H^1(\Omega)},$$

where the third term above equals $\langle S_{\mathrm{FEM}} v_h, v_h \rangle_{\partial\Omega}$. The rest of the proof follows from Lemma 1.56 and from Definition 1.41(ii). □

Remark 1.58. A similar result to that of Lemma 1.56 is the *discrete trace theorem*: for each function $v_h \in V^h(\partial\Omega)$ there exists a function $\tilde{v}_h \in V^h(\Omega)$

$$|\tilde{v}_h|^2_{H^1(\Omega_k)} \le C_t \, |v_h|^2_{H^{1/2}(\partial\Omega)},$$

for a constant $C_t > 0$ depending on the shape of Ω and the shape regularity constant of $\mathscr{T}^h(\Omega)$. For a proof using regularity theory see [TW05, Lemma 4.6].

1.2.6.3 Conditioning of the Schur Complement

Lemma 1.59. *Let $\Gamma_D \subset \partial\Omega$ have positive surface measure, let condition* (1.22) *hold, and assume that the triangulation $\mathscr{T}(\partial\Omega)$ is quasi-uniform. Then, for all vectors $\mathbf{v} \leftrightarrow v_h \in V^h(\partial\Omega)$, $v_{h|\Gamma_D} = 0$,*

$$c\, \alpha_{\min} \operatorname{diam}(\Omega)^{-1} h^{d-1} \, \|\mathbf{v}\|^2_{\ell^2} \le (\mathbf{S}\,\mathbf{v}, \mathbf{v})_{\ell^2} \le C\, \|\mathscr{A}\|_{L^\infty(\Omega)} \, h^{d-2} \, \|\mathbf{v}\|^2_{\ell^2}.$$

The constants c, C depend on the shape regularity and quasi-uniformity constants of $\mathscr{T}(\partial\Omega)$ and the Friedrichs constant $C_F(\partial\Omega, \Gamma_D)$ from Lemma 1.36.

Proof. Using condition (1.22) and Corollary 1.57, one shows

$$c\, \alpha_{\min} \, |v_h|^2_{H^{1/2}(\partial\Omega)} \le (\mathbf{S}\,\mathbf{v}, \mathbf{v})_{\ell^2} \le \|\mathscr{A}\|_{L^\infty(\Omega)} \, C\, |v_h|^2_{H^{1/2}(\partial\Omega)},$$

where c and C are generic constants. Thanks to an inverse inequality

$$|v_h|^2_{H^{1/2}(\partial\Omega)} \le C\, h^{-1} \, \|v_h\|^2_{L^2(\partial\Omega)},$$

see, e.g., [TW05, Lemma B.27]. Lemma 1.36 implies

$$|v_h|^2_{H^{1/2}(\partial\Omega)} \geq C_F(\partial\Omega, \Gamma_D)^{-1} \operatorname{diam}(\Omega)^{-1} \|v_h\|^2_{L^2(\partial\Omega)}.$$

The shape regularity and quasi-uniformity of the mesh implies that $\|v_h\|_{L^2(\partial\Omega)} \approx h^{(d-1)/2} \|\mathbf{v}\|_{\ell^2}$. This concludes the proof. □

Let $\overline{\mathbf{S}}$ be the submatrix of $\mathbf{S}$ corresponding to the non-Dirichlet DOFs. Lemma 1.59 implies that

$$\kappa(\overline{\mathbf{S}}) \leq C \frac{\|\mathscr{A}\|_{L^\infty(\Omega)}}{\alpha_{\min}} \frac{\operatorname{diam}(\Omega)}{h},$$

i.e., $\mathscr{O}(h^{-1})$ as $h \to \infty$. Later on, we will also need an estimate for functions which have a certain vanishing average.

Lemma 1.60. *(i) Let the triangulation $\mathscr{T}^h(\Omega)$ be quasi-uniform. Then*

$$\|v_h\|_{L^2(\Omega)} \leq C \operatorname{diam}(\Omega) \, |v_h|_{H^1(\Omega)} \qquad \forall \mathbf{v} \leftrightarrow v_h \in V^h(\Omega),\ (\mathbf{v}, \mathbf{1})_{\ell^2} = 0,$$

where $\mathbf{1}$ is the vector where all entries equal to one.
(ii) Let the triangulation $\mathscr{T}^h(\partial\Omega)$ be quasi-uniform. Then

$$\|v_h\|_{L^2(\partial\Omega)} \leq C \operatorname{diam}(\Omega)^{1/2} |v_h|_{H^{1/2}(\partial\Omega)} \qquad \forall \mathbf{v} \leftrightarrow v_h \in V^h(\partial\Omega),\ (\mathbf{v}, \mathbf{1})_{\ell^2} = 0.$$

In both estimates, the constant C is independent of h and just depends on the respective quasi-uniformity constant and on the Poincaré constants $C_P(\Omega)$, $C_P(\partial\Omega)$, respectively.

Proof. Let $\bar{\mathbf{v}} := (\mathbf{v}, \mathbf{1})_{\ell^2} / \|\mathbf{1}\|^2_{\ell^2}$ denote the algebraic average of $\mathbf{v}$. Then

$$\|v_h - \bar{\mathbf{v}}\|_{L^2(\Omega)} \simeq h^{d/2} \|\mathbf{v} - \bar{\mathbf{v}}\|_{\ell^2} = h^{d/2} \inf_{c \in \mathbb{R}} \|\mathbf{v} - c\|_{\ell^2} \simeq \inf_{c \in \mathbb{R}} \|v_h - c\|_{L^2(\Omega)}.$$

The first estimate is now implied by Lemma 1.27. The proof of the second estimate is analogous and uses Lemma 1.35. □

Corollary 1.61. *Let $\mathscr{T}^h(\partial\Omega)$ be quasi-uniform and let condition* (1.22) *hold. Then for all $\mathbf{v} \leftrightarrow v_h \in V^h(\partial\Omega)$ with $\overline{v_h}^{\partial\Omega} = 0$ or with $(\mathbf{v}, \mathbf{1})_{\ell^2} = 0$,*

$$c\, \alpha_{\min} \operatorname{diam}(\Omega)^{-1} h^{d-1} \|\mathbf{v}\|^2_{\ell^2} \leq (\mathbf{S}\mathbf{v}, \mathbf{v})_{\ell^2} \leq C \|\mathscr{A}\|_{L^\infty(\Omega)} h^{d-2} \|\mathbf{v}\|^2_{\ell^2}.$$

1.2.7 The Exterior Laplace Problem

Let $\Omega \subset \mathbb{R}^d$ be a bounded Lipschitz domain and define its *exterior*

$$\Omega^{\text{ext}} := \mathbb{R}^d \setminus \overline{\Omega},$$

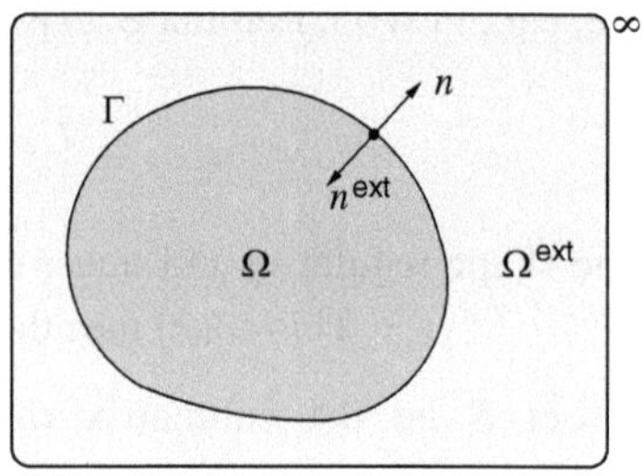

Fig. 1.5 Sketch of an exterior domain Ω^{ext}

and the *interface*

$$\Gamma := \partial\Omega = \partial\Omega^{\text{ext}} = \mathbb{R}^d \setminus (\Omega \cup \Omega^{\text{ext}}),$$

see Fig. 1.5. We consider the distributional Laplace equation

$$-\Delta u = 0 \qquad \text{in } \mathscr{D}(\Omega^{\text{ext}})^*,$$

i.e.,

$$\int_{\Omega^{\text{ext}}} \nabla u \cdot \nabla v \,\mathrm{d}x = 0 \qquad \forall u \in \mathscr{C}_0^\infty(\Omega^{\text{ext}}).$$

Unfortunately, we cannot use $H^1(\Omega^{\text{ext}})$ as the solution space: it holds that

$$\overline{\mathscr{C}_0^\infty(\overline{\Omega}^{\text{ext}})}^{|\cdot|_{H^1(\Omega^{\text{ext}})}} \not\subset H^1(\Omega^{\text{ext}}),$$

and so there is no hope for a coercive setting in any subspace of $H^1(\Omega^{\text{ext}})$.

Definition 1.62. The linear space $H^1_{\text{loc}}(\Omega^{\text{ext}})$ is given by

$$H^1_{\text{loc}}(\Omega^{\text{ext}}) := \left\{u \in \mathscr{D}^*(\overline{\Omega}^{\text{ext}}) : u \in H^1(\Omega^{\text{ext}} \cap B_R) \quad \forall R > 0\right\}, \tag{1.41}$$

where B_R is the open ball with center in the origin and radius R. Furthermore, we define

$$H^{1,\star}(\Omega^{\text{ext}}) := \overline{\mathscr{C}_0^\infty(\overline{\Omega}^{\text{ext}})}^{|\cdot|_{H^1(\Omega^{\text{ext}})}}.$$

The spaces $H^1_{\text{loc}}(\mathbb{R}^d)$ and $H^{1,\star}(\mathbb{R}^d)$ are defined analogously.

Note that $H^1_{\text{loc}}(\Omega^{\text{ext}})$ is much larger than $H^1(\Omega^{\text{ext}})$. It does not have any norm but a topology, cf. [SS11, Sect. 2.6]. Note also that functions from $\mathscr{C}_0^\infty(\overline{\Omega}^{\text{ext}})$ have compact support but do not necessarily vanish on Γ.

The following lemma characterizes the space $H^{1,\star}(\Omega^{\text{ext}})$ that we will use as a solution space for exterior problems.

Lemma 1.63. *The space*

$$H^{1,\star}(\Omega^{\text{ext}}) = \left\{v \in H^1_{\text{loc}}(\Omega^{\text{ext}}) : \|v\|_{H^{1,\star}(\Omega^{\text{ext}})} < \infty\right\}$$

equipped with norm

$$\|v\|_{H^{1,\star}(\Omega^{\text{ext}})} := \left(\int_{\Omega^{\text{ext}}} |\nabla v|^2 + \frac{|v|^2}{1+\rho(x)^2} \, \mathrm{d}x \right)^{1/2}$$

is a Hilbert space, where $\rho(x) := |x| \log |x|$ *if* $d = 2$, *and* $\rho(x) := |x|$ *if* $d = 3$. *The same holds true if the set* Ω^{ext} *is replaced by* $\mathbb{R}^d$.

Proof. For a proof see, e.g., [Lun06].

As a particularity of the Laplace problem, there is a principal difference between two and three space dimensions. For $d = 3$, one can show that functions $v \in H^{1,\star}(\Omega^{\text{ext}})$ fulfill the *radiation condition*

$$v(|x|) = \mathcal{O}(|x|^{-1}) \qquad \text{as } |x| \to \infty.$$

For $d = 2$, however,

$$1_{\Omega^{\text{ext}}} \in H^{1,\star}(\Omega^{\text{ext}}).$$

We could say that in three dimensions, $H^{1,\star}(\Omega^{\text{ext}})$ functions have to vanish at infinity, which is not the case in two dimensions.

Analogous to Theorem 1.23, the exterior Dirichlet trace operator

$$\gamma_0^{\text{ext}} : H^1_{\text{loc}}(\Omega^{\text{ext}}) \to H^{1/2}(\Gamma)$$

is well-defined. At least when restricted to $H^{1,\star}(\Omega^{\text{ext}})$, it is continuous and has a continuous right inverse. The generalized exterior normal derivative can be defined analogously to Theorem 1.40. Let n^{ext} be the unit normal vector outward to Ω^{ext}, see Fig. 1.5. Then the Neumann trace operator

$$\gamma_1^{\text{ext}} : \{u \in H^1_{\text{loc}}(\Omega^{\text{ext}}) : \Delta u = 0\} \to H^{-1/2}(\Gamma)$$

satisfies the relation

$$\langle \gamma_1^{\text{ext}} u, \gamma_0^{\text{ext}} v \rangle_\Gamma = \int_{\Omega^{\text{ext}}} \nabla u \cdot \nabla v \, \mathrm{d}x \qquad \forall v \in \mathcal{C}_0^\infty(\Omega^{\text{ext}}). \tag{1.42}$$

For a smooth function u, the functional $\gamma_1^{\text{ext}} u$ coincides with the classical normal derivative $\partial u / \partial n^{\text{ext}}$. Be aware that in the literature, γ_1^{ext} often has the opposite sign and equals $-\partial u / \partial n^{\text{ext}} = \partial u / \partial n$ for smooth functions u.

For $\Gamma_D \subset \Gamma$ (possibly empty) and $\Gamma_N := \Gamma \setminus \Gamma_D$, the variational problem reads: for given $g_D \in H^{1/2}(\Gamma_D)$, $g_N \in H_{00}^{-1/2}(\Gamma_N)$, find $u \in H^{1,\star}(\Omega^{\text{ext}})$, $u_{|\Gamma_D} = g_D$ such that

$$a_{\text{ext}}(u, v) = \langle g_N, v \rangle_{\Gamma_N} \qquad \forall v \in H^{1,\star}(\Omega^{\text{ext}}),\ v|_{\Gamma_D} = 0. \tag{1.43}$$

where

$$a_{\text{ext}}(u,v) := \int_{\Omega^{\text{ext}}} \nabla u \cdot \nabla v \, \mathrm{d}x.$$

Lemma 1.64. *For $d = 3$, the bilinear form $a_{\text{ext}}(\cdot,\cdot)$ is coercive on $H^{1,\star}(\Omega^{\text{ext}})$. Therefore, problem* (1.43) *is uniquely solvable for any choice of Γ_D, even the exterior pure Neumann problem ($\Gamma_D = \emptyset$, $\Gamma_N = \Gamma$).*

Proof. See, e.g., [Lun06].

The next two lemmas discuss the solvability of the two-dimensional exterior problem. Their statements will follow from the results in Sect. 1.3.3.

Lemma 1.65. *For $d = 2$ and let Γ_D have positive surface measure. Then problem* (1.43) *is uniquely solvable and there exists a constant $b \in \mathbb{R}$ such that*

$$u(x) = b + \mathcal{O}(|x|^{-1}) \qquad \text{as } |x| \to \infty.$$

Lemma 1.66. *For $d = 2$, the exterior pure Neumann problem (i.e. problem* (1.43) *with $\Gamma_D = \emptyset$) is solvable if and only if*

$$\langle g_N, 1 \rangle_\Gamma = 0.$$

Two solutions differ by a constant. There exists a particular solution $u \in H^{1,\star}(\Omega^{\text{ext}})$ that fulfills the radiation condition

$$u(x) = \mathcal{O}(|x|^{-1}) \qquad \text{as } |x| \to \infty.$$

The exterior Laplace problem will be further investigated in Sect. 1.3. Finally, we give the definitions of the exterior harmonic extension and the exterior Steklov-Poincaré operator.

Definition 1.67. (i) For $v \in H^{1/2}(\Gamma)$, the *exterior harmonic extension* $\mathcal{H}^{\text{ext}} v \in H^{1,\star}(\Omega^{\text{ext}})$ is the unique solution of

$$(\mathcal{H}^{\text{ext}} v)_{|\Gamma} = v, \quad \int_{\Omega^{\text{ext}}} \nabla(\mathcal{H}^{\text{ext}} v) \cdot \nabla w \, \mathrm{d}x = 0 \qquad \forall w \in H^{1,\star}(\Omega^{\text{ext}}),\ w_{|\Gamma} = 0,$$

i.e., $\mathcal{H}^{\text{ext}} v = \operatorname*{argmin}_{\substack{\tilde{v} \in H^{1,\star}(\Omega^{\text{ext}}) \\ \tilde{v}_{|\Gamma} = v}} |\tilde{v}|_{H^1(\Omega^{\text{ext}})}.$

(ii) The linear and continuous map $S^{\text{ext}} := \gamma_1^{\text{ext}} \mathcal{H}^{\text{ext}} : H^{1/2}(\Gamma) \to H^{-1/2}(\Gamma)$ is called *exterior Steklov-Poincaré operator*. We have

$$\langle S^{\text{ext}} v, w \rangle_\Gamma = \int_{\Omega^{\text{ext}}} \nabla(\mathscr{H}^{\text{ext}} v) \cdot \nabla(\mathscr{H}^{\text{ext}} w)\, \mathrm{d}x \qquad \forall v, w \in H^{1/2}(\Gamma),$$

$$\langle S^{\text{ext}} v, v \rangle_\Gamma = \min_{\substack{\tilde{v} \in H^{1,\star}(\Omega^{\text{ext}}) \\ \tilde{v}|_\Gamma = v}} |\tilde{v}|^2_{H^1(\Omega^{\text{ext}})} \qquad \forall v \in H^{1/2}(\Gamma).$$

In three dimensions, $\langle S^{\text{ext}} v, v \rangle_\Gamma^{1/2}$ is equivalent to the norm $\|v\|_{H^{1/2}(\Gamma)}$. In two dimensions, $\langle S^{\text{ext}} v, v \rangle_\Gamma^{1/2}$ is equivalent to the semi-norm $|v|_{H^{1/2}(\Gamma)}$ and we have the identity $S^{\text{ext}} 1_\Gamma = 0$.

1.3 Boundary Integral Equations

In this section, we briefly introduce the concept of boundary integral equations. Using fundamental solutions, the partial differential equation on a domain can be reformulated in terms of integral equations on the boundary. For more details, we refer to [HW08b, McL00, SS11, Ste08]. Additionally, we will discuss the exterior problem, and give representations of the (interior and exterior) Steklov-Poincaré operators in terms of boundary integral operators. Finally, we provide an overview on the boundary element method.

1.3.1 Surface Potentials and Representation Formulae

Let $\Omega \subset \mathbb{R}^d$ ($d = 2$ or 3) be a bounded Lipschitz domain with a simply connected boundary $\Gamma := \partial\Omega$. In the following we consider the interior $\Omega^{\text{int}} := \Omega$ and the exterior $\Omega^{\text{ext}} := \mathbb{R}^d \setminus \overline{\Omega}$ of Γ. Recall that n is the unit normal on Γ pointing outwards of Ω^{int} and $n^{\text{ext}} = -n$, see Fig. 1.5 on page 40. We consider the interior and exterior Laplace problem

$$-\Delta u = 0 \qquad \text{in } \mathscr{D}(\Omega^{\text{int}})^*, \tag{1.44}$$

$$-\Delta u = 0 \qquad \text{in } \mathscr{D}(\Omega^{\text{ext}})^*, \tag{1.45}$$

respectively. A fundamental solution to (1.44) and (1.45) is given by

$$U^*(x, y) := \begin{cases} -\dfrac{1}{2\pi} \log|x-y| & \text{for } d = 2, \\ \dfrac{1}{4\pi} \dfrac{1}{|x-y|} & \text{for } d = 3, \end{cases} \tag{1.46}$$

see, e.g., [Eva98] or [McL00, Theorem 8.1]. In fact, in two dimensions, we could use $1/(2\pi)\log(r/|x-y|)$ for any $r > 0$.

Using the fundamental solution we define the single and double layer potential

$$\widetilde{V} : H^{-1/2}(\Gamma) \to H^1_{\mathrm{loc}}(\mathbb{R}^d), \qquad \widetilde{W} : H^{1/2}(\Gamma) \to H^1_{\mathrm{loc}}(\Omega^{\mathrm{int}} \cup \Omega^{\mathrm{ext}}),$$

respectively (the double layer potential is discontinuous across Γ). For smooth functions v and w we have the integral representations

$$(\widetilde{V} w)(x) = \int_\Gamma U^*(x, y)\, w(y)\, \mathrm{d}s_y\,, \tag{1.47}$$

$$(\widetilde{W} v)(x) = \int_\Gamma \frac{\partial U^*}{\partial n_y}(x, y)\, v(y)\, \mathrm{d}s_y\,. \tag{1.48}$$

For the exact definitions involving tempered distributions see, e.g., [McL00]. In the following we use γ_0^{int} and γ_1^{int} for the *interior* Dirichlet and Neumann trace operator, respectively, in order to outrule misunderstandings.

Lemma 1.68. *Let $v \in H^{1/2}(\Gamma)$ and $w \in H^{-1/2}(\Gamma)$.*

(i) The surface potentials are weak solutions of the Laplace equation:

$$-\Delta\widetilde{V} w = 0 \quad \textit{and} \quad -\Delta\widetilde{W} v = 0 \qquad \textit{in } \mathscr{D}(\Omega^{\mathrm{int}} \cup \Omega^{\mathrm{ext}})^*.$$

(ii) Their behavior at infinity is as follows:

$$(\widetilde{V} w)(x) = \left.\begin{cases} -\frac{1}{2\pi}\langle w, 1\rangle_\Gamma \log|x| + \mathscr{O}(|x|^{-1}) & \textit{if } d = 2 \\ \mathscr{O}(|x|^{-1}) & \textit{if } d = 3 \end{cases}\right\} \quad \textit{as } |x| \to \infty,$$

$$(\widetilde{W} v)(x) = \mathscr{O}(|x|^{1-d}) \quad \textit{as } |x| \to \infty.$$

Consequently,

$$(\widetilde{W} v)_{|\Omega^{\mathrm{ext}}} \in H^{1,\star}(\Omega^{\mathrm{ext}}), \quad (\widetilde{V} w)_{|\Omega^{\mathrm{ext}}} \in H^{1,\star}(\Omega^{\mathrm{ext}}) \begin{cases} \textit{if } d = 2 \textit{ and } \langle w, 1\rangle_\Gamma = 0, \\ \textit{if } d = 3. \end{cases}$$

(iii) The following jump relations hold

$$\gamma_0^{\mathrm{ext}}\widetilde{V} w - \gamma_0^{\mathrm{int}}\widetilde{V} w = 0, \qquad \gamma_1^{\mathrm{ext}}\widetilde{V} w + \gamma_1^{\mathrm{int}}\widetilde{V} w = w,$$

$$\gamma_0^{\mathrm{ext}}\widetilde{W} v - \gamma_0^{\mathrm{int}}\widetilde{W} v = v, \qquad \gamma_1^{\mathrm{ext}}\widetilde{W} v + \gamma_1^{\mathrm{int}}\widetilde{W} v = 0.$$

Remark 1.69. In two dimensions, we see from Lemma 1.68(ii) that the single layer potential does not quite fit into the variational framework from Sect. 1.2.7. Indeed, it includes *algebraic* solutions of Laplace's equation that grow logarithmically towards infinity. These solutions do not have finite energy: if $\langle w, 1\rangle_\Gamma \neq 0$, then $|\widetilde{V} w|_{H^1(\Omega^{\mathrm{ext}})} = \infty$. Instead of the term $\log|x-y|$ in the fundamental solution

Table 1.2 Definitions and properties of four boundary integral operators

Operator, mapping properties	Relations	Integral representation ($x \in \Gamma$)
Single layer potential operator $V : H^{-1/2}(\Gamma) \to H^{1/2}(\Gamma)$	$\gamma_0^{\text{int}}\widetilde{V} = V$ $\gamma_0^{\text{ext}}\widetilde{V} = V$	for $w \in H^{-1/2}(\Gamma) \cap L^\infty(\Gamma)$: $(Vw)(x) = \int_\Gamma U^*(x, y)\, w(y)\, \mathrm{d}s_y$
Double layer potential operator $K : H^{1/2}(\Gamma) \to H^{1/2}(\Gamma)$	$\gamma_0^{\text{int}}\widetilde{W} = -\frac{1}{2}I + K$ $\gamma_0^{\text{ext}}\widetilde{W} = \frac{1}{2}I + K$	for $v \in H^{1/2}(\Gamma) \cap L^\infty(\Gamma)$: $(K\, v)(x) = \int_\Gamma \left[\frac{\partial U^*}{\partial n_y}(x, y)\right] v(y)\, \mathrm{d}s_y$
Adjoint double layer pot. op. $K^\top : H^{-1/2}(\Gamma) \to H^{-1/2}(\Gamma)$	$\gamma_1^{\text{int}}\widetilde{V} = \frac{1}{2}I + K^\top$ $\gamma_1^{\text{ext}}\widetilde{V} = \frac{1}{2}I - K^\top$	for $w \in H^{-1/2}(\Gamma) \cap L^\infty(\Gamma)$: $(K^\top w)(x) = \int_\Gamma \frac{\partial U^*}{\partial n_x}(x, y)\, v(y)\, \mathrm{d}s_y$
Hypersingular integral operator $D : H^{1/2}(\Gamma) \to H^{-1/2}(\Gamma)$	$\gamma_1^{\text{int}}\widetilde{W} = -D$ $\gamma_1^{\text{ext}}\widetilde{W} = D$	for $v \in H^{1/2}(\Gamma) \cap C(\Gamma)$: $(D\, v)(x) = \int_\Gamma \frac{\partial}{\partial n_x}\frac{\partial}{\partial n_y}\big[U^*(x, y)\big(v(x)$ $- v(y)\big)\big] ds_y$

(1.46), one can also use $\log(|x - y|/r)$ for any $r > 0$. This does not change the two surface potentials, except for the single layer potential evaluated at functions where $\langle w, 1\rangle_\Gamma \neq 0$; there we get different algebraic solutions with infinite energy. The single layer potential in two dimension is well studied in classical potential theory, see, e.g., [Hil87].

The two surface potentials can be used to reconstruct the solution of the interior (exterior) Laplace equation from the Cauchy data, i.e., from the Dirichlet and the Neumann data.

Lemma 1.70. *(i) Every weak solution $u \in H^1(\Omega^{\text{int}})$ of the interior Laplace problem* (1.44) *is given by the representation formula*

$$u = \widetilde{V}\gamma_1^{\text{int}}u - \widetilde{W}\gamma_0^{\text{int}}u \qquad \text{in } \Omega^{\text{int}}.$$

(ii) Let $d = 3$ (for $d = 2$, see Sect. 1.3.3). Then every weak solution $u \in H^{1,\star}(\Omega^{\text{ext}})$ of the exterior Laplace problem (1.45) *is given by the representation formula*

$$u = \widetilde{V}\gamma_1^{\text{ext}}u + \widetilde{W}\gamma_0^{\text{ext}}u \qquad \text{in } \Omega^{\text{ext}}.$$

1.3.2 Boundary Integral Operators and Equations

The traces of the surface potentials themselves have representations in terms of four boundary integral operators (V, K, $K^\top$, and D) as summarized in Table 1.2. Note that the representation for V involves a weakly singular surface integral, the other integral representations have to be understood in the sense of the Cauchy principal value. All four operators are linear and continuous, and $K^\top$ is indeed the adjoint of K. For details and proofs see, e.g., [McL00, Ste08].

Lemma 1.71. *(i) The Cauchy data* $(\gamma_0^{\text{int}}u, \gamma_1^{\text{int}}u)$ *of a solution* $u \in H^1(\Omega^{\text{int}})$ *of the interior Laplace problem* (1.44) *fulfills the two boundary integral equations of the* Calderón system

$$\begin{bmatrix} \gamma_0^{\text{int}}u \\ \gamma_1^{\text{int}}u \end{bmatrix} = \begin{bmatrix} \frac{1}{2}I - K & V \\ D & \frac{1}{2}I + K^\top \end{bmatrix} \begin{bmatrix} \gamma_0^{\text{int}}u \\ \gamma_1^{\text{int}}u \end{bmatrix}. \tag{1.49}$$

(ii) For $d = 3$, *the Cauchy data* $(\gamma_0^{\text{ext}}u, \gamma_1^{\text{ext}}u)$ *of a solution* $u \in H^{1,\star}(\Omega^{\text{ext}})$ *of the exterior Laplace problem* (1.45) *fulfills the Caldéron system*

$$\begin{bmatrix} \gamma_0^{\text{ext}}u \\ \gamma_1^{\text{ext}}u \end{bmatrix} = \begin{bmatrix} \frac{1}{2}I + K & V \\ D & \frac{1}{2}I - K^\top \end{bmatrix} \begin{bmatrix} \gamma_0^{\text{ext}}u \\ \gamma_1^{\text{ext}}u \end{bmatrix}. \tag{1.50}$$

For $d = 2$ *see Sect. 1.3.3.*

In the following, we discuss important properties of the boundary integral operators. The next assumption ensures that the single layer potential operator V is positive definite and can always be obtained by a simple scaling of the coordinates.

Assumption 1.72. If $d = 2$ then $\operatorname{diam}(\Omega^{\text{int}}) < 1$.

Lemma 1.73. *(i) Let Assumption 1.72 hold. Then the single layer potential operator* V *is positive definite, in particular, there exists a constant* $c_V > 0$ *such that*

$$\langle w, V\,w \rangle \geq c_V\, \|w\|^2_{H^{-1/2}(\Gamma)} \qquad \forall w \in H^{-1/2}(\Gamma).$$

It defines a norm $\|w\|_V := \langle w, Vw \rangle_\Gamma^{1/2}$ *on* $H^{-1/2}(\Gamma)$. *The inverse* $V^{-1} : H^{1/2}(\Gamma) \to H^{-1/2}(\Gamma)$ *is bounded and SPD, and* $\|v\|_{V^{-1}} := \langle V^{-1}v, v\rangle^{1/2}$ *defines a norm on* $H^{1/2}(\Gamma)$.

(ii) There is a unique functional $w_{\text{eq}} \in H^{-1/2}(\Gamma)$, *the* equilibrium density, *that fulfills*

$$V w_{\text{eq}} = \text{const}, \qquad \langle w_{\text{eq}}, 1 \rangle_\Gamma = 1. \tag{1.51}$$

Remark 1.74. In two dimensions, the constant $V w_{\text{eq}}$ might be negative or zero. Under Assumption 1.72, however, it is strictly positive. This particularity is linked to the exterior Neumann problem, see also Sect. 1.3.3. The quantity $e^{-2\pi V w_{\text{eq}}}$ is called *logarithmic capacity* of Γ and is well-studied in complex analysis (see, e.g., [Hil87]). Dahlberg [Dah77] shows that for a Lipschitz domain, the equilibrium density is in $L^2(\Gamma)$.

Definition 1.75.

$$H_*^{1/2}(\Gamma) := \{v \in H^{1/2}(\Gamma) : \langle w_{\text{eq}}, v \rangle_\Gamma = 0\}$$
$$H_*^{-1/2}(\Gamma) := \{w \in H^{-1/2}(\Gamma) : \langle w, 1 \rangle_\Gamma = 0\}$$

Table 1.3 Properties of the four boundary integral operators related to the splittings in Definition 1.75

Isomorphism property		Properties on complement	
$V:$	$H_*^{-1/2}(\Gamma) \leftrightarrow H_*^{1/2}(\Gamma)$	$V w_{\mathrm{eq}} = \mathrm{const}$	$V^{-1} 1_\Gamma = (V w_{\mathrm{eq}})^{-1} w_{\mathrm{eq}}$
$\frac{1}{2}I \pm K:$	$H_*^{1/2}(\Gamma) \leftrightarrow H_*^{1/2}(\Gamma)$	$(\frac{1}{2}I + K)1_\Gamma = 0$	$(\frac{1}{2}I - K)1_\Gamma = 1_\Gamma$
$\frac{1}{2}I \pm K^\top:$	$H_*^{-1/2}(\Gamma) \leftrightarrow H_*^{-1/2}(\Gamma)$	$(\frac{1}{2}I + K^\top) w_{\mathrm{eq}} = 0$	$(\frac{1}{2}I - K^\top) w_{\mathrm{eq}} = w_{\mathrm{eq}}$
$D:$	$H_*^{1/2}(\Gamma) \leftrightarrow H_*^{-1/2}(\Gamma)$	$D 1_\Gamma = 0$	

Apparently,

$$\begin{aligned} H^{1/2}(\Gamma) &= H_*^{1/2}(\Gamma) \ \oplus \ \mathrm{span}\{1_\Gamma\}, \\ H^{-1/2}(\Gamma) &= H_*^{-1/2}(\Gamma) \oplus \mathrm{span}\{w_{\mathrm{eq}}\}. \end{aligned}$$

Under Assumption 1.72, the first splitting is V^{-1}-orthogonal, the second one V-orthogonal.

Properties of the four boundary integral operators related to the splittings in Definition 1.75 are displayed in Table 1.3

Lemma 1.76. *The hypersingular integral operator D is self-adjoint and there exists a constant $c_D > 0$ such that*

$$\begin{aligned} \langle D v, v\rangle &\geq c_D \, \|v\|^2_{H^{1/2}(\Gamma)} \qquad \forall v \in H_*^{1/2}(\Gamma), \\ \langle D v, v\rangle &\geq c_D \, |v|^2_{H^{1/2}(\Gamma)} \qquad \forall v \in H^{1/2}(\Gamma), \end{aligned}$$

i.e., $\langle D v, v\rangle_\Gamma^{1/2}$ defines a semi-norm equivalent to $|v|_{H^{1/2}(\Gamma)}$.

Lemma 1.77. *Let Assumption 1.72 hold. Then the double layer potential operator K admits the contraction properties*

$$\begin{aligned} (1 - c_K)\|v\|_{V^{-1}} &\leq \|(\tfrac{1}{2}I \pm K)v\|_{V^{-1}} \leq c_K \, \|v\|_{V^{-1}} \qquad \forall v \in H_*^{1/2}(\Gamma), \\ (1 - c_K)\|v\|_{V^{-1}} &\leq \|(\tfrac{1}{2}I - K)v\|_{V^{-1}} \leq \|v\|_{V^{-1}} \qquad \forall v \in H^{1/2}(\Gamma), \\ 0 &\leq \|(\tfrac{1}{2}I + K)v\|_{V^{-1}} \leq c_K \, \|v\|_{V^{-1}} \qquad \forall v \in H^{1/2}(\Gamma), \end{aligned}$$

where the constants $c_0 \in (0, \frac{1}{4})$ and $c_K \in (\frac{1}{2}, 1)$, defined by

$$c_0 := \inf_{v \in H_*^{1/2}(\Gamma)} \frac{\langle D v, v\rangle}{\langle V^{-1} v, v\rangle}, \qquad c_K := \frac{1}{2} + \sqrt{\frac{1}{4} - c_0}\,,$$

depend only on the shape of $\partial\Omega$.

Remark 1.78. For details on the contraction properties see [SW01], as well as [Cos07] for a historical remark. Apparently, $c_0 \geq c_D\, c_V > 0$ with the constants c_V from Lemma 1.73 and c_D from Lemma 1.76. Explicit estimates for c_0 and c_K involving special norms can be found in [Pec12].

The two boundary integral equations appearing in the Calderón systems (1.49) and (1.50) can be used to solve Dirichlet, Neumann, and mixed boundary value problems in the sense that we determine the complete Cauchy data $(\gamma_0^{\text{int}} u,\ \gamma_1^{\text{int}} u)$. From Lemmas 1.73 and 1.76 one can conclude that the variational problems corresponding to the Dirichlet and to the Neumann problem are well-posed. For more details and for mixed problems see, e.g., [Ste08, Sect. 7].

1.3.3 *The Exterior Laplace Problem in Two Dimensions*

With the equilibrium density from Lemma 1.73, we are able to discuss the exterior Laplace problem in two dimensions.

Lemma 1.79. *For* $d = 2$, *every weak solution* $u \in H^{1,\star}(\Omega^{\text{ext}})$ *of the exterior Laplace problem* (1.45) *necessarily satisfies the compatibility condition* $\gamma_1^{\text{ext}} u \in H_*^{-1/2}(\Gamma)$ *and is represented by*

$$u = \widetilde{V}\gamma_1^{\text{ext}} u + \widetilde{W}\gamma_0^{\text{ext}} u + \langle w_{\text{eq}},\ \gamma_0^{\text{ext}} u\rangle_\Gamma \qquad \text{in } \Omega^{\text{ext}}.$$

In particular,

$$u(x) = \langle w_{\text{eq}},\ \gamma_0^{\text{ext}} u\rangle + \mathcal{O}(|x|^{-1}) \qquad \text{as } |x| \to \infty.$$

The Cauchy data $(\gamma_0^{\text{ext}} u,\ \gamma_1^{\text{ext}} u) \in H^{1/2}(\Gamma) \times H_*^{-1/2}(\Gamma)$ *fulfills the Caldéron system*

$$\begin{bmatrix} \gamma_0^{\text{ext}} u \\ \gamma_1^{\text{ext}} u \end{bmatrix} = \begin{bmatrix} \frac{1}{2}I + K & V \\ D & \frac{1}{2}I - K^\top \end{bmatrix} \begin{bmatrix} \gamma_0^{\text{ext}} u \\ \gamma_1^{\text{ext}} u \end{bmatrix} + \begin{bmatrix} \langle w_{\text{eq}},\ \gamma_0^{\text{ext}} u\rangle_\Gamma \\ 0 \end{bmatrix}. \tag{1.52}$$

Remark 1.80. In contrast to V itself, the operator $V_{|H_*^{-1/2}(\Gamma)}$ is invariant with respect to coordinate scalings. Together with Table 1.3, this implies that the entire system (1.52) is scaling invariant. We can think of V as of a regularization of $V_{|H_*^{-1/2}(\Gamma)}$. If $\widehat{V}$ denotes the extension by zero of $V_{|H_*^{-1/2}(\Gamma)}$ to $H^{-1/2}(\Gamma)$, then

$$\langle y,\ Vw\rangle \;=\; \langle y, \widehat{V}w\rangle + V w_{\text{eq}}\, \langle w,\ 1\rangle_\Gamma\, \langle y,\ 1\rangle_\Gamma \qquad \text{for } w,\ y \in H^{-1/2}(\Gamma).$$

The regularization parameter $V w_{\text{eq}}$ depends on the diameter $\operatorname{diam}(\Gamma)$ and can become zero or negative. Assumption 1.72 is sufficient that $V w_{\text{eq}}$ is positive.

Remark 1.81. Lemma 1.79 implies that for the two-dimensional exterior Neumann problem with $\gamma_1^{\text{ext}} u \in H_*^{-1/2}(\Gamma)$ given, the integral equation

$$\gamma_0^{\text{ext}} u = (\tfrac{1}{2} I + K)\gamma_0^{\text{ext}} u + V\gamma_1^{\text{ext}} u + b$$

selects the particular solution that decays to $b \in \mathbb{R}$ at infinity.

1.3.4 Representations of Steklov-Poincaré Operators

In the following, we denote by S^{int} the interior Steklov-Poincaré operator from Definition 1.41 and by S^{ext} the exterior Steklov-Poincaré operator from Definition 1.67. The next lemma shows that the two Steklov-Poincaré operators (mapping $H^{1/2}(\Gamma)$ to $H^{-1/2}(\Gamma)$) can be written in terms of the four boundary integral operators. Due to the principal difference between two and three space dimensions, let us define

$$S^{\text{corr}} := \begin{cases} \frac{1}{V w_{\text{eq}}} \langle w_{\text{eq}}, \cdot \rangle_\Gamma \, w_{\text{eq}} & \text{if } d = 2, \\ 0 & \text{if } d = 3. \end{cases} \tag{1.53}$$

Lemma 1.82. *Assume that V is invertible. Then the Steklov-Poincaré operators S^{int} and S^{ext} admit the representations*

$$\begin{aligned} S^{\text{int}} &= V^{-1}(\tfrac{1}{2} I + K) &&= D + (\tfrac{1}{2} I + K^\top)V^{-1}(\tfrac{1}{2} I + K), \\ S^{\text{ext}} &= V^{-1}(\tfrac{1}{2} I - K) - S^{\text{corr}} &&= D + (\tfrac{1}{2} I - K^\top)V^{-1}(\tfrac{1}{2} I - K) - S^{\text{corr}}. \end{aligned} \tag{1.54}$$

Moreover,

$$V^{-1} = S^{\text{int}} + S^{\text{ext}} + S^{\text{corr}}.$$

Proof. Follows from the Caldéron systems in Lemmas 1.71 and 1.79. □

Recall that

$$\ker(S^{\text{int}}) = \operatorname{span}\{1_\Gamma\}, \qquad \ker(S^{\text{ext}}) = \begin{cases} \operatorname{span}\{1_\Gamma\} & \text{if } d = 2, \\ \{0\} & \text{if } d = 3. \end{cases}$$

The following lemma summarizes spectral equivalence relations between V^{-1} and the two Steklov-Poincaré operators. See also [SW01, Cos07].

Lemma 1.83. *(i) Let c_K be the constant from Lemma 1.77. Then under Assumption 1.72,*

$$\begin{aligned}
(1-c_K)\,\|v\|_{V^{-1}}^2 \le \langle S^{\text{int}} v, v\rangle &\le c_K\,\|v\|_{V^{-1}}^2 && \forall v \in H_*^{1/2}(\Gamma),\\
0 \le \langle S^{\text{int}} v, v\rangle &\le c_K\,\|v\|_{V^{-1}}^2 && \forall v \in H^{1/2}(\Gamma),
\end{aligned}$$

$$\begin{aligned}
(1-c_K)\,\|v\|_{V^{-1}}^2 \le \langle S^{\text{ext}} v, v\rangle &\le c_K\,\|v\|_{V^{-1}}^2 && \forall v \in H_*^{1/2}(\Gamma),\\
0 \le \langle S^{\text{ext}} v, v\rangle &\le c_K\,\|v\|_{V^{-1}}^2 && \forall v \in H^{1/2}(\Gamma),\ \text{if } d=2,\\
(1-c_K)\,\|v\|_{V^{-1}}^2 \le \langle S^{\text{ext}} v, v\rangle &\le \|v\|_{V^{-1}}^2 && \forall v \in H^{1/2}(\Gamma),\ \text{if } d=3,
\end{aligned}$$

with the contraction constant $c_K > 0$ from Lemma 1.77, which only depends on the shape of Γ.

(ii) In three dimensions, the splitting $H^{1/2}(\Gamma) = H_*^{1/2}(\Gamma) \oplus \operatorname{span}\{1_\Gamma\}$ is orthogonal with respect to the $\langle S^{\text{ext}}\cdot,\cdot\rangle$-inner product.

Proof. Recall that for $d=3$, $S^{\text{int/ext}} = V^{-1}(\frac{1}{2}I \pm K)$. Property (ii) follows essentially from

$$V^{-1}1_\Gamma \;=\; V^{-1}(\tfrac{1}{2}I - K)1_\Gamma \;=\; S^{\text{ext}}1_\Gamma\,.$$

Property (i) follows from the Cauchy-Schwarz inequality and the contraction properties from Lemma 1.77: for $v \in H_*^{1/2}(\Gamma)$,

$$\begin{aligned}
\langle S^{\text{int/ext}} v, v\rangle &= \langle V^{-1}(\tfrac{1}{2}I \pm K)v, v\rangle \;\le\; \|(\tfrac{1}{2}I \pm K)v\|_{V^{-1}}\|v\|_{V^{-1}} \le c_K\|v\|_{V^{-1}}^2,\\
\langle S^{\text{int/ext}} v, v\rangle &= \langle V^{-1}(\tfrac{1}{2}I \pm K)v, v\rangle \;=\; \langle V^{-1}v, v\rangle - \langle V^{-1}(\tfrac{1}{2}I \mp K)v, v\rangle\\
&\ge \|v\|_{V^{-1}}^2 - c_K\|v\|_{V^{-1}}^2\,.
\end{aligned}$$

The other estimates are shown using the V^{-1}-orthogonal splitting from (ii) and the properties summarized in Table 1.3. □

Remark 1.84. From Lemma 1.83, one can conclude further estimates between the two Steklov-Poincaré operators, e.g., Poincaré's *fundamental theorem* (see [Cos07])

$$\langle S^{\text{ext}} v, v\rangle \;\le\; \frac{c_K}{1-c_K}\langle S^{\text{int}} v, v\rangle \qquad \begin{cases} \forall v \in H^{1/2}(\Gamma) & \text{if } d=2,\\ \forall v \in H_*^{1/2}(\Gamma) & \text{if } d=3,\end{cases}$$

$$\langle S^{\text{int}} v, v\rangle \;\le\; \frac{c_K}{1-c_K}\langle S^{\text{ext}} v, v\rangle \qquad \forall v \in H^{1/2}(\Gamma).$$

1.3.5 Newton Potentials

The following definition and lemma allow boundary integral equations for a Poisson problem too. See [Ste08] for a more general definition.

Definition 1.85. The Newton volume potential $\widetilde{N} : L^2(\Omega) \to H^1(\Omega^{\text{int}})$ is defined as

$$(\widetilde{N} f)(x) := \int_{\Omega^{\mathrm{int}}} U^*(x,y)\, f(y)\, \mathrm{d}y \qquad \text{for } x \in \Omega^{\mathrm{int}}.$$

Lemma 1.86. *Every weak solution $u \in H^1(\Omega^{\mathrm{int}})$ of the interior Poisson problem*

$$-\Delta u = f \qquad \text{in } \mathscr{D}^*(\Omega^{\mathrm{int}})$$

with $f \in L^2(\Omega^{\mathrm{int}})$ is represented by

$$u = \widetilde{V}\, \gamma_1^{\mathrm{int}} u - \widetilde{W}\, \gamma_0^{\mathrm{int}} u + \widetilde{N} f.$$

The Cauchy data $(\gamma_0^{\mathrm{int}} u,\ \gamma_1^{\mathrm{int}} u)$ fulfills

$$\begin{bmatrix} \gamma_0^{\mathrm{int}} u \\ \gamma_1^{\mathrm{int}} u \end{bmatrix} = \begin{bmatrix} \frac{1}{2} I - K & V \\ D & \frac{1}{2} I + K^\top \end{bmatrix} \begin{bmatrix} \gamma_0^{\mathrm{int}} u \\ \gamma_1^{\mathrm{int}} u \end{bmatrix} + \begin{bmatrix} \gamma_0^{\mathrm{int}} \widetilde{N} f \\ \gamma_1^{\mathrm{int}} \widetilde{N} f \end{bmatrix}. \tag{1.55}$$

Lemma 1.87. *We have*

$$\gamma_1^{\mathrm{int}} \widetilde{N} = V^{-1} \gamma_0^{\mathrm{int}} \widetilde{N}.$$

Furthermore,

$$N = V^{-1} \gamma_0^{\mathrm{int}} \widetilde{N}.$$

where N is the interior Newton potential for the Poisson problem from Definition 1.41.

Proof. The two identities follow directly from (1.55). □

1.3.6 Generalizations to the Potential Equation

If we replace the Laplace operator Δ by $\alpha\Delta$, where α is a constant coefficient, we just need to scale the fundamental solution and the normal derivatives $\partial/\partial n$, γ_1^{int}, γ_1^{ext} by α. Then, all statements from the previous subsections hold true. Effectively, only some of the operators will scale linearly with some power of α compared to the operators for $\alpha = 1$. For a constant SPD matrix $\mathscr{A}$, such a generalization is also possible, see [SS11].

However, a direct generalization to non-constant coefficients $\alpha(x)$ or $\mathscr{A}(x)$ is not possible because fundamental solutions cannot be constructed explicitly (see [HW08b, Sect. 6.2.3] for more details and an alternative approach). The framework in Chap. 2 will allow the treatment of problems with piecewise constant coefficient via the coupling of boundary integral equations on subdomains, or the coupling of boundary integral equations on one part of the domain (where the coefficient is piecewise constant) to the original variational formulation on the other part of the domain (where the coefficient is variable).

1.3.7 *The Boundary Element Method*

The Galerkin boundary element method is a special Galerkin finite element method applied to boundary integral equations. For a comprehensive introduction we refer to the monographs [HW08b, RS07, SS11, Ste08]. Throughout this section, let Assumption 1.72 hold.

1.3.7.1 BEM for the Dirichlet Problem

For simplicity, we consider the interior Laplace problem on Ω^{int}, but the techniques below can be transferred to the exterior problem without difficulties. The first equation of the Caldéron system (1.49) rewrites

$$\text{find } t \in H^{-1/2}(\Gamma): \quad \langle w, Vt\rangle_\Gamma = \langle w, (\tfrac{1}{2}I + K)\, g\rangle_\Gamma \qquad \forall w \in H^{-1/2}(\Gamma), \tag{1.56}$$

where $g_D = \gamma_0^{\text{int}} u \in H^{1/2}(\Gamma)$ is the given Dirichlet trace on Γ and $t = \gamma_1^{\text{int}} u \in H^{-1/2}(\Gamma)$ the unknown Neumann trace. This variational problem is uniquely solvable thanks to the Lax-Milgram theorem.

In the following, assume that Ω^{int} is a piecewise smooth Lipschitz domain, i.e., its boundary splits into disjoint parts $\{\Gamma_j\}_{j=1}^N$ that are images of $(d-1)$-dimensional parameter polytopes Q_j under smooth bijective mappings ζ_j. For each $j = 1, \ldots, N$, let $\mathscr{T}^h(Q_j)$ be a triangulation of the corresponding parameter polytope into triangles ($d = 3$) or line segments ($d = 2$), see e.g. [Ste08, Chap. 10] or [SS11, Sect. 4.1]. Let $\mathscr{T}^h(\Gamma)$ denote the associated subdivision of Γ. Next we define the boundary element space

$$Z^h(\Gamma) := \big\{z \in L^2(\Gamma) : (z \circ \zeta_j)_{|\tau} \in \mathscr{P}_0 \ \forall \tau \in \mathscr{T}^h(Q_j), \quad \forall j = 1, \ldots, N\big\} \tag{1.57}$$

of piecewise constant functions which is *conforming* in $H^{-1/2}(\Gamma)$. We now apply the Galerkin method to (1.56):

$$\text{find } t_h \in Z^h(\Gamma): \quad \langle w_h, Vt_h\rangle_\Gamma = \langle w_h, (\tfrac{1}{2}I + K)\, g_D\rangle_\Gamma \qquad \forall w_h \in Z^h(\Gamma)\,. \tag{1.58}$$

As we know from Céa's lemma (Lemma 1.5), the discretization error $t - t_h$ can be estimated in terms of the approximation error of the space $Z^h(\Gamma)$.

Definition 1.88. Let Ω^{int} be a Lipschitz polytope, then the boundary Γ splits into a number of disjoint flat faces ($d = 3$) or straight edges ($d = 2$): $\Gamma = \bigcup_{j=1}^J \overline{\Gamma}_j$. We define the space

$$H^s_{\mathrm{pw}}(\Gamma) := \{v \in L^2(\Gamma) : v_{|\Gamma_j} \in H^s(\Gamma_j)\} \qquad \text{for } s \in [0, 1]$$

of piecewise H^s functions, equipped with the norm

$$\|v\|_{H^s_{\mathrm{pw}}(\Gamma)} := \left(\|v\|^2_{L^2(\Gamma)} + \sum_{j=1}^{J} \|v\|^2_{H^s(\Gamma_j)}\right)^{1/2}.$$

Furthermore, we define

$$H^{-s}_{\mathrm{pw}}(\Gamma) := \prod_{j=1}^{J} H^s(\Gamma_j)^* \qquad \text{for } s \in (0, 1],$$

equipped with the norm $\|w\|_{H^{-s}_{\mathrm{pw}}(\Gamma)} := \sum_{j=1}^{J} \|w_j\|_{H^s(\Gamma_j)^*}$.

Lemma 1.89. *Let Γ the boundary of a Lipschitz polytope. Let $\sigma \in [-1, 0]$, $s \in [\sigma, 1]$ and $w \in H^s(\Gamma)$. Then there exists a constant C depending on Γ such that*

$$\inf_{w_h \in Z^h(\Gamma)} \|w - w_h\|_{H^\sigma(\Gamma)} \leq Ch^{s-\sigma} \|w\|_{H^s_{\mathrm{pw}}(\Gamma)} \qquad \forall w \in H^s_{\mathrm{pw}}(\Gamma).$$

Corollary 1.90. *If the exact Neumann trace $t = \gamma_1^{\mathrm{int}} u \in H^s_{\mathrm{pw}}(\Gamma)$ with $s \in [-\frac{1}{2}, 1]$, then we get the discretization error estimate*

$$\|t - t_h\|_{H^{-1/2}(\Gamma)} \leq C\, h^{s+\frac{1}{2}} \|t\|_{H^s_{\mathrm{pw}}(\Gamma)}.$$

In the optimal case $s = 1$, we get a convergence order of $\mathcal{O}(h^{3/2})$.

Since the normal vector is constant on each of the facets Γ_j, the regularity assumption $u \in H^{3/2+s}(\Omega)$ implies that $\nabla u \in (H^{1/2+s}(\Omega))^d$ and so $t = \nabla u \cdot n \in H^s_{\mathrm{pw}}(\Gamma)$ for $s \in (0, 1)$ due to the the trace theorem (Theorem 1.23). The last lemma discusses errors on the actual domain Ω^{int}.

Lemma 1.91. *Let t_h be the solution of* (1.58), *let u be the exact solution of the Laplace equation in Ω^{int} with $\gamma_0^{\mathrm{int}} u = g$, and let $\tilde{u}_h \in H^1(\Omega^{\mathrm{int}})$ be defined by*

$$\tilde{u}_h = \widetilde{V} t_h - \widetilde{W} g.$$

Then there is a constant C depending only on Ω^{int} such that

$$\|u - \tilde{u}_h\|_{H^1(\Omega^{\mathrm{int}})} \leq C\, \|t - t_h\|_{H^{-1/2}(\Gamma)}.$$

If Γ is the boundary of a Lipschitz polytope, then if $u \in H^{5/2-\varepsilon}(\Omega^{\mathrm{int}})$, we have $t \in H^{1-\varepsilon}_{\mathrm{pw}}(\Gamma)$ and so we get the convergence order of $\mathcal{O}(h^{3/2-\varepsilon})$.

For a more detailed presentation and estimates in other norms we refer, e.g., to [Ste08]. We note that other boundary integral equations can be used to get the Dirichlet data. Using the second equation in (1.49) leads to an integral equation of the second kind. Moreover, one could also follow an indirect approach (see, e.g., [SS11]).

1.3.7.2 BEM for the Neumann Problem

Again, we consider the interior Laplace problem. Let $g_N \in H^{-1/2}(\Gamma)$ be the given Neumann data and $u \in H^{1/2}(\Gamma)$ the unknown Dirichlet data. Then the second integral equation in (1.49) yields the variational formulation

$$\text{find } u \in H^{1/2}(\Gamma): \quad \langle D\,u, v\rangle_\Gamma = \langle (\tfrac{1}{2}I - K^\top)\,g_N, v\rangle_\Gamma \qquad \forall v \in H^{1/2}(\Gamma) \tag{1.59}$$

(other boundary integral equations could be used but are not considered here). From the properties of D and $K^\top$ (see Table 1.3), we see that this equation is only solvable if $\langle g_N, 1\rangle_\Gamma = 0$ and that the solution u is only unique up to an element from $\ker(D) = \text{span}\{1_\Gamma\}$.

Assume that Ω^{int} is a piecewise smooth Lipschitz domain as in Sect. 1.3.7.1. Analogously to (1.57), we define the space

$$V^h(\Gamma) := \{v \in \mathscr{C}(\Gamma) : (v \circ \zeta_j)_{|\tau} \in \mathscr{P}_1 \; \forall \tau \in \mathscr{T}^h(Q_j), \quad \forall j = 1, \ldots, N\}, \tag{1.60}$$

of piecewise linear functions which is conforming in $H^{1/2}(\Gamma)$. The Galerkin projection of (1.59) reads

$$\text{find } u_h \in V^h(\Gamma): \quad \langle D\,u_h, v_h\rangle_\Gamma = \langle (\tfrac{1}{2}I - K^\top)\,g_N, v_h\rangle_\Gamma \qquad \forall v_h \in V^h(\Gamma). \tag{1.61}$$

Lemma 1.92. *Let u be the solution of* (1.59) *and let $u_h \in V^h(\Gamma)$ be defined as in* (1.61)*. Furthermore, let Ω be a $\mathscr{C}^{k,1}$ domain (cf. Sect. 1.2.1.1). If the exact Dirichlet trace fulfills $u \in H^s(\Gamma)$ with $s \in (\frac{1}{2}, 2]$ and $s \le \max(3/2, k+1)$, then*

$$|u - u_h|_{H^{1/2}(\Gamma)} \le C\,h^{s-1/2}\,|u|_{H^s(\Gamma)}\,.$$

In the optimal case $s = 2$ (if the solution u lies in $H^{5/2}(\Omega^{\text{int}})$), we get a convergence order of $\mathscr{O}(h^{3/2})$.

Proof. Thanks to the properties of D and Céa's lemma, we have

$$|u - u_h|_{H^{1/2}(\Gamma)} \leq C \inf_{v_h \in V^h(\Gamma)} |u - v_h|_{H^{1/2}(\Gamma)} .$$

The approximation result from Lemma 1.44 can be transferred to the boundary,

$$\inf_{v_h \in V^h(\Gamma)} \|u - v_h\|_{H^{1/2}(\Gamma)} \leq C\, h^{s-1/2}\, |u|_{H^s(\Gamma)} ,$$

see, e.g., [Ste08, Theorem 10.9]. This concludes the proof. □

1.3.7.3 BEM Matrices and Data-Sparse Approximation

First, let us recall the Galerkin equations of the Dirichlet and the Neumann problem,

$$\begin{aligned} &\text{find } t_h \in Z^h(\Gamma): \quad \langle w_h, V t_h\rangle_\Gamma = \langle w_h, (\tfrac{1}{2}I + K)g_D\rangle_\Gamma \qquad \forall w_h \in Z^h(\Gamma),\\ &\text{find } u_h \in V^h(\Gamma): \quad \langle D\, u_h, v_h\rangle_\Gamma = \langle(\tfrac{1}{2}I - K^\top)g_N, v_h\rangle_\Gamma \qquad \forall v_h \in V^h(\Gamma). \end{aligned}$$

We assume that the Dirichlet data g_D in the first problem is in $V^h(\Gamma)$ (i.e., piecewise linear) and that the Neumann data g_N is in $Z^h(\Gamma)$ (i.e., piecewise constant). Otherwise, we project the given data (or we interpolate it if it is sufficiently smooth).

Introducing a numbering of the nodes $x_1, \ldots, x_{n_N}$ and the elements $\tau_1, \ldots, \tau_{n_T}$ of $\mathcal{T}^h(\Gamma)$, we obtain bases for the two approximation spaces:

$$V^h(\Gamma) = \operatorname{span}\{\varphi_i\}_{i=1}^{n_N}, \qquad Z^h(\Gamma) = \operatorname{span}\{\psi_k\}_{k=1}^{n_T}. \tag{1.62}$$

Here, φ_i is the nodal basis function being 1 at the i-th node and zero elsewhere, and ψ_k is 1 on the k-th element and zero elsewhere. These two bases define the boundary element matrices **D**, **V**, **K**, and **M** given by their entries

$$\left.\begin{aligned} D_{ij} &:= \langle D\,\varphi_j, \varphi_i\rangle_\Gamma \quad K_{ki} := \langle \psi_k, K\,\varphi_i\rangle_\Gamma \\ V_{kl} &:= \langle \psi_k, V\,\psi_l\rangle_\Gamma \quad M_{ki} := \langle \psi_k, \varphi_i\rangle_\Gamma = \int_\Gamma \psi_k\, \varphi_i \,\mathrm{d}s \end{aligned}\right\} \tag{1.63}$$

for $i, j = 1, \ldots, n_N$, and $k, l = 1, \ldots, n_T$.

Let $\mathbf{g}_D \leftrightarrow g_D \in V^h(\Gamma)$ and $\mathbf{g}_N \leftrightarrow g_N \in Z^h(\Gamma)$ be the vector representations with respect to the bases in (1.62), respectively. Then the problems from above read

$$\begin{aligned} \mathbf{V}\,\mathbf{t} &= (\tfrac{1}{2}\mathbf{M} + \mathbf{K})\mathbf{g}_D ,\\ \mathbf{D}\,\mathbf{u} &= (\tfrac{1}{2}\mathbf{M}^\top - \mathbf{K}^\top)\mathbf{g}_N . \end{aligned}$$

Except for the mass matrix **M**, which is apparently sparse, all the boundary element matrices in (1.63) are dense (despite the fact that the basis functions have local support). This is because of the non-local operators involving the fundamental solution; for assembling the matrices, double integrals have to be evaluated.

However, the fundamental solution $U^*(x, y)$ decays exponentially when moving away from the singularity at $x = y$.

The notion *fast boundary element methods* embraces techniques to design methods making use of this phenomenon in order to realize a fast (i.e., quasi-optimal) application of (approximate) BEM matrices, or even achieve data-sparse approximation of BEM matrices (i.e., quasi-optimal memory requirements). Only with fast BEM, quasi-optimal solvers of BEM systems become possible. For a quick overview we refer to [Ste08, RS07]. The probably most important techniques are (in alphabetical order)

- The *adaptive cross approximation* (ACA) [Beb00],
- The *fast multipole method* [Rok85, CGR99],
- The *panel clustering method* [HN89], and
- *Wavelet approximations* [Sch98a].

In these methods, the computational complexity of a matrix vector multiplication is typically $\mathscr{O}(N \log N)$ where N is the number of unknowns on the boundary.

The data structure for a data-sparse representation is called *hierarchical matrix* or *$\mathscr{H}$-matrix* using a hierarchy of low-rank approximations, where the maximum rank is bounded. The class of such matrices induces the so-called *$\mathscr{H}$-arithmetic*, cf. [Hac99, HK00], see also the recent monograph by Bebendorf [Beb08]. Within this arithmetic, one can compute an *$\mathscr{H}$-LU* factorization, cf. [Beb05], i.e., an approximate factorization of an $\mathscr{H}$-matrix, where the two factors are again $\mathscr{H}$-matrices. The factorization can also be done in a $\mathscr{O}(N \log N)$-complexity, although it additionally depends on the maximum rank. We mention two software packages:

HLib (hierarchical matrices)
http://www.hlib.org
See also [BGH06].

AHMED (another software library on hierarchical matrices for elliptic differential equations)
http://www.math.uni-leipzig.de/~bebendorf/AHMED.html

1.3.8 BEM Approximation of S^{int}, S^{ext}, and $\widetilde{N}$

In this section, we derive symmetric approximations of the two Steklov-Poincaré operators S^{int}, S^{ext} and of the Newton volume potential using discretized boundary integral equations.

1.3.8.1 BEM Approximation of S^{int}, S^{ext}

Assume that Ω^{int} is a Lipschitz polytope and consider a triangulation $\mathscr{T}^h(\Gamma)$ of its boundary Γ. We approximate the Dirichlet trace $\gamma_0^{\text{int}} u$ by a piecewise linear function $u_h \in V^h(\Gamma)$, see (1.60). The Neumann trace has two different kinds of approximations: one the one hand, we can approximate $\gamma_1^{\text{int}} u$ by a functional $t_h \in V^h(\Gamma)^* \subset H^{-1/2}(\Gamma)$, on the other hand, we can approximate it by a piecewise constant function $w_h \in Z^h(\Gamma) \subset H^{-1/2}(\Gamma)$, see (1.57).

From the symmetric representations in (1.54), we can conclude that

$$\begin{aligned} S^{\text{int}} v &= D v + (\tfrac{1}{2} I + K^\top) w^{\text{int}}(v), \\ S^{\text{ext}} v &= D v + (\tfrac{1}{2} I - K^\top) w^{\text{ext}}(v) - S^{\text{corr}} v, \end{aligned} \tag{1.64}$$

where $w^{\text{int}}(v), w^{\text{ext}}(v) \in H^{-1/2}(\Gamma)$ are given by

$$\begin{aligned} \langle z, V w^{\text{int}}(v) \rangle &= \langle z, (\tfrac{1}{2} I + K) v \rangle \qquad \forall z \in H^{-1/2}(\Gamma), \\ \langle z, V w^{\text{ext}}(v) \rangle_\Gamma &= \langle z, (\tfrac{1}{2} I - K) v \rangle \qquad \forall z \in H^{-1/2}(\Gamma). \end{aligned} \tag{1.65}$$

The Galerkin projection of (1.65) to $Z^h(\Gamma)$ reads: Find $w_h^{\text{int}}(v), w_h^{\text{ext}}(v) \in Z^h(\Gamma)$ such that

$$\begin{aligned} \langle z_h, V w_h^{\text{int}}(v) \rangle &= \langle z_h, (\tfrac{1}{2} I + K) v \rangle \qquad \forall z_h \in Z^h(\Gamma), \\ \langle z_h, V w_h^{\text{ext}}(v) \rangle &= \langle z_h, (\tfrac{1}{2} I - K) v \rangle \qquad \forall z_h \in Z^h(\Gamma). \end{aligned} \tag{1.66}$$

Replacing $w^{\text{int}}(v)$, $w^{\text{ext}}(v)$ in (1.64) by $w_h^{\text{int}}(v)$, $w_h^{\text{ext}}(v)$ respectively leads to the approximations $S_{\text{BEM}}^{\text{int}}, S_{\text{BEM}}^{\text{ext}} : H^{1/2}(\Gamma) \to H^{-1/2}(\Gamma)$ defined by

$$\begin{aligned} S_{\text{BEM}}^{\text{int}} v &:= D v + (\tfrac{1}{2} I + K^\top) w_h^{\text{int}}(v), \\ S_{\text{BEM}}^{\text{ext}} v &:= D v + (\tfrac{1}{2} I - K^\top) w_h^{\text{ext}}(v) - S_{\text{BEM}}^{\text{corr}} v, \end{aligned} \tag{1.67}$$

where for $d = 3$, $S_{\text{BEM}}^{\text{corr}} = 0$. For the two-dimensional case, see Sect. 1.3.8.4. According to [Ste03b] we have the quasi-optimal error estimate

$$\|(S^{\text{int}} - S_{\text{BEM}}^{\text{int}}) v\|_{H^{-1/2}(\Gamma)} \le C \inf_{z_h \in Z^h(\Gamma)} \|S^{\text{int}} v - z_h\|_{H^{-1/2}(\Gamma)} \qquad \forall v \in H^{1/2}(\Gamma),$$

and the analogous estimate for S^{ext} if $d = 3$. Hence, the error is controlled by the approximation property of the trial space $Z^h(\Gamma)$, see Lemma 1.89.

The Galerkin projections of $S_{\text{BEM}}^{\text{int}}$ and $S_{\text{BEM}}^{\text{ext}}$ to the space $V^h(\Gamma)$ can be represented in matrix form by

$$\begin{aligned} \mathbf{S}^{\text{int}} &= \mathbf{D} + (\tfrac{1}{2} \mathbf{M}^\top + \mathbf{K}^\top) \mathbf{V}^{-1} (\tfrac{1}{2} \mathbf{M} + \mathbf{K}), \\ \mathbf{S}^{\text{ext}} &= \mathbf{D} + (\tfrac{1}{2} \mathbf{M}^\top - \mathbf{K}^\top) \mathbf{V}^{-1} (\tfrac{1}{2} \mathbf{M} - \mathbf{K}), \end{aligned} \tag{1.68}$$

where the second formula only holds for $d = 3$. The approximations $S^{\text{int}}_{\text{BEM}}$ and $S^{\text{ext}}_{\text{BEM}}$ are self-adjoint and they are positive definite on the same subspaces as the original operators S^{int}, S^{ext}. In case of exact arithmetics we preserve the kernel of S^{int}, i.e.,

$$\ker(S^{\text{int}}_{\text{BEM}}) = \operatorname{span}\{1_\Gamma\}. \tag{1.69}$$

1.3.8.2 BEM Approximation of the Newton Potential

Recall the definition (Definition 1.85)

$$(\widetilde{N} f)(x) := \int_{\Omega^{\text{int}}} U^*(x, y)\, f(y)\, \mathrm{d}y \qquad \text{for } x \in \Omega^{\text{int}}.$$

of the Newton volume potential $\widetilde{N}$ and recall Lemma 1.86 from Sect. 1.3.5. Assume that we have an integral representation of $\gamma_0 \widetilde{N} f$, then we can derive a Galerkin formulation for the first integral equation in (1.55). Using the identity $N = \gamma_1^{\text{int}} \widetilde{N} = V^{-1} \gamma_0^{\text{int}} \widetilde{N}$, one can derive an approximation N_{BEM} of N analogously to Sect. 1.3.8.1.

However, only in special cases, it is possible to come up with an efficient representation of $\gamma_0 \widetilde{N} f \in H^{1/2}(\Gamma)$ that is comparable to the representations of the boundary integral operators, see [Ste03b] and [BHK08, Kho08b, Kho08a, OSU06, Urt08]. In the rest of this book we will use the boundary element method only on (sub)domains with vanishing volume sources.

1.3.8.3 Spectral Estimates of the Steklov-Poincaré Approximants

In this final section, we work out spectral equivalence relations between the exact Steklov Poincaré operators S^{int} and S^{ext} and their approximants from Sect. 1.3.8.1.

Lemma 1.93. *Let Ω^{int} be a Lipschitz polytope, let $\mathcal{T}^h(\Gamma)$ be any triangulation of Γ, and let S^{int}_{BEM}, S^{ext}_{BEM} be the approximants defined in* (1.66) *and* (1.67).

(i) The interior operators satisfy

$$\frac{c_0}{c_K} \langle S^{\text{int}} v, v\rangle \le \langle S^{\text{int}}_{BEM} v, v\rangle \le \langle S^{\text{int}} v, v\rangle \qquad \forall v \in H^{1/2}(\Gamma), \tag{1.70}$$

where c_0, c_K are the constants from Lemma 1.77.

(ii) If $d = 3$ then

$$\begin{aligned} \langle S^{\text{ext}}_{BEM} v, v\rangle &\le \langle S^{\text{ext}} v, v\rangle && \forall v \in H^{1/2}(\Gamma),\\ \tfrac{c_0}{c_K} \langle S^{\text{ext}} v, v\rangle &\le \langle S^{\text{ext}}_{BEM} v, v\rangle && \forall v \in H^{1/2}_*(\Gamma),\\ \tfrac{c_0}{c_K} \langle S^{\text{int}} v, v\rangle &\le \langle S^{\text{ext}}_{BEM} v, v\rangle && \forall v \in H^{1/2}(\Gamma). \end{aligned} \tag{1.71}$$

(iii) If $d = 3$ *then*

$$c^{\mathrm{ext}}\left[\langle S^{\mathrm{int}} v, v\rangle + \frac{1}{\mathrm{diam}(\Omega^{\mathrm{int}})}\|v\|^2_{L^2(\Gamma)}\right] \leq \langle S^{\mathrm{ext}}_{BEM} v, v\rangle \qquad \forall v \in H^{1/2}(\Gamma),$$

where the constant $c^{\mathrm{ext}} > 0$ *depends on* Γ *but not on its diameter.*

Proof. From (1.64) to (1.67), one can show that

$$\left.\begin{aligned}
\langle S^{\mathrm{int}} v, v\rangle &= \langle D\, v, v\rangle + \|w^{\mathrm{int}}(v)\|^2_V \\
\langle S^{\mathrm{int}}_{\mathrm{BEM}} v, v\rangle &= \langle D\, v, v\rangle + \|w^{\mathrm{int}}_h(v)\|^2_V \\
\langle S^{\mathrm{ext}} v, v\rangle &= \langle D\, v, v\rangle + \|w^{\mathrm{ext}}(v)\|^2_V - \langle S^{\mathrm{corr}}, v, v\rangle \\
\langle S^{\mathrm{ext}}_{\mathrm{BEM}} v, v\rangle &= \langle D\, v, v\rangle + \|w^{\mathrm{ext}}_h(v)\|^2_V - \langle S^{\mathrm{corr}}_{\mathrm{BEM}} v, v\rangle
\end{aligned}\right\} \quad \forall v \in H^{1/2}(\Gamma).$$

and due to Lemma 1.6,

$$\left.\begin{aligned}
\|w^{\mathrm{int}}_h(v)\|_V &\leq \|w^{\mathrm{int}}(v)\|_V \\
\|w^{\mathrm{ext}}_h(v)\|_V &\leq \|w^{\mathrm{ext}}(v)\|_V
\end{aligned}\right\} \quad \forall v \in H^{1/2}(\Gamma). \tag{1.72}$$

This implies that

$$\langle S^{\mathrm{int}}_{\mathrm{BEM}} v, v\rangle \leq \langle S^{\mathrm{int}} v, v\rangle, \qquad \langle S^{\mathrm{ext}}_{\mathrm{BEM}} v, v\rangle \leq \langle S^{\mathrm{ext}} v, v\rangle \qquad \forall v \in H^{1/2}(\Gamma),$$

where the latter inequality only holds for $d = 3$ (and $S^{\mathrm{corr}} = S^{\mathrm{corr}}_{\mathrm{BEM}} = 0$). This shows the upper bounds in (i) and (ii). We now turn to the lower bounds in (i) and (ii). First, let $\tilde{v} \in H^{1/2}_*(\Gamma)$ be arbitrary but fixed. The fact that V is SPD, Lemmas 1.77 and 1.83(i) yield

$$\begin{aligned}
\langle S^{\mathrm{int/ext}}_{\mathrm{BEM}} \tilde{v}, \tilde{v}\rangle &= \langle D\, \tilde{v}, \tilde{v}\rangle + \|w^{\mathrm{int/ext}}_h(\tilde{v})\|^2_V \\
&\geq \langle D\, \tilde{v}, \tilde{v}\rangle \geq c_0\langle V^{-1}\tilde{v}, \tilde{v}\rangle \geq \frac{c_0}{c_K}\langle S^{\mathrm{int/ext}} \tilde{v}, \tilde{v}\rangle,
\end{aligned} \tag{1.73}$$

where again, the estimate for the exterior operators only holds for $d = 3$. Secondly, for $v \in H^{1/2}(\Gamma)$, we can find $\tilde{v} \in H^{1/2}_*(\Gamma)$ and $v_0 \in \mathbb{R}$ such that $v = v_0 + \tilde{v}$. The same arguments as above yield

$$\langle S^{\mathrm{int/ext}}_{\mathrm{BEM}} v, v\rangle \geq \langle D\, v, v\rangle = \langle D\, \tilde{v}, \tilde{v}\rangle \geq \frac{c_0}{c_K}\langle S^{\mathrm{int}} \tilde{v}, \tilde{v}\rangle = \frac{c_0}{c_K}\langle S^{\mathrm{int}} v, v\rangle,$$

which concludes the proof of the lower bounds in (i) and (ii). Finally, to get (iii), we set

$$w^{\mathrm{ext}}_1(v) := \frac{\langle(\frac{1}{2}I - K)v, 1_\Gamma\rangle}{\langle V\, 1_\Gamma, 1_\Gamma\rangle},$$

which is equivalent to

$$w_1^{\text{ext}}(v) \in \operatorname{span}\{1_\Gamma\} : \quad \langle z, V\, w_1^{\text{ext}}(v)\rangle \;=\; \langle z, (\tfrac{1}{2}I - K)v\rangle_\Gamma \quad \forall z \in \operatorname{span}\{1_\Gamma\}.$$

Since $\operatorname{span}\{1_\Gamma\} \subset Z^h(\Gamma)$ for any triangulation of Γ, Lemma 1.6 implies that

$$\left\langle w_h^{\text{ext}}(v),\, V\, w_h^{\text{ext}}(v)\right\rangle \;\geq\; \left\langle w_1^{\text{ext}}(v),\, V\, w_1^{\text{ext}}(v)\right\rangle \;=\; \frac{\langle 1_\Gamma, (\frac{1}{2}I - K)v\rangle^2}{\|1_\Gamma\|_V^2} \;=\; |\psi(v)|^2,$$

where the linear functional $\psi : H^{1/2}(\Gamma) \to \mathbb{R}$ is defined as

$$\psi(v) \;:=\; \frac{\langle 1_\Gamma, (\frac{1}{2}I - K)v\rangle}{\|1_\Gamma\|_V}.$$

Using the above estimates and Definition 1.41, we find that

$$\langle S_{\text{BEM}}^{\text{ext}} v, v\rangle \;\geq\; \frac{c_0}{c_K}\,\langle S^{\text{int}} v, v\rangle + |\psi(v)|^2 \;=\; \frac{c_0}{c_K}\left(|\mathscr{H} v|^2_{H^1(\Omega^{\text{int}})} + |\psi(v)|^2\right), \tag{1.74}$$

where $\mathscr{H}$ denotes the harmonic extension from Γ into Ω^{int}. The functional $\psi(v)$ is bounded in $H^{1/2}(\Gamma)$ and a short calculation reveals that

$$\psi(1_\Gamma) \;=\; |\Gamma|\, \|1_\Gamma\|_V^{-1} \;\neq\; 0.$$

Since the corresponding normalized functional $\widetilde{\psi}(v) := \psi(v)/\psi(1_\Gamma)$ reproduces constants, Lemma 1.33 and Cauchy's inequality yield that

$$\begin{aligned}\tfrac{1}{2}\,\|\mathscr{H} v\|^2_{L^2(\Omega^{\text{int}})} \leq \left(C_P(\Omega^{\text{int}}) + \sqrt{1 + C_P(\Omega^{\text{int}})^2}\,\|\widetilde{\psi}\|\right)^2 \operatorname{diam}(\Omega^{\text{int}})^2\, |\mathscr{H} v|^2_{H^1(\Omega^{\text{int}})} \\ + |\Omega^{\text{int}}|\, |\widetilde{\psi}(v)|^2 .\end{aligned} \tag{1.75}$$

Finally, one can show that

$$\left.\begin{aligned} \|\widetilde{\psi}\| &\lesssim 1 \\ |\Omega^{\text{int}}|\, |\widetilde{\psi}(v)|^2 &\lesssim \operatorname{diam}(\Omega^{\text{int}})^2\, |\psi(v)|^2 \\ \frac{1}{\operatorname{diam}(\Omega^{\text{int}})}\, \|v\|^2_{L^2(\Gamma)} &\lesssim \frac{1}{\operatorname{diam}(\Omega^{\text{int}})^2}\, \|\mathscr{H} v\|^2_{L^2(\Omega^{\text{int}})} + |\mathscr{H} v|^2_{H^1(\Omega^{\text{int}})} \end{aligned}\right\} \quad \forall v \in H^{1/2}(\Gamma), \tag{1.76}$$

where the hidden constants only depend on the shape of Ω^{int}. Hence, we can conclude from (1.74) to (1.76) that

$$\langle S_{\mathrm{BEM}}^{\mathrm{ext}} v, v\rangle \gtrsim |\mathscr{H}v|^2_{H^1(\Omega^{\mathrm{int}})} + \mathrm{diam}(\Omega^{\mathrm{int}})^2 \, \|\mathscr{H}v\|^2_{L^2(\Omega^{\mathrm{int}})} \,.$$

The fact that $|\mathscr{H}v|^2_{H^1(\Omega^{\mathrm{int}})} = \langle S^{\mathrm{int}}v, v\rangle$ concludes the proof. □

Finally, we get a spectral equivalence between the FEM- and the BEM-approximant of S^{int}.

Corollary 1.94. *For the Laplace problem in Ω and for any shape regular triangulation $\mathscr{T}^h(\Omega)$, we have*

$$\frac{c_0}{c_K\, C_{SZ}} \langle S_{FEM}^{\mathrm{int}} v, v\rangle \le \langle S_{BEM}^{\mathrm{int}} v, v\rangle \le \langle S_{FEM}^{\mathrm{int}} v, v\rangle = |\mathscr{H}^h v|^2_{H^1(\Omega)} \quad \forall v \in V^h(\Gamma).$$

Proof. Follows immediately by combining Corollary 1.57 and Lemma 1.93. □

1.3.8.4 BEM Approximation of S^{ext} for $d = 2$

Let $d = 2$ and recall that

$$\ker(S^{\mathrm{ext}}) = \mathrm{span}\{1_\Gamma\}.$$

To preserve the kernel for the BEM approximant, we set

$$S_{\mathrm{BEM}}^{\mathrm{corr}} v := \frac{\langle (\frac{1}{2}I - K^\top) w_h^{\mathrm{ext}}(1_\Gamma), v\rangle}{\langle (\frac{1}{2}I - K^\top) w_h^{\mathrm{ext}}(1_\Gamma), 1_\Gamma\rangle} (\tfrac{1}{2}I - K^\top) w_h^{\mathrm{ext}}(1_\Gamma) \qquad \text{for } v \in H^{1/2}(\Gamma).$$

Then, by construction and since $\ker(D) = \mathrm{span}\{1_\Gamma\}$,

$$S_{\mathrm{BEM}}^{\mathrm{ext}} 1_\Gamma = (\tfrac{1}{2}I - K^\top)\, w_h^{\mathrm{ext}}(1_\Gamma) - S_{\mathrm{BEM}}^{\mathrm{corr}} 1_\Gamma = 0.$$

In vector notation, we first compute

$$\mathbf{z} := (\tfrac{1}{2}\mathbf{M} - \mathbf{K})\mathbf{V}^{-1}(\tfrac{1}{2}\mathbf{M} - \mathbf{K})\mathbf{1},$$

where $\mathbf{1}$ is the vector of ones. Note that we need the scaling condition from Assumption 1.72 to ensure the invertibility of $\mathbf{V}$. Finally, the matrix $\mathbf{S}^{\mathrm{ext}}$ corresponding to $S_{\mathrm{BEM}}^{\mathrm{ext}}$ is given by

$$\mathbf{S}^{\mathrm{ext}} = \mathbf{D} + (\tfrac{1}{2}\mathbf{M}^\top - \mathbf{K}^\top)\mathbf{V}^{-1}(\tfrac{1}{2}\mathbf{M} - \mathbf{K}) - \frac{\mathbf{z}\,\mathbf{z}^\top}{\mathbf{z}^\top \mathbf{1}},$$

which is a rank-one correction of the matrix $\mathbf{S}^{\mathrm{ext}}$ in (1.68).

Rigorous spectral equivalences between S^{ext} and $S_{\mathrm{BEM}}^{\mathrm{ext}}$ for $d = 2$ are not known to the author.

Chapter 2
One-Level FETI/BETI Methods

This chapter deals with tearing and interconnecting methods based on the finite element method (FEM) and the boundary element method (BEM). Here we allow a mixture of FEM and BEM within a single discretization of a PDE: the computational domain is partitioned ("torn") into several non-overlapping subdomains, and on each individual subdomain one may choose FEM or BEM as the local discretization. The coupling ("interconnecting") of these local discretizations is maintained by Lagrange multipliers. Additionally, the tearing and interconnecting framework is used to construct fast solvers for the resulting global system of equations.

The term "one-level" in the title of this chapter refers to a special treatment of the so-called *floating* subdomains that do not touch the Dirichlet boundary. An alternative treatment is used in the dual-primal methods, see Chap. 5.

There are two subclasses of one-level methods: the *classical* formulation, and the *total* or *all-floating* formulation. The *classical* finite element tearing and interconnecting (FETI) method was proposed by Farhat and Roux [FR91, FR92, FR94] as a solver for large-scale finite element systems. Note that some basic ideas can already be found in an earlier work by Glowinsky and Wheeler [GW88] on certain mixed methods. The FETI method was enhanced with the so-called Dirichlet preconditioner by Farhat, Mandel, and Roux [FMR94]. The latter method was first analyzed by Mandel and Tezaur [MT96], who showed that the condition number grows at most as $C\,(1 + \log(H/h))^3$ where H is the subdomain size and h the element size. See also [Tez98, Bre02, Bre03a] for further analyses. Klawonn and Widlund [KW01] proposed new preconditioners including an earlier algorithm by Rixen and Farhat [RF98a, RF99] and including ideas from balancing Neumann-Neumann methods (see Sect. 2.3). They also generalized the theory in several different respects (e.g., the case of three-dimensions and so-called fully redundant Lagrange multipliers) and could show the improved bound $C\,(1+\log(H/h))^2$. The boundary element tearing and interconnecting (BETI) and the coupled FETI/BETI methods were proposed and analyzed by Langer and Steinbach [LS03, LS05].

The *total* FETI method and the *all-floating* BETI method were introduced independently by Dostál, Horák, and Kučera [DHK06] and Of [Of06, Of08],

C. Pechstein, *Finite and Boundary Element Tearing and Interconnecting Solvers for Multiscale Problems*, Lecture Notes in Computational Science and Engineering 90,
DOI 10.1007/978-3-642-23588-7_2, © Springer-Verlag Berlin Heidelberg 2013

respectively (see also [OS09]), and the corresponding preconditioner was analyzed in [Pec08b]. Note that these methods have been successfully generalized to mechanical contact problems, see e.g. [BDS08, DKV+10].

The remainder of this chapter is organized as follows. In Sect. 2.1 we work out a (discrete) skeleton formulation using the (approximate) Steklov-Poincaré operators from Chap. 1. Section 2.2 describes the classical FETI/BETI and the all-floating (total) FETI/BETI method in detail and discusses implementation issues. Section 2.3 briefly introduces the related balancing Neumann-Neumann method. The analysis of one-level FETI/BETI is performed in two steps. In Sect. 2.4 we analyze the unpreconditioned method (which turns out to be sub-optimal) and we begin to analyze the so-called *scaled Dirichlet preconditioner* on an abstract level. After providing a set of technical tools (Sect. 2.5) we will conclude the analysis in Sect. 2.6. Finally, we provide some numerical results in Sect. 2.7, and we discuss generalizations to other equations and/or discretization spaces in Sect. 2.8.

2.1 Skeleton Formulations

FETI and BETI are iterative substructuring methods based on a non-overlapping decomposition of the computational domain Ω. A good starting point for these methods, especially for BETI, is a discrete skeleton formulation, which is derived from a continuous one.

2.1.1 Continuous Skeleton Formulation

Let $\Omega \subset \mathbb{R}^d$ ($d = 2$ or 3) be a bounded Lipschitz domain whose boundary $\partial\Omega$ consists of a Dirichlet boundary $\Gamma_D = \overline{\Gamma_D}$ with positive surface measure and a Neumann boundary $\Gamma_N = \partial\Omega \setminus \Gamma_D$. The outward unit normal vector to $\partial\Omega$ is denoted by n. We consider the weak form of the potential equation (1.21): find $u \in H^1(\Omega)$, $u_{|\Gamma_D} = g_D$ such that

$$\underbrace{\int_\Omega \mathscr{A}\, \nabla u \cdot \nabla v \,\mathrm{d}x}_{=:\, a(u,v)} = \underbrace{\int_\Omega f_\Omega\, v \,\mathrm{d}x + \int_{\Gamma_N} g_N\, v \,\mathrm{d}s}_{=:\, \langle f, v\rangle} \qquad \forall v \in H_D^1(\Omega), \tag{2.1}$$

where $H_D^1(\Omega) := \{v \in H^1(\Omega) : v_{|\Gamma_D} = 0\}$ and

$$f_\Omega \in L^2(\Omega), \qquad g_N \in L^2(\Gamma_N), \qquad g_D \in H^{1/2}(\Gamma_D) \tag{2.2}$$

are given. We assume that the coefficient $\mathscr{A}$ fulfills Condition (1.22) from Lemma 1.39 (p. 20), such that we have unique solvability.

Table 2.1 Geometric sets and quantities associated to the subdomain decomposition

$H_i := \operatorname{diam}(\Omega_i)$	Subdomain diameter
$\Gamma_{ij} := (\partial\Omega_i \cap \partial\Omega_j) \setminus \Gamma_D$	Subdomain interface
$\Gamma := \bigcup_{i \neq j} \Gamma_{ij}$	Interface
$\Gamma_S := \bigcup_i \partial\Omega_i$	Skeleton

Let $\{\Omega_i\}_{i=1}^s$ be a non-overlapping partition of Ω into open Lipschitz domains Ω_i (called *subdomains* or *substructures*) such that

$$\overline{\Omega} = \bigcup_{i=1}^{s} \overline{\Omega}_i, \qquad \Omega_i \cap \Omega_j = \emptyset \qquad \text{for } i \neq j. \tag{2.3}$$

Furthermore, we introduce a couple of geometric quantities summarized in Table 2.1 (recall that Γ_D is closed!).

Thanks to the assumptions on f_Ω and g_N, we have the splitting property

$$a(u, v) = \sum_{i=1}^{s} \underbrace{\int_{\Omega_i} \mathscr{A}\nabla u \cdot \nabla v \, \mathrm{d}x}_{=: \, a_i(u_{|\Omega_i}, v_{|\Omega_i})}, \qquad \langle f, v \rangle = \sum_{i=1}^{s} \underbrace{\Big(\int_{\Omega_i} f_\Omega \, v \, \mathrm{d}x + \int_{\partial\Omega_i \cap \Gamma_N} g_N \, v \, \mathrm{d}s \Big)}_{=: \, \langle f_i, v_{|\Omega_i} \rangle} \tag{2.4}$$

with $a_i : H^1(\Omega_i) \times H^1(\Omega_i) \to \mathbb{R}$ and $f_i \in H^1(\Omega_i)^*$.

Remark 2.1. We can also allow for a general functional $f \in H^1(\Omega)^*$ (not necessarily of the form (2.1)), provided that we have a splitting into subdomain functionals $f_i \in H^1(\Omega_i)^*$ as in (2.4).

For each subdomain Ω_i, let $S_i : H^{1/2}(\partial\Omega_i) \to H^{-1/2}(\partial\Omega_i)$ denote the Steklov-Poincaré operator corresponding to the bilinear form $a_i(\cdot,\cdot)$ and $N_i : H^1(\Omega_i)^* \to H^{-1/2}(\partial\Omega_i)$ the corresponding Newton potential, see Definition 1.41. Furthermore, we define the skeletal spaces

$$H^{1/2}(\Gamma_S) := \{ v \in L^2(\Gamma_S) : \exists \tilde{v} \in H^1(\Omega) : v = \tilde{v}_{|\Gamma_S} \}, \tag{2.5}$$

$$H_D^{1/2}(\Gamma_S) := \{ v \in H^{1/2}(\Gamma_S) : v_{|\Gamma_D} = 0 \}. \tag{2.6}$$

Lemma 2.2. *The variational formulation* (2.1) *is equivalent to finding $u \in H^1(\Omega)$ with $u_{|\Gamma_D} = g_D$ such that*

$$\sum_{i=1}^{s} \langle S_i \, u_{|\partial\Omega_i}, v_{|\partial\Omega_i} \rangle = \sum_{i=1}^{s} \langle N_i \, f_i, v_{|\partial\Omega_i} \rangle \qquad \forall v \in H_D^{1/2}(\Gamma_S),$$

$$a_i(u_{|\Omega_i}, v_0) = \langle f_i, v_0 \rangle_{\Omega_i} \qquad \forall v_0 \in H_0^1(\Omega_i) \quad \forall i = 1, \ldots, s.$$

Proof. The equations in the second line follow immediately from (2.1). With these local equations fulfilled, Lemma 1.42(ii) implies that

$$a_i(u_{|\Omega_i}, v_{|\Omega_i}) - \langle f_i, v_{|\Omega_i}\rangle = \langle S_i\, u_{|\partial\Omega_i}, v_{|\partial\Omega_i}\rangle - \langle N_i\, f_i, v_{|\partial\Omega_i}\rangle \qquad \forall v \in H^1_D(\Omega).$$

This proves the equivalence. □

The system in Lemma 2.2 can be seen as an algorithm:

1. Determine $u_S \in H^{1/2}(\Gamma_S)$ with $u_{S|\Gamma_D} = g_D$ such that

$$\sum_{i=1}^{s} \langle S_i\, u_{S|\partial\Omega_i}, v_{|\partial\Omega_i}\rangle = \sum_{i=1}^{s} \langle N_i\, f_i, v_{|\partial\Omega_i}\rangle \qquad \forall v \in H_D^{1/2}(\Gamma_S). \tag{2.7}$$

2. Determine the local functions $u_i = u_{|\Omega_i}$ as the solution of the variational problems, find $u_i \in H^1(\Omega_i)$ with $u_{i|\partial\Omega_i} = u_{S|\partial\Omega_i}$ such that

$$a_i(u_i, v_0) = \langle f_i, v_0\rangle_{\Omega_i} \qquad \forall v_0 \in H^1_0(\Omega_i). \tag{2.8}$$

Problem (2.7) is called *skeletal variational formulation*. Under the assumptions of Lemma 1.39, it is straightforward to show (with Theorem 1.1) that Problem (2.7) is well-posed. Should we only be interested in the trace of the solution u on Γ_S, we can stop after step 1.

Sometimes, the following homogeneous version of (2.7) is convenient. Find $\tilde{u} \in H_D^{1/2}(\Gamma_S)$ such that

$$\sum_{i=1}^{s} \langle S_i\, \tilde{u}_{|\partial\Omega_i}, v_{|\partial\Omega_i}\rangle = \sum_{i=1}^{s} \langle N_i\, f_i - S_i\, \tilde{g}_{D|\partial\Omega_i}, v_{|\partial\Omega_i}\rangle \qquad \forall v \in H_D^{1/2}(\Gamma_S), \tag{2.9}$$

where $\tilde{g}_D \in H^{1/2}(\Gamma_S)$ fulfills $\tilde{g}_{D|\Gamma_D} = g_D$ (see Lemma 1.21). Then $u = \tilde{g}_D + \tilde{u}$ solves Problem (2.7).

2.1.2 Discrete Skeleton Formulations

Discrete skeleton formulations are obtained by applying a Galerkin method to (2.7) or (2.9).

Step 1. We project the equation to a finite-dimensional space. To this end, we consider a shape regular triangulation $\mathscr{T}^h(\Gamma_S)$ of the skeleton Γ_S into line segments (if $d = 2$) or triangles (if $d = 3$). We require that the subdomain boundaries $\partial\Omega_i$ are unions of elements from $\mathscr{T}^h(\Gamma_S)$.

The $H^{1/2}(\Gamma_S)$-conforming skeletal finite element space is defined as

$$V^h(\Gamma_S) := \{v \in \mathscr{C}(\Gamma_S) : v_{|\tau} \in \mathscr{P}_1 \quad \forall \tau \in \mathscr{T}^h(\Gamma_S)\}. \tag{2.10}$$

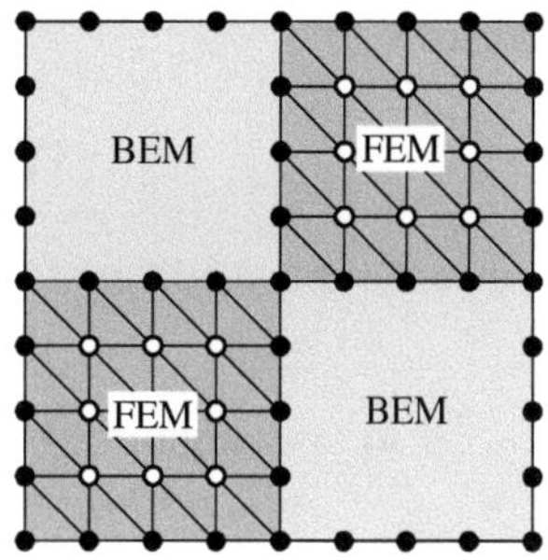

Fig. 2.1 Example of a skeleton mesh for a decomposition into four subdomains, extended to local meshes in two of the subdomains(• skeleton nodes, ∘ interior FE nodes)

For simplicity, we assume that the Dirichlet data g_D is piecewise linear too.[1] Hence, there is a unique function $\tilde{g}_D \in V^h(\Gamma_S)$ with $\tilde{g}_{D|\Gamma_D} = g_D$ which vanishes on all nodes except those on Γ_D.

Step 2. We use an *approximate* bilinear form and right hand side. Let $\mathscr{I}_{\mathrm{BEM}} \subset \{1, \ldots, s\}$ be an index set and assume that

$$\mathscr{A}_{|\Omega_i} = \alpha_i \, I, \qquad \forall i \in \mathscr{I}_{\mathrm{BEM}},$$

with constants $\alpha_i > 0$. In each subdomain, the restriction of $\mathscr{T}^h(\Gamma_S)$ to $\partial\Omega_i$ is a triangulation of $\partial\Omega_i$, simply denoted by $\mathscr{T}^h(\partial\Omega_i)$. For $i \notin \mathscr{I}_{\mathrm{BEM}}$, we extend $\mathscr{T}^h(\partial\Omega_i)$ to a shape regular triangulation $\mathscr{T}^h(\Omega_i)$ of Ω_i. For an example see Fig. 2.1. We now replace the local Steklov-Poincaré operators S_i and Newton potentials N_i by the FEM- and BEM-approximations from Sects. 1.2.6 and 1.3.8.1. With $\mathscr{I}_{\mathrm{FEM}} := \{1, \ldots, s\} \setminus \mathscr{I}_{\mathrm{BEM}}$, we set

$$S_{i,h} := \left\{ \begin{array}{l} S_{i,\mathrm{BEM}} \text{ if } i \in \mathscr{I}_{\mathrm{BEM}} \\ S_{i,\mathrm{FEM}} \text{ if } i \in \mathscr{I}_{\mathrm{FEM}} \end{array} \right\}, \qquad N_{i,h} := \left\{ \begin{array}{l} N_{i,\mathrm{BEM}} \text{ if } i \in \mathscr{I}_{\mathrm{BEM}} \\ N_{i,\mathrm{FEM}} \text{ if } i \in \mathscr{I}_{\mathrm{FEM}} \end{array} \right\}.$$

To ensure the invertibility of the discretized single layer potential operators occurring in $S_{i,\mathrm{BEM}}$, we assume that in two dimensions, $\mathrm{diam}(\Omega_i) < 1$ for all $i \in \mathscr{I}_{\mathrm{BEM}}$, cf. Assumption 1.72, p. 46. A sufficient and practicable condition for this is of course $\mathrm{diam}(\Omega) < 1$, which can be achieved by a simple scaling of the coordinates.

The resulting *discrete skeleton variational problem* reads: find $u_h \in V^h(\Gamma_S)$ with $u_{h|\Gamma_D} = g_D$ such that

$$\sum_{i=1}^{s} \langle S_{i,h} u_{h|\partial\Omega_i}, v_{h|\partial\Omega_i} \rangle = \sum_{i=1}^{s} \langle N_{i,h} f_i, v_{h|\partial\Omega_i} \rangle \qquad \forall v_h \in V_D^h(\Gamma_S), \tag{2.11}$$

[1] Otherwise, we can use an interpolation (if the data is continuous) or an L^2-orthogonal projection of the Dirichlet data to $V^h(\Gamma_D)$.

where

$$V_D^h(\Gamma_S) := V^h(\Gamma_S) \cap H_D^{1/2}(\Gamma_S). \tag{2.12}$$

Since the above bilinear form is symmetric, Lemma 1.3 implies

$$u_h = \underset{\substack{v_h \in V^h(\Gamma_S) \\ v_{h|\Gamma_D} = g_D}}{\operatorname{argmin}} \sum_{i=1}^{s} \Big(\tfrac{1}{2} \langle S_{i,h} v_{h|\partial\Omega_i}, v_{h|\partial\Omega_i} \rangle - \langle N_{i,h} f_i, v_{h|\partial\Omega_i} \rangle \Big). \tag{2.13}$$

The homogeneous version reads: find $\tilde{u}_h \in V_D^h(\Gamma_S)$ such that

$$\sum_{i=1}^{s} \langle S_{i,h} \tilde{u}_{h|\partial\Omega_i}, v_{h|\partial\Omega_i} \rangle = \sum_{i=1}^{s} \langle N_{i,h} f_i - S_{i,h} \tilde{g}_{D|\partial\Omega_i}, v_{h|\partial\Omega_i} \rangle \qquad \forall v_h \in V_D^h(\Gamma_S). \tag{2.14}$$

Then $u_h = \tilde{g}_D + \tilde{u}_h$ solves (2.11). The minimization problem equivalent to (2.14) reads

$$\tilde{u}_h = \underset{v_h \in V_D^h(\Gamma_S)}{\operatorname{argmin}} \sum_{i=1}^{s} \Big(\tfrac{1}{2} \langle S_{i,h} v_{h|\partial\Omega_i}, v_{h|\partial\Omega_i} \rangle - \langle N_{i,h} f_i - S_{i,h} \tilde{g}_{D|\partial\Omega_i}, v_{h|\partial\Omega_i} \rangle \Big). \tag{2.15}$$

Remark 2.3. For a pure FE formulation ($\mathscr{I}_{\mathrm{FEM}} = \{1, \ldots, s\}$), system (2.11) is the Schur complement system (cf. Sect. 1.2.6) of the global FE system

$$\text{find } u_h^{\mathrm{FE}} \in V^h(\Omega),\ u_{h|\Gamma_D}^{\mathrm{FE}} = g_D : \quad a(u_h^{\mathrm{FE}}, v_h) = \langle f, v_h \rangle \qquad \forall v_h \in V^h(\Omega),\ v_{h|\Gamma_D} = 0,$$

and $u_h = u_{h|\Gamma_S}^{\mathrm{FE}}$.

Remark 2.4. For the case that both FEM and BEM are included, the present form of the skeleton problem is also known as the *symmetric coupling* of FEM and BEM [Cos87]. For the advantages of this coupling and for other types of couplings see e.g. [ZKB77, ZKB79, BJ79, BJN78, JN80, CS90, Lan94, HHKL97, Hip02, KS02, Ste03b, Ste11]. See also the related works [CKL98, HSW00, HW91] on pure boundary element domain decomposition.

Remark 2.5. Note that on a FEM subdomain Ω_i, the Schur complement matrix may be defined differently by eliminating non-coupling DOFs on the Neumann boundary *together* with the interior DOFs, cf. [TW05, Chap. 4]. The associated operator then maps from $V^h(\partial\Omega_i \cap (\Gamma \cup \Gamma_D))$ to its dual. Correspondingly, one can define a modified Steklov-Poincaré operator $S_i : H^{1/2}(\partial\Omega_i \setminus \Gamma_N) \to H_{00}^{-1/2}(\partial\Omega_i \setminus \Gamma_N)$, and show the analogous statements of Sect. 1.2.6 for the two modified operators.

This approach is very natural in implementations, since one only needs to know whether a DOF is a coupling one (or a Dirichlet DOF) rather than if it is on the subdomain boundary. On the contrary, for BEM subdomains this approach is not natural, since all DOFs in $V^h(\partial\Omega_i)$ are already on the subdomain boundary. Hence, for a unified presentation, we have chosen the separation into "true" boundary DOFs and interior DOFs for all subdomains. However, generalizations of the statements below to the modified operators can be proved without major effort.

2.1.3 Error Analysis of the Discrete Skeleton Formulation

The discrete skeleton formulation (2.14) introduces variational crimes when compared to (2.9). In this short section, we provide an a priori error analysis for the simplified case that

$$g_D = 0 \qquad \text{and} \qquad f_i = 0 \quad \forall i \in \mathscr{I}_{\mathrm{BEM}}.$$

In this case, the solution u of (2.1) lies in the space

$$V_{S,D} := \left\{ v \in H^1_D(\Omega) : \forall i \in \mathscr{I}_{\mathrm{BEM}} : v_{|\Omega_i} = \mathscr{H}_i(v_{|\partial\Omega_i}) \right\},$$

where $\mathscr{H}_i$ denotes the harmonic extension from $H^{1/2}(\partial\Omega_i) \to H^1(\Omega_i)$. From the relation

$$a_i(v_{|\Omega_i}, w_{|\Omega_i}) = \langle S_i v_{|\partial\Omega_i}, w_{|\partial\Omega_i} \rangle, \qquad \forall v, w \in V_{S,D},\ i \in \mathscr{I}_{\mathrm{BEM}},$$

it is straightforward to show that (2.9) is then equivalent to finding $u \in V_{S,D}$ such that

$$\sum_{i \in \mathscr{I}_{\mathrm{FEM}}} a_i(u_{|\Omega_i}, v_{|\Omega_i}) + \sum_{i \in \mathscr{I}_{\mathrm{BEM}}} \langle S_i\, u_{|\partial\Omega_i}, v_{|\partial\Omega_i} \rangle = \sum_{i \in \mathscr{I}_{\mathrm{FEM}}} \langle f_i, v_{|\Omega_i} \rangle \qquad \forall v \in V_{S,D}\,. \tag{2.16}$$

We now define the space

$$V^h_{S,D} := \left\{ v \in V_{S,D} : \forall i \in \mathscr{I}_{\mathrm{BEM}} : v_{|\partial\Omega_i} \in V^h(\partial\Omega_i),\ \forall i \in \mathscr{I}_{\mathrm{FEM}} : v_{|\Omega_i} \in V^h(\Omega_i) \right\}.$$

The discrete problem (2.14) is equivalent to finding $u_h \in V^h_{S,D}$ such that

$$\sum_{i \in \mathscr{I}_{\mathrm{FEM}}} a_i(u_{h|\Omega_i}, v_{h|\Omega_i}) + \sum_{i \in \mathscr{I}_{\mathrm{BEM}}} \langle S_{i,h}\, u_{h|\partial\Omega_i}, v_{h|\partial\Omega_i} \rangle = \sum_{i \in \mathscr{I}_{\mathrm{FEM}}} \langle f_i, v_{h|\Omega_i} \rangle \quad \forall v_h \in V^h_{S,D}\,. \tag{2.17}$$

Comparing (2.16) and (2.17), we see that the variational crime is located in the BEM subdomains only, as we replace S_i by $S_{i,h}$ for $i \in \mathscr{I}_{\mathrm{BEM}}$ (for the case $\mathscr{I}_{\mathrm{BEM}} = \emptyset$

see also Remark 2.3). Following the proof of [HLP10, Lemma 4.2], which involves a Strang lemma (cf. [Cia87]), we obtain that there exists a constant C depending on the domain Ω, the coefficient $\mathscr{A}$, and on the BEM subdomains Ω_i, $i \in \mathscr{I}_{\mathrm{BEM}}$ such that

$$\begin{aligned}\|u - u_h\|_{H^1(\Omega)} \leq C & \inf_{v_h \in V^h_{S,D}} \|u - v_h\|_{H^1(\Omega)} \\ & + C \Big(\sum_{i \in \mathscr{I}_{\mathrm{BEM}}} \inf_{z_h \in Z_h(\partial\Omega_i)} \|t_i(u) - z_h\|^2_{H^{-1/2}(\partial\Omega_i)} \Big)^{1/2},\end{aligned}$$

where $t_i(u) = S_i \, u_{|\partial\Omega_i}$ is the generalized conormal derivative of the solution u on $\partial\Omega_i$ for $i \in \mathscr{I}_{\mathrm{BEM}}$. Alternatively, one can employ an error estimate for the associated mixed setting, cf. [Hof11]. Using the minimizing property of the harmonic extension (Lemma 1.54), the FE approximation estimates from Lemma 1.44, and the BE approximation estimates from Lemma 1.89, we finally get the a priori error estimate

$$\|u - u_h\|_{H^1(\Omega)} \leq C\, h^s \Big(|u|^2_{H^{1+s}(\Omega)} + \sum_{i \in \mathscr{I}_{\mathrm{BEM}}} \|t_i(u)\|^2_{H_{\mathrm{pw}}^{-1/2+s}(\partial\Omega_i)} \Big)^{1/2},$$

provided that $u \in H^{1+s}(\Omega)$ with $s \in (0, 1]$. In terms of h, this is the same error behavior as for a pure FEM discretization. Note, however, that the constant C depends on the subdomain partition. Finding an estimate which is explicit in the subdomains is non-trivial, but should be possible under the regularity assumptions that we will introduce in Sect. 2.5.2. For the limit case $H_i \to h$ see [HLP10].

2.1.4 Conditioning of the Skeleton Problem

Lemma 2.6. *Let $\mathscr{T}^H(\Omega)$ be a shape regular and quasi uniform triangulation of Ω, and let each subdomain Ω_i be a union of a few coarse elements of $\mathscr{T}^H(\Omega)$, such that the number of elements per subdomain is uniformly bounded. Furthermore, let the triangulation $\mathscr{T}^h(\Gamma_S)$ be quasi-uniform. Then there exist uniform constants c_1, $c_2 > 0$ such that for all $v \in V^h_D(\Gamma_S)$,*

$$c_1\, \alpha_{\min}\, H\, \|v\|^2_{L^2(\Gamma_S)} \leq \underbrace{\sum_{i=1}^{s} \langle S_{i,h}\, v_{|\partial\Omega_i},\, v_{|\partial\Omega_i} \rangle}_{=:\langle \mathsf{S}_h v, v\rangle} \leq c_2\, \|\mathscr{A}\|_{L^\infty(\Omega)}\, h^{-1}\, \|v\|^2_{L^2(\Gamma_S)},$$

where $\alpha_{\min} > 0$ is the constant from (1.22) *(p. 21). Hence, the conditioning of the skeleton problems* (2.11) *and* (2.14) *is given by*

$$\kappa(\mathsf{S}_h) = \mathscr{O}\Big(\frac{\|\mathscr{A}\|_{L^\infty(\Omega)}}{\alpha_{\min}}\, H^{-1} h^{-1} \Big).$$

Proof. The following proof is identical to [TW05, Lemma 4.11]. To simplify its presentation, we assume that $\operatorname{diam}(\Omega) = 1$, and we extend v from Γ_S to Ω by setting $v_{|\Omega_i} := \mathscr{H}_i v$. By Theorem 1.23 and a scaling argument we get that

$$\begin{aligned} H_i \, \|v\|^2_{L^2(\partial\Omega_i)} &\leq H_i^2 \left(|v|^2_{H^{1/2}(\partial\Omega_i)} + H_i^{-1} \, \|v\|_{L^2(\partial\Omega_i)} \right) \\ &\lesssim H_i^2 \left(|v|^2_{H^1(\Omega_i)} + H_i^{-2} \, \|v\|^2_{L^2(\Omega_i)} \right). \end{aligned}$$

Above, $\lesssim$ stands for $\leq C$, where C is a generic, uniform constant. Summing over $i = 1, \ldots, s$, using that $H_i \eqsim H \leq 1$, Friedrichs' inequality, as well as the spectral equivalence relations from Sect. 1.3.8.3, we obtain

$$\begin{aligned} \alpha_{\min} \, H \, \|v\|^2_{L^2(\Gamma_S)} \lesssim \alpha_{\min} \, \|v\|^2_{H^1(\Omega)} &\lesssim \alpha_{\min} \, |v|^2_{H^1(\Omega)} \\ &\leq \sum_{i=1}^{s} \langle S_i \, v, \, v \rangle_{\partial\Omega_i} \lesssim \sum_{i=1}^{s} \langle S_{i,h} \, v, \, v \rangle_{\partial\Omega_i} \, . \end{aligned}$$

This shows the lower bound. For the upper bound, note first that with the same notation as above,

$$\sum_{i=1}^{s} \langle S_{i,h} \, v, \, v \rangle_{\partial\Omega_i} \lesssim \sum_{i=1}^{s} \langle S_i \, v, \, v \rangle_{\partial\Omega_i} \lesssim \|\mathscr{A}\|^2_{L^\infty(\Omega)} \sum_{i=1}^{s} |v|^2_{H^{1/2}(\partial\Omega_i)} \, .$$

To conclude the proof, we apply the inverse inequality

$$|w|_{H^{1/2}(\partial\Omega_i)} \lesssim h^{-1/2} \, \|w\|_{L^2(\Omega_i)} \qquad \forall w \in V^h(\partial\Omega_i),$$

which can derived using interpolation theory (cf. [TW05, Lemma B.27]). □

Remark 2.7. Preconditioners for the skeleton problem are also called Schur complement preconditioners. Examples are the BPS type and wirebasket preconditioners (see [TW05, Sect. 5] and the pioneering papers [BPS86, BPS87, BPS88, BPS89]), the Neumann-Neumann preconditioners (Sect. 2.3) and the BDDC preconditioner (Sect. 5.1.4). See also the related papers [HL92, HLM91a, HLM91b, CKL98, HS01, Ste03a, KL04] as well as the work of Nepomnyaschikh [Nep91b, Nep07].

2.2 Formulation of One-Level FETI/BETI Methods

In the following Sect. 2.2.1, we derive the classical FETI/BETI methods starting from the homogenized minimization problem (2.14). Section 2.2.2 deals with the all-floating (total) FETI/BETI formulation starting from the non-homogenized

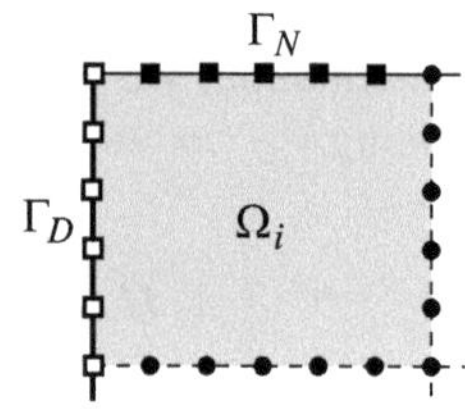

Fig. 2.2 Nodes of the set $\partial\Omega_i^h$ for a single subdomain Ω_i (● interface nodes in $\Gamma^h \cap \partial\Omega_i^h$, □ Dirichlet nodes, ■ remaining Neumann nodes)

minimization problem (2.11). In Sect. 2.2.3, we give an interpretation of the Lagrange multipliers involved in both formulations. Section 2.2.4 introduces various preconditioners, and Sect. 2.2.5 discusses implementation issues.

Throughout the remainder of this chapter, we work exclusively in the discrete setting. To simplify the notation, we drop the subscript h in the operators $S_{i,h}$ and $N_{i,h}$. That means from now on, S_i, N_i refer to discrete operators

$$S_i : V^h(\partial\Omega_i) \to V^h(\partial\Omega_i)^*, \quad N_i : H^1(\Omega_i)^* \to V^h(\partial\Omega_i)^*. \tag{2.18}$$

Moreover, we need a few definitions concerning the skeletal triangulation. Let Γ_S^h be the set of nodes on Γ_S. Analogously, $\partial\Omega_i^h$, Γ_D^h, Γ^h, Γ_{ij}^h are the sets of nodes on the respective parts of the skeleton, see also Fig. 2.2. A typical node will be denoted by x^h.

Remark 2.8. Note that, for all FEM subdomains, we can define the discrete operators S_i, N_i differently by eliminating non-coupling Neumann DOFs *together* with the interior DOFs, which is much more practicable in implementations, cf. Remark 2.5 and [TW05, Chap. 4]. This results in operators

$$S_i : V^h(\partial\Omega_i \setminus \Gamma_N) \to V^h(\partial\Omega_i \setminus \Gamma_N)^*, \qquad N_i : H^1(\Omega_i) \to V^h(\partial\Omega_i \setminus \Gamma_N)^*. \tag{2.19}$$

This is in fact done in the original FETI method, but for a unified presentation of FETI/BETI, we have chosen the setting in (2.18). The theory below, however, can be generalized without major difficulties to the case of (2.19).

2.2.1 Formulation of Classical FETI/BETI

In this section we derive the classical formulation of the FETI method (as introduced in [FR94]) as well as the BETI and coupled FETI/BETI methods (as introduced in [LS03, LS05]). Our presentation mainly follows [KW01, LS05], and [TW05, Sect. 6.3]. Our starting point is the homogeneous skeleton problem (2.14).

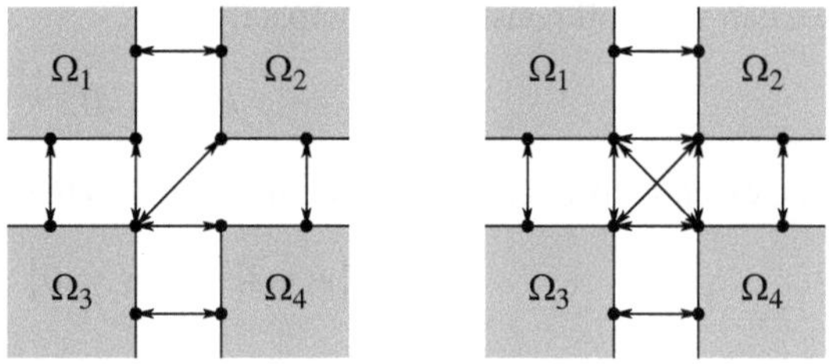

Fig. 2.3 Sketch of non-redundant constraints (*left*) and fully redundant constraints (*right*) for a node that is shared by four subdomains

2.2.1.1 Tearing and Interconnecting

The idea of "tearing" is to introduce separate unknowns for $\tilde{u}_{|\partial\Omega_i} \in V_D^h(\partial\Omega_i)$ on the subdomain boundaries. To this end we define the spaces[2]

$$W_i := V_D^h(\partial\Omega_i) \qquad \text{and} \qquad W := \prod_{i=1}^{s} W_i\,. \tag{2.20}$$

We denote the components of $w \in W$ by w_i and write

$$w := [w_i]_{i=1}^s \in W. \tag{2.21}$$

A Jump Operator. Functions in the product space W are typically discontinuous across subdomain interfaces. Continuity ("interconnecting") is enforced by constraints of the form

$$w_i(x^h) - w_j(x^h) = 0 \qquad \text{for } x^h \in \Gamma_{ij}^h, \quad i > j. \tag{2.22}$$

In this book, we restrict ourselves to *fully redundant* constraints, i.e., we impose *all* constraints of the form (2.22), see [TW05, Sect. 6.3.3] and Fig. 2.3, right. They turn out to be advantageous in implementations due to the full symmetry. In the non-redundant case (see [TW05, Sect. 6.3.2] and Fig. 2.3, left) a minimal number of necessary constraints is used. For other variants, such as orthogonal constraints, see e.g. [FP03, FP04].

Definition 2.9. Let us assume a numbering of the (fully redundant) constraints (2.22) with M being the total number of constraints. The entry of a vector $\mu \in \mathbb{R}^M$ corresponding to the constraint (2.22) at the node $x^h \in \Gamma_{ij}^h$ is denoted by $\mu_{ij}(x^h)$.

Introducing the (linear) jump operator $B : W \to \mathbb{R}^M$, given by

$$(B\,w)_{ij}(x^h) = w_i(x^h) - w_j(x^h) \qquad \text{for } x^h \in \Gamma_{ij}^h,\ i > j, \tag{2.23}$$

[2] For the setting (2.19), $W_i = V_D^h(\partial\Omega_i \setminus \Gamma_N)$.

we can write all constraints compactly as

$$B\,w \;=\; 0.$$

The essential property of B is (and should be) that the subspace

$$\widehat{W} \;:=\; \big\{w \in W : B\,w = 0\big\} \;=\; \ker(B) \tag{2.24}$$

can be identified with $V_D^h(\Gamma_S)$, in short: $V_D^h(\Gamma_S) \equiv \widehat{W}$.

Remark 2.10. With respect to the standard nodal basis, the operator B is represented by signed Boolean matrix (with entries 0, 1 or -1). For other discretization spaces we refer to Sect. 2.8.

Saddle point formulation. In the sequel, to avoid cumbersome notation, we will regard the approximate Steklov-Poincaré approximants S_i as operators mapping from W_i to W_i^*, with the only exception of $S_i\,\tilde{g}_{D|\partial\Omega_i}$. In addition, we define $S : W \to W^*$ by

$$\langle S v,\, w\rangle \;:=\; \sum_{i=1}^{s} \langle S_i\, v_i,\, w_i\rangle \qquad \text{for } v,\, w \in W, \tag{2.25}$$

in short $S := \operatorname{diag}(S_i)_{i=1}^s$, and the linear functional $g \in W^*$ by

$$\langle g,\, w\rangle \;:=\; \sum_{i=1}^{s} \langle g_i,\, w_i\rangle \;:=\; \sum_{i=1}^{s} \langle N_i\, f_i - S_i\, \tilde{g}_{D|\partial\Omega_i},\, w_i\rangle \qquad \text{for } w \in W, \tag{2.26}$$

in short $g := [g_i]_{i=1}^s = [N_i\, f_i - S_i\, \tilde{g}_{D|\partial\Omega_i}]_{i=1}^s$. Using this notation we can write the minimization problem (2.15) equivalently as the constrained minimization problem

$$\tilde{u} \;=\; \operatorname*{argmin}_{\substack{w\in W\\ B\,w=0}} \tfrac{1}{2}\langle S\, w,\, w\rangle - \langle g,\, w\rangle. \tag{2.27}$$

For simplicity, we use the symbol $\tilde{u}$ simultaneously for the solution of (2.14) and the solution of the above problem. The equivalent saddle point problem reads as follows (see also the following Sect. 2.2.1.2). Find $(\tilde{u}, \lambda) \in W \times U$:

$$\begin{bmatrix} S & B^\top \\ B & 0 \end{bmatrix} \begin{bmatrix} \tilde{u} \\ \lambda \end{bmatrix} = \begin{bmatrix} g \\ 0 \end{bmatrix}, \tag{2.28}$$

where $U = \mathbb{R}^M$ is the space of Lagrange multipliers. The second equation ($B\,\tilde{u} = 0$) ensures that $\tilde{u} \in \widehat{W} \equiv V_D^h(\Gamma_S)$.

Lemma 2.11. *Under the assumptions made in Sect. 2.1,* $\ker(S) \cap \ker(B) = \{0\}$. *Therefore, Problem* (2.28) *is uniquely solvable up to adding elements from* $\ker B^\top$ *to* λ. *The solution* $\tilde{u} \in \widehat{W} \equiv V_D^h(\Gamma_S)$ *is the unique solution of* (2.14).

Proof. Since $\ker(B) = \widehat{W} \equiv V_D^h(\Gamma_S)$ and the bilinear form $\langle S\cdot,\cdot\rangle$ is coercive on $V_D^h(\Gamma_S)$, it follows that $\ker(S) \cap \ker(B) = \{0\}$. The classical Brezzi theory (cf. Lemma 1.16) implies the solvability of (2.28). Finally, $\tilde{u}$ solves the constrained miminization problem (2.27), which is equivalent to (2.15) and (2.14). □

For reasons that will become clear later on, we agree that the Lagrange multipliers are in U, but $B\,w$ is in its dual:

$$B : W \to U^*, \qquad B^\top : U \to W^*.$$

Of course, $U^* = \mathbb{R}^M$ too, and the duality pairing in $U^* \times U$ is nothing else than the Euclidean inner product, i.e., $\langle \mu, \lambda\rangle = (\mu, \lambda)_{\ell^2}$ for $\lambda \in U$, $\mu \in U^*$. Nevertheless, we will keep track of the subtle difference between U and U^* for this has several advantages.

The jump operator B can be decomposed into local operators $B_i : W_i \to U^*$, $i = 1, \ldots, s$ such that

$$\langle B\,w, \lambda\rangle = \sum_{i=1}^{s} \langle B_i\,w_i, \lambda\rangle \qquad \forall \lambda \in U.$$

2.2.1.2 Interpretation of the Lagrange Multipliers

This subsection may be skipped by readers who are mainly interested in the derivation of the FETI/BETI method. It contains an interpretation of the Lagrange multipliers as normal fluxes, and it discusses the adjoint $B^\top$ in more detail.

The saddle point formulation (2.28) can also be derived directly from the discrete skeleton formulation (2.14) without a detour via the minimization problem.[3] We show this alternative derivation because it provides more insight on the Lagrange multipliers. Let $\tilde{u} \in V_D^h(\Gamma_S)$ be the solution of (2.14) and let

$$t_i := S_i\,\tilde{u}_{|\partial\Omega_i} - g_i$$

denote the discrete (generalized) conormal derivative[4] and set $t = [t_i]_{i=1}^s \in W^*$. Substituting this formula into (2.14) and using that $V_D^h(\Gamma_S) \equiv \widehat{W} = \ker(B)$, we obtain

$$\langle t, w\rangle = 0 \qquad \forall w \in \ker(B). \tag{2.29}$$

[3]This way, FETI formulations can be derived for non-symmetric or indefinite problems.

[4]Note that the given Neumann data g_N is already incorporated in g_i; if $\tilde{u}$ is the solution, t_i vanishes on all the interior nodes of the local Neumann boundary.

This is equivalent to

$$t \in \ker(B)^\circ = \operatorname{range}(B^\top),$$

cf. Lemma 1.8. We now

- Parametrize the solution $\tilde{u} \in V_D^h(\Gamma_S)$ by $\tilde{u} \in W$ with the side condition $B\,\tilde{u} = 0$,
- Parametrize the conormal derivative $t \in W^*$ by $-B^\top\lambda$ with $\lambda \in U$, therefore fulfilling condition (2.29) automatically.

From the definition of t and g, we get the equation $-B^\top\lambda \;=\; S\,\tilde{u}-g$. Together with $B\,\tilde{u} = 0$, this yields exactly (2.28). Under this perspective, the Lagrange multipliers themselves can be interpreted as normal fluxes. More precisely, they *parametrize* the normal fluxes of the solution $\tilde{g}_D + \tilde{u}$ of (2.11):

$$-B_i^\top\lambda \;=\; S_i(\,\tilde{g}_{D|\partial\Omega_i} + \tilde{u}_i) - N_i\, f_i\,.$$

For an interpretation in a mechanical context see [RF98a, RF99].

Definition 2.12. For $x^h \in \Gamma_S^h$ we define

$$\mathcal{N}_{x^h} \;:=\; \{k = 1,\ldots,s : x^h \in \partial\Omega_k^h\},$$

i.e., the index set of the subdomains sharing the node x^h. Furthermore, we set

$$\mathcal{N}_i \;:= \{k = 1,\ldots,s : \partial\Omega_i \cap \partial\Omega_k \neq \emptyset\}.$$

Definition 2.13. For $i = 1,\ldots,s$ and $x^h \in \partial\Omega_i^h$, let $\varphi_{i,x^h} \in V^h(\partial\Omega_i)$ be the nodal basis function corresponding to node x^h. For $t \in W^*$, we set

$$t_{i,x^h} \;:=\; \langle t_i,\, \varphi_{i,x^h}\rangle_{\partial\Omega_i}\,.$$

Lemma 2.14. *For $\mu \in U^*$ and $\lambda \in U$,*

$$\langle \mu,\, \lambda\rangle \;=\; \sum_{x^h\in\Gamma^h}\; \sum_{\substack{i,\,j\in\mathcal{N}_{x^h}\\ i>j}} \mu_{ij}(x^h)\,\lambda_{ij}(x^h).$$

The adjoints $B^\top : U \to W^*$ and $B_i^\top : U \to W_i^*$ fulfill

$$(B^\top\lambda)_{i,x^h} \;=\; \langle B_i^\top\lambda,\, \varphi_{i,x^h}\rangle_{\partial\Omega_i} \;=\; \sum_{j\in\mathcal{N}_{x^h}\setminus\{i\}} \operatorname{sign}(i-j)\,\lambda_{ij}(x^h) \qquad \textit{for } x^h \in \partial\Omega_i^h \cap \Gamma^h.$$

Proof. The proof is straightforward.

2.2.1.3 Eliminating Local Variables

The first set of equations in (2.28) read

$$S_i \, \tilde{u}_i \;=\; g_i - B_i^\top \lambda \qquad \forall i = 1, \ldots, s.$$

The goal of this subsection is to obtain an explicit formula for $\tilde{u}_i$, which is not straightforward as some of the operators S_i are singular.

In the usual theory of iterative substructuring methods, a *floating subdomain* is defined as a subdomain Ω_i whose boundary $\partial\Omega_i$ does not intersect the Dirichlet boundary Γ_D. Here, we use a more abstract form.

Definition 2.15 (floating subdomain). A subdomain Ω_i is called a *floating subdomain* if $S_i : W_i \to W_i^*$ is singular, otherwise it is a *non-floating subdomain*. The index set corresponding to the floating subdomains is denoted by $\mathscr{I}_{\text{float}}$.

In the case of the potential equation, the solution of the pure Neumann problem on a floating subdomain is only unique up to an additive constant, and so

$$\ker(S_i) \;=\; \begin{cases} \text{span}\{1_{\partial\Omega_i}\} & \text{if } i \in \mathscr{I}_{\text{float}}, \\ \{0\} & \text{else.} \end{cases}$$

In a non-floating subdomain,

$$\tilde{u}_i \;=\; S_i^{-1}(g_i - B_i^\top \lambda) \qquad \forall i \notin \mathscr{I}_{\text{float}}.$$

For the remaining subdomains, we need the solvability conditions

$$g_i - B_i^\top \lambda \;\in\; \text{range}(S_i) \qquad \forall i \in \mathscr{I}_{\text{float}}.$$

Choosing an injective operator

$$R_i : \mathbb{R}^{\dim(\ker(S_i))} \to W_i, \qquad \text{range}(R_i) \;=\; \ker(S_i), \tag{2.30}$$

the local solution u_i can be represented by

$$\tilde{u}_i \;=\; S_i^\dagger (g_i - B_i^\top \lambda) + R_i \, \xi_i, \tag{2.31}$$

where $S_i^\dagger$ is a pseudo inverse of S_i (see Sect. 1.1.3.4), and $\xi_i \in \mathbb{R}^{\dim(\ker(S_i))}$. Actually, formula (2.31) is valid for all $i = 1, \ldots, s$. In our setting, we choose $R_i \xi_i = \xi_i$ if $i \in \mathscr{I}_{\text{float}}$ and 0 otherwise. Since $\text{range}(S_i) = \ker(S_i)^\circ = \text{range}(R_i)^\circ = \ker(R_i^\top)$ (see Lemma 1.8), the compatibility conditions rewrite as

$$R_i^\top (g_i - B_i^\top \lambda) \;=\; 0 \qquad \forall i = 1, \ldots, s, \tag{2.32}$$

where for $i \notin \mathscr{I}_{\text{float}}$, this condition is trivial. Introducing

$$Z := \prod_{i=1}^{s} \mathbb{R}^{\dim(\ker(S_i))}$$

and the operators $R = \text{diag}(R_i)_{i=1}^s : Z \to W$, and $S^\dagger := \text{diag}(S_i^\dagger) : W^* \to W$, we see from (2.31) and (2.32) that

$$\tilde{u} = S^\dagger(g - B^\top \lambda) + R\xi \tag{2.33}$$

for some $\xi \in Z$, provided that

$$R^\top B^\top \lambda = R^\top g. \tag{2.34}$$

Eliminating $\tilde{u}$ from (2.28) using (2.33) yields

$$BS^\dagger(g - B^\top \lambda) + BR\xi = 0.$$

After reordering the terms above and adding (2.34) to the set of equations, we obtain the system

$$\begin{aligned} BS^\dagger B^\top \lambda - BR\xi &= BS^\dagger g, \\ R^\top B^\top \lambda &= R^\top g. \end{aligned}$$

By defining the operators

$$F := BS^\dagger B^\top, \qquad G := BR, \tag{2.35}$$

we see that the above system has saddle point structure. In the following, we briefly discuss the solvability of this problem and its relation to (2.28).

Lemma 2.16. *The problem of finding* $(\lambda, \xi) \in U \times Z$ *such that*

$$\begin{bmatrix} F & -G \\ G^\top & 0 \end{bmatrix} \begin{bmatrix} \lambda \\ \xi \end{bmatrix} = \begin{bmatrix} BS^\dagger g \\ R^\top g \end{bmatrix} \tag{2.36}$$

is uniquely solvable up to adding an element from $\ker(B^\top)$ *to* λ. *With*

$$\tilde{u} = S^\dagger(g - B^\top \lambda) + R\xi,$$

$(\tilde{u}, \lambda)$ *solves* (2.28), *and* $\tilde{u} \in \widehat{W} \equiv V_D^h(\Gamma_S)$ *solves* (2.14).

Proof. First note that the operators $F, G^\top$ vanish on $\ker(B^\top)$. Therefore, F and $G^\top$ are well-defined on the factor space $U_{/\ker(B^\top)}$. We apply Lemma 1.16 (with $V \mapsto U_{/\ker(B^\top)}$, $Q \mapsto Z$). The assumptions hold because $\ker(F) \cap \ker(G^\top) = \ker(B^\top)$.

One easily shows that $\ker(G) = \ker(B_{|\mathrm{range}(R)}) = \ker(B) \cap \ker(S) = \{0\}$, which implies uniqueness. The rest of the proof is straightforward. □

Remark 2.17. For a pure FEM, i.e., $\mathcal{I}_{\mathrm{BEM}} = \emptyset$, the use of Schur complement operators in the derivation can be circumvented. For each $i = 1, \ldots, s$, let $X_i := V_D^h(\Omega_i)$ be the FE space on Ω_i with respect to $\mathcal{T}^h(\Omega_i)$ and set $X := \prod_{i=1}^s X_i$. Moreover, let $A_i : X_i \to X_i^*$ and $\ell_i \in X_i^*$ be the operator and functional corresponding to the local stiffness matrix and load vector (after the homogenization). Let $\overline{B}_i : X_i \to U^*$ be the jump operator defined by $\overline{B}_i v_i := B_i v_{i|\partial\Omega_i}$, then the global problem is identical to find $(u, \lambda) \in V \times U$:

$$\begin{bmatrix} A & \overline{B}^\top \\ \overline{B} & 0 \end{bmatrix} \begin{bmatrix} u \\ \lambda \end{bmatrix} = \begin{bmatrix} \ell \\ 0 \end{bmatrix},$$

where $A = \mathrm{diag}(A_i)_{i=1}^s$, and $\ell = [\ell_i]_{i=1}^s \in V^*$. Having at hand pseudo inverses $A_i^\dagger$ and injective operators $\overline{R}_i : \mathbb{R}^{\dim(\ker(A_i))} \to X_i$ such that $\ker(A_i) = \mathrm{range}(\overline{R}_i)$, we can reduce this system as in Sect. 2.2.1.3. The resulting system is identical to (2.36), i.e.,

$$F = BS^\dagger B^\top = \overline{B} A^\dagger \overline{B}^\top, \quad BS^\dagger g = \overline{B} A^\dagger \ell, \quad G = BR = \overline{B}\,\overline{R}, \quad R^\top g = \overline{R}^\top \ell,$$

cf. [RFTM99, Sect. 2.1.3].

Remark 2.18. The result of Remark 2.17 also holds for arbitrary combinations of FEM and BEM if we reformulate the equations in a mixed setting. For $i \in \mathcal{I}_{\mathrm{BEM}}$, we use the local space $X_i := V_D^h(\partial\Omega_i) \times Z^h(\partial\Omega_i)$ and define $A_i : X_i \to X_i^*$, $\overline{B}_i : X_i \to U^*$, and $\ell_i \in X_i^*$ by

$$A_i := \begin{bmatrix} D_i & \frac{1}{2}I + K_i^\top \\ \frac{1}{2}I + K_i & -V_i \end{bmatrix}, \quad \overline{B}_i \begin{bmatrix} v_i \\ t_i \end{bmatrix} := B_i v_i, \quad \ell_i := \begin{bmatrix} N_i f_i - D_i \tilde{g}_{D|\partial\Omega_i} \\ -(\frac{1}{2}I + K_i)\tilde{g}_{D|\partial\Omega_i} \end{bmatrix}.$$

2.2.1.4 A Projection Method

Note that the saddle point problem (2.36) is different in its structure from the saddle point problem (2.28) because the variables ξ lives in the space Z of small dimension:

$$\dim(Z) < s \ll \dim(U).$$

Therefore, an inversion of a sparse system on Z is acceptable. Since the lower right block of System (2.36) is zero, we can use a *projection method*, cf. [FR91, FR94].

Following [FCM95], we define the space of *admissible Lagrange increments*

$$U_{\mathrm{ad}} := \ker(G^\top) = \{\lambda \in U : B^\top \lambda \in \mathrm{range}(S)\}. \tag{2.37}$$

The Case $R^\top g = 0$. Assume for a moment that $R^\top g = 0$. Then the solution λ of (2.36) lies in U_{ad}. We can always reach this subspace by an orthogonal projection (see Sect. 1.1.3.3). Let the self-adjoint linear operator

$$Q : U^* \to U$$

be positive definite on range(G); specific choices of Q will be given later on. The operator

$$P := I - Q\,G\,(G^\top Q\,G)^{-1} G^\top \tag{2.38}$$

is a projection from U onto U_{ad}. Note that $G^\top Q\,G$ is SPD because of the assumptions on Q and because $\ker(G) = \{0\}$, see the proof of Lemma 2.16. Furthermore, for a suitable choice of Q, the matrix representing $G^\top Q\,G$ is sparse, cf. Sect. 2.2.5.

Hence, if $R^\top g = 0$, it suffices to test the first line of (2.36) (rewritten as a variational problem) with test functions from $U_{\text{ad}} = \text{range}(P)$, i.e.,

$$\text{find } \lambda \in U_{\text{ad}} : \quad P^\top F\,\lambda = P^\top B\,S^\dagger g.$$

The Case $R^\top g \neq 0$. In the general case, we can decompose $\lambda = \lambda_0 + \widetilde{\lambda}$, where $G^\top \lambda_0 = R^\top g$ and $\widetilde{\lambda} \in U_{\text{ad}}$. Apparently, the choice

$$\lambda_0 = Q\,G(G^\top Q\,G)^{-1} R^\top g \tag{2.39}$$

fulfills these requirements, and so we can homogenize the equation. Summarizing, we obtain the problem

$$\text{find } \widetilde{\lambda} \in U_{\text{ad}} : \quad P^\top F\,\widetilde{\lambda} = P^\top \underbrace{(B\,S^\dagger g - F\,\lambda_0)}_{=B\,S^\dagger(g - B^\top \lambda_0)}. \tag{2.40}$$

We will discuss this equation in detail in Sect. 2.2.1.5. Before, we need to see how to recover the variable ξ from λ_0 and $\widetilde{\lambda}$. Testing the first line of (2.36) (rewritten as a variational problem) with test functions from range($I - P$), we obtain

$$(I - P^\top) F\,\lambda - \underbrace{(I - P^\top) G\,\xi}_{=G\,\xi} = (I - P^\top) B\,S^\dagger g$$

$$\iff \qquad G\,\xi = \underbrace{(I - P^\top)}_{G(G^\top Q\,G)^{-1} G^\top Q} \quad \underbrace{(F\,\lambda - B\,S^\dagger g)}_{B\,S^\dagger(B^\top \lambda - g)}.$$

Applying $(G^\top Q\,G)^{-1} G^\top Q$ to the last equation, we obtain the formula

$$\xi = -(G^\top Q\,G)^{-1}\,G^\top Q\,B\,S^\dagger (g - B^\top \lambda). \tag{2.41}$$

Algorithm 2: Classical FETI/BETI method based on PCG

$g = [N_i f_i - S_i \tilde{g}_{D|\partial\Omega_i}]_{i=1}^s$
$\lambda_0 = Q\, G(G^\top Q\, G)^{-1} R^\top g$
$d = P^\top B\, S^\dagger (g - B^\top \lambda_0)$
solve $P^\top F\, \widetilde{\lambda} = d$ with PCG and initial value $\widetilde{\lambda}^{(0)} = 0$ and stop after k iterations
$\lambda^{(k)} = \lambda_0 + \widetilde{\lambda}^{(k)}$
$\xi^{(k)} = -(G^\top Q\, G)^{-1} G^\top Q\, B\, S^\dagger (g - B^\top \lambda^{(k)})$
$\tilde{u}^{(k)} = S^\dagger (g - B^\top \lambda^{(k)}) + R\, \xi^{(k)}$

Table 2.2 Overview on the spaces and operators involved in FETI/BETI

Spaces	
$U = \mathbb{R}^M$	Space of Lagrange multipliers
$U_{\text{ad}} = \ker(G^\top) \subset U$	Space of admissible Lagrange increments
$\widetilde{U}_{\text{ad}} = U_{\text{ad}/\ker(B^\top)}$	Factor space modulo redundancies
$V_D^h(\Gamma_S)$	Skeletal FE space with homogeneous Dirichlet conditions
$W_i = \begin{cases} V_D^h(\partial\Omega_i) & \text{(classical)} \\ V^h(\partial\Omega_i) & \text{(all-floating)} \end{cases}$	Local spaces
$W = \prod_{i=1}^s W_i$	Product space ("discontinuous" functions)
$\widehat{W} = \ker(B) \subset W$	Subspace of continuous functions in W, identifiable with $V_D^h(\Gamma_S)$
$Z = \prod_{i=1}^s \mathbb{R}^{\dim(\ker(S_i))}$	Parameter space of $\ker(S)$
Operators	
$B : W \to U^*$	Jump operator
$F : U \to U^*$	$F = B\, S^\dagger B^\top$
$G : Z \to U^*$	$G = B\, R$
$P : U \to U_{\text{ad}}$	Projection, $P = I - Q\, G(G^\top Q\, G)^{-1} G^\top$
$P^\top : U^* \to U_{\text{ad}}^*$	Projection, $P^\top = I - G(G^\top Q\, G)^{-1} G^\top Q$
$Q : U^* \to U$	Self-adjoint operator, SPD on range(G)
$R : Z \to W$	Injective operator, range$(R) = \ker(S)$
$S : W \to W^*$	Block operator of local approximate Steklov-Poincaré operators
$S^\dagger : W^* \to W$	Pseudo inverse of S

The complete method is summarized in Algorithm 2, p. 81. An overview on the spaces and operators involved is given in Table 2.2, p. 81.

2.2.1.5 The Central FETI/BETI Equation

Herein, we discuss the solvability and a solution algorithm of Problem (2.40). Since the Lagrange multipliers are only unique modulo $\ker(B^\top)$, we introduce the factor space

$$\widetilde{U}_{\text{ad}} := U_{\text{ad}/\ker(B^\top)}. \tag{2.42}$$

As mentioned in the proof of Lemma 2.16, F is well-defined on $\widetilde{U}_{\text{ad}}$ because it vanishes on $\ker(B^\top)$. Recall from Sect. 1.1.3.3 that $\text{range}(P^\top)$ is a realization of the dual of U_{ad}. Moreover, one can show that $\text{range}(P^\top) \cap \text{range}(B)$ is a realization of the dual of $\widetilde{U}_{\text{ad}}$. In short,

$$\widetilde{U}^*_{\text{ad}} = \text{range}(P^\top) \cap \text{range}(B) = \{\mu \in \text{range}(B) : \langle B\, z,\, Q\,\mu\rangle = 0 \quad \forall z \in \ker(S)\}. \tag{2.43}$$

Lemma 2.19. *The operator $P^\top F$ maps $\widetilde{U}_{\text{ad}}$ to $\widetilde{U}^*_{\text{ad}}$ and $P^\top F_{|\widetilde{U}_{\text{ad}}}$ is SPD.*

Proof. By definition, F maps U_{ad} to $\text{range}(B)$. From the definition of $P^\top$ we see that $P^\top(\text{range}(B)) \subset \text{range}(B) \cap \text{range}(P^\top) = \widetilde{U}^*_{\text{ad}}$.

Since F is self-adjoint and positive semi-definite and P is a projection onto U_{ad}, it follows immediately that $P^\top F_{|\tilde{U}_{\text{ad}}}$ is self-adjoint and positive semi-definite. It remains to show the definiteness. Assume that for $\lambda \in \widetilde{U}_{\text{ad}}$,

$$0 = \langle P^\top F\,\lambda,\, \lambda\rangle = \langle S^\dagger B^\top \lambda,\, B^\top \lambda\rangle.$$

Since $\lambda \in \widetilde{U}_{\text{ad}}$ implies $B^\top\lambda \in \text{range}(S)$, we conclude from the properties of the pseudo inverse $S^\dagger$ that $B^\top\lambda = 0$. Since λ is in the factor space modulo $\ker(B^\top)$, this means $\lambda = 0$ and concludes the proof. □

As a consequence of Lemma 2.19, Problem (2.40) can be solved using a PCG method, see Sect. 1.2.4.2 and Corollary 1.50. Recall that we have left the choice of $Q : U^* \to U$ open yet, and that we need a preconditioner (at least a formal one) mapping from $\widetilde{U}^*_{\text{ad}}$ back to $\widetilde{U}_{\text{ad}}$.

The unpreconditioned case. We choose $Q = I$ (recall that $U = \mathbb{R}^M = U^*$). Then $P^\top = P$ and $U^*_{\text{ad}} = U_{\text{ad}}$. Hence, $P^\top F_{|\tilde{U}_{\text{ad}}}$ maps to $U_{\text{ad}} \cap \text{range}(B)$, which is naturally embedded in the factor space $\widetilde{U}_{\text{ad}}$. Summarizing, the formal preconditioner can be chosen as the identity, i.e., we take *no* preconditioner.

The preconditioned case. All preconditioners under our consideration have the form

$$P M^{-1} \tag{2.44}$$

where $M^{-1} : U^* \to U$. The projection P makes sure that the preconditioner maps back to the space U_{ad}, which is embedded in $\widetilde{U}_{\text{ad}}$. Of course, $(P M^{-1})_{|\tilde{U}^*_{\text{ad}}}$ must be

SPD in order to make PCG applicable. The choice $M^{-1} = I$ and $Q = I$ gives the unpreconditioned case.

The complete FETI/BETI method for the classical formulation is summarized in Algorithm 2, p. 81. Note that in that algorithm, we can substitute g, B, R, S, and $S^\dagger$ by ℓ, $\overline{B}$, $\overline{R}$, A, and $A^\dagger$ from Remarks 2.17 and 2.18, respectively (and omit the first line), see also Sect. 2.2.5.

Remark 2.20. By simple linear algebra, one shows the residual identity

$$d - P^\top F \widetilde{\lambda}^{(k)} \;=\; B\, \tilde{u}^{(k)},$$

(see e.g. [FR94, FCRR98]) i.e., the residual in the CG algorithm controls the jump of the approximant $\tilde{u}^{(k)}$, and equivalently the jump of $u^{(k)} = [\tilde{g}_{D|\partial\Omega_i}]_{i=1}^s + \tilde{u}^{(k)}$. We note that the entire method can be rewritten in terms of the variables $t^{(k)} := -B^\top (\lambda_0 + \widetilde{\lambda}^{(k)})$ and $\tilde{u}^{(k)}$, see Sects. 2.2.1.2 and 2.2.3.

Remark 2.21. In Algorithm 2, we have chosen $\widetilde{\lambda}^{(0)} = 0$, but any value in U_{ad} would be suitable. Formally, the PCG runs in the factor space $\widetilde{U}_{\text{ad}}$, but in a standard implementation, we just use vectors in U (projected to U_{ad}). Working in factor spaces might often be dangerous in practice, but not in the current case: the components in $\ker(B^\top)$ of the iterates cannot blow up. If an iterate $\widetilde{\lambda}^{(k)}$ should have a non-zero contribution from $\ker(B^\top)$, the next iterate $\widetilde{\lambda}^{(k+1)}$ does not depend on this contribution, see also Sect. 2.2.3.

Remark 2.22. If a preconditioner M^{-1} is SPD on the whole of U^*, one can use $Q = M^{-1}$. In that case, $P\,Q = Q\,P^\top$ (cf. formula (1.5)) and so

$$PM^{-1}P^\top F \;=\; M^{-1}P^\top F,$$

i.e., we can leave out P in (2.44), see also Sect. 2.2.5.

We leave it up to the reader to follow the subsequent presentation linearly or not. Here is a guide for "nonlinear" readers.

Conditioning. For the case $Q = M^{-1} = I$, the convergence of the CG method is determined by the condition number $\kappa(P^\top F_{|\widetilde{U}_{\text{ad}}})$, cf. Lemma 1.49, which we will analyze in Sect. 2.4.1. In case of a global quasi-uniform mesh $\mathcal{T}^h(\Gamma_S)$ with mesh parameter h and suitable assumptions on the subdomains, it can be shown that

$$\kappa(P^\top F_{|\widetilde{U}_{\text{ad}}}) \;\le\; C\, \frac{\|\mathcal{A}\|_{L^\infty(\Omega)}}{\alpha_{\min}}\, \frac{H}{h},$$

where H denotes the maximal subdomain diameter (see Theorem 2.38). The constant C depends on the shape of the subdomains but is independent of $\mathcal{A}$, h, and H, in particular independent of the number of subdomains. Note that the robustness with respect to the number of subdomains comes from the projection P which plays the role of a coarse solve, cf. [FMR94].

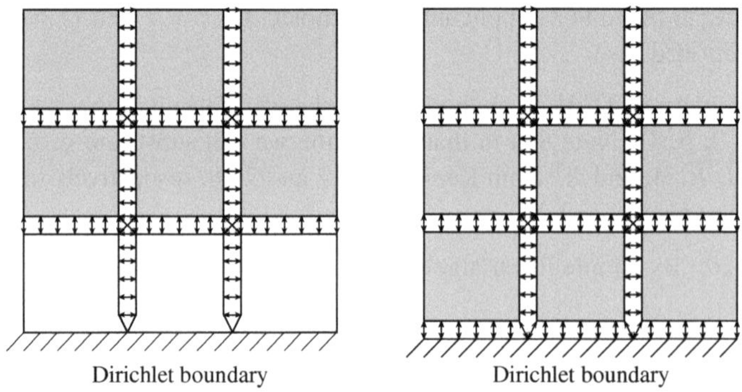

Fig. 2.4 Illustration of a classical formulation (*left*) and the all-floating (or total) formulation (*right*). Floating subdomains are *dark-shaded*

Preconditioning is (in general) required if

- H/h gets large (there is a large number of local unknowns in each subdomain), or
- $\|\mathscr{A}\|_{L^\infty(\Omega)}/\alpha_{\min}$ gets large (the coefficient $\mathscr{A}$ varies over several orders of magnitude).

For the definition of preconditioners for piecewise constant coefficients see Sect. 2.2.4, for their analysis see Sect. 2.6. The case of highly varying (multiscale) coefficients is subject of Chap. 3.

All-floating (Total) FETI/BETI. An important variant of the classical FETI/BETI method introduced in the current subsection is the *all-floating (total) FETI/BETI method*, see Sect. 2.2.2, where additional Lagrange multipliers are used to enforce the Dirichlet boundary conditions. This simplifies the method in a certain sense.

Implementation and Parallelization of the classical and the all-floating FETI/BETI method are discussed in Sect. 2.2.5.

2.2.2 All-Floating (Total) FETI/BETI

The all-floating method is a variant of the classical FETI/BETI method, where one introduces additional Lagrange multipliers that enforce Dirichlet conditions. For an illustration see Fig. 2.4.

Saddle Point Formulation. We start with the inhomogeneous skeleton formulation (2.11). In contrast to Sect. 2.2.1, the working spaces are chosen as[5]

[5]For the setting (2.19), $W_i = V^h(\partial\Omega_i \setminus \Gamma_N)$.

$$W_i := V^h(\partial\Omega_i) \qquad \text{and} \qquad W := \prod_{i=1}^{N} W_i, \tag{2.45}$$

i.e., we do not incorporate any Dirichlet boundary conditions. In this subsection, we regard S_i as a mapping from W_i to W_i^* (opposed to Sect. 2.2.1).

In addition to the (fully redundant) interface constraints

$$w_i(x^h) - w_j(x^h) = 0 \qquad \text{for } x^h \in \Gamma_{ij}^h, \quad i > j, \tag{2.46}$$

(cf. (2.22)) we also require

$$w_i(x^h) \;=\; \tilde{g}_D(x^h) \qquad \text{for } x^h \in \Omega_i^h \cap \Gamma_D\,. \tag{2.47}$$

Definition 2.23. Let us assume a numbering of the constraints (2.46) and (2.47) with M being the total number of constraints. The entry of a vector $\mu \in \mathbb{R}^M$ corresponding to constraint (2.46) is denoted by $\mu_{ij}(x^h)$, the entry corresponding to constraint (2.47) is denoted by $\mu_{iD}(x^h)$.

We define $B : W \to U^*$ (where $U = \mathbb{R}^M$) by

$$\left.\begin{aligned} (B\,w)_{ij}(x^h) &= w_i(x^h) - w_j(x^h) \text{ for } x^h \in \Gamma_{ij}^h, \quad i > j, \\ (B\,w)_{iD}(x^h) &= w_i(x^h) \qquad\qquad\quad \text{for } x^h \in \partial\Omega_i^h \cap \Gamma_D. \end{aligned}\right\} \tag{2.48}$$

Recall that we have chosen $\tilde{g}_D \in V^h(\Gamma_S)$ such that it vanishes on all nodes in $\Gamma_S^h \setminus \Gamma_D$. Of course, we can identify $\tilde{g}_D$ with a function in W, and (for simplicity) denote it again by $\tilde{g}_D$. With these considerations, the condition

$$B\,w \;=\; B\,\tilde{g}_D$$

requires that w is continuous across the subdomain interfaces, and that $w_{|\Gamma_D} = g_D$. Hence, the space

$$\widehat{W} \;:=\; \ker(B) \;\equiv\; V_D^h(\Gamma_S)$$

coincides with that from Sect. 2.2.1. Analogously to Sect. 2.2.1.2, we derive a saddle point formulation from (2.11). We define

$$t_i \;:=\; S_i\,u_{|\partial\Omega_i} - N_i\,f_i, \qquad t = [t_i]_{i=1}^s \in W^*.$$

Equation (2.11) implies that

$$\langle t, v\rangle \;=\; 0 \qquad \forall v \in \widehat{W},$$

and so $t \in \ker(B)^\circ = \operatorname{range}(B^\top)$. Parametrizing t by $-B^\top\lambda$ for $\lambda \in U$ and representing the solution as $u \in W$ with $B\,u = B\,\tilde{g}_D$ yields the saddle point problem:

Algorithm 3: All-floating FETI/BETI method based on PCG

$g = [N_i f_i]_{i=1}^s$
$\lambda_0 = Q\,G(G^\top Q\,G)^{-1} R^\top g$
$d = P^\top B\big[S^\dagger(g - B^\top \lambda_0) - \tilde{g}_D\big]$
solve $P^\top F\,\widetilde{\lambda} = d$ with PCG and initial value $\widetilde{\lambda}^{(0)} = 0$ and stop after k iterations
$\lambda^{(k)} = \lambda_0 + \widetilde{\lambda}^{(k)}$
$\xi^{(k)} = -(G^\top Q\,G)^{-1} G^\top Q\,B\big[S^\dagger(g - B^\top \lambda^{(k)}) - \tilde{g}_D\big]$
$u^{(k)} = S^\dagger(g - B^\top \lambda^{(k)}) + R\,\xi^{(k)}$

find $(u, \lambda) \in W \times U$ such that

$$\begin{bmatrix} S & B^\top \\ B & 0 \end{bmatrix} \begin{bmatrix} u \\ \lambda \end{bmatrix} = \begin{bmatrix} [N_i f_i]_{i=1}^s \\ B\,\tilde{g}_D \end{bmatrix}. \tag{2.49}$$

This problem has the same properties as (2.28): the solution u is unique, and λ is unique modulo $\ker(B^\top)$, cf. Lemma 2.11. Note that $-B^\top\lambda$ coincides with the discrete (generalized) conormal derivative of the solution u (also on the Dirichlet boundary).

Dual Saddle Point Formulation. Going through the same steps as in Sect. 2.2.1.3, we derive the dual saddle point formulation. The spaces and operators are slightly different because (in our setting)

$$\ker(S_i) = \operatorname{span}\{1_{\partial\Omega_i}\} \qquad \forall i = 1, \dots, s,$$

and so *all subdomains are floating subdomains* in the sense of Definition 2.15. Hence, $Z = \mathbb{R}^s$ and $R : Z \to \ker(S)$ with $(R\xi)_i = \xi_i$. With the notations from Table 2.2, p. 81, the dual saddle point problem reads as follows. Find $(\lambda, \xi) \in U \times Z$ such that

$$\begin{bmatrix} F & -G \\ G^\top & 0 \end{bmatrix} \begin{bmatrix} \lambda \\ \xi \end{bmatrix} = \begin{bmatrix} B(S^\dagger[N_i f_i]_{i=1}^s - \tilde{g}_D) \\ R^\top [N_i f_i]_{i=1}^s \end{bmatrix}.$$

We see that the structure of this system is identical to (2.40). The projection method is applied analogously to Sect. 2.2.1.4 and leads to an SPD problem for the variable $\widetilde{\lambda}$, see Lemma 2.24 below. The complete all-floating FETI/BETI method is summarized in Algorithm 3. Also here, we can substitute g, B, R, S, and $S^\dagger$ by ℓ, $\overline{B}$, $\overline{R}$, A, and $A^\dagger$ from Remarks 2.17 and 2.18, respectively (and omit the first line), see also Sect. 2.2.5.

Lemma 2.24. *In the all-floating formulation, the operator $P^\top F$ maps $\widetilde{U}_{\text{ad}}$ to $\widetilde{U}^*_{\text{ad}}$ and $P^\top F_{|\widetilde{U}_{\text{ad}}}$ is SPD.*

Proof. The proof is identical to that of Lemma 2.19.

Remark 2.25. Similarly to Remark 2.20, one can show the residual identity

$$d - P^\top F \widetilde{\lambda}^{(k)} \;=\; B\,(u^{(k)} - \tilde{g}_D),$$

where above, $\tilde{g}_D$ is interpreted as an element in W. Hence, in the all-floating method, the residual in the CG method controls the jump of the approximant $u^{(k)}$ as well as the error in the Dirichlet conditions.

Lemma 2.26. *In the all-floating formulation, for $\mu \in U^*$ and $\lambda \in U$,*

$$\langle \mu, \lambda \rangle \;=\; \sum_{x^h \in \Gamma^h} \sum_{\substack{i,j \in \mathcal{N}_{x^h} \\ i>j}} \mu_{ij}(x^h)\,\lambda_{ij}(x^h) + \sum_{i=1}^{s} \sum_{x^h \in \partial\Omega_i^h \cap \Gamma_D} \mu_{iD}(x^h)\,\lambda_{iD}(x^h).$$

The adjoints $B^\top : U \to W^*$ and $B_i^\top : U \to W_i^*$ fulfill

$$(B^\top \lambda)_{i,x^h} \;=\; \langle B_i^\top \lambda, \varphi_{i,x^h} \rangle_{\partial\Omega_i} \;=\; \begin{cases} \sum\limits_{j \in \mathcal{N}_{x^h} \setminus \{i\}} \operatorname{sign}(i-j)\,\lambda_{ij}(x^h) & \text{if } x^h \in \partial\Omega_i^h \cap \Gamma^h, \\ \lambda_{iD}(x^h) & \text{if } x^h \in \partial\Omega_i^h \cap \Gamma_D, \\ 0 & \text{else.} \end{cases}$$

Proof. The proof is straightforward.

2.2.3 FETI/BETI in Terms of Fluxes and Traces

Algorithm 4 displays the FETI/BETI method (both for the classical and the all-floating formulation) including explicitly the PCG method. Note that in order to unify the two formulations, we have used that in the classical formulation, $B\,\tilde{g}_D = 0$ and

$$u^{(k)} \;=\; \tilde{u}^{(k)} + [\tilde{g}_{D|\partial\Omega_i}]_{i=1}^s \;=\; P_Z\,S^\dagger([N_i\,f_i]_{i=1}^s - B^\top \lambda^{(k)}),$$

where

$$P_Z := I - R(G^\top Q\,G)^{-1} G^\top Q\,B. \tag{2.50}$$

We note that this particular projection is also treated in [FP03, SM08]. One can now show that for *both* formulations,

$$\tilde{u}^{(k)} \;=\; u^{(k)} - \tilde{g}_D \;=\; P_Z\big(S^\dagger(g - B^\top \lambda^{(k)}) - \tilde{g}_D\big).$$

We have seen in Sect. 2.2.1.2 that the Lagrange multipliers are temporary variables that parametrize fluxes. Also, due to Remarks 2.20 and 2.25,

$$d^{(k)} \;=\; B\,\tilde{u}^{(k)}.$$

Algorithm 4: FETI/BETI method in terms of Lagrange multipliers ($\lambda^{(k)}$) and jumps ($d^{(k)}$) including PCG

$$g = \begin{cases} [N_i f_i - S_i \tilde{g}_{D|\partial\Omega_i}]_{i=1}^s & \text{in classical formulation} \\ [N_i f_i]_{i=1}^s & \text{in all-floating formulation} \end{cases}$$

$\lambda^{(0)} = Q\,G(G^\top Q\,G)^{-1} R^\top g$

$d^{(0)} = P^\top B\left(S^\dagger(g - B^\top \lambda^{(0)}) - \tilde{g}_D\right)$

$k = 0$

repeat

$z^{(k)} = P M^{-1} d^{(k)}$ (in the unpreconditioned case: $z^{(k)} = d^{(k)}$)

$q^{(k)} = z^{(k)} + \beta_{k-1}\, q^{(k-1)}$ where $\beta_{-1} = 0,\ \beta_{k-1} = \dfrac{(d^{(k)}, z^{(k)})_{\ell^2}}{(d^{(k-1)}, z^{(k-1)})_{\ell^2}}$

$\lambda^{(k+1)} = \lambda^{(k)} + \alpha^{(k)}\, q^{(k)}$ where $\alpha_k = \dfrac{(d^{(k)}, z^{(k)})_{\ell^2}}{(P^\top F\, q^{(k)}, q^{(k)})_{\ell^2}}$

$d^{(k+1)} = d^{(k)} - \alpha_k\, P^\top F\, q^{(k)} \quad = B(u^{(k)} - \tilde{g}_D)$

$k = k + 1$

until *stopping criterion fulfilled for* $d^{(k)}$

$u^{(k)} = \tilde{g}_D + P_Z\left(S^\dagger(g - B^\top \lambda^{(k)}) - \tilde{g}_D\right)$

Introducing the additional variables

$$\begin{aligned} t^{(k)} &:= -B^\top \lambda^{(k)}, \\ s^{(k)} &:= B^\top z^{(k)} = B^\top P M^{-1} d^{(k)} = B^\top P M^{-1} B\, \tilde{u}^{(k)}, \\ p^{(k)} &:= B^\top q^{(k)}, \end{aligned}$$

we can rewrite the whole algorithm in terms of the fluxes $t^{(k)}$, the (discontinuous) Dirichlet traces $\tilde{u}^{(k)}$, the preconditioned "residual" $s^{(k)}$, and the search directions $p^{(k)}$ in the flux space, see Algorithm 5. In that algorithm we have to use the projection operators P_Z and $P_Z^\top$ rather than P and $P^\top$. We have

$$\begin{aligned} \text{range}(P_Z) &= \ker(S)^{\perp_{B^\top Q B}}, & P^\top B &= B\, P_Z \\ \text{range}(P_Z^\top) &= \text{range}(S) & B^\top P &= P_Z^\top B^\top. \end{aligned}$$

We notice that Algorithm 5 is formally a PCG algorithm (cf. Algorithm 1) with the operator

$$P_Z S^\dagger : Y \to Y^*$$

and the preconditioner

$$P_Z^\top B^\top M^{-1} B : Y^* \to Y,$$

where the flux space Y and the Dirichlet trace space Y^* are given by

---

Algorithm 5: FETI/BETI method including PCG rewritten in terms of fluxes ($t^{(k)}$) and traces ($\tilde{u}^{(k)}$)

$$g = \begin{cases} [N_i f_i - S_i\, \tilde{g}_{D|\partial\Omega_i}]_{i=1}^s & \text{in classical formulation} \\ [N_i f_i]_{i=1}^s & \text{in all-floating formulation} \end{cases}$$

$t^{(0)} = -(I - P_Z^\top) g$

$\tilde{u}^{(0)} = P_Z\big(S^\dagger(g + t^{(0)}) - \tilde{g}_D\big)$

$k = 0$

repeat

$\quad s^{(k)} = P_Z^\top B^\top M^{-1} B\, \tilde{u}^{(k)}$ (in the unpreconditioned case: $s^{(k)} = B^\top B\, \tilde{u}^{(k)}$)

$\quad p^{(k)} = s^{(k)} + \beta_{k-1}\, p^{(k-1)}$ where $\beta_{-1} = 0,\ \beta_{k-1} = \dfrac{(\tilde{u}^{(k)},\, s^{(k)})_{\ell^2}}{(\tilde{u}^{(k-1)},\, s^{(k-1)})_{\ell^2}}$

$\quad t^{(k+1)} = t^{(k)} - \alpha_k\, p^{(k)}$ where $\alpha_k = \dfrac{(\tilde{u}^{(k)},\, s^{(k)})_{\ell^2}}{(P_Z S^\dagger p^{(k)},\, p^{(k)})_{\ell^2}}$

$\quad \tilde{u}^{(k+1)} = \tilde{u}^{(k)} - \alpha_k\, P_Z S^\dagger p^{(k)} \qquad = P_Z\big(S^\dagger(g + t^{(k)}) - \tilde{g}_D\big)$

$\quad k = k + 1$

until *stopping criterion fulfilled for* $B\, \tilde{u}^{(k)}$

$u^{(k)} = \tilde{g}_D + \tilde{u}^{(k)}$

$$Y := \operatorname{range}(B^\top) \cap \operatorname{range}(S),$$

$$Y^* := \{w \in \ker(S)^{\perp_{B^\top Q B}} : S\, w \in \operatorname{range}(B^\top)\} = \{w \in \ker(S)^{\perp_{B^\top Q B}} \cap \ker(B)^{\perp_S}\}$$

(it can be argued that Y^* above is a realization of the dual of Y). If $Q = M^{-1} = I$, then the operator $P_Z^\top$ above can be left out (see Remark 2.22). These spaces make perfect sense because the solution $\tilde{u}$ must fulfill $B\, \tilde{u} = 0$, thus it must lie in $\ker(S)^{\perp_{B^\top Q B}}$. On the other hand, its flux must vanish on $\ker(B)$, i.e., the flux must lie in $\ker(B)^\circ = \operatorname{range}(B^\top)$. During the algorithm, the flux iterates are indeed in equilibrium:

$$t^{(k)} - t^{(0)} \in \ker(B)^\circ .$$

2.2.4 Preconditioning

As announced in Sect. 2.2.1.5, all preconditioners for $P^\top F_{|\widetilde{U}_{\text{ad}}}$ under our consideration have the form

$$P M^{-1} : \widetilde{U}_{\text{ad}}^* \to \widetilde{U}_{\text{ad}},$$

where $M^{-1} : U^* \to U$.

2.2.4.1 The Dirichlet Preconditioner

The *Dirichlet preconditioner* proposed by Farhat, Mandel, and Roux [FMR94] is given by

$$M_{\mathrm{D}}^{-1} = B S B^{\top} . \tag{2.51}$$

It was first analyzed by Mandel and Tezaur [MT96] who showed that in two dimensions, the condition number of FETI with the Dirichlet preconditioner and with $Q = I$ is bounded by

$$C \max_{i=1}^{s} (1 + \log(H_i / h_i))^{\beta},$$

where the constant C is independent of h_i (which is the local mesh size of subdomain Ω_i), H_i, and the number of subdomains. In general, $\beta = 3$. In some situations, $\beta = 2$, see [MT96, Lemma 3.8 and Remark 3.9]. With their pioneering article, Mandel and Tezaur paved the ground for all the refined FETI type analyses that appeared later. Note also that Tezaur [Tez98] showed that a method by Park, Justino, and Felippa [PJF97] is equivalent to the method in [FR91], see also [RFTM99].

Note that in general, the constant C above depends on the coefficient $\mathcal{A}$ including possible jumps. Also, classical primal substructuring methods (see [BPS86] and [TW05, Chap. 5]) are known to have a condition number involving just two powers of the logarithmic term. To get rid of the third power in FETI and to address coefficient jumps, one has to use a scaling of the jump operator in the preconditioner.

2.2.4.2 The Scaled Dirichlet Preconditioner

The scaled Dirichlet preconditioner has its roots in the following works. Rixen and Farhat [RF98a] provided a derivation using an energy-minimizing smoothing procedure. Klawonn and Widlund [KW01] used the so-called *weighted counting functions*, which are a basic ingredient of balancing Neumann-Neumann methods [DL91, Man93, MB96] (see also Sect. 2.3) and of related methods, see [DSW94, DW95, DSW96, Sar93, Sar94, Sar97]. A special choice of scalings leads to the method in [RF98a]. Moreover, Klawonn and Widlund [KW01] gave a rigorous analysis (covering also the three-dimensional case), showing that FETI with the scaled Dirichlet preconditioner results in a condition number of

$$C \max_{i=1}^{s} (1 + \log(H_i / h_i))^{2},$$

where the constant C is independent of the local mesh sizes h_i, the diameters H_i, the number of subdomains, and of jumps in the coefficients (provided that $\mathcal{A}$ is isotropic and piecewise constant on each of the subdomains). We will present this analysis in detail in Sect. 2.6 below.

Scalings and Weighted Counting Functions. Firstly, we need scalar weights

$$\rho_i(x^h) \; > \; 0 \tag{2.52}$$

for each $i = 1, \ldots, s$ and for each node $x^h \in \partial\Omega_i^h$. We will discuss several choices below. Secondly, for each $j \in \{1, \ldots, s\}$ and $x^h \in \Gamma_S^h$, we define the weight

$$\delta_j^\dagger(x^h) := \begin{cases} \dfrac{\rho_j(x^h)^\gamma}{\sum\limits_{k \in \mathcal{N}_{x^h}} \rho_k(x^h)^\gamma} & \text{for } x^h \in \partial\Omega_j^h, \\ 0 & \text{for } x^h \in \Gamma_S^h \setminus \partial\Omega_j^h, \end{cases} \tag{2.53}$$

where $\mathcal{N}_{x^h} = \{i = 1, \ldots, s : x^h \in \partial\Omega_i\}$, cf. Definition 2.12, and $\gamma \in [1/2, \infty)$ is a fixed exponent. A default choice for γ is one. The resulting piecewise linear functions $\delta_j^\dagger \in V^h(\Gamma_S)$ for $j = 1, \ldots, s$, are called *weighted counting functions*. The union of these functions forms a partition of unity on the skeleton, i.e.,

$$\sum_{j=1}^{s} \delta_j^\dagger(x^h) \; = \; 1 \qquad \forall x^h \in \Gamma_S^h \, .$$

Note finally that $\delta_j^\dagger(x^h) = 1$ for all $x^h \in \partial\Omega_j \setminus \Gamma^h$.

Remark 2.27. The limit case $\gamma \to \infty$ corresponds to the choice

$$\delta_j^\dagger(x^h) := \begin{cases} 1/m & \text{if } x^h \in \partial\Omega_j^h \text{ and } \rho_j(x^h) = \max\limits_{k \in \mathcal{N}_{x^h}} \rho_k(x^h), \\ 0 & \text{else,} \end{cases} \tag{2.54}$$

where $m = \#\big\{k \in \mathcal{N}_{x^h} : \rho_k(x^h) = \max\limits_{\ell \in \mathcal{N}_{x^h}} \rho_\ell(x^h)\big\}$ is the number of times the maximal coefficient is attained. For an early domain decomposition method with this scaling see [WX94].

The following scalings are common.

Multiplicity Scaling. If there is no (or only very little) variation in the coefficient $\mathcal{A}$, one usually chooses

$$\rho_i(x^h) \; = \; 1.$$

Then $\delta_j^\dagger(x^h)$ simply equals the reciprocal of the multiplicity of the node x^h.

Coefficient Scaling. In case of coefficient variation, the weight $\rho_i(x^h)$ should in a way resemble the coefficient $\mathcal{A}$ in subdomain Ω_i *around* x^h. If $\mathcal{A}_{|\Omega_i} = \alpha_i\, I$, we can set $\rho_i(x^h) = \alpha_i$. The case of a varying coefficient is treated in Chap. 3. If $\mathcal{A}$ is isotropic and globally constant, coefficient and multiplicity scaling are equivalent.

Stiffness Scaling. For a pure FETI method, $\rho_i(x^h)$ is often chosen as the diagonal entry of the subdomain stiffness matrix corresponding to the node x^h, see also Sects. 2.6.4 and 3.3.2.

Note that the coefficient scaling is frequently called *ρ-scaling* in the literature (where the scalar coefficient itself is denoted by ρ), see e.g. [KRW08]. The stiffness scaling was first proposed by Rixen and Farhat [RF98a, RF99], there called *superlumped smoothing*. For more scalings see Sect. 2.6.4 and the recent paper [DW12b].

A Weighted Jump Operator. Recall from Definition 2.13 that

$$t_{i,x^h} = \langle t_i, \varphi_{i,x^h} \rangle_{\partial\Omega_i} \qquad \text{for } t \in W,$$

where $\varphi_{i,x^h} \in V^h(\partial\Omega_i)$ is the nodal basis function associated to node x^h. The weighted jump operator $B_D : W^* \to U$ is given by

$$(B_D\, t)_{ij}(x^h) = \delta_j^\dagger(x^h)\, t_{i,x^h} - \delta_i^\dagger(x^h)\, t_{j,x^h} \qquad \text{for } x^h \in \Gamma_{ij}^h,\ i > j. \tag{2.55}$$

In the all-floating formulation, we set

$$(B_D\, t)_{iD}(x^h) = t_{i,x^h} \qquad \text{for } x^h \in \partial\Omega_i^h \cap \Gamma_D. \tag{2.56}$$

Lemma 2.28. *The adjoint $B_D^\top : U^* \to W$ is given by*

$$(B_D^\top \mu)_i(x^h) = \begin{cases} \sum_{j \in \mathcal{N}_{x^h} \setminus \{i\}} \delta_j^\dagger(x^h)\, \text{sign}(i-j)\, \mu_{ij}(x^h) & \text{if } x^h \in \Gamma^h \cap \partial\Omega_i, \\ \mu_{iD}(x^h) & \text{if } x^h \in \partial\Omega_i^h \cap \Gamma_D \text{ (in the all-floating formulation)}, \\ 0 & \text{else.} \end{cases}$$

Proof. The proof is straightforward.

The Scaled Dirichlet Preconditioner is finally given by

$$M_{\text{sD}}^{-1} = B_D\, S\, B_D^\top. \tag{2.57}$$

Remark 2.29. Each block S_i in S appearing in the preconditioner (2.57) may be replaced by the hypersingular operator on $\partial\Omega_i$ (then called *scaled hypersingular BETI preconditioner*, cf. [LS03, LS05]), or by any other operator that is spectrally equivalent to S_i.

Remark 2.30. For non-redundant Lagrange multipliers, the preconditioner takes the same form but with

$$B_D = (BD^{-1}B^\top)^{-1}BD^{-1},$$

where $D : W^* \to W$ is a diagonal scaling operator where the entry corresponding to node x^h in subdomain Ω_i is $\delta_i^\dagger(x^h)$, cf. [KW01] and [TW05, Sect. 6.3.2]. For an efficient algorithmic construction of B_D see also [Of06, Sect. 5.5.2].

2.2.4.3 Lumped Preconditioners

For a pure FETI method, "lumped" preconditioners (see [FR91, Sect. 5] and [FMR94, Sect. 6]) are constructed by replacing the FE Schur complements $\mathbf{S}_i$ in (2.51) or (2.57) by the block $\mathbf{K}_{i,BB}$ of the stiffness matrix corresponding to the boundary/interface unknowns. In other words, instead of solving for the PDE-harmonic extension in each subdomain, we simply extend by zero.

The application of a lumped preconditioner is more economic because one does not need the factorizations and solves appearing in the operators S_i. In some situations, the overall CPU time of a FETI method with a lumped preconditioner can be shorter than with a Dirichlet preconditioner, even though the condition number of the preconditioned system might be larger. This is also due to a superconvergence effect, cf. [FMR94, Sect. 7].

Remark 2.31. The hypersingular BETI preconditioner could be seen as a BEM equivalent of the lumped preconditioner, because there is no additional solving involved. However, this operator has a corresponding dense matrix and thus needs an effective implementation. Note also that the hypersingular preconditioner is still quasi-optimal.

For the remainder of this book, we will only treat the scaled Dirichlet preconditioner, because it turns out to be quasi-optimal with respect to the condition number.

2.2.4.4 The Operator Q

If there is no (or only little) coefficient variation, one commonly chooses

$$Q = I,$$

which is supported by the analysis, see Remark 2.111. In case of large coefficient variation, one often chooses

$$Q = M_{\mathrm{sD}}^{-1}.$$

where M_{sD}^{-1} is the scaled Dirichlet preconditioner, cf. [BDF$^+$00]. Another practically successful option is to set Q to the lumped Dirichlet preconditioner. These choice originate from [FR94] and were further investigated in [Rix97]. There is also a diagonal choice

$$Q = Q_{\mathrm{diag}}$$

(first suggested and analyzed in [KW01]) which mimics the choice $Q = M_{\mathrm{sD}}^{-1}$ under certain regularity assumptions and which simplifies the implementation, see

Sects. 2.6.3 and 3.3.5.4. Note all these different choices of Q lead to a sparse coarse matrix $\mathbf{G}^\top \mathbf{Q}\mathbf{G}$, but to different sparsity patterns, cf. [FR94].

2.2.5 *Implementation Issues*

In this subsection, we discuss implementation issues of the classical and the all-floating FETI/BETI methods described above (cf. also [LP98, RF98b, Kam00]). In particular, we address the issue of parallelization.

2.2.5.1 The Basic Input Data

To make an implementation of FETI/BETI possible, one needs the vector representations for the local spaces $V^h(\Omega_i)$ or $V^h(\Gamma_i)$ with respect to the standard nodal basis ($V^h_D(\Omega_i)$ or $V^h_D(\Gamma_i)$ in the classical formulation) and a local numbering of the degrees of freedom (DOFs). Correspondingly, for each FEM subdomain we need the local stiffness matrix (for a floating subdomain the "Neumann" matrix) and the load vector associated to each of these spaces. For each BEM subdomain we need (data-sparse) matrix approximations of the boundary integral operators (for matrix-free methods see also Sect. 2.2.6). For the classical formulation, we assume that the system is homogenized, i.e., the prescribed Dirichlet values are already contained in the load vector, and there are either no Dirichlet DOFs, or these are decoupled from the remaining DOFs.

2.2.5.2 Interconnecting

In order to do interconnecting, each DOF on the interface must have a local and a global index. With this information, one can set up the Lagrange multipliers on the interface: each multiplier (numbered by an index) is described by two subdomain indices and two local DOF indices, cf. (2.22). The global index is used for identification only. In the all-floating formulation, one additionally needs to know which DOFs are on the Dirichlet boundary in order to set up the additional multipliers there, cf. (2.47).

If the subdomain decomposition is generated from a mesh partitioner such as METIS [KK98], the local to global mappings come together with the output (at least for scalar P^1-elements).

2.2.5.3 Additional Input Data

For the scaled Dirichlet preconditioner, one needs the scalings $\rho_i(x^h)$ corresponding to each interface DOF. For the coefficient scaling (cf. Sect. 2.2.4.2), the coefficient is required, whereas for the stiffness scaling and the multiplicity scaling no additional information is needed. See however Sect. 2.6.4.2 for undesired effects that can occur with the stiffness scaling and/or with varying coefficients.

Most importantly, one needs a description of the local kernels, i.e., one needs matrices $\mathbf{R}_i$ corresponding to R_i from (2.30). If $i \in \mathscr{I}_{\text{FEM}}$, it is advantageous to set up $\mathbf{R}_i$ such that the columns of $\mathbf{R}_i$ span the kernel of the local stiffness matrix $\mathbf{K}_i$ (rather than that of the Schur complement $\mathbf{S}_i$ which should never be formed).

If the kernel is not known a priori, one can run a singular value decomposition or try to get low-frequent eigenpairs of $\mathbf{K}_i$ by an inverse power method (see e.g. [GV96]) or the LOBPCG method [Kny01], but this is in general expensive. Here lies a true advantage of the all-floating formulation: for the most widely used types of PDEs, especially for the potential equation or linear elasticity (see also Sect. 2.8), the kernel is known a priori in all the subdomains and the matrix $\mathbf{R}_i$ can be given explicitly.

For the diagonal choice $Q = Q_{\text{diag}}$ due to Klawonn and Widlund, which we will expose in Sect. 2.6.3, additional mesh information (H_i and h_i) is required. Furthermore, one needs to know whether a coupling DOF is associated to a subdomain vertex, or if it is in the interior of a subdomain edge or face. However, this can be figured out from the Lagrange multipliers by combinatorial means (cf. Sect. 2.48).

2.2.5.4 Implementation of the Underlying Operators

Having R and $R^\top$ at our disposal, and assuming that $Q \in \{I, M_{\text{sD}}^{-1}, Q_{\text{diag}}\}$, we can reduce Algorithms 2 and 3 to the applications of B, $B^\top$, B_D, $B_D^\top$, S, $S^\dagger$, and $(G^\top Q\, G)^{-1}$, which are discussed in the sequel.

Jump Operators. The operators B, $B^\top$, B_D and $B_D^\top$ need not be stored but are encoded as routines which perform their application to vectors. These routines mainly use the description of the Lagrange multipliers and the scalings $\rho_i(x^h)$.

Local FEM Neumann Problems. For each $i \in \mathscr{I}_{\text{FEM}}$, the action $v_i = S_i^\dagger f_i$ for a given $f_i \in \text{range}(S_i)$ is performed as follows. Let $\mathbf{v}_B$ and $\mathbf{f}_B$ denote the vector representations of v_i and f_i, respectively. Then, with the analogous notation as in Sect. 1.2.6, the equation $\mathbf{S}_i\, \mathbf{v}_B = \mathbf{f}_B$ is equivalent to

$$\begin{bmatrix} \mathbf{K}_{i,BB} & \mathbf{K}_{i,BI} \\ \mathbf{K}_{i,IB} & \mathbf{K}_{i,II} \end{bmatrix} \begin{bmatrix} \mathbf{v}_B \\ \mathbf{v}_I \end{bmatrix} = \begin{bmatrix} \mathbf{f}_B \\ 0 \end{bmatrix},$$

i.e., we need to solve a local Neumann problem, cf. Sect. 1.2.5. The most convenient way is to regularize the matrix $\mathbf{K}_i$ (if Ω_i is floating) and store its factorization. For the setting (2.19) (see also Remark 2.5, p. 68), the set "I" above includes interior DOFs as well as non-coupling Neumann DOFs.

Local BEM Neumann Problems. For each $i \in \mathscr{I}_{\text{BEM}}$, the action $v_i = S_i^\dagger f_i$ for a given $f_i \in \text{range}(S_i)$ is performed as follows. Let $\mathbf{v}$ and $\mathbf{f}$ denote the vector representations of v_i and f_i, respectively. Let $\mathbf{D}_i$, $\mathbf{K}_i$, and $\mathbf{V}_i$ denote the matrix representations of the hypersingular operator, the double layer potential, and the single layer potential, respectively, and let $\mathbf{M}_i$ be the mass matrix from Sect. 1.3.7. Then instead of solving $\mathbf{S}_i\mathbf{v} = \mathbf{f}$ with $\mathbf{S}_i = \mathbf{D}_i + (\frac{1}{2}\mathbf{M}_i^\top + \mathbf{K}_i^\top)\mathbf{V}_i^{-1}(\frac{1}{2}\mathbf{M}_i + \mathbf{K}_i)$, we solve the equivalent saddle point problem

$$\begin{bmatrix} \mathbf{D}_i & \frac{1}{2}\mathbf{M}_i^\top + \mathbf{K}_i^\top \\ \frac{1}{2}\mathbf{M}_i + \mathbf{K}_i & -\mathbf{V}_i \end{bmatrix} \begin{bmatrix} \mathbf{v} \\ \mathbf{t} \end{bmatrix} = \begin{bmatrix} \mathbf{f} \\ \mathbf{0} \end{bmatrix},$$

which is a standard BEM problem corresponding to the local Neumann problem. Recall that for $d = 2$, we should ensure that $\operatorname{diam}(\Omega_i) < 1$ (e.g. by scaling all the coordinates of Ω) in order to ensure the invertibility of $\mathbf{V}_i$.

Again, if Ω_i is a floating subdomain, we can regularize this problem by regularizing the hypersingular operator $\mathbf{D}_i$ similar as in Sect. 1.2.5. As briefly described in Sect. 1.3.7, the matrices $\mathbf{D}_i$, $\mathbf{K}_i$, and $\mathbf{V}_i$ can be approximated in data-sparse form using $\mathcal{H}$-matrices. Thus, also the matrix corresponding to the above (possibly regularized) saddle point problem can be represented by an $\mathcal{H}$-matrix, and for each BEM subdomain its $\mathcal{H}$-*LU* factorization can be built and stored in quasi-optimal time and memory complexity in the preprocessing phase. For matrix-free fast BEM, such as the fast multipole methods, in connection with FETI/BETI methods see Sect. 2.2.6.

Local Dirichlet Problems. The action of S_i for $i \in \mathcal{I}_{\text{FEM}}$, corresponds essentially to solving a system of the form

$$\mathbf{K}_{i,II}\,\mathbf{v}_I = -\mathbf{K}_{i,IB}\,\mathbf{v}_B\,.$$

For the input $\mathbf{v}_B$, the output is given by $\mathbf{S}_i\mathbf{v}_B = \mathbf{K}_{i,BB}\,\mathbf{v}_B + \mathbf{K}_{i,BI}\,\mathbf{v}_I$. Hence, it is most convenient to build and store a factorization of $\mathbf{K}_{i,II}$ in the preprocessing phase. Again, for the setting (2.19) (see also Remark 2.5, p. 68), the set "I" above includes interior DOFs as well as non-coupling Neumann DOFs. For $i \in \mathcal{I}_{\text{BEM}}$, the action of S_i (see Sect. 1.3.8.1) involves the inverse of the single layer potential, which can be realized by $\mathcal{H}$-*LU* factorization.

The Coarse Problem. Let $\mathbf{G}$ and $\mathbf{Q}$ denote the matrix representations of G and Q. As discussed in Remark 2.108 below, the matrix $\mathbf{G}^\top\mathbf{Q}\,\mathbf{G}$ is sparse, and its sparsity pattern is determined by the connectivity graph of the subdomain partition where each floating subdomain is a node of that graph. Once $\mathbf{G}^\top\mathbf{Q}\,\mathbf{G}$ is assembled (cf. [RF98b, Sect. 3.2]), its factorization can be computed efficiently during the preprocessing phase, as long as the number of subdomains is not very large. If we set $Q = M_{\text{sD}}^{-1}$, an efficient assembly of $\mathbf{G}^\top\mathbf{Q}\,\mathbf{G}$ is possible but tricky (see Remark 2.36). Note that the extra cost of applying Q during the FETI/BETI algorithm (see Algorithm 2 and (2.38)) involves the solution of additional local Dirichlet problems. However, within each step of the iterative solver, this is compensated by the fact that we can leave out the action of P (cf. Remark 2.22). Nevertheless, from the implementation point of view, it is more attractive to use a diagonal choice of Q if the context allows to do so; see Sects. 2.6.3 and 3.3.5.4 below.

Remark 2.32. We warn the reader that if the coefficient α varies extremely within a single FEM subdomain (see also Chap. 3), the matrices $\mathbf{K}_{i,II}$ and $\mathbf{K}_i$ (the latter possibly regularized) can become very ill-conditioned. For example if the coefficient varies of between 1 and 10^9, even direct solvers may run into stability problems. The same can happen with the coarse matrix $\mathbf{G}^\top \mathbf{Q}\,\mathbf{G}$ if the coefficient varies a lot throughout the global domain Ω.

2.2.5.5 Parallelization

The FETI/BETI algorithm is very suitable for multiprocessor machines with shared and especially with distributed memory. The coding should follow the MIMD (multiple instruction multiple data) paradigm, cf. [DHL03, Haa99, SBG96]. In the following, let the processors be numbered from 1 to p. For software supporting parallelization (on different levels), we refer e.g. to the following frameworks.

- MPI (message-passing interface) standard [MPI09]
- *PETSc*: http://www-unix.mcs.anl.gov/petsc/petsc-as/
- *Hypre*: http://acts.nersc.gov/hypre/
- *DUNE*: http://www.dune-project.org/
- *Parallel toolbox:* http://paralleltoolbox.sourceforge.net/

For further literature see also [Bas96, SBG96, Zum03].

Since the main work are subdomain solves that are independent of each other, we assign each subdomain Ω_i to a processor $p_i \in \{1, \ldots, p\}$. Hence, each processor handles one or several subdomains. There are two kinds of global objects involved in the algorithm that need to be parallelized:

(i) Lagrange multipliers, i.e., vectors from U, U^*,
(ii) Coarse vectors representing elements from Z, Z^*.

The remaining variables are elements from the spaces $W = \Pi_{i=1}^s W_i$ and W^* which are parallel by construction.

We will first show how to deal with parallelizing the Lagrange multipliers (kind (i)) using the concept of accumulated and distributed vectors, see e.g. [DHL03, Haa99].

Accumulated and Distributed Vectors

Definition 2.33. For $\lambda \in \mathbb{R}^M$ and $q = 1, \ldots, p$, we denote by $\lambda_{\text{acc}}^{(q)} \in \mathbb{R}^{M_q}$ the vector of those entries $\lambda_{ij}(x^h)$ where

$$p_i = q \;\text{ or }\; p_j = q \quad \text{and} \quad x^h \in \Gamma_{ij}^h,$$

i.e., the entries *shared* by processor q. As for λ itself, the local numbering within $\lambda_{\mathrm{acc}}^{(q)}$ can be arbitrary. To be general, we denote the entries of $\lambda_{\mathrm{acc}}^{(q)}$ again by $\lambda_{\mathrm{acc},ij}^{(q)}(x^h)$. For the all-floating formulation, the vector $\lambda_{\mathrm{acc}}^{(q)}$ additionally contains those entries $\lambda_{iD}(x^h)$ where $p_i = q$ and $x^h \in \partial\Omega_i^h \cap \Gamma_D^h$. The parallel vector $\lambda_{\mathrm{acc}} := [\lambda_{\mathrm{acc}}^{(q)}]_{q=1}^p$ is called *accumulated* realization of λ. It has the property that $\lambda_{\mathrm{acc},ij}^{(q)}(x^h) = \lambda_{ij}(x^h)$ for all processors q.

Definition 2.34. A parallel vector $\lambda_{\mathrm{dist}} = [\lambda_{\mathrm{dist}}^{(q)}]_{q=1}^p$ with $\lambda_{\mathrm{dist}}^{(q)} \in \mathbb{R}^{M_q}$ is called *distributed* realization of $\lambda \in \mathbb{R}^M$, if

$$\lambda_{ij}(x^h) = \sum_{q\in\{p_i,p_j\}} \lambda_{\mathrm{dist},ij}^{(q)}(x^h).$$

In the implementation, we will use both representations, but the vectors λ_{acc}, λ_{dist} are never formed, but only their local components $\lambda_{\mathrm{acc}}^{(q)}$, $\lambda_{\mathrm{dist}}^{(q)}$ on each processor q.

It is immediate, that for scalars $\alpha, \beta \in \mathbb{R}$ and vectors $\lambda, \mu \in U$ or U^*, the accumulated (or distributed) representation of $\alpha\lambda + \beta\mu$ is simply $\alpha\lambda_{\mathrm{acc}} + \beta\mu_{\mathrm{acc}}$ (or $\alpha\lambda_{\mathrm{dist}} + \beta\mu_{\mathrm{dist}}$, respectively), i.e., these operations can be performed fully in parallel.

The next lemma shows that computing the scalar product between a distributed and an accumulated vector can be performed by computing the local scalar products and then just communicate the (scalar) results between all processors and add them up. In the MPI standard, this is done by the `allreduce` command, see [MPI09].

Lemma 2.35. *For $\lambda \in U$ and $\mu \in U^*$,*

$$\langle\mu, \lambda\rangle = (\mu, \lambda)_{\ell^2} = \sum_{q=1}^p (\mu_{\mathrm{dist}}^{(q)}, \lambda_{\mathrm{acc}}^{(q)})_{\ell^2}.$$

Proof. The proof is straightforward.

With a slight abuse of notation, we will write $\langle\mu_{\mathrm{dist}}, \lambda_{\mathrm{acc}}\rangle = \langle\mu, \lambda\rangle$ in the sequel, which indicates that the calculation of the inner product is based on the result of Lemma 2.35.

Let U_{acc}, U_{dist} and U_{acc}^*, U_{dist}^* denote the (formal) spaces of accumulated and distributed vectors representing elements in U, U^*, respectively. Let A denote the *accumulation operator* such that for $\lambda \in U$ or in U^*, $\mathsf{A}\,\lambda_{\mathrm{dist}} = \lambda_{\mathrm{acc}}$. Moreover, let D denote the *distribution operator* under whose action entries will be distributed by simply dividing them by their multiplicity. We have $\mathsf{A}\,\mathsf{D} = I$ but in general $\mathsf{D}\mathsf{A} \neq I$. Note also that $\mathsf{D} \neq \mathsf{A}^\top$.

When we want to compute the (global) Euclidean norm of a vector λ from its accumulated representation λ_{acc}, we can calculate $\sqrt{(\mathsf{D}\lambda_{\mathrm{acc}}, \lambda_{\mathrm{acc}})_{\ell^2}}$, without major communication. If only λ_{dist} is available, we can calculate $\sqrt{(\lambda_{\mathrm{dist}}, \mathsf{A}\,\lambda_{\mathrm{dist}})_{\ell^2}}$, which involves more communication due to the accumulation operator.

Setting Up Parallel Lagrange Multipliers

The most convenient starting point is a setting where each local DOF (in a subdomain) is associated to a global DOF. This can be easily achieved for continuous P^1 elements, if one has global indices for each vertex of the mesh (for high order elements, one additionally needs global indices for the edges/facets of the mesh). For simplicity, we continue with the P^1 elements for the scalar elliptic equation, but the concept can be generalized straightforwardly.

For each DOF, it should be known which subdomains are formally sharing it. If this information is not known a priori, it can be got from the associated global indices by sorting and communicating (see, e.g., [Lie06]). The communication can be reduced if the next neighbors of each processor are known.

Each multiplier $\lambda_{ij}(x^h)$ is identified by the triple $(i,\ j,\ g_{x^h})$ where g_{x^h} is the global index of node x^h and $i > j$. When it comes to sending/receiving entries to/from another processor, we order the corresponding entries with respect to the global index triple. This ensures that the entries are sent/received in the correct order.

Parallelization of Z and Z^*

Since each entry of a vector $\xi \in Z$ or in Z^* corresponds to a *unique* subdomain, parallelization is easier. For $\xi \in Z$ and $q = 1, \ldots, p$, let $\xi_{\text{loc}}^{(q)}$ be the local vector of entries ξ_i such that $p_i = q$, and let $\xi_{\text{loc}} := [\xi_{\text{loc}}^{(q)}]_{q=1}^p$ be the corresponding parallel vector. Moreover, let Z_{loc}, Z_{loc}^* be the spaces of such parallel vectors and let

$$\mathsf{A} : Z_{\text{loc}}^* \to Z^*, \qquad \mathsf{D} : Z \to Z_{\text{loc}}$$

denote the accumulation and distribution operator, respectively. Opposed to the global vectors of Lagrange multipliers, here the operator A creates a global vector *on each processor* by collecting the vectors from all processors. Note that the size of this global vector is of the same order as the number of subdomains. For the distribution operator, we have the identity $\mathsf{D} = \mathsf{A}^\top$.

Parallelization of the FETI/BETI Operators

We will now replace the operators occurring in Algorithm 4 by operators involving the parallel spaces defined above. The following operators arise naturally.
If $M^{-1} = I$, we have to set $\mathsf{M}^{-1} := \mathsf{A}$ as an operator mapping U_{dist}^* to U_{acc}.
The implementation of $\mathsf{Q} : U_{\text{dist}}^* \to U_{\text{acc}}$ depends on the choice of Q:

- If $Q = I$ or $Q = Q_{\text{diag}}$, we first implement local diagonal operators $\mathsf{Q}^{(q)}$. Applying them in parallel leads to the operator $\mathsf{Q}_{\text{loc}} : U_{\text{dist}}^* \to U_{\text{dist}}$. Finally, we set $\mathsf{Q} := \mathsf{A}\,\mathsf{Q}_{\text{loc}}$.
- If $Q = M_{\text{sD}}^{-1}$, we set $\mathsf{Q} := \mathsf{M}_{\text{sD}}^{-1}$.

Operator		Adjoint	
B	$: W \to U^*_{\text{dist}}$	$\mathsf{B}^\top$	$: U_{\text{acc}} \to W^*$
B_D	$: W^* \to U_{\text{dist}}$	$\mathsf{B}_D^\top$	$: U^*_{\text{acc}} \to W$
R	$: Z_{\text{loc}} \to W$	$\mathsf{R}^\top$	$: W^* \to Z^*_{\text{loc}}$
$\mathsf{G} := \mathsf{B}\,\mathsf{R}\,\mathsf{D}$	$: Z \to U^*_{\text{dist}}$	$\mathsf{G}^\top := \mathsf{A}\mathsf{R}^\top\mathsf{B}^\top$	$: U_{\text{acc}} \to Z^*$
$\mathsf{F} := \mathsf{B}\,S^\dagger\,\mathsf{B}^\top$	$: U_{\text{acc}} \to U^*_{\text{dist}}$		–
$\mathsf{M}_{\text{sD}}^{-1} := \mathsf{A}\mathsf{B}_D\,S\,\mathsf{B}_D^\top\mathsf{A}$	$: U^*_{\text{dist}} \to U_{\text{acc}}$		–

The parallel operators corresponding to the projections P, $P^\top$ are as follows:

$$\mathsf{P} := I - \mathsf{Q}\,\mathsf{G}(\mathsf{G}^\top\mathsf{Q}\mathsf{G})^{-1}\mathsf{G}^\top : U_{\text{acc}} \to U_{\text{acc}},$$
$$\mathsf{P}^\top := I - \mathsf{G}(\mathsf{G}^\top\mathsf{Q}\mathsf{G})^{-1}\mathsf{G}^\top\,\mathsf{Q} : U^*_{\text{dist}} \to U^*_{\text{dist}}.$$

Note that the matrix corresponding to the coarse operator $(\mathsf{G}^\top\mathsf{Q}\,\mathsf{G}) : Z \to Z^*$ can be assembled and made available on each processor, and its factorization can be built and stored. Recall from Remark 2.22 that if $Q = M_{\text{sD}}^{-1}$, we can leave out P.

Remark 2.36. A parallely efficient assembly of $\mathbf{G}^\top\mathbf{Q}\,\mathbf{G}$ for the case $Q = M_{\text{sD}}^{-1}$ is tricky but possible. One possible way of implementation is described in [FR94], see also [BDF+00] for a parallel speed-up of the factorization. Another recipe is the following. In Sect. 2.4.2.1, we will see that there is an averaging operator $E_D : W \to \widehat{W}$ such that $G^\top M_{\text{sD}}^{-1}\,G = R^\top(I - E_D)^\top S\,(I - E_D)\,R = R^\top E_D^\top S\,E_D\,R$, where the latter identity holds true because $S\,R = 0$. From this formula, we see that averaging each kernel function (by E_D) and employing suitable neighbor communication, we can figure out the corresponding row of the matrix $G^\top M_{\text{sD}}^{-1} G$.

The final FETI/BETI algorithm (including PCG) is displayed in Algorithm 6.

2.2.6 *Inexact FETI/BETI Methods*

Recall that the action of $S_i^\dagger$ requires to solve a system on subdomain Ω_i. For a FEM subdomain, the system matrix is the (regularized) stiffness matrix, for a BEM subdomain, the system matrix has the form

$$\begin{bmatrix} \widetilde{\mathbf{D}}_i & \frac{1}{2}\mathbf{M}_i^\top + \mathbf{K}_i^\top \\ \frac{1}{2}\mathbf{M}_i + \mathbf{K}_i & -\mathbf{V}_i \end{bmatrix},$$

where $\widetilde{\mathbf{D}}_i$ is the matrix corresponding to the hypersingular operator, regularized if Ω_i is floating, see Sect. 2.2.5.4. We already mentioned that this matrix can be factorized using $\mathscr{H}$-LU (or Cholesky) factorization. If one uses the fast multipole method, only the fast *application* of the above matrix to a vector is available. In [LOSZ07b, LOSZ07a], (see also [KW00] for the original idea in a FETI setting), an

Algorithm 6: Parallel FETI/BETI method including PCG

$$g = \begin{cases} [N_i f_i - S_i \tilde{g}_{D|\partial\Omega_i}]_{i=1}^s & \text{in classical formulation} \\ [N_i f_i]_{i=1}^s & \text{in all-floating formulation} \end{cases}$$

$$\lambda_{\text{acc}}^{(0)} = \mathsf{Q}\,\mathsf{G}(\mathsf{G}^\top \mathsf{Q}\,\mathsf{G})^{-1}\mathsf{A}\,\mathsf{R}^\top g$$

$$d_{\text{dist}}^{(0)} = \mathsf{P}^\top \mathsf{B}\left(S^\dagger(g - \mathsf{B}^\top \lambda_{\text{dist}}^{(0)}) - \tilde{g}_D\right)$$

$k = 0$

repeat

$$z_{\text{acc}}^{(k)} = \mathsf{P}\,\mathsf{M}^{-1} d_{\text{dist}}^{(k)} \qquad \text{(in the unpreconditioned case: } z_{\text{acc}}^{(k)} = \mathsf{A}\, d_{\text{dist}}^{(k)})$$

$$q_{\text{acc}}^{(k)} = z_{\text{acc}}^{(k)} + \beta_{k-1}\, q_{\text{acc}}^{(k-1)} \qquad \text{where} \quad \beta_{-1} = 0,$$

$$\beta_{k-1} = \frac{\langle d_{\text{dist}}^{(k)}, z_{\text{acc}}^{(k)} \rangle}{\langle d_{\text{dist}}^{(k-1)}, z_{\text{acc}}^{(k-1)} \rangle}$$

$$\lambda_{\text{acc}}^{(k+1)} = \lambda_{\text{acc}}^{(k)} + \alpha^{(k)}\, q_{\text{acc}}^{(k)} \qquad \text{where} \quad \alpha_k = \frac{\langle d_{\text{dist}}^{(k)}, z_{\text{acc}}^{(k)} \rangle}{\langle \mathsf{P}^\top \mathsf{F}\, q_{\text{acc}}^{(k)}, q_{\text{acc}}^{(k)} \rangle}$$

$$d_{\text{dist}}^{(k+1)} = d_{\text{dist}}^{(k)} - \alpha_k\, \mathsf{P}^\top \mathsf{F}\, q_{\text{acc}}^{(k)} \qquad = \mathsf{B}(u^{(k)} - \tilde{g}_D)$$

$k = k + 1$

until *stopping criterion fulfilled for* $d_{\text{dist}}^{(k)}$

$$u^{(k)} = \tilde{g}_D + \left(I - \mathsf{R}\,\mathsf{D}(\mathsf{G}^\top \mathsf{Q}\,\mathsf{G})^{-1}\mathsf{G}^\top \mathsf{Q}\,\mathsf{B}\right)\left(S^\dagger(g - \mathsf{B}^\top \lambda_{\text{acc}}^{(k)}) - \tilde{g}_D\right)$$

inexact BETI method has been developed which circumvents this factorization. The BETI system is rewritten as the threefold saddle point system

$$\begin{bmatrix} -V & \frac{1}{2}I + K & 0 \\ \frac{1}{2}I + K^\top & D & B^\top \\ 0 & B & \end{bmatrix} \begin{bmatrix} t \\ u \\ \lambda \end{bmatrix} = \begin{bmatrix} 0 \\ f \\ 0 \end{bmatrix}, \tag{2.58}$$

where $V = \text{diag}(V_i)$, $D = \text{diag}(D_i)$, and $K = \text{diag}(K_i)$, see also Remark 2.18. The authors use and extend the classical theory of saddle point preconditioners, see [BP88, Zul02, Zul11]. Their full preconditioner is based on the scaled hypersingular BETI preconditioner (M_{sD}^{-1} with S_i replaced by D_i) and further preconditioners for the local single layer potential operators V_i. For the latter, several choices have been proposed in the literature, see e.g. [FS97, Ste03a]. The same technique applies to hybrid FETI/BETI methods, as outlined in [LP08].

2.3 Balancing Neumann-Neumann Methods

In this section, we define the balancing Neumann-Neumann method. Its close connection to FETI will be subject of Sect. 2.4.2.4. The Neumann-Neumann methods were first developed by Bourgat, Glowinski, Le Tallec, and Vidrascu, as well as De Roeck, see [BGLV89, De 91, DL91]. They were considerably improved by adding a second coarse level, see Mandel and Brezina [Man93, MB96], Le Tallec [Le 94], as well as Dryja and Widlund [DW95]. See also [TW05, Sect. 6.2] for more details on the method. For an alternative with a different coarse space see [BS02, Sect. 7.7]. The connection between balancing Neumann-Neumann and FETI methods was demonstrated in [KW01], and supported with an underpinning theory in [SM08]. Note also that the Neumann-Neumann methods have been further developed by Dohrmann and Mandel, leading to the BDDC methods, see Sect. 5.1.4.

Assume for simplicity that $g_D = 0$ and let W, $\widehat{W}$, $S : W \to W^*$, and $g \in W^*$ be defined either as in the classical or as in the all-floating formulation. Let $V_D^h(\Gamma_S)$ be as in (2.12) and recall that $V_D^h(\Gamma_S) \equiv \widehat{W}$. Let $A^\top : V_D^h(\Gamma_S) \to W$ denote the natural embedding and $A : W^* \to V_D^h(\Gamma_S)^*$ its adjoint (the latter can be seen as an assembling operator). Furthermore, we define $\widehat{S} := A\, S\, A^\top$ and $\hat{g} := A\, g$. The equation

$$\text{find } u \in V_D^h(\Gamma_S): \qquad \widehat{S}\, u = \hat{g} \tag{2.59}$$

is then identical to the skeleton formulation (2.11).

The balancing Neumann-Neumann preconditioner can be classified as a symmetric, hybrid two-level Schwarz preconditioner for $\widehat{S}$ (cf. [TW05, Sect. 2.2]; here *hybrid* indicates the combination of Schwarz projectors in both an additive and multiplicative way). First, we define the "coarse" level. Let the averaging operator $\widehat{E}_D : W \to V_D^h(\Gamma_S)$ be defined by

$$(\widehat{E}_D)(x^h) := \begin{cases} \sum_{j \in \mathcal{N}_{x^h}} \delta_j^\dagger(x^h)\, w_j(x^h) & \text{for } x^h \in \partial\Omega_i^h \setminus \Gamma_D, \\ 0 & \text{for } x^h \in \partial\Omega_i^h \cap \Gamma_D, \end{cases} \tag{2.60}$$

where $\delta_j^\dagger$ are the weighted counting functions from (2.53) (for a fixed choice of scalings ρ_j). Let $r^{(i)} \in W$ be such that $\operatorname{span}\{r_i^{(i)}\} = \ker(S_i)$ and $r_j^{(i)} = 0$ for all $j \neq i$. If Ω_i is non-floating, then $r^{(i)} = 0$. Moreover, we define the subspace

$$\widehat{W}_0 := \operatorname{span}\{\widehat{E}_D\, r^{(i)} : i \in \mathcal{I}_{\text{float}}\} \subset V_D^h(\Gamma_S),$$

with the basis $\{\widehat{E}_D\, r^{(i)}\}_{i \in \mathcal{I}_{\text{float}}}$. Let $\widehat{R}_0^\top : \widehat{W}_0 \to V_D^h(\Gamma_S)$ denote the natural embedding and $\widehat{R}_0$ its adjoint ($\widehat{R}_0$ is a restriction and $\widehat{R}_0^\top$ the corresponding prolongation). Let $\widehat{S}_0 := \widehat{R}_0\, \widehat{S}\, \widehat{R}_0^\top$ be the corresponding projection of $\widehat{S}$ (which is still SPD). Note that the matrix $\widehat{\mathbf{S}}_0$ corresponding to $\widehat{S}_0$ with respect to the basis $\{\widehat{E}_D\, r^{(i)}\}_{i \in \mathcal{I}_{\text{float}}}$ is sparse. Moreover, it can be shown that for the choice $Q = M_{\text{sD}}^{-1}$, the matrix $\widehat{\mathbf{S}}_0$ is identical to $\mathbf{G}^\top \mathbf{Q}\, \mathbf{G}$ from the FETI method, cf. Remark 2.36.

Let the Schwarz projector $\widehat{P}_0 : V_D^h(\Gamma_S) \to \widehat{W}_0$ be defined by

$$\widehat{P}_0 := \widehat{R}_0^\top \widehat{S}_0^{-1}\, \widehat{R}_0 \widehat{S}, \tag{2.61}$$

see also Sect. 1.1.3.3. Finally, the balancing Neumann-Neumann preconditioner $M_{\mathrm{BNN}}^{-1} : V_D^h(\Gamma_S)^* \to V_D^h(\Gamma_S)$ is given by

$$M_{\mathrm{BNN}}^{-1} := (I - \widehat{P}_0)\, \widehat{E}_D\, S^\dagger\, \widehat{E}_D^\top (I - \widehat{P}_0^\top) + \widehat{R}_0^\top \widehat{S}_0^{-1} \widehat{R}_0 . \tag{2.62}$$

Note that the operator $\widehat{E}_D^\top : V_D^h(\Gamma_S)^* \to W^*$ distributes (global) residuals to the (local) subdomains using the weights $\delta_j^\dagger$. The process of replacing a residual $r \in \widehat{S}$ by $(I - \widehat{P}_0^\top) r$ when applying the first part of the preconditioner is called *balancing*, cf. [Man93, Sect. 2].

Let $P_Z : W \to W$ be defined as in (2.50). With Lemma 2.39 in Sect. 2.4.2.1 below, one can show that

$$(I - \widehat{P}_0)\, \widehat{E}_D = \widehat{E}_D\, P_Z . \tag{2.63}$$

Since $\mathrm{range}(P_Z^\top) = \mathrm{range}(S)$, this identity shows that the application of the pseudo inverse $S^\dagger$ in (2.62) is valid and the output is always consistent, cf. Definition 1.13.

We can apply the preconditioner (2.62) either to the original Schur complement problem (2.59) or to an auxiliary problem (2.64) that we will introduce below. In the latter case, many simplifications can be made in the algorithm. Furthermore, it can be shown that the two algorithms are equivalent.

In a first step, we use $\widehat{P}_0$ to project Eq. (2.59) to a subspace. Each element $w \in V_D^h(\Gamma_S)$ can be decomposed as

$$w = w_0 + \tilde{w}, \qquad \text{where } w_0 \in \widehat{W}_0,\ \tilde{w} \in \mathrm{range}(I - \widehat{P}_0),\ \langle \widehat{S}\, w_0, \tilde{w} \rangle = 0.$$

Thanks to the $\widehat{S}$-orthogonality of $\widehat{W}_0$ and $\mathrm{range}(I - \widehat{P}_0)$, the part u_0 of the solution u of (2.59) is given by

$$u_0 = \widehat{R}_0^\top \widehat{S}_0^{-1} \widehat{R}_0\, \hat{g}.$$

It is easily seen that $\hat{g} - \widehat{S}\, u_0 = (I - \widehat{P}_0^\top)\hat{g}$. The orthogonal part $\tilde{u}$ is given as the solution of the following problem:

$$\text{find } \tilde{u} \in \mathrm{range}(I - \widehat{P}_0) : \quad (I - \widehat{P}_0^\top)\widehat{S}\, \tilde{u} = (I - \widehat{P}_0^\top)\hat{g}. \tag{2.64}$$

Then $u = u_0 + \tilde{u}$ is the solution of (2.59). We now solve (2.64) using PCG with preconditioner M_{BNN}^{-1}. Two simplifications can then be made. Firstly, when applying M_{BNN}^{-1} to this equation, the second term $\widehat{R}_0^\top \widehat{S}_0^{-1} \widehat{R}_0$ in M_{BNN}^{-1} can be left out, as well as the term $(I - P_0^\top)$. To explain the second simplification, we first rewrite the whole procedure as a method in the original variables $u = u_0 + \tilde{u}$ (assuming that we have chosen the initial value 0 for $\tilde{u}^{(0)}$). The resulting algorithm is displayed in Algorithm 7. Due to the properties of $u^{(0)}$, all the residuals are in $\mathrm{range}(I - \widehat{P}_0^\top)$.

Algorithm 7: Balancing Neumann-Neumann algorithm based on PCG

$u^{(0)} = \widehat{R}_0^\top \widehat{S}_0^{-1} \widehat{R}_0\, \hat{g}$
$r^{(0)} = \hat{g} - \widehat{S}\, u^{(0)}$
$k = 0$
repeat
$z^{(k)} = \widehat{E}_D\, S^\dagger\, \widehat{E}_D^\top r^{(k)}$
$s^{(k)} = (I - \widehat{P}_0) z^{(k)}$
$p^{(k)} = s^{(k)} + \beta_{k-1}\, p^{k-1}$ where $\beta_{-1} = 0, \quad \beta_{k-1} = \frac{\langle r^{(k)}, s^{(k)}\rangle}{\langle r^{(k-1)}, s^{(k-1)}\rangle}$
for $k > 0$
$u^{(k)} = u^{(k)} + \alpha_k\, p^{(k)}$ where $\alpha_k = \frac{\langle r^{(k)}, s^{(k)}\rangle}{\langle \widehat{S}\, p^{(k)}, p^{(k)}\rangle}$
$r^{(k+1)} = r^{(k)} - \alpha_k\, \widehat{S}\, p^{(k)}$
$k = k + 1$
until *stopping criterion fulfilled for* $r^{(k)}$

Hence, the projection step $(I - \widehat{P}_0^\top)$ in the operator $(I - \widehat{P}_0^\top)\widehat{S}$ can be omitted as well (and is not included in Algorithm 7).

A brief analysis of the balancing Neumann-Neumann preconditioner will be given in Sect. 2.4.2.4, where we will also see a close connection to FETI/BETI.

2.4 Introduction to the Analysis of FETI/BETI

In this section, we first investigate the conditioning of the FETI/BETI operator itself (Sect. 2.4.1). Second, we introduce an abstract framework for analyzing the condition number of the preconditioned operator (Sect. 2.4.2).

2.4.1 The Unpreconditioned Case

Following [FMR94, Sect. 3], we analyze the convergence of Algorithms 2 and 3 for the case $M^{-1} = Q = I$. Thanks to Lemma 1.49, it suffices to find an upper bound for the condition number of the corresponding operator. Since $Q = I$, the operator $P^\top F_{|\widetilde{U}_{\text{ad}}}$ maps to $U_{\text{ad}}^* = U_{\text{ad}}$, and so, as a formal preconditioner we can choose the natural embedding of U_{ad} into the factor space $\widetilde{U}_{\text{ad}}$. Since each element in the factor space has a unique representative in range(B), we can use $U_{\text{ad}} \cap \text{range}(B)$ instead of $\widetilde{U}_{\text{ad}}$ and obtain

$$\kappa(P^\top F_{|\tilde{U}_{\text{ad}}}) = \frac{\sup\limits_{\lambda \in U_{\text{ad}} \cap \text{range}(B)} \dfrac{\langle F\lambda, \lambda\rangle}{\|\lambda\|_{\ell^2}^2}}{\inf\limits_{\lambda \in U_{\text{ad}} \cap \text{range}(B)} \dfrac{\langle F\lambda, \lambda\rangle}{\|\lambda\|_{\ell^2}^2}}. \tag{2.65}$$

In the following let $a \lesssim b$ be a short hand for $a \leq C\, b$ and $a \eqsim b$ for $a \lesssim b$ and $b \lesssim a$, where C is a generic constant.

Lemma 2.37. *On $U_{\text{ad}} \cap \text{range}(B)$, the operator $B^\top$ is injective. Let $\mathbf{B}_i \in \mathbb{R}^{M \times \dim(W_i)}$ be the matrix representation of B_i with respect to the nodal basis of W_i. Then*

$$\sum_{i=1}^{s} \|\mathbf{B}_i^\top \lambda\|_{\ell^2}^2 \eqsim \|\lambda\|_{\ell^2}^2 \qquad \forall \lambda \in U_{\text{ad}} \cap \text{range}(B),$$

where the equivalence constants only depend on the maximal number of subdomains that share a single node $x^h \in \Gamma_S^h$.

Proof. In the classical formulation, we have

$$\sum_{i=1}^{s} \|\mathbf{B}_i^\top \lambda\|_{\ell^2}^2 = \sum_{x^h \in \Gamma^h} \underbrace{\sum_{i \in \mathcal{N}_{x^h}} \Big| \sum_{j \in \mathcal{N}_{x^h} \setminus \{i\}} \text{sign}(i-j)\, \lambda_{ij}(x^h)\Big|^2}_{=: b_1(\lambda, x^h)}$$

(see Lemma 2.14) and

$$\|\lambda\|_{\ell^2}^2 = \sum_{x^h \in \Gamma^h} \underbrace{\sum_{i,j \in \mathcal{N}_{x^h}, i>j} |\lambda_{ij}(x^h)|^2}_{=: b_2(\lambda, x^h)}.$$

For $\lambda \in \text{range}(B)$, the expressions $\sqrt{b_1(\lambda, x^h)}$ and $\sqrt{b_2(\lambda, x^h)}$ are norms on a space of dimension $\#(\mathcal{N}_{x^h})$ and as such equivalent. Since there are only a bounded number of topologically different configurations of a shared node x^h, the equivalence in the statement of the lemma is indeed uniform. The treatment of the additional terms in the all-floating formulation is straightforward. □

Theorem 2.38. *Assume that*

(i) The triangulation $\mathcal{T}^h(\Gamma_S)$ is shape regular,
(ii) The local triangulations $\mathcal{T}^h(\partial\Omega_i)$ are quasi-uniform with mesh parameter h_i,
(iii) The intersection $\partial\Omega_i \cap \Gamma_D$ is either empty or has positive surface measure,
(iv) The number of subdomains sharing a node $x^h \in \Gamma_S^h$ is uniformly bounded.

Then

$$\kappa(P^\top F_{|\tilde{U}_{\rm ad}}) \;\leq\; C\,\frac{\|\mathscr{A}\|_{L^\infty(\Omega)}}{\alpha_{\min}}\,\frac{\max_{i=1}^s H_i\,h_i^{1-d}}{\min_{i=1}^s h_i^{2-d}},$$

where the constant C is independent of H_i, h_i, $\mathscr{A}$, and the number of subdomains, but it depends on the subdomain shapes. If $\mathscr{T}^h(\Gamma_S)$ is quasi-uniform, then

$$\kappa(P^\top F_{|\tilde{U}_{\rm ad}}) \;=\; \mathscr{O}\Big(\frac{H}{h}\Big) \qquad \textit{as } h \to \infty,$$

where $H := \max_{i=1}^s H_i$.

Proof. We define the subspaces

$$W_i^\perp := \begin{cases} \{w_i \in W_i : (\mathbf{w}_i,\,\mathbf{1})_{\ell^2} = 0\} & \text{if } \Omega_i \text{ floating},\\ W_i & \text{else,}\end{cases}$$

where $\mathbf{w}_i \leftrightarrow w_i$ and $\mathbf{1}$ is the vector of ones. Furthermore, we set

$$|w_i|_{S_i} \;:=\; \langle S_i\,w_i,\,w_i\rangle \qquad \text{for } w_i \in W_i\,.$$

Throughout the proof, let $\lambda \in U_{\rm ad} \cap \operatorname{range}(B)$ be arbitrary but fixed. Due to Lemma 1.14,

$$\langle F\,\lambda,\,\lambda\rangle \;=\; \langle S^\dagger B^\top\lambda,\,B^\top\lambda\rangle \;=\; \sum_{i=1}^{s}\sup_{w_i\in W_i^\perp}\frac{\langle B_i^\top\lambda,\,w_i\rangle^2}{|w_i|_{S_i}^2}.$$

Assume first that $\mathscr{I}_{\rm BEM} = \emptyset$. Corollary 1.61 implies that

$$\alpha_{\min}\,H_i^{-1}\,h_i^{d-1}\,\|\mathbf{w}_i\|_{\ell^2}^2 \;\lesssim\; |w_i|_{S_i}^2 \;\lesssim\; \|\mathscr{A}\|_{L^\infty(\Omega_i)}\,h_i^{d-2}\,\|\mathbf{w}_i\|_{\ell^2}^2 \qquad \forall \mathbf{w}_i \leftrightarrow w_i \in W_i^\perp\,.$$

If $\mathscr{I}_{\rm BEM} \neq \emptyset$ we can use the spectral equivalence in Corollary 1.94, then the equivalence constants depend additionally on the constant $c_0(\partial\Omega_i)$ defined as in Lemma 1.77. Moreover, since $B_i^\top\lambda \in \operatorname{range}(S_i)$ implies $(\mathbf{B}_i^\top\lambda,\,\mathbf{1})_{\ell^2} = 0$, we obtain

$$\sup_{\mathbf{w}_i\leftrightarrow w_i\in W_i^\perp}\frac{\langle B_i^\top\lambda,\,w_i\rangle^2}{\|\mathbf{w}_i\|_{\ell^2}^2} \;=\; \sup_{\mathbf{w}_i\leftrightarrow w_i\in W_i^\perp}\frac{(\mathbf{B}_i^\top\lambda,\,\mathbf{w}_i)_{\ell^2}^2}{\|\mathbf{w}_i\|_{\ell^2}^2} \;=\; \|\mathbf{B}_i^\top\lambda\|_{\ell^2}^2\,.$$

Combining the above estimate with Lemma 2.37 yields

$$\langle F\lambda,\lambda\rangle \lesssim \sum_{i=1}^{s} \alpha_{\min}^{-1} H_i\, h_i^{1-d}\, \|\mathbf{B}_i^\top \lambda\|_{\ell^2}^2 \lesssim \alpha_{\min}^{-1} \max_{i=1}^{s}(H_i\, h_i^{1-d})\, \|\lambda\|_{\ell^2}^2,$$

$$\langle F\lambda,\lambda\rangle \gtrsim \sum_{i=1}^{s} \|\mathscr{A}\|_{L^\infty(\Omega)}^{-1} h_i^{2-d}\, \|\mathbf{B}_i^\top \lambda\|_{\ell^2}^2 \gtrsim \|\mathscr{A}\|_{L^\infty(\Omega)}^{-1} \min_{i=1}^{s} h_i^{2-d}\, \|\lambda\|_{\ell^2}^2 .$$

Together with (2.65) this concludes the proof. □

2.4.2 Abstract Framework for the Preconditioned Case

In this subsection, we collect abstract results, whose proofs can be performed on the operator level. Later on, we will have to make rather strong assumptions on the coefficient and the subdomains and use technical tools for finite element functions.

2.4.2.1 The Projection Operators P_D and E_D

An important role in the analysis of the scaled Dirichlet preconditioner (2.57) is played by the projection operator

$$P_D := B_D^\top B, \tag{2.66}$$

whose properties are summarized in the following lemma. The result for classical FETI methods was proved by Klawonn and Widlund [KW01], see also [TW05].

Lemma 2.39. *The operator $P_D : W \to W$ defined in* (2.66) *satisfies the identities*

$$B\, M_{\mathrm{sD}}^{-1}\, B^\top = P_D^\top\, S\, P_D, \tag{2.67}$$

$$B\, P_D = B, \tag{2.68}$$

for both the classical and the all-floating formulation. Furthermore, $E_D := I - P_D$ is a projection onto the subspace $\widehat{W}$, and can be evaluated by

$$(E_D\, w)_i(x^h) = \begin{cases} \sum_{j\in\mathscr{N}_{x^h}} \delta_j^\dagger(x^h)\, w_j(x^h) & \text{for } x^h \in \partial\Omega_i^h \setminus \Gamma_D, \\ 0 & \text{for } x^h \in \partial\Omega_i^h \cap \Gamma_D . \end{cases} \tag{2.69}$$

Proof. Identity (2.67) follows from the definitions of M_{sD}^{-1} and P_D. Recall that

$$(B\,w)_{ij}(x^h) = \operatorname{sign}(i-j)\,\big(w_i(x^h) - w_j(x^h)\big) \qquad \text{for } x^h \in \Gamma_{ij}^h,$$

$$(B\,w)_{iD}(x^h) = w_i(x^h) \qquad \text{for } x^h \in \partial\Omega_i^h \cap \Gamma_D,$$

where the last formula holds only in the all-floating formulation. From Lemma 2.28, we immediately obtain that

$$(P_D\, w)_i(x^h) \;=\; \begin{cases} \sum\limits_{j \in \mathcal{N}_{x^h}} \delta_j^\dagger(x^h)\,\big(w_i(x^h) - w_j(x^h)\big) & \text{for } x \in \partial\Omega_i^h \cap \Gamma^h, \\ w_i(x^h) & \text{for } x^h \in \partial\Omega_i^h \cap \Gamma_D, \\ 0 & \text{else.} \end{cases} \tag{2.70}$$

This implies formula (2.69), and we see that $E_D\, w \in \widehat{W}$. Hence $B\, E_D = 0$, which implies (2.68). Therefore, P_D is a projection, and so is E_D. □

Corollary 2.40. *For each $\mu \in \operatorname{range}(B)$ we can find a function $w \in \operatorname{range}(P_D)$ such that $\mu = B\, w$.*

Proof. Lemma 2.39 implies that $\operatorname{range}(B) = \operatorname{range}(B\, P_D)$. □

Remark 2.41. The operator E_D is a weighted averaging operator and equals the operator $\widehat{E}_D$ from Sect. 2.3, up to identification of $\widehat{W}$ and $V_D^h(\Gamma_S)$, cf. [KW01]. For the multiplicity scaling $\rho_i(x^h) = 1$, $(E_D\, w)_i(x^h)$ is the algebraic mean value of $\{w_j(x^h)\}_{j \in \mathcal{N}_{x^h}}$.

2.4.2.2 Positivity of the Preconditioner

For a diagonal choice of Q (see [KW01]), the operator Q is SPD per definition, and so the projections P and $P^\top$ are well-defined. As the following lemma shows, P and $P^\top$ are also well-defined if $Q = M_{\mathrm{sD}}^{-1}$.

Lemma 2.42. *The scaled Dirichlet preconditioner M_{sD}^{-1} is SPD on $\operatorname{range}(G)$. Consequently, if $Q = M_{\mathrm{sD}}^{-1}$, then the projections P and $P^\top$ are well-defined.*

Proof. From (2.57) it is immediate that M_{sD}^{-1} is positive semi-definite. To show the definiteness on $\operatorname{range}(G)$, assume that $\langle B\, z,\; M_{\mathrm{sD}}^{-1}\, B\, z\rangle = 0$ for some $z \in \ker(S)$. Due to identity (2.67) we obtain $|P_D\, z|_S^2 = 0$ which implies that $P_D\, z = z - E_D\, z \in \ker(S)$ and consequently, $E_D\, z \in \ker(S)$. However, $E_D\, z \in \ker(B)$ and $\ker(S) \cap \ker(B) = \{0\}$ imply that $E_D\, z = 0$. This means that the function z, which is piecewise constant on the subdomains, is continuous across the subdomain interfaces and vanishes on the Dirichlet boundary. Since the domain Ω is connected, there is no other possibility than $z = 0$, which shows the definiteness. □

The next lemma discusses the positivity of $P\, M_{\mathrm{sD}}^{-1}$ on $\widetilde{U}_{\mathrm{ad}}^*$.

Lemma 2.43. *If Q is SPD on $\operatorname{range}(G)$, then $P\, M_{\mathrm{sD}}^{-1}$ is SPD on $\widetilde{U}_{\mathrm{ad}}^*$.*

Proof. From (2.57) it is immediate that M_{sD}^{-1} is positive semi-definite. To show the definiteness on $\widetilde{U}_{\mathrm{ad}}^*$, assume that $\langle \mu,\; M_{\mathrm{sD}}^{-1}\mu\rangle = 0$ for some $\mu \in \widetilde{U}_{\mathrm{ad}}^*$. Due to

Corollary 2.40, there exists a function $w \in \text{range}(P_D)$ with $\mu = B\,w$, and so

$$0 = \langle \mu, M_{\text{sD}}^{-1}\,\mu\rangle = |P_D\,w|_S^2 = |w|_S^2 = 0.$$

Hence, $w \in \ker(S)$ and $\mu = B\,w \in \text{range}(G)$. The definition (2.43) of $\widetilde{U}_{\text{ad}}^*$ yields

$$\langle \mu, Q\,\mu\rangle = \langle \mu, Q\,B\,w\rangle = 0.$$

Since Q is SPD on range(G), it follows that $\mu = 0$. □

2.4.2.3 An Abstract Condition Number Estimate for FETI/BETI

With the following two lemmas, we can reduce the condition number estimate of the preconditioned FETI/BETI method to a single stability estimate of the P_D operator. For the original proofs see [KW01] and also [MT96].

Lemma 2.44. *Let Q be SPD on* range(G)*. Then, for any $w \in W$, then there exists a unique element $z_w \in \ker(S)$ such that $B(w + z_w) \in \widetilde{U}_{\text{ad}}^*$, given by*

$$z_w = -R\,(G^\top Q\,G)^{-1}G^\top Q\,B\,w = -(I - P_Z)w,$$

where P_Z is defined in (2.50)*. Moreover, if Q is SPD on $\widetilde{U}^*$, then*

$$z_w = \operatorname*{argmin}_{z\in\ker S} \|B\,(w + z)\|_Q, \qquad \textit{and} \qquad \|B\,z_w\|_Q \le \|B\,w\|_Q,$$

where $\|\mu\|_Q := \langle \mu, Q\,\mu\rangle^{1/2}$.

Proof. The statements follow from Sect. 1.1.3.3 and Lemma 1.3. □

Lemma 2.45. *Assume that $P\,M_{\text{sD}}^{-1}$ is SPD on $\widetilde{U}_{\text{ad}}^*$ and let ω be a parameter such that*

$$|P_D(w + z_w)|_S^2 \le \omega\,|w|_S^2 \qquad \forall w \in W,$$

where z_w is the unique element from Lemma 2.44. Then

$$\kappa(P\,M_{\text{sD}}^{-1}\,P^\top F_{|\widetilde{U}_{\text{ad}}}) \le \omega.$$

Proof. Let $M_{\text{sD}} : \widetilde{U}_{\text{ad}} \to U_{\text{ad}}^*$ be the inverse of $(P\,M_{\text{sD}}^{-1})_{|\widetilde{U}_{\text{ad}}^*}$. From Corollary 1.50, we see that

$$\kappa(P\,M_{\text{sD}}^{-1}\,P^\top F_{|\widetilde{U}_{\text{ad}}}) = \frac{\displaystyle\sup_{\lambda\in\widetilde{U}_{\text{ad}}} \frac{\langle F\,\lambda, \lambda\rangle}{\langle M_{\text{sD}}\,\lambda, \lambda\rangle}}{\displaystyle\inf_{\lambda\in\widetilde{U}_{\text{ad}}} \frac{\langle F\,\lambda, \lambda\rangle}{\langle M_{\text{sD}}\,\lambda, \lambda\rangle}}. \tag{2.71}$$

In order to estimate the numerator in (2.71), we bound F in terms of M_{sD}. Let $\lambda \in \widetilde{U}_{\mathrm{ad}}$ arbitrary but fixed. Lemma 1.14 and our assumptions imply

$$\langle F\,\lambda,\,\lambda\rangle \;=\; \sup_{w\in W}\frac{\langle B\,w,\,\lambda\rangle^2}{|w|_S^2} \;\le\; \omega\,\sup_{w\in W}\frac{\langle B\,w,\,\lambda\rangle^2}{|P_D\,(w+z_w)|_S^2}\,.$$

From to the definition (2.37) of U_{ad} and from $\widetilde{U}_{\mathrm{ad}} = U_{\mathrm{ad}/\ker(B^\top)}$, we can conclude that

$$\langle B\,w,\,\lambda\rangle \;=\; \langle B\,(w+z_w),\,\lambda\rangle.$$

Together with the above it follows from (2.67) and Lemma 2.44 that

$$\langle F\lambda,\lambda\rangle \le \omega\,\sup_{w\in W}\frac{\langle B(w+z_w),\lambda\rangle^2}{\langle B(w+z_w),M_{\mathrm{sD}}^{-1}B(w+z_w)\rangle} \le \omega\,\sup_{\mu\in\widetilde{U}^*_{\mathrm{ad}}}\frac{\langle\mu,\lambda\rangle^2}{\langle\mu,M_{\mathrm{sD}}^{-1}\mu\rangle} = \omega\langle M_{\mathrm{sD}}\lambda,\lambda\rangle.$$

We now turn to the denominator in (2.71). Let $\lambda \in \widetilde{U}_{\mathrm{ad}}$ be fixed and $\mu \in \widetilde{U}^*_{\mathrm{ad}}$ arbitrary. Thanks to Corollary 2.40, there exists $w \in \mathrm{range}(P_D)$ such $B\,w = \mu$. Hence,

$$\langle F\,\lambda,\,\lambda\rangle \;\ge\; \frac{\langle B\,w,\,\lambda\rangle^2}{|w|_S^2} \;=\; \frac{\langle B\,w,\,\lambda\rangle^2}{|P_D w|_S^2} \;=\; \frac{\langle B\,w,\,\lambda\rangle^2}{\langle B\,w,\,M_{\mathrm{sD}}^{-1}\,B\,w\rangle} \;=\; \frac{\langle\mu,\,\lambda\rangle^2}{\langle\mu,\,M_{\mathrm{sD}}^{-1}\,\mu\rangle}.$$

Since $\mu \in \widetilde{U}^*_{\mathrm{ad}}$ was arbitrary, we get that

$$\langle F\,\lambda,\,\lambda\rangle \;\ge\; \langle M_{\mathrm{sD}}\,\lambda,\,\lambda\rangle \qquad \forall\lambda\in\widetilde{U}_{\mathrm{ad}}\,.$$

Combining the two bounds concludes the proof. □

In Sect. 2.6 we will work out in detail the missing bound

$$|P_D(w+z_w)|_S^2 \;\le\; \omega\,|w|_S^2 \qquad \forall w\in W,$$

which requires a series of assumptions and technical tools that we present in Sect. 2.5.

2.4.2.4 An Estimate for the Balancing Neumann-Neumann Method

For the original analysis of the balancing Neumann-Neumann preconditioner we refer to [Man93, MB96], see also [DW95] and [TW05, Sect. 6.2.3]. Here, we provide an analysis which makes use of an abstract framework evolved from the above works on Neumann-Neumann as well as from the analysis of the more recent BDDC methods (see Sect. 5.1.4). We note that the following theorem (cf. [MS07a, Theorem 2] and [Sou10, Theorem 2.5]) has certain relations to the *ficticious space lemma* by Nepomnyaschikh, cf. [Nep91a, Nep92, Nep07].

Theorem 2.46. *Let $\widehat{X} \subset X$ be two Hilbert spaces, $a(\cdot,\cdot) : X \times X \to \mathbb{R}$ a symmetric positive definite bilinear form, and let $A : X \to X^*$ and $\widehat{A} : \widehat{X} \to \widehat{X}^*$ denote the corresponding operator and its restriction to $\widehat{X}$. Furthermore, let $\widehat{Q} : X \to \widehat{X}$ be a projector. The abstract BDDC preconditioner $\widehat{B}^{-1} : \widehat{X}^* \to \widehat{X}$ is defined by*

$$\widehat{B}^{-1} : \widehat{Q}\, A^{-1} \widehat{Q}^\top .$$

Let $X^M := \{v \in X : \forall z \in X : \widehat{Q}\, v = \widehat{Q}\, z \implies \|v\|_a \le \|z\|_a\}$, where $\|\cdot\|_a$ is the norm associated to $a(\cdot,\cdot)$, and let ω be the minimal constant such that

$$\|\widehat{Q}\, v\|_a^2 \;\le\; \omega\, \|v\|_a^2 \qquad \forall v \in X^M .$$

Then

$$\kappa(\widehat{B}^{-1}\widehat{A}) \;\le\; \omega \;=\; \sup_{v \in X} \frac{\|\widehat{Q}\, v\|_a^2}{\|v\|_a^2} .$$

Proof. For completeness we display the proof from [MS07a, Sou10]. Let $\widehat{G} : \widehat{X} \to X$ be defined by

$$\widehat{G}\, v := \operatorname*{argmin}_{w \in X^M :\; v = \hat{Q}\, w} a(w, w).$$

A short computation reveals that

$$\widehat{G} \;=\; A^{-1}\, \widehat{Q}^\top \widehat{B}, \qquad \widehat{Q}\, \widehat{G} \;=\; I, \qquad \widehat{G}^\top A\, \widehat{G}^\top \;=\; \widehat{B}, \tag{2.72}$$

where $\widehat{B} : \widehat{X} \to \widehat{X}^*$ is the inverse of $\widehat{B}^{-1}$. Using (2.72), the definition of $\widehat{G}$, and the fact that $\widehat{Q}$ is a projection, we obtain

$$\langle \widehat{B}\, v, v\rangle \;=\; a(\widehat{G}\, v, \widehat{G}\, v) \;\le\; a(v, v) \qquad \forall v \in \widehat{X},$$

which shows that $\lambda_{\min}(\widehat{B}^{-1}\widehat{A}) \ge 1$. Now, we conclude from (2.72) and the defining property of ω that for any $v \in \widehat{\widehat{W}}$,

$$\|v\|_a^2 \;=\; \|\widehat{Q}\, \widehat{G}\, v\|_a^2 \;\le\; \omega\, \|\widehat{G}\, v\|_a^2 \;=\; \omega\, \langle \widehat{B}\, v, v\rangle.$$

The last estimate implies that $\lambda_{\max}(\widehat{B}^{-1}\,\widehat{A}) \le \omega$. The alternative characterization of ω is trivial. □

The following lemma makes use of the above abstract theorem and shows that a similar estimate as in Lemma 2.45 implies a bound for the condition number of the balancing Neumann-Neumann preconditioner.

Lemma 2.47. *For each $i = 1, \ldots, s$, let $W_i^\perp$ be a subspace of W_i such that the sum $W_i = \ker(S_i) \oplus W_i^\perp$ is direct, and let $W^\perp := \prod_{i=1}^s W_i^\perp$. Then a bound of the form*

$$|E_D\, w|_S^2 \;\le\; \omega\, |w|_S^2 \qquad \forall w \in W^{\perp} \tag{2.73}$$

implies that

$$\kappa(M_{\mathrm{BNN}}^{-1} \widehat{S}_{|\mathrm{range}(I-\widehat{P}_0)}) \;\le\; \omega.$$

Proof. Firstly, we convince ourselves that

$$M_{\mathrm{BNN}}^{-1} \widehat{S}_{|\mathrm{range}(I-\widehat{P}_0)} \;=\; (I - \widehat{P}_0)\widehat{E}_D\, S^{\dagger}\, \widehat{E}_D^{\top}(I - \widehat{P}_0^{\top})\, \widehat{S}_{|\mathrm{range}(I-\widehat{P}_0)}\,.$$

Using identity (2.63) and the fact that $P_Z\, S^{\dagger}{}_{|\mathrm{range}(S)}$ is the inverse of the restriction of S to $\mathrm{range}(P_Z)$, it can be shown that the balancing Neumann-Neumann preconditioner fits into the framework of Theorem 2.46 with

$$X := \mathrm{range}(P_Z), \quad \widehat{X} := \mathrm{range}(I - \widehat{P}_0), \quad \widehat{Q} := (I - \widehat{P}_0)\widehat{E}_D = \widehat{E}_D\, P_Z\,.$$

Hence, Theorem 2.46 implies that

$$\kappa(M_{\mathrm{BNN}}^{-1} \widehat{S}_{|\mathrm{range}(I-\widehat{P}_0)}) \;\le\; \sup_{w \in \mathrm{range}(P_Z)} \frac{|E_D\, w|_S^2}{|w|_S^2} \;=\; \sup_{w \in W^{\perp}} \frac{|E_D\, P_Z\, w|_S^2}{|w|_S^2},$$

where in the last step, we have used the same argumentation as in Lemma 2.103 below. Finally, the statement of the lemma follows since $\widehat{E}_D\, P_Z = (I - \widehat{P}_0)\, E_D$ and $(I - \widehat{P}_0)$ is an $\widehat{S}$-orthogonal projector with its $\widehat{S}$-norm less or equal than one. □

The connection between FETI and balancing Neumann-Neumann methods can firstly be seen from the ingredients of the two methods, which are related by the two operators S and $S^{\dagger}$, by $B_D^{\top} B = P_D = I - E_D$, by the fact that the coarse matrices coincide for the case $Q = M_{\mathrm{sD}}^{-1}$, cf. Sect. 2.3, and by many common parts of their analyses (see also [KW01]). Secondly, since the norm of a non-trivial projection in a finite-dimensional Hilbert space depends only on the angle between its kernel and range [IM95], the bound (2.73) implies

$$|P_D\, w|_S^2 \;\le\; \omega\, |w|_S^2 \qquad \forall w \in W^{\perp}, \tag{2.74}$$

and vice versa. Moreover, Sousedik and Mandel [SM08] proved that the spectra of the FETI preconditioner (with $Q = M_{\mathrm{sD}}^{-1}$) and the balancing Neumann-Neumann preconditioner (with corresponding scalings) are identical except for the eigenvalue of one, i.e., $\sigma(M_{\mathrm{BNN}}^{-1}\, \widehat{S}_{|\mathrm{range}(I-\hat{P}_0)}) \setminus \{1\} = \sigma(M_{\mathrm{sD}}^{-1}\, F_{|\widetilde{U}_{\mathrm{ad}}}) \setminus \{1\}$.

2.5 Technical Tools

In this section, we first define a partition of the skeleton into so-called *globs* and introduce associated cut-off functions. These will be used to split the contributions in the operator P_D that involve different neighbors. To estimate the effect of the cut-off, we need regularity assumptions of the subdomains (Sect. 2.5.2) and a series of technical estimates (Sects. 2.5.3–2.5.6). Readers interested in the main line of the analysis may initially bypass the technical estimates, continue with Sect. 2.6, and return to the necessary lemmas at a later stage.

2.5.1 Globs and Cut-Off Functions

When looking to formula (2.70), we see that the operator P_D involves different subsets of the functions $\{w_i\}_{i=1}^N$ on different parts of the skeleton Γ_S. This is the main reason for the following definition.

Definition 2.48. (i) For $x \in \Gamma_S$ we define the set of subdomains that share x:

$$\mathcal{N}_x := \{i = 1, \ldots, s : x \in \partial\Omega_i\}.$$

(ii) The interface Γ decomposes into equivalence classes of the relation

$$x \sim y \iff \mathcal{N}_x = \mathcal{N}_y \,.$$

Each *connected* component of such an equivalence class is called a *glob* (cf. [Mat08]). For a glob $\mathcal{G}$, we simply denote by $\mathcal{N}_{\mathcal{G}}$ the set of sharing subdomains ($\mathcal{N}_{\mathcal{G}} = \mathcal{N}_x$ for all $x \in \mathcal{G}$).

(iii) In three dimensions, a glob $\mathcal{G}$ on the interface Γ is called

- *Subdomain vertex* if it consists of a single point,
- *Subdomain face* or *subdomain facet* if it is shared by exactly two subdomains,
- *Subdomain edge* otherwise.

In two dimensions, a glob $\mathcal{G}$ on the interface Γ is called

- *Subdomain edge* or *subdomain facet* if it is shared by exactly two subdomains,
- *Subdomain vertex* otherwise (being a single point).

Note that subdomain edges may include one of their endpoints (see the definition of the interface Γ, Table 2.1, p. 65). Similarly, there are subdomain faces which include a part of the Neumann boundary.

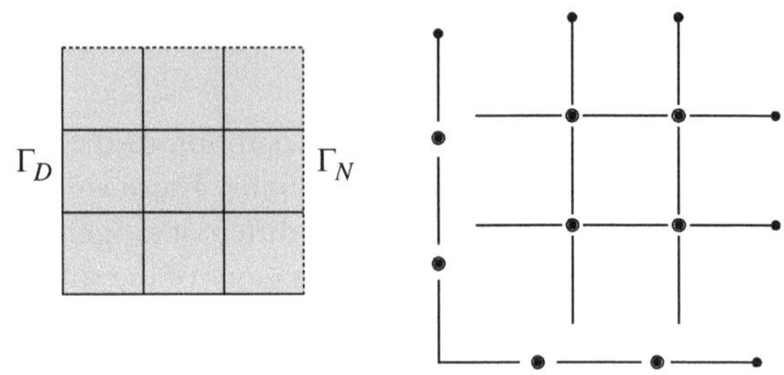

Fig. 2.5 Example of a decomposition of a two-dimensional domain with the corresponding globs on $\Gamma \cup \Gamma_D$,subdomain vertices indicated by ◉, subdomain edges indicated by — and —•

(iv) We decompose the Dirichlet boundary Γ_D into equivalence classes of the same type as above, and each connected component of such an equivalence class is called a *Dirichlet glob*.

(v) In three dimensions, a Dirichlet glob $\mathscr{G}$ is called

- *Subdomain vertex* if it consists of a single point,
- *Subdomain face* or *subdomain facet* if it belongs to exactly one subdomain,
- *Subdomain edge* otherwise.

In two dimensions, we have again subdomain vertices consisting of a single point, all other globs are called subdomain edges or also subdomain facets.

Figure 2.5 illustrates the decomposition of Γ and Γ_D into globs for a two-dimensional example.

Remark 2.49. Our definition of subdomain faces, edges, and vertices slightly differs from the "common" one in [TW05, Definition 4.1] because we will need to treat all-floating methods that operate on the Dirichlet boundary as well.

Notation. We denote subdomain facets, edges, and vertices on Γ that are shared by (at least) two subdomains Ω_i and Ω_j generically by $\mathscr{F}_{ij}$, $\mathscr{E}_{ij}$, $\mathscr{V}_{ij}$, respectively. We would like to point out that an index pair does not necessarily specify a subdomain facet/edge/vertex uniquely. Subdomain facets, edges, and vertices on $\partial\Omega_i$ (possibly shared by Ω_i and the outer boundary) are denoted generically by $\mathscr{F}_i$, $\mathscr{E}_i$, and $\mathscr{V}_i$, respectively.

Definition 2.50. For a subdomain face $\mathscr{F}$, let $\mathscr{F}^h$ be the set of nodes contained in $\mathscr{F}$, and for a subdomain edge $\mathscr{E}$, let $\mathscr{E}^h$ be the set of nodes contained in $\mathscr{E}$. For a subdomain vertex $\mathscr{V}$ we set $\mathscr{V}^h := \{\mathscr{V}\}$.

In order to separate the contributions of P_D on the different globs, we define the following cut-off functions, according to [TW05, Sect. 4.6]. Let $V^h(\partial\Omega_i)$ denote the restriction of $V^h(\Omega_i)$ to the boundary, cf. Sect. 1.2.6.2.

Definition 2.51 (Finite element cut-off functions).

- For a subdomain vertex $\mathscr{V}_i$, the piecewise linear function $\theta_{\mathscr{V}_i} \in V^h(\partial\Omega_i)$ equals one at the vertex $\mathscr{V}_i$, and zero on all other nodes.
- For a subdomain edge $\mathscr{E}_i$, the piecewise linear function $\theta_{\mathscr{E}_i} \in V^h(\partial\Omega_i)$ equals one at all nodes contained in $\mathscr{E}_i$, and zero at all other nodes.

- For a subdomain face $\mathscr{F}_i$, the piecewise linear function $\theta_{\mathscr{F}_i} \in V^h(\partial\Omega_i)$ equals one at the nodes contained in $\mathscr{F}_i$, and zero at all other nodes.

Extending these functions by zero, we have $\theta_{\mathscr{G}_i} \in V^h(\Gamma_S)$.

Definition 2.52. Let I^h denote the nodal interpolator that interpolates continuous functions to $V^h(\Omega_i)$ or $V^h(\partial\Omega_i)$.

Notation. By writing

$$\sum_{\mathscr{G}_i} I^h(\theta_{\mathscr{G}_i} w_i), \qquad \sum_{\mathscr{G}_i \subset \Gamma} I^h(\theta_{\mathscr{G}_i} w_i), \quad \text{and} \quad \sum_{\mathscr{G}_i \subset \Gamma_D} I^h(\theta_{\mathscr{G}_i} w_i),$$

we mean that we sum over all globs on $\partial\Omega_i$, $\partial\Omega_i \cap \Gamma$, and $\partial\Omega_i \cap \Gamma_D$, respectively.

The cut-off functions from Definition 2.51 provide a partition of unity in the sense that

$$\sum_{\mathscr{G}_i} I^h(\theta_{\mathscr{G}_i} v) = v \qquad \forall v \in V^h(\partial\Omega_i),\ v_{|\partial\Omega_i \cap \Gamma_N^h} = 0, \tag{2.75}$$

where

$$\Gamma_N^h := \Gamma_S^h \cap (\Gamma_N \setminus \Gamma) \tag{2.76}$$

denotes the set of non-coupling Neumann nodes.

Within the following sections, we work out tools in order to estimate the effect of a single cut-off in the H^1-energy norm.

2.5.2 *Regularity Assumptions on the Subdomains*

In this subsection we collect all the regularity assumptions that we need for the technical tools exposed in subsequent sections. Most of the assumptions can be relaxed (while of course complicating the theory) and we will indicate that at the corresponding places.

Assumption 2.53. The subdomain triangulations $\mathscr{T}^h(\Omega_i)$, $i \in \mathscr{I}_{\text{FEM}}$ and $\mathscr{T}^h(\partial\Omega_i)$, $i \in \mathscr{I}_{\text{BEM}}$ are quasi-uniform with mesh parameter h_i.

Assumption 2.54. There is a shape regular simplicial coarse triangulation $\mathscr{T}^H(\Omega)$ of Ω, such that each subdomain Ω_i is the union of coarse elements and the number of coarse elements per subdomain is uniformly bounded.

Under Assumption 2.54, the subdomain diameter H_i is equivalent to the local mesh parameter of $\mathscr{T}^H(\Omega_i)$. Furthermore, we can extend the triangulations $\mathscr{T}^h(\partial\Omega_i)$, $i \in \mathscr{I}_{\text{BEM}}$ to auxiliary triangulations $\mathscr{T}^h(\Omega_i)$ that are quasi-uniform with mesh parameter h_i. We will use these auxiliary triangulations in the analysis.

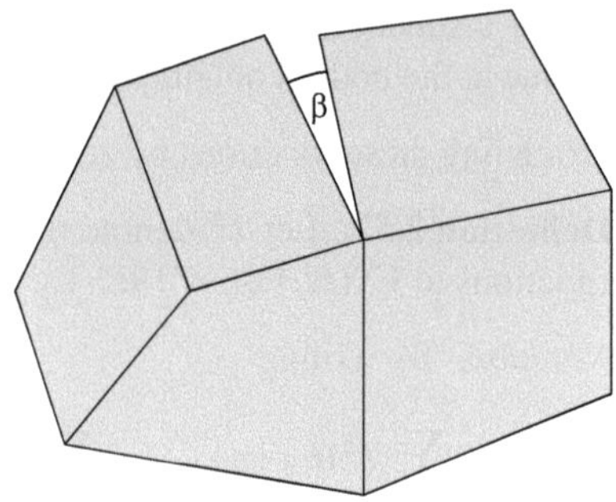

Fig. 2.6 Non-convex domain with abritrarily small exterior angle β. The displayed subdomain decomposition satisfies Assumptions 2.54 and 2.55 with shape regularity constants uniform in β

The next assumption essentially states that the exterior angles of the BEM subdomains must be bounded from below. Note that even for a pure BETI method, this does not necessarily prohibit the global domain Ω from having bad exterior angles, cf. Fig. 2.6.

Assumption 2.55. For each subdomain Ω_i, $i \in \mathscr{I}_{\text{BEM}}$, there exists a neighborhood $\Omega_i' \supset \overline{\Omega_i}$ and a shape regular simplicial coarse triangulation $\mathscr{T}^H(\Omega_i')$, such that the shape regularity constants and the number of coarse elements in $\mathscr{T}^H(\Omega_i')$ are uniformly bounded.

Assumption 2.56. The Dirichlet boundary Γ_D is the union of facets (faces/edges) of the coarse triangulation $\mathscr{T}^H(\Omega)$ from Assumption 2.54.

Thanks to Assumptions 2.54 and 2.56, each glob is the union of a few vertices, edges, and faces of the coarse triangulation $\mathscr{T}^H(\Omega)$.

Notation. We write $a \lesssim b$ if there is a constant C depending only on the shape regularity and quasi-uniformity constants from Assumptions 2.53–2.55 such that $a \leq C\, b$. The notation $a \eqsim b$ is a short hand for $a \lesssim b$ and $b \lesssim a$.

2.5.3 *An Explicit Poincaré Inequality*

In this subsection, we provide a uniform bound of the Poincaré constant $C_P(\Omega_i)$ (see Lemma 1.27) under Assumption 2.54. The following lemma is taken from [PS12b, Lemma A.1]. For similar approaches see also [BH70, DS80, DW06, VV09].

Lemma 2.57. *Let T be a triangle ($d = 2$) or tetrahedron ($d = 3$) and let F be one of its facets (edge if $d = 2$ and triangular face if $d = 3$). Then*

$$\|u - \overline{u}^F\|_{L^2(T)} \leq \text{diam}(T)\, |u|_{H^1(T)} \qquad \forall u \in H^1(T).$$

Proof. Due to Veeser and Verfürth (see [VV09, Sect. 4, Remark 4.6, formula (2.3), and Corollary 4.5]), for all $v \in H^1(T)$:

$$\frac{1}{\text{meas}_{d-1}(F)} \|v\|^2_{L^2(F)} \leq \frac{1}{\text{meas}_d(T)} \|v\|^2_{L^2(T)} + \frac{2\,\text{diam}(T)}{d\,\text{meas}_d(T)} \|v\|_{L^2(T)} |v|_{H^1(T)}, \tag{2.77}$$

Due to [PW60, Beb03],

$$\|u - \overline{u}^T\|_{L^2(T)} \leq \frac{\text{diam}(T)}{\pi} |u|_{H^1(T)} \qquad \forall u \in H^1(T), \tag{2.78}$$

because T is convex. With the triangle inequality and Cauchy's inequality,

$$\begin{aligned} \|u - \overline{u}^F\|_{L^2(T)} &\leq \|u - \overline{u}^T\|_{L^2(T)} + \sqrt{\text{meas}_d(T)}\, |\overline{u}^T - \overline{u}^F| \\ &\leq \|u - \overline{u}^T\|_{L^2(T)} + \frac{\sqrt{\text{meas}_d(T)}}{\sqrt{\text{meas}_{d-1}(F)}} \|u - \overline{u}^T\|_{L^2(F)} \end{aligned}$$

Using (2.77) and (2.78) in the estimate above yields

$$\begin{aligned} \|u - \overline{u}^F\|_{L^2(T)} &\leq \tfrac{\text{diam}(T)}{\pi} |u|_{H^1(T)} + \sqrt{\|u - \overline{u}^T\|^2_{L^2(T)} + \tfrac{2\,\text{diam}(T)}{d} \|u - \overline{u}^T\|_{L^2(T)} |u|_{H^1(T)}} \\ &\leq \tfrac{\text{diam}(T)}{\pi} |u|_{H^1(T)} + \sqrt{\tfrac{\text{diam}(T)^2}{\pi^2} |u|^2_{H^1(T)} + \tfrac{2\,\text{diam}(T)}{d} \tfrac{\text{diam}(T)}{\pi} |u|^2_{H^1(T)}} \\ &= \left(\tfrac{1}{\pi} + \sqrt{\tfrac{1}{\pi^2} + \tfrac{2}{d\,\pi}}\right) \text{diam}(T)\, |u|_{H^1(T)} \end{aligned}$$

Since $d \geq 2$, the factor in the parentheses is $\leq 0.96609936 \leq 1$. □

Definition 2.58. Let $\mathcal{T}^H(\Omega)$ be a simplicial triangulation. An open set $Y \subset \Omega$ is called (d-dimensional) *agglomerate* of $\mathcal{T}^H(\Omega)$ (in short $\mathcal{T}^H$*-agglomerate*) if Y is a connected union of elements from $\mathcal{T}^H(\Omega)$.

Definition 2.59. Let Y be an agglomerate of $\mathcal{T}^H(\Omega)$ with $\overline{Y} = \bigcup_{\ell=1}^{L} \overline{T}_\ell$. We call the region $P_{\ell_1,\ell_s} = \text{interior}(\overline{T}_{\ell_1} \cup \overline{T}_{\ell_2} \cup \dots \cup \overline{T}_{\ell_s})$, $1 \leq \ell_1, \dots, \ell_s \leq L$, a *path* from T_{ℓ_1} to T_{ℓ_s} of *length* s if for each $i = 1, \dots, s-1$, the elements $\overline{T}_{\ell_i}$ and $\overline{T}_{\ell_{i+1}}$ share a common facet F_i.

The following two lemmas lift the result of Lemma 2.57 from a single simplex to an agglomerate, cf. [PS11c, Lemmas 2 and 4]. We will revisit the same approach in Sect. 3.4.4.

Lemma 2.60. *Let Y be an agglomerate of $\mathscr{T}^H(\Omega)$, let $P_{\ell_1,\ell_s} = \text{interior}(\overline{T}_{\ell_1} \cup \overline{T}_{\ell_2} \cup \cdots \cup \overline{T}_{\ell_s})$ be a path and let F_i be the facet shared by $\overline{T}_{\ell_i}$ and $\overline{T}_{\ell_{i+1}}$ for $i = 1, \ldots, s-1$. Furthermore, let F_s be any facet of T_{ℓ_s}. Then*

$$\|u - \overline{u}^{F_s}\|^2_{L^2(T_{\ell_1})} \le 4 \left(\sum_{i=1}^{s} \frac{\text{meas}_d(T_{\ell_1})\,\text{diam}(T_{\ell_i})^2}{\text{meas}_d(T_{\ell_i})} \right) |u|^2_{H^1(P_{\ell_1,\ell_2})} \qquad \forall u \in H^1(Y).$$

Proof. By a telescoping argument we have

$$\|u - \overline{u}^{F_s}\|_{L^2(T_{\ell_1})} \le \|u - \overline{u}^{F_1}\|_{L^2(T_{\ell_1})} + \sum_{i=2}^{s} \sqrt{\text{meas}_d(T_{\ell_1})}\,\left|\overline{u}^{F_{i-1}} - \overline{u}^{F_i}\right|.$$

With Lemma 2.57,

$$\|u - \overline{u}^{F_1}\|_{L^2(T_{\ell_1})} \le \text{diam}(T_{\ell_1})\,|u|_{H^1(T_{\ell_1})}.$$

For fixed $2 \le i \le s$ the same lemma implies that

$$\begin{aligned} \left|\overline{u}^{F_{i-1}} - \overline{u}^{F_i}\right|^2 &\le \frac{2}{\text{meas}_d(T_{\ell_i})} \left(\|\overline{u}^{F_{i-1}} - u\|^2_{L^2(T_{\ell_i})} + \|u - \overline{u}^{F_i}\|^2_{L^2(T_{\ell_i})} \right) \\ &\le \frac{4}{\text{meas}_d(T_{\ell_i})}\,\text{diam}(T_{\ell_i})^2\,|u|^2_{H^1(T_{\ell_i})}. \end{aligned}$$

Combining the three estimates above, Cauchy's inequality (in $\mathbb{R}^s$) yields

$$\|u - \overline{u}^{F_s}\|^2_{L^2(T_{\ell_1})} \le \left[\sum_{i=1}^{s} \frac{4\,\text{meas}_d(T_{\ell_1})\,\text{diam}(T_{\ell_i})^2}{\text{meas}_d(T_{\ell_i})} \right] |u|^2_{H^1(P_{\ell_1,\ell_s})},$$

which completes the proof. □

Lemma 2.61. *Let Y an agglomerate of $\mathscr{T}^H(\Omega)$ consisting of at most L elements. Then*

$$C_P(Y) \le C,$$

where C only depends on L and on the shape regularity constant of $\mathscr{T}^H(\Omega)$.

Proof. One easily shows that

$$\text{meas}_d(T) \ge c\,\text{meas}_d(Y), \qquad \text{diam}(T) \ge c\,\text{diam}(Y)$$

for all elements T contained in Y, where the constant c only depends on L and on the shape regularity constant of $\mathscr{T}^H(\Omega)$. Let F be an arbitrary facet of Y. For each element T in Y there exists a path from T to the element containing F of maximal length L. Summing the estimate in Lemma 2.60 over all elements in Y yields

$$\inf_{c \in \mathbb{R}} \|u - c\|^2_{L^2(Y)} \le \|u - \overline{u}^F\|^2_{L^2(Y)} \le C\,\text{diam}(Y)^2\,|u|^2_{H^1(Y)}.$$

for all $u \in H^1(Y)$. □

Corollary 2.62. *Under Assumption 2.54, the Poincaré constants of all the subdomains are uniformly bounded.*

Remark 2.63. Without Assumption 2.54, one can instead assume that the isoperimetric constants of the subdomains are uniformly bounded, see [DKW08b, DKW08a, KRW08]. Note that ragged boundaries alone do not make these constants blow up.

2.5.4 Trace and Poincaré Inequalities for FE Functions

In this subsection, we elaborate on inequalities for finite element functions, which do (in general) not hold in the Sobolev space H^1. Therefore, we obtain expressions depending on the mesh parameter h. All the estimates can e.g. be found in [TW05, Sect. 4.6] as well as in early papers [BPS86, BPS87, BPS88, BPS89]. Here, we work out the common pattern and the dependence of the constants on the geometry, using the assumptions of Sect. 2.5.2.

Definition 2.64. For $0 \leq m \leq d-1$, the *m-facets* of a simplicial (coarse) triangulation $\mathscr{T}^H(\Omega)$ are

- The vertices of $\mathscr{T}^H(\Omega)$ if $m = 0$,
- The edges if $m = 1$,
- The triangular faces if $m = 2$.

Definition 2.65. For $0 \leq m \leq d-1$, a set $X \subset \overline{\Omega}$ is called m-dimensional *agglomerate* of $\mathscr{T}^H(\Omega)$ (in short $\mathscr{T}^H$*-agglomerate*) if X is a union of m facets. The dimension of X is denoted by $d_X := m$.

Apparently, a zero-dimensional agglomerate consists of finitely many (isolated) vertices. We see that under Assumption 2.54, the globs from Definition 2.48 are m-agglomerates of the coarse triangulation $\mathscr{T}^H(\Omega)$ with $0 \leq m \leq d-1$.

If X is a $\mathscr{T}^H$-agglomerate of dimension $d_X > 0$, the d_X-dimensional Lebesgue measure properly defines the space $L^2(X)$ and the average

$$\overline{v}^X := \frac{1}{\text{meas}_{d_X}(X)} \int_X v \, ds \qquad \text{for } v \in L^2(X).$$

The case $d_X = 0$ is treated in the following definition.

Definition 2.66. Let $X = \{p_j : j = 1, \ldots, J\}$ be a zero-dimensional agglomerate of $\mathscr{T}^H(\Omega)$, consisting of J different points. For a function $v : X \to \mathbb{R}$, we define

$$\text{meas}_0(X) := J, \qquad \int_X v \, ds := \sum_{j=1}^{J} v(p_j), \qquad \|v\|_{L^2(X)} := \Big(\int_X |v|^2 ds\Big)^{1/2}$$

and the average

$$\overline{v}^X := \frac{1}{\operatorname{meas}_0(X)} \int_X v \, ds = J^{-1} \sum_{j=1}^{J} v(p_j).$$

Let $\{\mathcal{T}^h(\Omega)\}_h$ be a family of triangulations that are refinements of $\mathcal{T}^H(\Omega)$ and let $V^h(\Omega)$ denote the finite element space of continuous and piecewise linear functions with respect to $\mathcal{T}^h(\Omega)$. For any d-dimensional agglomerate Y of $\mathcal{T}^H(\Omega)$, we denote by $\mathcal{T}^h(Y)$ and $V^h(Y)$ the restrictions of $\mathcal{T}^h(\Omega)$ and $V^h(\Omega)$ to Y.

Definition 2.67. Let Y be a d-dimensional $\mathcal{T}^H$-agglomerate.

(i) For a $\mathcal{T}^H$-agglomerate $X \subset \overline{Y}$ of dimension $0 \le d_X \le d$, let $C_P(Y, X; h)$ denote the smallest parameter such that

$$\|v - \overline{v}^X\|_{L^2(Y)} \le C_P(Y, X; h) \operatorname{diam}(Y) \, |v|_{H^1(Y)} \qquad \forall v \in V^h(Y).$$

For $d_X \ge d - 1$ let $C_P(Y, X)$ denote the smallest constant such that the above inequality holds in $H^1(Y)$.

(ii) For $\mathcal{T}^H$-agglomerates $X, W \subset \overline{Y}$ of dimensions $d_X, d_W \in [0, d]$, let $C_P(Y, X, W; h)$ denote the smallest parameter such that

$$\|v - \overline{v}^X\|_{L^2(W)} \le C_P(Y, X, W; h) \operatorname{diam}(Y) \sqrt{\frac{\operatorname{meas}_{d_W}(W)}{\operatorname{meas}_d(Y)}} \, |v|_{H^1(Y)} \quad \forall v \in V^h(Y).$$

For $d_X \ge d - 1$ and $d_W \ge d - 1$ let $C_P(Y, X, W)$ denote the smallest constant such that the above inequality holds in $H^1(Y)$.

By definition, for $d_W = d$ and $d_X \ge d - 1$, we have

$$C_P(D) \le C_P(D, X), \qquad C_P(D, X, W) \le C_P(D, X).$$

Definition 2.68. For an integer $j \ge 0$, we define the indicator function

$$\sigma^j(x) := \begin{cases} 1 & \text{if } j \le 1, \\ 1 + \log(x) & \text{if } j = 2, \\ x^{j-2} & \text{if } j \ge 3. \end{cases}$$

The following lemma states a discrete trace inequality with explicit dependence on the geometric parameters.

Lemma 2.69 (discrete trace inequality). *Let Assumption 2.54 hold, let Y and W be $\mathcal{T}^H$-agglomerates of dimensions $d_Y = d$ and $0 \le d_W \le d$ with $W \subset \overline{Y}$. Furthermore, let $\mathcal{T}^h(Y)$ be quasi-uniform. Then*

$$\|v\|^2_{L^2(W)} \leq C\,\sigma^{d-d_W}\big(\tfrac{H_Y}{h}\big)\,\frac{\mathrm{meas}_{d_W}(W)}{\mathrm{meas}_d(Y)}\,\Big(H_Y^2\,|v|^2_{H^1(Y)} + \|v\|^2_{L^2(Y)}\Big) \quad \forall v \in V^h(Y),$$

where $H_Y = \mathrm{diam}(Y)$*. The constant* C *only depends on the shape regularity and quasi-uniformity constants of* $\mathcal{T}^H(\Omega)$ *and* $\mathcal{T}^h(Y)$*, and on the number of coarse elements contained in* Y*.*

Proof. For $d_W = d-1$, the estimate follows from the trace theorem and a scaling argument. For the case $d=2$, $d_W = 0$, see [TW05, Lemma 4.15] and also [BX91]. For $d=3$, $d_W=0$ we combine the embedding of $H^1(D)$ in $L^{2d/(d-2)}(D)$ and the inverse inequality $\|v\|_{L^\infty(D)} \lesssim h^{-(d-2)/2}\|v\|_{H^1(D)}$ on a domain of unit diameter. Finally, the case $d=3$, $d_W=1$ is treated by integrating the two-dimensional L^∞-estimate along the edges forming W, see also [TW05] and [PS12b, Lemma A.4]. □

The statement of the following lemma can be found (in different form) in [TW05, Sect. 4.6], see also the references therein.

Lemma 2.70 (discrete Poincaré type inequality). *Let Assumption 2.54 hold, let* Y *be an agglomerate and let* X*,* W *be agglomerates of dimensions* $d_X, d_W \in [0, d]$*. Furthermore, let* $\mathcal{T}^h(Y)$ *be quasi-uniform. Then*

$$\|v - \overline{v}^X\|^2_{L^2(W)} \leq C\,\sigma^{d-\min(d_X,d_W)}\big(\tfrac{H}{h}\big)\,\frac{\mathrm{meas}_{d_W}(W)}{\mathrm{meas}_d(Y)}\,H^2\,|v|^2_{H^1(Y)} \quad \forall v \in V^h(Y),$$

where $H = \mathrm{diam}(Y)$*. The constant* C *only depends on the shape regularity and quasi-uniformity constants of* $\mathcal{T}^H(\Omega)$ *and* $\mathcal{T}^h(Y)$*, respectively, and on the number of elements that* Y *contains.*

Proof. We have

$$\begin{aligned}
\tfrac{1}{2}\,\|v - \overline{v}^X\|^2_{L^2(W)} &\leq \|v - \overline{v}^Y\|^2_{L^2(W)} + \mathrm{meas}_{d_W}(W)\,|\overline{v}^Y - \overline{v}^X|^2 \\
&\leq \|v - \overline{v}^Y\|^2_{L^2(W)} + 2\,\frac{\mathrm{meas}_{d_W}(W)}{\mathrm{meas}_{d_X}(X)}\Big(\|v - \overline{v}^Y\|^2_{L^2(X)} + \|v - \overline{v}^X\|^2_{L^2(X)}\Big).
\end{aligned}$$

Due to the Ritz minimum principle (cf. Lemma 1.3),

$$\|v - \overline{v}^X\|^2_{L^2(X)} = \inf_{c\in\mathbb{R}} \|v - c\|^2_{L^2(X)} \leq \|v - \overline{v}^Y\|^2_{L^2(X)}.$$

Hence, applying Lemma 2.69 for the manifolds X and W, we obtain

$$\begin{aligned}
\tfrac{1}{2}\,\|v - \overline{v}^X\|^2_{L^2(W)} &\leq \|v - \overline{v}^Y\|_{L^2(W)} + 4\,\frac{\mathrm{meas}_{d_W}(W)}{\mathrm{meas}_{d_X}(X)}\,\|v - \overline{v}^Y\|^2_{L^2(X)} \\
&\leq \Big(\sigma^{d-d_W}\big(\tfrac{H}{h}\big) + 4\,\sigma^{d-d_X}\big(\tfrac{H}{h}\big)\Big)\,\frac{\mathrm{meas}_{d_W}(W)}{\mathrm{meas}_d(Y)}\,\Big(|v|^2_{H^1(Y)} + \mathrm{diam}(Y)^{-2}\|v - \overline{v}^Y\|^2_{L^2(Y)}\Big).
\end{aligned}$$

Applying Poincaré's inequality on Y and taking Lemma 2.61 into account concludes the proof. □

Corollary 2.71. *Under the assumptions of Lemma 2.70,*

$$C_P(Y, X, W; h) \leq C \sqrt{\sigma^{d-\min(d_X, d_W)}\left(\tfrac{H}{h}\right)},$$

$$C_P(Y, X; h) \leq C \sqrt{\sigma^{d-d_X}\left(\tfrac{H}{h}\right)}.$$

2.5.5 Cut-Off Estimates

In this subsection, we estimate the effect of a cut-off in the energy norm. As for Sect. 2.5.4, the following estimates can be found in [TW05]. They mainly stem from the early works [BPS86, BX91, Dry87, DW94, DW95] on iterative substructuring.

Definition 2.72. Let $\mathscr{F}_i$ be a subdomain facet (i.e., a subdomain face if $d = 3$, a subdomain edge if $d = 2$) on $\partial\Omega_i$. We set

$$\vartheta_{\mathscr{F}_i} := \mathscr{H}_i^h(\theta_{\mathscr{F}_i}) \in V^h(\Omega_i).$$

Here, the space $V^h(\Omega_i)$ and the discrete harmonic extension $\mathscr{H}_i^h$ are defined with respect to the (possibly auxiliary) triangulation $\mathscr{T}^h(\Omega_i)$. For any $\mathscr{G}_i$ on $\partial\Omega_i$ that is not a subdomain facet, we define $\vartheta_{\mathscr{G}_i} \in V^h(\Omega_i)$ as the extension of $\theta_{\mathscr{G}_i} \in V^h(\partial\Omega_i)$ that vanishes at all interior nodes of Ω_i.

Remark 2.73. For simplicity, we did not indicate the subdomain index in the definition above. For any interface glob $\mathscr{G}$ we may regard $\vartheta_{\mathscr{G}}$ as a global function in $V^h(\Omega)$ only supported in the subdomains that share the glob $\mathscr{G}$.

For any $v \in V^h(\Omega_i)$, the function $I^h(\vartheta_{\mathscr{G}_i} v)$ is an extension of $I^h(\theta_{\mathscr{G}_i} v_{|\partial\Omega})$ and so by the minimizing property of $\mathscr{H}_i^h$ (cf. Definition 1.55),

$$|\mathscr{H}_i^h(\theta_{\mathscr{G}_i} v)|_{H^1(\Omega_i)} \leq |I^h(\vartheta_{\mathscr{G}_i} v)|_{H^1(\Omega_i)}.$$

Notation. Above and in the following, with a small abuse of notation, whenever v is a function with well-defined values at the nodes contained in $\mathscr{G}_i$, by

$$\mathscr{H}_i^h(\theta_{\mathscr{G}_i} v) \tag{2.79}$$

we mean that we first interpret the expression $\theta_{\mathscr{G}_i} v$ as a function in $V^h(\partial\Omega_i)$ (by interpolating at the nodes contained in $\mathscr{G}_i$, and choosing zero at all other nodes) and then apply $\mathscr{H}_i^h$ to it.

The following lemmas give further bounds in terms of the original function v. First, we discuss the case of lower-dimensional globs.

Lemma 2.74. *Let $\mathscr{T}^h(\Omega_i)$ be quasi-uniform. Then for any glob $\mathscr{G}_i$ of dimension $0 \le m < d-1$ (a subdomain vertex or edge if $d = 3$, a subdomain vertex if $d = 2$),*

$$|I^h(\vartheta_{\mathscr{G}_i} v)|^2_{H^1(\Omega_i)} \le C\, h_i^{d-2-m} \|v\|^2_{L^2(\mathscr{G}_i)} \qquad \forall v \in V^h(\Omega_i),$$

where $\|v\|_{L^2(\mathscr{V}_i)} = |v(\mathscr{V}_i)|$. In particular,

$$|\vartheta_{\mathscr{V}_i}|^2_{H^1(\Omega_i)} \le C\, h_i^{d-2}.$$

For $d = 3$, let $\mathscr{E}_i$ be a subdomain edge such that $\mathscr{V}_i \in \overline{\mathscr{E}_i}$. Then

$$|I^h(\vartheta_{\mathscr{V}_i} v)|^2_{H^1(\Omega_i)} \le C\, \|v\|^2_{L^2(\mathscr{E}_i)} \qquad \forall v \in V^h(\Omega_i).$$

If $d = 3$ and if Assumption 2.54 holds then

$$|\vartheta_{\mathscr{E}_i}|^2_{H^1(\Omega_i)} \le C\, H_i\,.$$

The constant C only depends on the shape regularity constants of $\mathscr{T}^h(\Omega_i)$ (and on the shape regularity of $\mathscr{T}^H(\Omega)$ in the last estimate).

Proof. Let $v \in V^h(\Omega_i)$ be arbitrary but fixed and note that $I^h(\vartheta_{\mathscr{G}_i} v)$ vanishes on all nodes of $\mathscr{T}^h(\Omega_i)$ except those on $\mathscr{G}_i$. Then, by a an inverse inequality,

$$|I^h(\vartheta_{\mathscr{G}_i} v)|^2_{H^1(\Omega_i)} \le C \sum_{x^h \in \mathscr{G}_i} h_i^{d-2}\, |v(x^h)|^2 \le C\, h_i^{d-2-m} \sum_{x^h \in \overline{\mathscr{G}_i}} h_i^m\, |v(x^h)|^2.$$

The sum in the last expression is equivalent to $\|v\|^2_{L^2(\mathscr{G}_i)}$. For $d = 3$, we obtain from the above that $|I^h(\vartheta_{\mathscr{V}_i} v)|^2_{H^1(\Omega_i)} \le C\, h_i\, |v(\mathscr{V}_i)|^2 \le C\, \|v\|^2_{L^2(\mathscr{E}_i)}$. The last estimate follows from $\|1\|^2_{L^2(\mathscr{E}_i)} = \text{meas}_1(\mathscr{E}_i) \le C\, H_i$. □

Combining Lemma 2.74 with the trace inequality from Lemma 2.69, we obtain the following result.

Lemma 2.75. *Let Assumptions 2.53 and 2.54 hold. Then for any glob $\mathscr{G}_i$ of dimension $0 \le m < d-1$ (a subdomain vertex or edge if $d = 3$, a subdomain vertex if $d = 2$),*

$$|I^h(\vartheta_{\mathscr{G}_i} v)|^2_{H^1(\Omega_i)} \lesssim \left(1 + \log\left(\tfrac{H_i}{h_i}\right)\right) \left(|v|^2_{H^1(\Omega_i)} + H_i^{-2}\, \|v\|^2_{L^2(\Omega_i)}\right) \quad \forall v \in V^h(\Omega_i).$$

Proof. First, we select a glob $\mathscr{X}_i$ of dimension at least $d-2$. If $d = 3$ and $m = 0$, we choose $\mathscr{X}_i$ as a subdomain edge touching $\mathscr{G}_i$, otherwise we set $\mathscr{X}_i = \mathscr{G}_i$. Thanks to Lemma 2.74,

$$|I^h(\vartheta_{\mathcal{G}_i} v)|^2_{H^1(\Omega_i)} \lesssim \|v\|^2_{L^2(\mathcal{X}_i)}.$$

Lemma 2.69 yields

$$\|v\|^2_{L^2(\mathcal{X}_i)} \lesssim \sigma^2\big(\tfrac{H_i}{h_i}\big) \frac{\operatorname{meas}_{d-2}(\mathcal{X}_i)\, H_i^2}{\operatorname{meas}_d(\Omega_i)} \Big(|v|^2_{H^1(\Omega_i)} + H_i^{-2}\, \|v\|^2_{L^2(\Omega_i)}\Big).$$

Since $\operatorname{meas}_{d-2}(\mathcal{X}_i) \eqsim H_i^{d-2}$ and $\operatorname{meas}_d(\Omega_i) \eqsim H_i^d$, this concludes the proof. □

We now turn to the case of a subdomain facet (i.e., a subdomain face if $d = 3$, a subdomain edge if $d = 2$).

Lemma 2.76. *Let Assumptions 2.53 and 2.54 hold. Then for any subdomain facet* $\mathcal{F}_i$*,*

$$|I^h(\vartheta_{\mathcal{F}_i} v)|^2_{H^1(\Omega_i)} \lesssim \big(1 + \log\big(\tfrac{H}{h}\big)\big)^2 \Big(|v|^2_{H^1(\Omega_i)} + H_i^{-2}\, \|v\|^2_{L^2(\Omega_i)}\Big) \quad \forall v \in V^h(\Omega_i),$$

and

$$|\vartheta_{\mathcal{F}_i}|^2_{H^1(\Omega_i)} \lesssim \big(1 + \log\big(\tfrac{H}{h}\big)\big)\, H_i^{d-2}.$$

Proof. For a proof in three dimensions, we refer to [TW05, Lemmas 4.24 and 4.25]. For the two-dimensional case, see [MB96] as well as the pioneering paper [BPS86]. Sharpness of the estimates is shown in [BS00].

Remark 2.77. On a subdomain facet $\mathcal{F}_i$, the term $|\mathcal{H}_i^h(\theta_{\mathcal{F}_i} v)|_{H^1(\Omega_i)}$ realizes the $H^{1/2}_{00}(\mathcal{F}_i)$-norm of $v \in V^h(\mathcal{F}_i)$, see also [TW05, Lemma 4.26] and Sect. 2.5.7.

Combining the cut-off lemmas and the discrete Poincaré inequality straightforwardly may create several powers of the term $(1 + \log(H_i/h_i))$. The following corollaries help to keep a power of two.

Corollary 2.78. *Let Assumptions 2.53 and 2.54 hold, and let* $\mathcal{G}_i$ *be a glob. Then*

$$|I^h(\vartheta_{\mathcal{G}_i} v)|^2_{H^1(\Omega_i)} \lesssim \big(1 + \log\big(\tfrac{H_i}{h_i}\big)\big)^2\, |v|^2_{H^1(\Omega_i)} + \big(1 + \log\big(\tfrac{H_i}{h_i}\big)\big)\, H_i^{-2}\, \|v\|^2_{L^2(\Omega_i)}$$

for all $v \in V^h(\Omega_i)$.

Proof. If the dimension of $\mathcal{G}_i$ is less than $d - 1$, the statement follows from Lemma 2.74. If $\mathcal{G}_i = \mathcal{F}_i$ is a subdomain facet, we have

$$|I^h(\vartheta_{\mathcal{F}_i} v)|^2_{H^1(\Omega_i)} \lesssim |I^h(\vartheta_{\mathcal{F}_i}(v - \overline{v}^{\Omega_i}))|^2_{H^1(\Omega_i)} + |\vartheta_{\mathcal{F}_i}|^2_{H^1(\Omega_i)}\, |\overline{v}^{\Omega_i}|^2.$$

By Cauchy's inequality (cf. (1.13)) and the estimates from Lemma 2.76, we get

$$\begin{aligned}|I^h(\vartheta_{\mathcal{F}_i} v)|^2_{H^1(\Omega_i)} &\lesssim \big(1 + \log\big(\tfrac{H_i}{h_i}\big)\big)^2 \Big(|v|^2_{H^1(\Omega_i)} + H_i^{-2}\, \|v - \overline{v}^{\Omega_i}\|^2_{L^2(\Omega_i)}\Big) \\ &\quad + \big(1 + \log\big(\tfrac{H_i}{h_i}\big)\big)\, H_i^{d-2}\, |\Omega_i|^{-1}\, \|v\|^2_{L^2(\Omega_i)}\end{aligned}$$

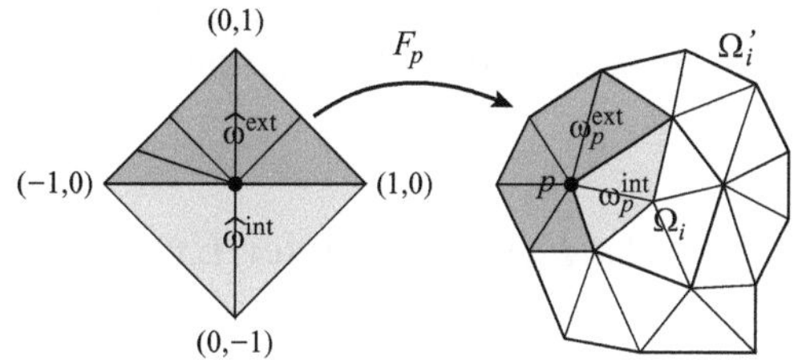

Fig. 2.7 Mapping of a node patch ω_p in two dimensions

Poincaré's inequality (Lemma 2.61) and using that $|\Omega_i| \eqsim H_i^d$ concludes the proof. □

Corollary 2.79. *Let Assumptions 2.53 and 2.54 hold, and let $\mathcal{G}_i$, $\mathcal{X}_i$ be two globs of Ω_i, such that the dimension of $\mathcal{X}_i$ is at least $d-2$. Then*

$$|I^h(\vartheta_{\mathcal{G}_i}(v - \overline{v}^{\mathcal{X}_i}))|^2_{H^1(\Omega_i)} \lesssim \left(1 + \log\left(\tfrac{H_i}{h_i}\right)\right)^2 |v|^2_{H^1(\Omega_i)} \qquad \forall v \in V^h(\Omega_i).$$

Proof. The statement follows immediately from Corollary 2.78 and Lemma 2.70. □

Summing the estimate from Corollary 2.78 over all globs of a subdomain, we obtain the following stable decomposition (recall our convention (2.79) from p. 122),

$$\sum_{\mathcal{G}_i} |\mathcal{H}_i^h(\theta_{\mathcal{G}_i} v)|^2_{H^1(\Omega_i)} \lesssim \left(1 + \log\left(\tfrac{H_i}{h_i}\right)\right)^2 |v|^2_{H^1(\Omega_i)} \tag{2.80}$$
$$+ \left(1 + \log\left(\tfrac{H_i}{h_i}\right)\right) H_i^{-2} \|v\|^2_{L^2(\Omega_i)} \qquad \forall v \in V^h(\Omega_i).$$

The following lemma is kind of inverse to (2.80).

Lemma 2.80. *Let Assumption 2.54 hold. Then*

$$|\mathcal{H}_i^h v|^2_{H^1(\Omega_i)} \leq C \sum_{\mathcal{G}_i} |\mathcal{H}_i^h(\theta_{\mathcal{G}_i} v)|^2_{H^1(\Omega_i)} \qquad \forall v \in V^h(\partial\Omega_i),\ v_{|\partial\Omega_i \cap \Gamma_N^h} = 0,$$

where the constant C only depends on the shape regularity constant of Assumption 2.54.

Proof. Since the functions $\{\theta_{\mathcal{G}_i}\}$ provide a partition of unity on $\partial\Omega_i \setminus \Gamma_N$, the function $\sum_{\mathcal{G}_i} \mathcal{H}_i^h(\theta_{\mathcal{G}_i} v)$ is a discrete extension of v from $\partial\Omega_i$ to Ω_i. Therefore, the result follows from Lemma 1.54, the triangle inequality, and the fact that the number of globs of Ω_i is uniformly bounded (see Assumption 2.54). See also Lemma 3.21, p. 169. □

Remark 2.81. Almost all proofs of the cut-off estimates in this section rely on Assumption 2.54 and so does Lemma 2.69. An extension to less regular subdomains (so-called John domains that can have ragged boundaries) for two dimensions can be found in [KRW08]. The three-dimensional case seems to be open yet.

2.5.6 An Explicit Sobolev Extension

In this subsection, we define a Sobolev extension operator for Lipschitz polytopes in the spirit of Stein [Ste70] and provide an explicit estimate in terms of shape regularity constants only.

For a Lipschitz polytope Ω_i, let $\Omega_i' \supset \overline{\Omega_i}$ be a neighborhood such that both domains are resolved by a shape regular coarse triangulation $\mathscr{T}^H(\Omega_i')$ consisting of a bounded number of elements, cf. Assumption 2.55. For an illustration see Fig. 2.7 (right). Let $\partial\Omega_i^H$ denote the set of coarse vertices of $\mathscr{T}^H(\Omega_i')$ that lie on $\partial\Omega_i$. For each vertex $p \in \partial\Omega_i^H$, we define the open vertex patch ω_p by

$$\overline{\omega}_p = \bigcup \{\overline{T} : T \in \mathscr{T}^H(\Omega'),\ p \in \overline{T}\},$$

and

$$\omega_p^{\text{int}} := \omega_p \cap \Omega_i, \qquad \omega_p^{\text{ext}} := \omega_p \cap (\Omega_i' \setminus \overline{\Omega}_i),$$

cf. Fig. 2.7 (right). Without loss of generality, we assume that ω_p^{int} and ω_p^{ext} each contain at least one coarse vertex that does not lie on $\partial\Omega_i$. This condition can always be fulfilled by formally subdividing some of the coarse elements.

We define the open reference patch $\widehat{\omega}$ by

$$\overline{\widehat{\omega}} = \begin{cases} \text{conv}(\{(-1,0),\ (1,0),\ (0,1),\ (0,-1)\}) & \text{if } d = 2, \\ \text{conv}(\{(-1,0,0),\ (1,1,0),\ (1,-1,0),\ (0,0,1),\ (0,0,-1)\}) & \text{if } d = 3, \end{cases}$$

where conv(S) denotes the convex hull of the set S. Furthermore, we define the subsets

$$\widehat{\omega}^{\text{int}} := \widehat{\omega} \cap \{x : x_d < 0\}, \qquad \widehat{\omega}^{\text{ext}} := \widehat{\omega} \cap \{x : x_d > 0\},$$

where x_d refers to the d-th component of x.

Let $\mathscr{T}_p(\widehat{\omega})$ be a shape regular simplicial triangulation of $\widehat{\omega}$ such that there exists a bijective continuous mapping $F_p : \widehat{\omega} \to \omega_p$ with the following properties.

- For each element $T \in \mathscr{T}_p(\widehat{\omega})$, the restricted mapping $F_{p|T}$ is affine linear,
- $F_p(0) = p$,
- $F_p(\widehat{\omega} \cap \{x : x_d = 0\}) = \omega_p \cap \partial\Omega_i$,
- $F_p(\widehat{\omega}^{\text{int}}) = \omega_p^{\text{int}}$ and $F_p(\widehat{\omega}^{\text{ext}}) = \omega_p^{\text{ext}}$,
- For each element $T \in \mathscr{T}_p(\widehat{\omega})$,

$$c_1 H_i^d \le \det(F'_{p|T}) \le c_2 H_i^d,$$

$$\|F'_{p|T}\|_{\ell^2} \le c_3 H_i, \qquad \|(F'_{p|T})^{-1}\|_{\ell^2} \le c_4 H_i^{-1},$$

where the constants c_1, c_2, c_3, and c_4 only depend on the shape regularity constants of $\mathcal{T}^H(\Omega_i')$.

For an illustration in two dimensions, see Fig. 2.7. Under the conditions on $\mathcal{T}^H(\Omega_i)$ stated in Assumption 2.55, such a triangulation and mapping exists for every coarse vertex $p \in \partial\Omega_i^H$.

On the reference patch, we define the linear operator

$$\widehat{E} : \mathcal{C}^\infty(\overline{\widehat{\omega}^{\text{int}}}) \to \mathcal{C}^\infty(\overline{\widehat{\omega}^{\text{ext}}}), \qquad (\widehat{E}w)(x_1, \ldots, x_d) := w(x_1, \ldots, x_{d-1}, -x_d),$$

i.e., the reflection of w across the hyperplane $\{x : x_d = 0\}$. By construction we have $\|\widehat{E}w\|_{H^1(\widehat{\omega}^{\text{ext}})} = \|w\|_{H^1(\widehat{\omega}^{\text{int}})}$. Since the $\mathcal{C}^\infty$-functions are dense in H^1, it follows that $\widehat{E} : H^1(\widehat{\omega}^{\text{int}}) \to H^1(\widehat{\omega}^{\text{ext}})$ is linear and continuous. For each coarse node $p \in \partial\Omega_i^H$, we define the linear operator

$$E^{(p)} : H^1(\omega_p^{\text{int}}) \to H^1(\omega_p^{\text{ext}}), \qquad E^{(p)}v := \left(\widehat{E}(v \circ F_p)\right) \circ F_p^{-1}.$$

Since F_p is continuous and piecewise affine linear, $E^{(p)}$ maps indeed into H^1 and is linear and continuous. Furthermore, one easily shows that

$$(E^{(p)}v)_{|\omega_p \cap \partial\Omega_i} = v_{|\omega_p \cap \partial\Omega_i}.$$

Finally, we define the extension operator

$$E_i : H^1(\Omega_i) \to H^1(\Omega_i'), \quad E_i\, v := \begin{cases} v & \text{in } \Omega_i, \\ \displaystyle\sum_{p \in \partial\Omega_i^H} \varphi_p\, E^{(p)}v & \text{in } \Omega_i' \setminus \Omega_i, \\ 0 & \text{else.} \end{cases} \tag{2.81}$$

where φ_p is the nodal finite element basis function on $\mathcal{T}^H(\Omega_i')$ associated with the coarse node p.

Lemma 2.82. *Let Ω_i be a subdomain and $\Omega_i' \supset \overline{\Omega_i}$ a Lipschitz domain that fully contains Ω_i and that is resolved by a shape regular coarse triangulation $\mathcal{T}^H(\Omega_i')$. Then the extension operator E_i as defined above maps into $H^1(\mathbb{R}^d)$. Furthermore, there exists a constant C_{E_i} depending only on the number of coarse elements in Ω_i and on the shape regularity constants of $\mathcal{T}^H(\Omega_i')$ such that*

$$|E_i\, v|^2_{H^1(\mathbb{R}^d)} + H_i^{-2}\, \|E_i\, v\|_{L^2(\mathbb{R}^d)} \le C_{E_i} \left(|v|^2_{H^1(\Omega_i)} + H_i^{-2}\, \|v\|_{L^2(\Omega_i)}\right) \quad \forall v \in H^1(\Omega_i).$$

Proof. Let $v \in H^1(\Omega_i)$ be arbitrary but fixed. For each coarse vertex p_j, $j = 1, \ldots, J$, the function $\varphi_{p_j} E^{(p_j)} v$ vanishes on $\mathbb{R}^d \setminus (\overline{\Omega}_i \cup \omega_p^{\text{ext}})$. Hence,

$$(E_i\, v)_{|\mathbb{R}^d \setminus \overline{\Omega}_i} \;\in\; H^1(\mathbb{R}^d \setminus \overline{\Omega}_i).$$

Thanks to the partition of unity property

$$\sum_{j=1}^{J} \varphi_{p_j}(x) \;=\; 1 \qquad \forall x \in \partial\Omega_i,$$

we can conclude that $(E_i v)_{|\partial\Omega_i} = v_{|\partial\Omega_i}$. Since in addition $(E_i v)_{|\partial\Omega_i'} = 0$, it follows that $E_i v \in H^1(\mathbb{R}^d)$. With standard finite element techniques (see e.g. [BS02, Cia87]), one shows that for each coarse node p on $\partial\Omega_j$,

$$|E^{(p)} v|_{H^1(\omega_p^{\text{ext}})} \;\le\; C\, |v|_{H^1(\omega_p^{\text{int}})}, \qquad \|E^{(p)} v\|_{L^2(\omega_p^{\text{ext}})} \;\le\; C\, \|v\|_{L^2(\omega_p^{\text{int}})}.$$

The constant C is uniform because there are only a small number of different triangulations $\mathscr{T}_p(\widehat{\omega})$. Since $\|\varphi_p\|_{L^\infty} = 1$, it follows from the above that

$$\|\varphi_p E^{(p)} v\|_{L^2(\omega_p^{\text{ext}})} \;\le\; C\, \|v\|_{L^2(\omega_p^{\text{int}})}.$$

Summing over $p \in \partial\Omega_i^H$, we obtain

$$\|E_i\, v\|_{L^2(\Omega_i' \setminus \Omega_i)} \;\le\; \sum_{p \in \partial\Omega_i^H} C\, \|v\|_{L^2(\omega_p^{\text{int}})} \;\le\; C\, \|v\|_{L^2(\Omega_i)}, \tag{2.82}$$

where we have used that the number of coarse nodes on $\partial\Omega_i$ is uniformly bounded.

We now turn to the H^1-seminorm. Since $\|\nabla\varphi_p\|_{L^\infty} \le C\, H_i^{-1}$, we can conclude from the product rule and the local L^2-estimate from above that

$$\begin{aligned} |\varphi_p(E^{(p)} v)|^2_{H^1(\omega_p^{\text{ext}})} &\le C\, \Big(|E^{(p)} v|^2_{H^1(\omega_p^{\text{ext}})} + H_i^{-2}\, \|E^{(p)} v\|^2_{L^2(\omega_p^{\text{ext}})} \Big) \\ &\le C\, \Big(|v|^2_{H^1(\omega_p^{\text{int}})} + H_i^{-2}\, \|v\|^2_{L^2(\omega_p^{\text{int}})} \Big). \end{aligned}$$

Summing over all coarse nodes $p \in \partial\Omega_i^H$, we get

$$|E_i\, v|^2_{H^1(\Omega_i' \setminus \Omega_i)} \;\le\; C\, \Big(|v|^2_{H^1(\Omega_i)} + H_i^{-2}\, \|v\|^2_{L^2(\Omega_i)} \Big) \tag{2.83}$$

Using the definition of E_i and combining (2.82) and (2.83) concludes the proof. □

Remark 2.83. The relaxation of Assumptions 2.54 and 2.55 is possible by using Jones' extension operator introduced in [Jon81], which is defined for the so-called *ε-δ domains* or *uniform domains*, and which employs a bound of the same form as in Lemma 2.82. The constant can be made explicit in a geometric parameter linked to the domain, see also [DKW08b, DKW08a, KRW08]. As for Poincaré's constant, ragged boundaries alone do not make the extension constant blow up. For a degree-independent extension operator see also [Rog06].

Remark 2.84. The extension operator E_i is also well-defined and bounded from $L^2(\Omega_i)$ to $L^2(\mathbb{R}^d)$, and by classical interpolation theory (see [AF03, Sect. 7.22f]) it follows that it is bounded from $H^s(\Omega_i)$ to $H^s(\mathbb{R}^d)$ for $s \in [0,1]$, cf. [Ste70].

Lemma 2.85. *The operator* $\widetilde{E}_i : H^1(\Omega_i) \to H^1(\Omega_i')$ *defined by*

$$\widetilde{E}_i\, v := \begin{cases} v & \text{in } \Omega_i, \\ \bar{v}^{\Omega_i} + \sum\limits_{p \in \partial\Omega_i^H} \varphi_p\, E^{(p)}(v - \bar{v}^{\Omega_i}) & \text{in } \Omega_i' \setminus \Omega_i \end{cases}$$

(cf. (2.81)*) is linear and continuous, and it preserves constants, i.e.,*

$$\widetilde{E}_i\, c = c \qquad \forall c \in \mathbb{R}. \tag{2.84}$$

Furthermore, we have the separate stability estimates

$$\left.\begin{aligned} \|\widetilde{E}_i\, v\|_{L^2(\Omega_i')} &\le C\, \|v\|_{L^2(\Omega_i)} \\ |\widetilde{E}_i\, v|_{H^1(\Omega_i')} &\le C\, |v|_{H^1(\Omega_i)} \end{aligned}\right\} \qquad \forall v \in H^1(\Omega_i),$$

where the constant C *depends only on the number of coarse elements in* Ω_i *and on the shape regularity constants of* $\mathscr{T}^H(\Omega_i')$.

Proof. The proof of the L^2-stability follows from (2.82), from Cauchy's inequality, and the fact that $\text{meas}_d(\Omega_i' \setminus \Omega_i) \eqsim \text{meas}_d(\Omega_i)$. The H^1-stability immediately follows from Poincaré's inequality (Lemma 2.61). □

2.5.7 A Subdomain Transfer Operator for FE Functions

In this subsection, we construct an extension operator that transfers finite element functions from a subdomain to its neighboring subdomains. This technique was first used in [KRW08, Lemma 4.5], see also the references therein. Note that the original analysis of FETI [KW01] was performed using trace norms, which we can circumvented by the transfer operators. Such, one can make the dependence on the shapes of the subdomains more explicit.

Definition 2.86. For any glob $\mathscr{G}$, the *glob patch* $\mathscr{U}_{\mathscr{G}}$ is given by

$$\overline{\mathscr{U}_{\mathscr{G}}} := \bigcup_{k \in \mathscr{N}_{\mathscr{G}}} \overline{\Omega}_k\,.$$

Lemma 2.87. *Let Assumption 2.54 hold. Then, for any glob $\mathcal{G}_i$ there exists a discrete extension operator $E^h_{i,\mathcal{G}_i} : V^h(\Omega_i) \to V^h(\mathcal{U}_{\mathcal{G}_i})$ such that $(E^h_{i,\mathcal{G}_i} v)_{|\Omega_i} = v$ and*

$$\left.\begin{aligned} |E^h_{i,\mathcal{G}_i} v|_{H^1(\mathcal{U}_{\mathcal{G}_i})} &\le C\, |v|_{H^1(\Omega_i)} \\ \|E^h_{i,\mathcal{G}_i} v\|_{L^2(\mathcal{U}_{\mathcal{G}_i})} &\le C\, \|v\|_{L^2(\Omega_i)} \end{aligned}\right\} \quad \forall v \in V^h(\Omega_i),$$

where the constant C only depends on the shape regularity constant of $\mathcal{T}^H(\Omega)$. In particular, $E^h_{i,\mathcal{G}_i}$ preserves the (fine) nodal values on $\overline{\mathcal{G}_i}$ and it preserves constants.

Proof. Assume that $\mathcal{G}_i \subset \Gamma$ (otherwise $\mathcal{U}_{\mathcal{G}_i} = \Omega_i$ and the extension is trivial). Therefore, $\mathcal{U}_{\mathcal{G}_i} \setminus \Omega_i$ contains at least one coarse element and $\text{meas}_d(\mathcal{U}_{\mathcal{G}_i} \setminus \Omega_i) \eqsim \text{meas}_d(\Omega_i)$. Let $\widetilde{E}_{i,\mathcal{G}_i} : H^1(\Omega_i) \to H^1(\mathcal{U}_{\mathcal{G}_i})$ be defined analogously to Lemma 2.85, with the only modification that we work on the coarse mesh on $\mathcal{U}_{\mathcal{G}_i}$ instead of Ω_i' and that we sum only the reflections corresponding to the coarse nodes on $\overline{\mathcal{G}_i}$. Doing so, we obtain

$$\left.\begin{aligned} \|\widetilde{E}_{i,\mathcal{G}_i} v\|_{L^2(\mathcal{U}_{\mathcal{G}_i})} &\le C\, \|v\|_{L^2(\Omega_i)} \\ |\widetilde{E}_{i,\mathcal{G}_i} v|_{H^1(\mathcal{U}_{\mathcal{G}_i})} &\le C\, |v|_{H^1(\Omega_i)} \end{aligned}\right\} \quad v \in H^1(\Omega_i). \tag{2.85}$$

Note that, again, the extension preserves constants: $\widetilde{E}_{i,\mathcal{G}_i} c = c$ for all $c \in \mathbb{R}$. Let $\Pi^h : H^1(\mathcal{U}_{\mathcal{G}_i} \setminus \Omega_i) \to V^h(\mathcal{U}_{\mathcal{G}_i} \setminus \Omega_i)$ denote the Scott-Zhang operator, which preserves piecewise linear data on the boundary. We set

$$E^h_{i,\mathcal{G}_i} v := \begin{cases} v & \text{in } \Omega_i, \\ \Pi^h \widetilde{E}_{i,\mathcal{G}_i} v & \text{in } \mathcal{U}_{\mathcal{G}_i} \setminus \Omega_i\,. \end{cases}$$

Indeed, $E^h_{i,\mathcal{G}_i} v$ is continuous across $\partial\Omega_i \cap \overline{\mathcal{U}_{\mathcal{G}_i}}$ and the stability estimates follow immediately from (2.85) and Lemma 1.45. □

Remark 2.88. We note that the essential assumption in Lemma 2.87 is that the *target* domain $\mathcal{U}_{\mathcal{G}_i}$ obeys a shape regular coarse triangulation consisting of a bounded number of coarse elements.

2.5.8 Uniform Bounds for the Constants c_0 and c_K

Let V_i denote the single layer potential operator on $\partial\Omega_i$ and D_i the corresponding hypersingular integral operator. Moreover, let $c_{0,i}$ and $c_{K,i}$ denote the respective constants from Lemma 1.77. In this subsection, we show that under the assumptions made in Sect. 2.5.2, at least in three dimensions, we can bound $c_{0,i}$ from below in terms of the shape regularity constants of $\mathscr{T}^H(\Omega)$ and $\mathscr{T}^H(\Omega_i')$. The following presentation is based on [Pec12].

To get a bound for c_0, we introduce special trace norms.

Definition 2.89. For $v \in H^{1/2}(\partial\Omega_i)$ we define

$$\|v\|_{\star,H^{1/2}(\partial\Omega_i)} := \left(|\mathscr{H}_i v|^2_{H^1(\Omega_i)} + H_i^{-2}\, \|\mathscr{H}_i v\|^2_{L^2(\Omega_i)} \right)^{1/2}$$

and

$$\|w\|_{\star,H^{-1/2}(\partial\Omega_i)} := \sup_{v \in H^{1/2}(\partial\Omega)} \frac{\langle w, v\rangle}{\|v\|_{\star,H^{1/2}(\partial\Omega_i)}}.$$

According to Definition 1.41 and Sect. 1.2.1.6, these norms are equivalent to $\|\cdot\|_{H^{1/2}(\partial\Omega_i)}$ and $\|\cdot\|_{H^{-1/2}(\partial\Omega_i)}$.

Definition 2.90. Assume that V_i is coercive and let $c^\star_{V_i}$ and $c^\star_{D_i}$ be the largest constants and $C^\star_{P,i}$ the smallest constant such that

$$\begin{aligned} \langle w, V_i w\rangle &\geq c^\star_{V_i}\, \|w\|^2_{\star,H^{-1/2}(\partial\Omega_i)} && \forall w \in H^{-1/2}(\partial\Omega_i), \\ \langle D_i v, v\rangle &\geq c^\star_{D_i}\, |\mathscr{H}_i v|^2_{H^1(\Omega_i)} && \forall v \in H^{1/2}(\partial\Omega_i), \\ \|\mathscr{H}_i v\|_{L^2(\Omega_i)} &\leq C^\star_{P,i}\, H_i\, |\mathscr{H}_i v|_{H^1(\Omega_i)} && \forall v \in H^{1/2}_*(\partial\Omega_i). \end{aligned}$$

Lemma 2.91. *We have that*

$$c_{0,i} \geq \frac{c^\star_{V_i}\, c^\star_{D_i}}{1 + (C^\star_{P,i})^2}.$$

Proof. With a standard duality argument, it follows that

$$\langle V_i^{-1} v, v\rangle \leq (c^\star_{V_i})^{-1}\, \|v\|^2_{\star,H^{1/2}(\partial\Omega_i)} \qquad \forall v \in H^{1/2}(\partial\Omega_i).$$

Using the definition of $c_{0,i}$, Definition 2.90, and the above boundedness result, we obtain

$$c_{0,i} = \inf_{v \in H^{1/2}_*(\partial\Omega_i)} \frac{\langle D_i\, v, v\rangle}{\langle V_i^{-1}\, v, v\rangle} \leq \inf_{v \in H^{1/2}_*(\partial\Omega_i)} \frac{c^\star_{D_i}\, |\mathscr{H}_i v|^2_{H^1(\Omega_i)}}{(c^\star_{V_i})^{-1}\, (|\mathscr{H}_i v|^2_{H^1(\Omega_i)} + H_i^{-2}\, \|\mathscr{H}_i v\|^2_{L^2(\Omega_i)})}.$$

Using the definition of $C^\star_{P,i}$ concludes the proof. □

Lemma 2.92. *Let Assumption 2.55 holds. Then in three dimensions,*

$$c^\star_{V_i} \geq \frac{1}{2\, C^2_{E_i}},$$

where C_{E_i} is the constant from Lemma 2.82. Hence $c^\star_{V_i}$ can be bounded from below in terms of the shape regularity constants of $\mathcal{T}^H(\Omega_i')$.

Proof. For a proof see [Pec12, Corollary 6.2]. It follows basically the line of the standard coercivity proof (see e.g. [Ste08, Sect. 6.6.1]), but uses the carefully chosen norms and the extension operator.

To bound the constant for the hypersingular integral operator, we need another extension operator which extends functions in the annulus $\Omega_i' \setminus \overline{\Omega}_i$ back to Ω_i. By changing the roles of Ω_i and the annulus in the construction of Lemma 2.85, we get the extension operator

$$\widetilde{E}_i' : H^1(\Omega_i' \setminus \overline{\Omega}_i) \to H^1(\Omega_i')$$

which preserves constants and obeys the following stability estimates.

Lemma 2.93. *There exists a constant $C_{\widetilde{E}_i'}$ depending only on the maximal number of coarse elements per subdomain and on the shape regularity constants of $\mathcal{T}^H(\Omega_i')$ such that for each subdomain Ω_i,*

$$\left.\begin{aligned} |\widetilde{E}_i' v|_{H^1(\Omega_i')} &\leq C_{\tilde{E}_i'} \, |v|_{H^1(\Omega_i' \setminus \overline{\Omega}_i)} \\ \|\widetilde{E}_i' v\|_{L^2(\Omega_i')} &\leq C_{\tilde{E}_i'} \, \|v\|_{L^2(\Omega_i' \setminus \overline{\Omega}_i)} \end{aligned}\right\} \quad \forall v \in H^1(\Omega_i' \setminus \overline{\Omega}_i).$$

Lemma 2.94. *Let Assumption 2.55 hold. Then*

$$c^\star_{D_i} \;\geq\; \frac{1}{2\, C^2_{\tilde{E}_i'}},$$

where $C_{\tilde{E}_i'^2}$ is the constant from Lemma 2.93. Hence $c^\star_{D_i}$ can be bounded from below in terms of the shape regularity constants of $\mathcal{T}^H(\Omega_i')$.

Proof. For a proof see [Pec12, Lemma 6.4].

Lemma 2.95. *Let Assumption 2.55 holds. Then in three dimensions,*

$$C^\star_{P,i} \;\leq\; \left[2\, C_P(\Omega_i)^2 + \frac{1 + C_P(\Omega_i)^2}{C^2_{E_i}}\right]^{1/2}.$$

Hence, $C^\star_{P,i}$ can be bounded from above in terms of the shape regularity constants from $\mathcal{T}^H(\Omega_i')$.

Proof. For a proof see [Pec12, Lemma 6.7].

Corollary 2.96. *Let Assumption 2.55 holds. Then in three dimensions, the constant* $c_{0,i}$ *can be bounded from below and* $c_{K,i}$ *from above in terms of the shape regularity constants of* $\mathcal{T}^H(\Omega_i')$.

Remark 2.97. The generalization of these uniformity results to the case of two dimensions is not yet known, mainly due to the particularity of the two-dimensional exterior problem, see Sect. 1.3.3, or one might say due to the logarithm in the fundamental solution. Ideas towards such estimates are given in [Pec12, Remark 4].

2.5.9 An Elementary Inequality

The following lemma provides an important inequality involving the weighted counting functions from (2.53). See e.g. [DW95, MB96, Sar94] for early works where this result has been used.

Lemma 2.98. *For* $x^h \in \Gamma_S^h$, *let* $\{\rho_k(x^h)\}_{k\in\mathcal{N}_{x^h}}$ *be arbitrary positive weights and let* $\{\delta_k^\dagger(x^h)\}_{k\in\mathcal{N}_{x^h}}$ *be defined as in* (2.53). *Then*

$$\rho_i(x^h)\,\big(\delta_j^\dagger(x^h)\big)^2 \;\le\; \min(\rho_i(x^h),\, \rho_j(x^h)) \qquad \forall i,\, j \in \mathcal{N}_{x^h}$$

for any choice of the exponent $\gamma \in [1/2, \infty)$. *The same estimate holds for the choice* (2.54)*, which corresponds to* $\gamma \to \infty$.

Proof. For simplicity, we drop the dependence on x^h during the proof. Recall that

$$\delta_j^\dagger \;=\; \frac{\rho_j^\gamma}{\sum_{k\in\mathcal{N}} \rho_k^\gamma}\,.$$

Assume that $\#(\mathcal{N}) \ge 2$ (otherwise the inequality is trivial). Since $0 < \delta_j^\dagger \le 1$, we immediately get that $\rho_i\,\big(\delta_j^\dagger\big)^2 \le \rho_i$. Secondly,

$$\rho_i\,(\delta_j^\dagger)^2 \;\le\; \frac{\rho_i\,\rho_j^{2\gamma}}{(\rho_i+\rho_j)^{2\gamma}} \;=\; \frac{\rho_i}{(\rho_i+\rho_j)}\,\frac{\rho_j^{2\gamma-1}}{(\rho_i+\rho_j)^{2\gamma-1}}\,\rho_j\,.$$

The first factor on the right-hand side is less than one. Since the function $y \mapsto y^{2\gamma-1}$ is monotonically non-decreasing for $\gamma \in [1/2, \infty)$, the second factor is less than one as well. This implies $\rho_i\,\big(\delta_j^\dagger\big)^2 \le \rho_j$. The proof of the estimate for the choice (2.54) (corresponding to $\gamma \to \infty$) is trivial. □

2.6 Preconditioner Analysis for Subdomain Resolved Coefficients

In this section, we give the complete convergence analysis of FETI/BETI (both classical and all-floating) for the case that the diffusion coefficient $\mathscr{A}$ is isotropic and piecewise constant in each subdomain, and with two further assumptions stated below.

Assumption 2.99. In each subdomain Ω_i, there exists a constant $\alpha_i > 0$ such that

$$\mathscr{A}_{|\Omega_i} = \alpha_i \, I .$$

Assumption 2.100. The *coefficient scaling* is used, i.e., $\rho_i(x^h) = \alpha_i$, cf. Sect. 2.2.4.2.

Remark 2.101. Note that the theory below can be carried over to matrix-valued coefficients with mild anisotropy that are piecewise constant in each subdomain. In such a case, the value of $\rho_i(x^h)$ should be chosen as the maximal eigenvalue of the matrix coefficient on Ω_i.

Assumption 2.102. For the classical formulation of FETI/BETI in three dimensions only, we assume that $\Gamma_D \cap \partial\Omega_i$ is either empty or contains at least a subdomain edge (i.e., it should not collapse to a subdomain vertex).

Recall the missing estimate of the operator P_D from Lemma 2.45,

$$|P_D(w + z_w)|_S^2 \le \omega \, |w|_S^2 \qquad \forall w \in W.$$

Lemma 2.103. *For each $i = 1, \ldots, s$, let $W_i^\perp$ be a subspace of W_i such that the sum*

$$W_i = \ker(S_i) \oplus W_i^\perp$$

is direct. Let $W^\perp := \prod_{i=1}^s W_i^\perp$. Then

$$|P_D(w + z_w)|_S^2 \le \omega \, |w|_S^2 \qquad \forall w \in W^\perp$$

implies

$$|P_D(w + z_w)|_S^2 \le \omega \, |w|_S^2 \qquad \forall w \in W.$$

Proof. Let $w \in W^\perp$ be arbitrary but fixed. First, Lemma 2.44 states that the mapping $v \mapsto z_v$ (for $v \in W$) is linear and that $z_y = -y$ for $y \in \ker(S)$. Therefore, we have the invariants

$$\left. \begin{aligned} w + z_w &= (w + y) + (z_{w+y}) \\ |w + y|_S &= |w|_S \end{aligned} \right\} \quad \forall y \in \ker(S).$$

Since $W = W^\perp \oplus \ker(S)$ the second inequality follows immediately. □

Throughout this section, we choose

$$W_i^\perp := \left\{ w_i \in W_i : \overline{\mathscr{H}_i^h w_i}^{\Omega_i} = 0 \right\} \quad \text{if } i \in \mathscr{I}_{\text{float}}, \tag{2.86}$$

and $W_i^\perp := W_i$ else. The strategy is to show bounds for $P_D\, w$ and $P_D\, z_w$ separately, for $w \in W^\perp$.

2.6.1 An Energy Estimate of P_D

The following lemma is essential for the condition number estimate and goes back to [KW01, Lemma 4.7], see also [MT96, Sect. 3.3] and [TW05, Lemma 6.3].

Lemma 2.104. *Let Assumptions 2.53–2.56 and 2.99–2.102 hold. Then*

$$|P_D\, w|_S^2 \lesssim \max_{i=1,\dots,s} (1 + \log(H_i/h_i))^2\, |w|_S^2 \qquad \forall w \in W^\perp.$$

Proof. Let $w \in W^\perp$ and $i = 1, \dots, s$ be fixed. Recall the characterization (2.70) of the P_D operator (p. 108), which reveals that there is no contribution from Γ_N^h (the non-coupling Neumann nodes). Recall also that $\mathscr{N}_{x^h} = \mathscr{N}_{\mathscr{G}_i}$ for all $x^h \in \mathscr{G}_i$. Using the BEM-FEM spectral equivalence from Corollary 1.94 and the cut-off result from Lemma 2.80 (with convention (2.79) from p. 122) we obtain

$$\begin{aligned} |(P_D\, w)_i|_{S_i}^2 &\lesssim \alpha_i\, |\mathscr{H}_i^h (P_D\, w)_i|_{H^1(\Omega_i)}^2 \qquad (2.87)\\ &\lesssim \underbrace{\alpha_i \sum_{\mathscr{G}_i \subset \Gamma} \Big| \mathscr{H}_i^h \Big(\theta_{\mathscr{G}_i} \sum_{j \in \mathscr{N}_{\mathscr{G}_i}} \delta_j^\dagger (w_i - w_j) \Big) \Big|_{H^1(\Omega_i)}^2}_{=: \Upsilon_i} + \alpha_i \sum_{\mathscr{G}_i \subset \Gamma_D} |\mathscr{H}_i^h (\theta_{\mathscr{G}_i} w_i)|_{H^1(\Omega_i)}^2 . \end{aligned}$$

Since $\rho_i(x^h) = \alpha_i$, the functions $\delta_j^\dagger$ are constant on each glob $\mathscr{G}_i$. Using the fact that each glob is shared by a uniformly bounded number of subdomains and the elementary inequality from Lemma 2.98, we obtain

$$\begin{aligned} \Upsilon_i &\lesssim \sum_{\mathscr{G}_i \subset \Gamma} \sum_{j \in \mathscr{N}_{\mathscr{G}_i}} \alpha_i\, \big(\delta_{j\,|\mathscr{G}_i}^\dagger\big)^2\, |\mathscr{H}_i^h (\theta_{\mathscr{G}_i} (w_i - w_j))|_{H^1(\Omega_i)}^2 \\ &\leq \sum_{\mathscr{G}_i \subset \Gamma} \sum_{j \in \mathscr{N}_{\mathscr{G}_i}} \min(\alpha_i, \alpha_j)\, |\mathscr{H}_i^h (\theta_{\mathscr{G}_i} (w_i - w_j))|_{H^1(\Omega_i)}^2 \end{aligned}$$

Inserting the estimate for Υ_i into (2.87) and using the triangle inequality yields

$$
\begin{aligned}
&|(P_D\, w)_i|_{S_i}^2 && (2.88)\\
&\lesssim \sum_{\mathcal{G}_i \subset \Gamma} \sum_{j \in \mathcal{N}_{\mathcal{G}_i}} \min(\alpha_i, \alpha_j)\, |\mathcal{H}_i^h(\theta_{\mathcal{G}_i}(w_i - w_j))|_{H^1(\Omega_i)}^2 + \sum_{\mathcal{G}_i \subset \Gamma_D} \alpha_i\, |\mathcal{H}_i^h(\theta_{\mathcal{G}_i} w_i)|_{H^1(\Omega_i)}^2 \\
&\lesssim \sum_{\mathcal{G}_i \subset \Gamma \cup \Gamma_D} \sum_{j \in \mathcal{N}_{\mathcal{G}_i}} \alpha_j\, |\mathcal{H}_i^h(\theta_{\mathcal{G}_i} w_j)|_{H^1(\Omega_i)}^2 . && (2.89)
\end{aligned}
$$

Note that the function w_j from Ω_j is cut down to $\mathcal{G}_{ij}$ and harmonically extended to the (possibly different) subdomain Ω_i. In order to estimate the energy of the extension in terms of the energy of the original function, we use the transfer operator from Sect. 2.5.7. Let $\tilde{w}_j = \mathcal{H}_j^h w_j \in V^h(\Omega_j)$. Then the function

$$
I^h(\vartheta_{\mathcal{G}_i} E_{j,\mathcal{G}_i}^h \tilde{w}_j)
$$

extends $\theta_{\mathcal{G}_i} w_j$ to Ω_i. Using Corollary 2.78 and Lemma 2.87, we get

$$
\begin{aligned}
|\mathcal{H}_i^h(\theta_{\mathcal{G}_i} w_j)|_{H^1(\Omega_i)}^2 &\leq |I^h(\vartheta_{\mathcal{G}_i} E_{j,\mathcal{G}_i}^h \tilde{w}_j)|_{H^1(\Omega_i)}^2 \\
&\lesssim (1 + \log(H_i/h_i))^2\, |E_{j,\mathcal{G}_i}^h \tilde{w}_j|_{H^1(\Omega_i)}^2 + H_i^{-2}\,(1 + \log(H_i/h_i))\, \|E_{j,\mathcal{G}_i}^h \tilde{w}_j\|_{L^2(\Omega_i)}^2 \\
&\lesssim (1 + \log(H_i/h_i))^2\, |\mathcal{H}_j^h w_j|_{H^1(\Omega_j)}^2 + H_i^{-2}\,(1 + \log(H_i/h_i))\, \|\mathcal{H}_j^h w_j\|_{L^2(\Omega_j)}^2 .
\end{aligned}
$$

If Ω_j is floating, $w_j \in W_j^\perp$, and we can eliminate the L^2-term using Poincaré's inequality. If Ω_j is non-floating, we know that w_j vanishes at least on a glob of dimension $d-2$ (see Assumption 2.102). Hence, we can eliminate the L^2-term at the cost of another factor of $(1 + \log(H_i/h_i))$ using Lemma 2.70. In either case, what we obtain is

$$
|\mathcal{H}_i^h(\theta_{\mathcal{G}_i} w_j)|_{H^1(\Omega_i)}^2 \lesssim (1 + \log(H_i/h_i))^2\, |\mathcal{H}_j^h w_j|_{H^1(\Omega_j)}^2 . \qquad (2.90)
$$

Combining (2.89) with (2.90) and using that the number of globs and neighbors per subdomain is uniformly bounded yields

$$
\sum_{i=1}^{s} |(P_D w)_i|_{S_i}^2 \lesssim \max_{i=1}^{s}(1 + \log(H_i/h_i))^2 \sum_{j=1}^{s} \alpha_j\, |\mathcal{H}_j^h w_j|_{H^1(\Omega_j)}^2 .
$$

The proof is concluded by the fact that $\alpha_j\, |\mathcal{H}_j^h w_j|_{H^1(\Omega_j)}^2 = |w_j|_{S_j}^2$ for $j \in \mathcal{I}_{\text{FEM}}$ and $\alpha_j\, |\mathcal{H}_j^h w_j|_{H^1(\Omega_j)}^2 \lesssim |w_j|_{S_j}^2$ for $j \in \mathcal{I}_{\text{BEM}}$ (by Corollary 1.94). □

2.6.2 The Case $Q = M_{\mathrm{sD}}^{-1}$

To complete the estimate from Lemma 2.45, we still need to bound $|P_D\, z_w|_S$. The next lemma stems from [KW01, Lemma 4.8].

Lemma 2.105. *Let z_w be defined as in Lemma 2.44 and let $Q = M_{\mathrm{sD}}^{-1}$. Then*

$$|P_D\, z_w|_S \;\le\; |P_D\, w|_S \qquad \forall w \in W.$$

Proof. Using identity (2.67), i.e., $P_D^\top S\, P_D \;=\; B^\top M_{\mathrm{sD}}^{-1} B \;=\; B^\top Q\, B$, and Lemma 2.44 we have $|P_D\, z_w|_S^2 \;=\; \|B\, z_w\|_Q^2 \;\le\; \|B\, w\|_Q^2 \;=\; |P_D\, w|_S^2$. □

The next theorem estimates the condition number of classical an all-floating FETI/BETI.

Theorem 2.106. *Let Assumptions 2.53–2.56 and 2.99–2.102 hold. Then, for the classical or the all-floating FETI/BETI method with the scaled Dirichlet preconditioner M_{sD}^{-1} and with the choice $Q = M_{\mathrm{sD}}^{-1}$,*

$$\kappa(P\, M_{\mathrm{sD}}^{-1} P^\top F_{|\tilde{U}_{\mathrm{ad}}}) \;\le\; C\; \max_{i=1}^{s}(1 + \log(H_i / h_i))^2,$$

where the constant C depends only on the uniform constants from the mentioned assumptions. If the subdomain meshes are not quasi-uniform (but still shape-regular), the analogous bound holds but then h_i has to be replaced by the minimal element diameter of $\mathcal{T}^h(\Omega_i)$.

Proof. The estimate follows immediately by combining Lemmas 2.45 and 2.103–2.105. □

2.6.3 Diagonal Choice of Q

An implementation of the FETI/BETI method with Q a diagonal matrix is of course much easier. The following choice, proposed and analyzed by Klawonn and Widlund [KW01], still gives a robust method with respect to coefficient jumps.

Definition 2.107. For each node $x^h \subset \Gamma^h \cup \Gamma_D^h$ we define

$$q_i(x^h) := \begin{cases} (1 + \log(H_i / h_i))\, \dfrac{h_i^{d-1}}{H_i} & \text{if } x^h \text{ lies on a subdomain facet,} \\ h_i^{d-2} & \text{else.} \end{cases}$$

Furthermore, we define the operator $Q_{\mathrm{diag}} : U^* \to U$ by

$$(Q_{\mathrm{diag}}\, \mu)_{ij}(x^h) := \min(\rho_i(x^h), \rho_j(x^h))\, q_{ij}(x^h)\, \mu_{ij}(x^h) \qquad \text{for } \mu \in U^*, \quad (2.91)$$

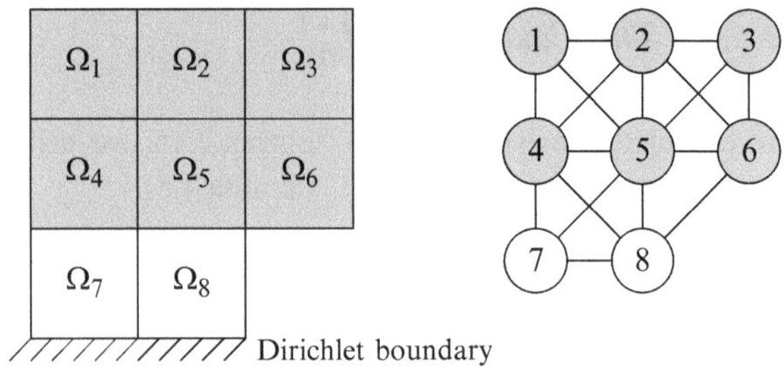

Fig. 2.8 *Left*: Subdomains with Dirichlet boundary. *Right*: Corresponding connectivity graph

where $q_{ij}(x^h) = \min(q_i(x^h), q_j(x^h))$ and (in case of the all-floating formulation)

$$(Q_{\mathrm{diag}}\,\mu)_{iD}(x^h) \;:=\; \rho_i(x^h)\,q_i(x^h)\,\mu_{iD}(x^h) \qquad \text{for } \mu \in U^*, \tag{2.92}$$

cf. [KW01, (4.14)] and [Of06, Pec08b].

Note that if $H_i \precsim H_j$ and $h_i \precsim h_j$ for neighboring subdomains Ω_i and Ω_j, then we have also $q_i(x^h) \precsim q_j(x^h)$. The operator Q_{diag} mimics the action of M_{sD}^{-1} when restricted to range(G), and it will be better understood in the proof of Lemma 2.109 below, where we analyze FETI/BETI with $Q = Q_{\mathrm{diag}}$. Note that if the coefficient is globally constant, we may also choose $Q = I$, see Remark 2.111.

Remark 2.108. For $Q = Q_{\mathrm{diag}}$, let us investigate the structure of the matrix $G^\top Q\,G$ that appears in the projections P and $P^\top$. Consider the connectivity graph whose nodes correspond to the subdomains Ω_i with an edge between two nodes whenever the corresponding subdomains are neighboring, cf. Fig. 2.8. Recall that $G = B\,R$ and that $R : Z \to \ker S$ with

$$Z \;=\; \prod_{i=1}^{s} \mathbb{R}^{\dim(\ker(S_i))}.$$

We can think of elements from Z as discrete functions on the nodes of the connectivity graph which satisfy homogeneous boundary conditions at the nodes which correspond to the non-floating subdomains, cf. Fig. 2.8. Using the definition of the jump operator B, we find that

$$\langle G^\top Q\,G\,y, z\rangle \;=\; \sum_{\substack{i>j\\ \Gamma_{ij}\neq\emptyset}} (y_i - y_j)\underbrace{\Big(\min(\alpha_i,\alpha_j)\sum_{x^h\in\Gamma_{ij}^h} q_{ij}(x^h)\Big)}_{=:\,\beta_{ij}}(z_i - z_j) \qquad \forall y, z \in Z.$$

In the all-floating formulation, we have to add

$$\sum_{i=1}^{s} y_i \underbrace{\left(\alpha_i \sum_{x^h \in \partial\Omega_i^h \cap \Gamma_D} q_i(x^h)\right)}_{=:\beta_{iD}} z_i \, .$$

This bilinear form corresponds to the (sparse) matrix induced by the graph Laplacian (see e.g., [Fie73]) where we assign each edge between node i and j in $\mathcal{G}$ the weight β_{ij}. As the subsequent analysis will show, the operator $(G^\top Q\, G)^{-1}$, as such solving a discrete Laplace problem on the connectivity graph, acts as a coarse problem for the FETI/BETI algorithm. In three dimensions, under Assumption 2.53, each subdomain face $\mathcal{F}_i$ contains $\mathcal{O}((H_i/h_i)^2)$ nodes and each subdomain edge $\mathcal{E}_i$ contains $\mathcal{O}(H_i/h_i)$ nodes. Hence,

$$\beta_{ij} \eqsim \begin{cases} \min(\alpha_i, \alpha_j)\,(1+\log(H_i/h_i))\, H_i & \text{if } \Omega_i \text{ and } \Omega_j \text{ share a subdomain face,} \\ \min(\alpha_i, \alpha_j)\, H_i & \text{if } \Omega_i \text{ and } \Omega_j \text{ share only a subd. edge,} \\ \min(\alpha_i, \alpha_j)\, h_i & \text{if } \Omega_i \text{ and } \Omega_j \text{ share only a subd. vertex.} \end{cases}$$

In two dimensions,

$$\beta_{ij} \eqsim \begin{cases} \min(\alpha_i, \alpha_j)\,(1+\log(H_i/h_i)) & \text{if } \Omega_i \text{ and } \Omega_j \text{ share a subdomain edge,} \\ \min(\alpha_i, \alpha_j) & \text{if } \Omega_i \text{ and } \Omega_j \text{ share only a subd. vertex.} \end{cases}$$

We observe that vertex connections in three dimensions are weighted weaker than others, and that connections between subdomains with large coefficients are in general weighted stronger than others.

The following lemma is essentially [KW01, Lemma 4.10] (there stated for the classical FETI method).

Lemma 2.109. *Let Assumptions 2.53–2.56 and 2.99–2.102 hold. Then, for $Q = Q_{\text{diag}}$,*

$$|P_D\, z_w|_S^2 \lesssim \max_{i=1}^{s}(1+\log(H_i/h_i))^2\, |w|_S^2 \qquad \forall w \in W^\perp .$$

Proof. Note that z_w is constant on each subdomain and vanishes on the non-floating subdomains. We denote the components by z_i. Using inequality (2.88) from the proof of Lemma 2.104 we obtain

$$\begin{aligned} |P_D\, z_w|_S^2 &\lesssim \sum_{\mathcal{G}_i \subset \Gamma} \sum_{j \in \mathcal{N}_{\mathcal{G}_i}} \min(\alpha_i, \alpha_j) |\mathcal{H}_i^h(\theta_{\mathcal{G}_i}(z_i - z_j))|_{H^1(\Omega_i)}^2 + \sum_{\mathcal{G}_i \subset \Gamma_D} \alpha_i |\mathcal{H}_i^h(\theta_{\mathcal{G}_i} z_i)|_{H^1(\Omega_i)}^2 \\ &\lesssim \sum_{\mathcal{G}_i \subset \Gamma} \sum_{j \in \mathcal{N}_{\mathcal{G}_i}} \min(\alpha_i, \alpha_j) |\mathcal{H}_i^h \theta_{\mathcal{G}_i}|_{H^1(\Omega_i)}^2 |z_i - z_j|^2 + \sum_{\mathcal{G}_i \subset \Gamma_D} \alpha_i |\mathcal{H}_i^h \theta_{\mathcal{G}_i}|_{H^1(\Omega_i)}^2 |z_i|^2 . \end{aligned}$$

By the subdomain facet estimate from Lemma 2.76, Definition 2.107, and the fact that a subdomain facet contains $\mathcal{O}((H_i/h_i)^{d-1})$ nodes, we can conclude that

$$|\mathcal{H}_i^h \theta_{\mathcal{F}_i}|^2_{H^1(\Omega_i)} \lesssim (1 + \log(H_i/h_i))\, H_i^{d-2} \lesssim \sum_{x^h \in \mathcal{F}_i} q_i(x^h).$$

For the remaining globs, we can conclude from Lemma 2.74, Definition 2.107, and the fact that a subdomain edge contains $\mathcal{O}(H_i/h_i)$ nodes, we can conclude that

$$|\mathcal{H}_i^h \theta_{\mathcal{V}_i}|^2_{H^1(\Omega_i)} \lesssim h_i^{d-2} \lesssim \sum_{x^h \in \mathcal{V}_i} q_i(x^h)$$

$$|\mathcal{H}_i^h \theta_{\mathcal{E}_i}|^2_{H^1(\Omega_i)} \lesssim H_i \lesssim \sum_{x^h \in \mathcal{E}_i} q_i(x^h) \qquad \text{if } d = 3.$$

Since $q_i(x^h) \eqsim q_j(x^h)$ for $x^h \in \mathcal{G}_{ij}$, and since

$$(B\, z_w)_{ij}(x^h) = \pm|z_i - z_j|, \qquad (B\, z_w)_{iD}(x^h) = z_i,$$

we obtain (comparing with Definition 2.107) that

$$|P_D\, z_w|^2_S \lesssim \|B\, z_w\|^2_{Q_{\text{diag}}} \leq \|B\, w\|^2_{Q_{\text{diag}}}, \tag{2.93}$$

where in the last step we have used Lemma 2.44. The particular choice of Q_{diag} actually stems from the estimates above.

In order to bound $\|B\, w\|^2_{Q_{\text{diag}}}$ in terms of $|w|^2_S$, we sort the contributions with respect to the globs. Using the definition of Q_{diag} and the quasi-uniformity of $\mathcal{T}^h(\Omega_i)$ we obtain

$$\begin{aligned}
&\|B\, w\|^2_{Q_{\text{diag}}} \\
&\lesssim \sum_{\mathcal{G}_i \subset \Gamma} \sum_{j \in \mathcal{N}_{\mathcal{G}_i}} \min(\alpha_i, \alpha_j) \sum_{x^h \in \mathcal{G}_i{}^h} q_{ij}(x^h) |w_i(x^h) - w_j(x^h)|^2 + \sum_{\mathcal{G}_i \subset \Gamma_D} \alpha_i \sum_{x^h \in \mathcal{G}_i{}^h} q_i(x^h) |w_i(x^h)|^2 \\
&\lesssim \sum_{\mathcal{G}_i \subset \Gamma \cup \Gamma_D} \alpha_i \sum_{x^h \in \mathcal{G}_i{}^h} q_i(x^h)\, |w_i(x^h)|^2 \\
&\lesssim \sum_{\mathcal{G}_i \subset \Gamma \cup \Gamma_D} \alpha_i\, q_{i|\mathcal{G}_i{}^h}\, h_i^{-d_{\mathcal{G}_i}}\, \|w\|^2_{L^2(\mathcal{G}_i)},
\end{aligned} \tag{2.94}$$

where $\mathcal{G}_i{}^h$ is as in Definition 2.50 and $q_{i|\mathcal{G}_i{}^h}$ denotes the constant value of q_i on the nodes on $\mathcal{G}_i$. Note that in the classical formulation, the globs on Γ_D can be left out. Combining this estimate with the discrete trace inequality from Lemma 2.69 yields

$$\|B\, w\|^2_{Q_{\text{diag}}} \lesssim \sum_{\mathcal{G}_i \subset \Gamma \cup \Gamma_D} \alpha_i\, r_{\mathcal{G}_i}(H_i, h_i) \Big(|\mathcal{H}_i^h w_i|^2_{H^1(\Omega_i)} + H_i^{-2}\, \|\mathcal{H}_i^h w_i\|^2_{L^2(\Omega_i)} \Big), \tag{2.95}$$

where

$$r_{\mathcal{G}_i}(H_i, h_i) := q_{i|\mathcal{G}_i^h} h_i^{-d_{\mathcal{G}_i}} \sigma^{d-d_{\mathcal{G}_i}}\left(\tfrac{H_i}{h_i}\right) H_i^{2+d_{\mathcal{G}_i}-d} .$$

In the all-floating formulation, we obtain the desired bound

$$|P_D w|_S^2 \lesssim \|B w\|_{Q_{\text{diag}}}^2 \lesssim \max_{i=1}^{s}(1+\log(H_i/h_i))^2 |w|_S^2 \qquad \forall w \in W^\perp$$

(by Poincaré's inequality from Lemma 2.61) if

$$r_{\mathcal{G}_i}(H_i, h_i) \lesssim (1+\log(H_i/h_i))^2. \tag{2.96}$$

For quasi-uniform fine and coarse triangulations, condition (2.96) is even necessary. A short computation reveals that (2.96) holds if and only if

$$\begin{aligned}
(1+\log(\tfrac{H_i}{h_i}))\, \frac{h_i^{d-1}}{H_i} \lesssim q_i(x^h) &\lesssim (1+\log(\tfrac{H_i}{h_i}))^2\, \frac{h_i^{d-1}}{H_i} && \text{if } x^h \in \mathcal{F}_i, \\
h_i \lesssim q_i(x^h) &\lesssim (1+\log(\tfrac{H_i}{h_i}))\, h_i && \text{if } x^h \in \mathcal{E}_i,\ d=3, \\
h_i^{d-2} \lesssim q_i(x^h) &\lesssim (1+\log(\tfrac{H_i}{h_i}))^{d-1}\, h_i^{d-2} && \text{if } x^h = \mathcal{V}_i,
\end{aligned}$$

where we have included the lower bounds for $q_i(x^h)$ that we have used in the beginning of the proof. In the classical formulation, we need to be able to apply a discrete Poincaré inequality in each non-floating subdomain (cf. Assumption 2.56) at the cost of a factor of $(1 + \log(H_i/h_i))$. Hence, this factor must be taken away from the upper bounds above unless all non-floating subdomains have a subdomain face in common with the Dirichlet boundary. In any case, we see that the respective bounds for $q_i(x^h)$ hold for the choice made in Definition 2.107, which concludes the proof. □

Theorem 2.110. *Let Assumptions 2.53–2.56 and 2.99–2.102 hold. Then, for the classical and the all-floating FETI/BETI method with the scaled Dirichlet preconditioner M_{sD}^{-1} and with the choice $Q = Q_{\text{diag}}$ (cf. Definition 2.107),*

$$\kappa(P M_{\text{sD}}^{-1} P^\top F_{|\tilde{U}_{\text{ad}}}) \le C \max_{i=1}^{s}(1+\log(H_i/h_i))^2 .$$

where the constant C depends only on the uniform constants from the mentioned assumptions.

Proof. The estimate follows immediately by combining Lemmas 2.45, 2.103, 2.104, and 2.109. □

Opposed to Theorem 2.106, the statement of Theorem 2.110 does not generalize to the case of non quasi-uniform meshes, as we heavily used the quasi-uniformity in the proof of Lemma 2.109.

Remark 2.111. Let us suppose that the coefficients are locally quasi-monotone, i.e., for each glob $\mathscr{G}_{ij}$ we can find an admissible face path $\mathscr{F}_{k_1,k_2}, \mathscr{F}_{k_2,k_3}, \ldots, \mathscr{F}_{k_{m-1},k_m}$ with

$$\min(\alpha_i, \alpha_j) = \alpha_{k_1} \leq \alpha_{k_2} \leq \ldots \leq \alpha_{k_m} = \max(\alpha_i, \alpha_j).$$

In that situation, the weights $q_{ij}(x^h)$ for subdomain vertices (and edges if $d = 3$) on the interface Γ can be decreased arbitrarily, see also the remarks after Theorems 4.11 and 5.7 in [KW01]. As a consequence, if the coefficient is globally constant (or at least $\alpha_i \approx \alpha_j$ for all neighboring subdomains Ω_i, Ω_j), and if the subdomain decomposition and the global mesh are quasi-uniform ($h_i \approx h$ and $H_i \approx H$), then the matrix Q can be chosen as the identity matrix or any multiple thereof.

2.6.4 *Alternative Scalings*

In this short section we discuss other scalings than the coefficient scaling, which is assumed in the theory above (cf. Assumption 2.100). Note a further scalings (not discussed below) based on eigensolves can be found in [DW12b].

2.6.4.1 Multiplicity Scaling

If we use the multiplicity scaling (which also effects Q_{diag}, cf. Definition 2.107), the statements of Theorems 2.106 and 2.110 remain true under the stated assumptions, but the condition number bounds have to be multiplied by a factor of

$$\max_{i=1}^{s} \max_{j \in \mathscr{N}_i} \frac{\alpha_i}{\alpha_j},$$

i.e., the maximal jump *between* subdomains, where $\mathscr{N}_i$ contains the indices of the subdomains neighboring Ω_i, cf. Definition 2.12, This is seen from the proofs of Lemmas 2.104 and 2.109 by using that $\delta_j^\dagger(x^h) \leq 1$ as well as the simple fact that

$$\alpha_i \leq \left(\max_{i=1}^{s} \max_{k \in \mathscr{N}_i} \frac{\alpha_i}{\alpha_k} \right) \alpha_j$$

whenever $j \in \mathscr{N}_i$.

2.6.4.2 On the Stiffness Scaling and on Effects of Varying Coefficients

If one has no a priori knowledge on the coefficient α_i, a common choice in FETI is the stiffness scaling, where $\rho_i(x^h)$ is set to the diagonal entry of the stiffness matrix $\mathbf{K}_i$ corresponding to the node x^h. Assuming an isotropic coefficient $\mathscr{A} = \alpha I$, the stiffness scaling is given by

$$\rho_i(x^h) = \int_{\omega_{i,x^h}} \alpha\, |\nabla \varphi_{i,x^h}|^2\, \mathrm{d}x,$$

where ω_{i,x^h} is the union of all elements τ from $\mathscr{T}^h(\Omega_i)$ such that $x^h \in \overline{\tau}$. Similar choices are the local average

$$\rho_i(x^h) = \frac{1}{\mathrm{meas}_{\omega_{i,x^h}}} \int_{\omega_{i,x^h}} \alpha\, \mathrm{d}x$$

and the local maximum

$$\rho_i(x^h) = \operatorname*{ess.sup}_{x \in \omega_{i,x^h}} \alpha(x).$$

If $\alpha_{|\Omega_i} = \alpha_i$ is constant, the latter two choices reproduce the coefficient scaling $\rho_i(x^h) = \alpha_i$. For the stiffness scaling, $\rho_i(x^h) \approx h_i^{d-2}\alpha_i$ (if $\mathscr{T}^h(\Omega_i)$ is quasi-uniform) which may look promising. However, in the presence of rough (or ragged) interfaces as they appear in METIS partitions, the stiffness scaling leads to extremely poor convergence of FETI type methods as it has been demonstrated in [KRW08, Sect. 5] (the term *ρ-scaling* therein corresponds to the *coefficient scaling* in this book).

In the analysis, we used that (for the coefficient scaling) $\delta_j^\dagger$ is constant at the nodes of each glob, which is not anymore true for the stiffness scaling. In the case of rough interfaces, the function $\rho_i \in V^h(\partial\Omega_i)$ becomes *oscillatory*, i.e.,

$$\frac{\rho_i(x^h) - \rho_i(y^h)}{\rho_i(x^h)} \gg \frac{|x^h - y^h|}{H_i}.$$

Typically, the weighted counting functions $\delta_j^\dagger$ have the same property, which is at high probability the reason behind the poor convergence.

Even if the interfaces are smooth, the phenomenon of an oscillatory function $\delta_j^\dagger$ can occur when the coefficient $\mathscr{A} = \alpha I$ is mildly varying in Ω_i. By this we mean that $\mathrm{ess.sup}_{x,y\in\Omega_i} \frac{\alpha(x)}{\alpha(y)}$ is relatively small, but the coefficient may change from element to element.

Summarizing, in case of rough interfaces or a mildly varying (isotropic) coefficient, provided that the subdomain meshes are quasi-uniform, a good choice is

$$\rho_i(x^h) = \max_{x^h \in \partial\Omega_i^h} k_{i,x^h} \qquad \text{or} \qquad \rho_i(x^h) = \max_{x^h \in \Omega_i^h} k_{i,x^h},$$

where k_{i,x^h} denotes the diagonal entry of the stiffness matrix $\mathbf{K}_i$ at node x^h. For an analysis with mildly varying coefficients (but smooth interfaces) see Sect. 3.2. For a sound theory of FETI-DP methods for the case of rough interfaces in two dimensions (using the coefficient scaling) see [KRW08].

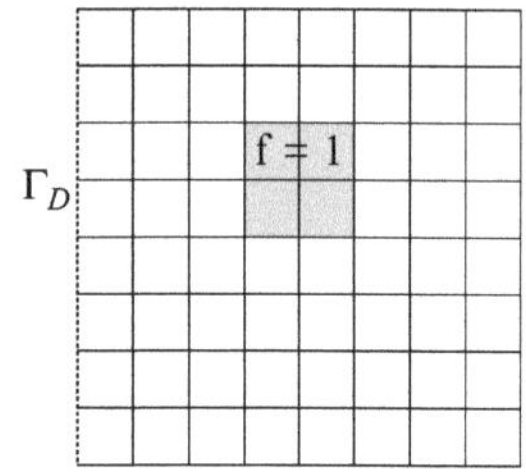

Fig. 2.9 Setup for the experiments on the unit square with a homogeneous coefficient

2.7 Numerical Results

In this section, we present some numerical results for one-level FETI/BETI methods for a two-dimensional model problem with and without coefficient jumps. This is mainly to give the reader an impression on how small the condition numbers and the number of PCG iterations actually are. The main implementation was done in C^{++}. The FEM stiffness matrices and the coarse matrix were factorized using PARDISO [PAR05, SG04, SG06]. For the boundary element method we have used Olaf Steinbach's Fortran package OSTBEM [Ste00]. The condition numbers are estimated using the Lanczos method, see Remark 1.51. Mainly interested in verifying the theoretical results of this chapter, we have not used any data-sparse approximation of the boundary element matrices.

Note that computational results for FETI methods (including the case of linear elasticity) can be found in [FR91, FR91, LP98, RFTM99, Rhe02]; for all-floating BETI methods see [Of06, Of08].

Unit Square – Homogeneous Coefficient. We consider the unit square $\Omega = (0, 1)^2$, subdivided into 64 equally-sized square-shaped subdomains, with homogeneous Dirichlet boundary conditions on the left side Γ_D, and homogeneous Neumann boundary conditions on the rest of $\partial\Omega$. The source term f is chosen to be zero except for the four shaded subdomains in Fig. 2.9, and the coefficient α is set uniformly to one.

Tables 2.3 and 2.4 show the results for FETI and FETI/BETI, respectively. There, the column entitled "Lagr. mult." indicates number of Lagrange multipliers (additional multipliers enforcing the Dirichlet boundary conditions in the all-floating method are not counted). For simplicity, H denotes the height/width of the subdomain. In the columns entitled "PCG" we give the number of PCG steps needed to get a reduction of $\varepsilon = 10^{-8}$ in the residual, and the columns entitled "cond." show the estimated condition number using the Lanczos method. We see that the condition numbers of the preconditioned systems behave as predicted by the theory. From the first column in the two tables one can observe the reduction in the global DOFs when using the boundary element method. In Table 2.5 we demonstrate the scalability, i.e., the robustness with respect to the number of subdomains.

Table 2.3 Unit square with homogeneous coefficient; classical one-level vs. all-floating FETI method; 64 subdomains

				Std. one-level		All-floating	
Global DOFs	Lagr. mult.	Local DOFs	H/h	PCG	Cond.	PCG	Cond.
289	406	9	2	9	1.67	8	1.40
1,089	630	25	4	11	2.20	10	1.88
4,225	1,078	81	8	13	2.97	12	2.43
16,641	1,974	289	16	16	3.92	14	3.15
66,049	3,766	1,089	32	18	5.05	16	4.05
263,169	7,350	4,225	64	21	6.33	18	5.12
1,050,625	14,518	16,641	128	23	7.77	19	6.36
4,198,403	28,854	66,049	256	24	9.38	21	7.76
16,785,409	57,526	263,169	512	25	11.15	23	9.33

Table 2.4 Unit square with homogeneous coefficient; classical one-level vs. all-floating FETI/BETI method; 64 subdomains (60 BEM, 4 FEM)

					Std. one-level		All-floating	
Global DOFs	Lagr. mult.	FEM loc. DOFs	BEM loc. DOFs	H/h	PCG	Cond.	PCG	Cond.
229	406	9	8	2	9	1.65	9	1.64
549	630	25	16	4	10	1.91	9	1.67
1,285	1,078	81	32	8	13	2.58	11	2.08
3,141	1,974	289	64	16	15	3.44	13	2.72
8,389	3,766	1,089	128	32	18	4.48	16	3.54
25,029	7,350	4,225	256	64	20	5.68	18	4.54
82,885	14,518	16,641	512	128	23	7.03	20	5.71
296,901	28,854	66,049	1,024	256	24	8.55	22	7.05
1,118,149	57,526	263,169	2,048	512	25	10.24	23	8.55

Table 2.5 Unit square with homogeneous coefficient; classical vs. all-floating FETI method; fixed ratio $H/h = 32$; fixed number 1,089 of local FEM DOFs, varying number of subdomains

		Std. one-level		All-floating	
Number of subdomains	Global DOFs	PCG	Cond.	PCG	Cond.
64	66,049	18	5.049	16	4.045
256	263,169	18	5.055	16	4.064
1,024	591,361	18	5.055	16	4.064
4,096	1,050,625	18	5.053	15	4.057

Unit Square – Heterogeneous Coefficient. In this example we consider the unit square $(0, 1)^2$ with the same partitioning as in Fig. 2.9, but we choose the coefficient α and the source f according to Fig. 2.10, left. The Dirichlet boundary conditions read

$$u(x_1, x_2) = 8\,x_2(1 - 8\,x_2) \qquad \text{for } (x_1, x_2) \in \Gamma_D = \{(0, x_2) : x_2 \in (0, 1)\}.$$

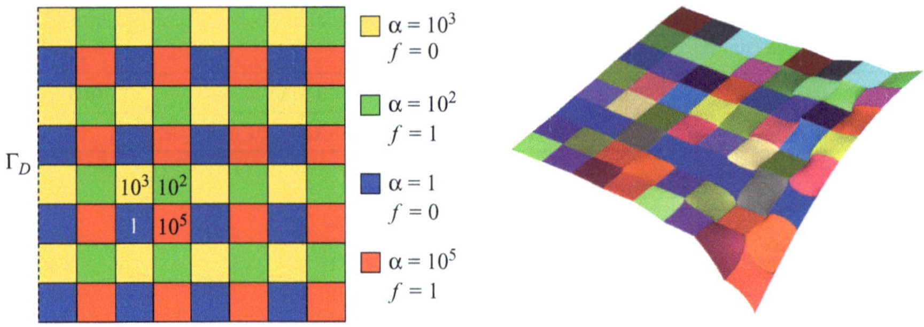

Fig. 2.10 *Left*: Setup of unit square with heterogeneous coefficient. *Right*: Visualization of the solution u via the graph $(x_1, x_2, u(x_1, x_2))$; different colors indicate different subdomains

On the remainder of $\partial\Omega$ we impose homogeneous Neumann boundary conditions.

Tables 2.6 and 2.7 show the number of PCG steps and the estimated condition number for the classical one-level and all-floating FETI and FETI/BETI method, respectively. In the second case, the BEM subdomains are exactly those where $f = 0$. The numbers in the tables demonstrate the robustness with respect to the heterogeneous coefficient, which would not be the case without the careful scalings in B_D and Q. Figure 2.10, right displays the solution u to the problem. We see that in the areas with large coefficients the solution is relatively flat.

2.8 Other PDEs and Other Discretization Spaces

The theory of the preceding sections carries over immediately from P^1 to Q^1 elements, i.e., bilinear quadrilateral or trilinear hexahedral elements. In this section, however, we give a brief overview (mostly in form of references) on FETI/BETI and related methods for PDEs other than the potential equation, and for discretizations other than piecewise linear FEM/BEM. For DD methods other than substructuring methods, we refer to [Mat08, TW05]. In Sect. 2.8.1 we discuss the influence of the discretization on the coupling (interconnecting), whereas Sect. 2.8.2 treats the change of the PDE (possibly implying specific discretizations).

More developments than mentioned below are and will be documented in the proceedings of the international conferences on domain decomposition methods, see http://www.ddm.org/.

2.8.1 *Other Discretizations Spaces for H^1-Problems*

Let Ω be the computational domain, and $V(\Omega)$ a space of functions on Ω related to the problem, e.g., $V = H^1(\Omega)$. First, we treat the case of a conforming finite

Table 2.6 Unit square with heterogeneous coefficient; classical one-level vs. all-floating FETI, 64 subdomains

			Std. one-level		All-floating	
Global DOFs	Local DOFs	H/h	PCG	Cond.	PCG	Cond.
289	9	2	10	2.23	8	1.54
1,089	25	4	11	2.69	11	2.12
4,225	81	8	13	3.18	13	2.90
16,641	289	16	15	3.84	15	3.84
66,049	1,089	32	17	4.91	17	4.91
263,169	4,255	64	19	6.10	19	6.10
1,050,625	16,641	128	21	7.42	22	7.43
4,198,401	66,049	256	23	8.91	24	8.88
16,785,409	263,169	512	26	10.60	25	10.47

Table 2.7 Unit square with heterogeneous coefficient; classical one-level vs. all-floating FETI/BETI, 32 FEM, 32 BEM subdomains

				Std. one-level		All-floating	
Global DOFs	FEM loc. DOFs	BEM loc. DOFs	H/h	PCG	Cond.	PCG	Cond.
257	9	8	2	7	1.68	7	1.64
801	25	16	4	10	2.24	9	2.24
2,657	81	32	8	11	3.09	12	3.09
9,441	289	64	16	13	4.08	13	4.08
35,297	1,089	128	32	14	5.21	15	5.21
136,161	4,255	256	64	16	6.46	17	6.46
534,497	16,641	512	128	17	7.83	18	7.83
2,117,601	66,049	1,024	256	18	9.33	20	9.33
8,429,537	263,169	2,048	512	19	10.96	21	10.96

element space $V^h(\Omega) \subset V$ that fulfills Assumption 2.112 below. Other types of discretization spaces will be treated in Sects. 2.8.1.3–2.8.1.5.

Assumption 2.112. (i) The FE space $V^h(\Omega)$ is based on a triangulation $\mathscr{T}^h(\Omega)$ of Ω, i.e., to each mesh element $T \in \mathscr{T}^h$ there is an associated space V_T of shape functions and a set $\mathscr{N}_T \subset V_T^*$ of DOFs (nodal variables), cf. [BS02, Definition 3.1.1] and [Cia87].

(ii) Each DOF is associated to a (fine) vertex, edge, face, or element of the mesh.

(iii) The global space $V^h(\Omega)$ is composed of the local spaces V_T, where corresponding DOFs on the same entity (vertex, edge, face) are globally identified.

Assumption 2.112 holds for the case for high order H^1-conforming spaces (either of hierarchical nature as in Sect. 1.2.3.7, or for spectral elements; see e.g., [BM97, KS99, Sch98b]). Note that Assumption 2.112 holds as well for the $H(\text{curl})$-conforming Nédélec edge elements, the $H(\text{div})$-conforming Raviart-Thomas (Nédélec face) elements of any order, see, e.g., [RT77, Néd80, Néd86, Mon03, Zag06], and the mixed elements in [Sin08, PS11a, PS12a].

With Assumption 2.112 being fulfilled, we can define the interface (or skeleton) DOFs as those DOFs associated to (fine) vertices, edges, or faces lying on the (geometrical) interface Γ (or the skeleton Γ_S, respectively) and eliminate the remaining (interior) DOFs. This yields a discrete skeleton formulation. Furthermore, we can define the local restrictions to Ω_i or $\partial\Omega_i$, which leads to the local spaces W_i and the (discontinuous) product space W. Reinstalling the original continuity follows the finite element construction above. Let $\psi_{i,k}$ and $\psi_{j,\ell}$ be two DOFs on subdomain Ω_i and Ω_j, respectively, that are globally identified. Then the correct jump condition simply reads

$$\psi_{i,k}(w_i) - \psi_{j,\ell}(w_j) \;=\; 0,$$

and all these conditions together define the jump operator $B : W \to U^* = \mathbb{R}^M$. If we represent B as a matrix $\mathbf{B}$ with respect to the nodal FE basis, then $\mathbf{B}$ is again signed Boolean (see, e.g., [TK01, TV03]).

Corresponding boundary element spaces on the skeleton Γ_S can be easily derived by restricting $V^h(\Omega)$ to the skeleton. This restricted space is parameterized by the DOFs associated to Γ_S, and the coupling procedure is identical to the above one.

The following two sections deal with specific H^1-conforming discretizations.

2.8.1.1 H^1-Conforming Interface Concentrated FETI/BETI

The interface concentrated FETI method was introduced by Beuchler, Eibner, and Langer [BEL08] as a solver for the scalar potential equation in H^1, see also [LP08] for a generalization to interface concentrated FETI/BETI. The main idea is to employ a *boundary concentrated* FEM (cf. [KM03] and Sect. 1.2.3.8) in each FEM subdomain, such that the polynomial degree equals one on each subdomain boundary. See Fig. 2.11 for an illustration. Obviously, the coupling is the same as for the low-order FETI/BETI methods. Since it can be shown that the corresponding FEM Schur complement in each subdomain is spectrally equivalent to the Steklov-Poincaré operator (cf. [BEL08, Theorem 3.13]), provided that the boundary mesh is quasi uniform, all the theoretical results of Sect. 2.6 carry over immediately to the interface concentrated FETI/BETI. Numerical results can be found in [BEL08, LP08], and also in [Pec08b, Sect. 2.3].

2.8.1.2 High Order H^1-Conforming Spaces

There are two classes of high order elements. Spectral elements [BM97, KS99] have uniform polynomial degree throughout the domain. The associated DOFs are point evaluations at Gauss-Lobatto-Legendre points. Hierarchical high order elements (see e.g. [Dem07, Zag06]) allow for variable polynomial degree, but the DOFs are associated to vertices, edges, faces, or elements and not necessarily point evaluations, but typically averages.

Fig. 2.11 Example for an interface concentrated mesh for a partition of the unit square into four subdomains

For spectral elements of degree k, it was shown that Neumann-Neumann and FETI type methods (in two and three dimensions) lead to condition number bounds of the form

$$\kappa \leq C\,(1 + \log(k))^2,$$

cf. [TW05, Sect. 7.2] as well as [Pav97, Pav07, KPR08]. However, in three dimensions these bounds do not carry over to the hierarchical high order elements, even if spanning the same polynomial space, see [TW05, Sect. 7.5] (this is because the coupling relies on the underlying DOFs, which are different for the two approaches). Indeed, the convergence can slow down for the hierarchichal case. For numerical studies of Neumann-Neumann and FETI methods see also [TV03, TV04, TV06].

Remark 2.113. Instead of coupling directly the DOFs, one could also introduce a suitable set of interpolation points and couple function values there. This is possible for any basis of a high order H^1-conforming FE space. In that case the structure of B is more involved.

2.8.1.3 Mortar Discretizations

Mortar discretizations (see e.g. [BMP94]) are FEM or BEM on non-conforming meshes. The coupling is done by integrating the jumps on the coupling interface against test functions that act as Lagrange multipliers. Thus, mortar methods contain already the needed interconnecting. A FETI like jump operator can be derived using the mass matrices on the coupling interface. See [SK98, Ste01, KL05, Kim07, KDW08, Kim08a, Kim08c, Kim08b, KT09] for FETI and balancing type methods for such mortar discretizations. Moreover, see [Rou09] for a related method called FETI-2LM.

2.8.1.4 Discontinuous Galerkin Method

Discontinuous Galerkin methods (see e.g. [DE12, HW08a, Riv08]) typically work with spaces of discontinuous functions across elements. The coupling is performed by suitable jump terms that are added to the bilinear form. Balancing type methods for DG discretizations on conforming and non-conforming meshes have been thoroughly investigated by Dryja, Galvis, and Sarkis [DGS07, DGS08, DGS11, DGS12].

2.8.1.5 Isogeometric Analysis

For the discretizations in *isogeometric analysis* (cf, [HCB05, BBC$^+$06, CHB09]), Assumption 2.112 is not necessarily fulfilled. This is because for C^k-elements, the NURBS basis functions cannot be localized to individual elements, and thus it is difficult to associate the DOFs to a *geometric* interface. See however [BSEH11] for a localization by a change to the Bernstein basis. The recently introduced IETI method, cf. [KPJT12], covers the case of C^0-continuity across subdomain interfaces, which makes the coupling more natural. The case of C^k-continuity with $k > 0$ has been studied recently in [BCPS12], using the concept of *fat interfaces*.

2.8.2 *FETI and Balancing Type Methods for General PDEs*

This section consists mainly of an (incomplete) list of references for FETI and balancing type methods for a variety of different PDEs. Note that many of the references are on FETI-DP and BDDC methods, which will be dealt with later on in Chap. 5. In the following, we briefly touch advection-diffusion problems, and problems in continuum mechanics, acoustics, and electromagnetics.

2.8.2.1 Dynamic Problems and Advection-Diffusion Problems

In this section, we consider the modified PDE

$$-\text{div}(\alpha\,\nabla u) + \boldsymbol{\beta}\cdot\nabla u + \gamma\, u \;=\; f, \tag{2.97}$$

The Case $\boldsymbol{\beta} = \mathbf{0}$ and $\boldsymbol{\gamma} > \mathbf{0}$. Assume that $\boldsymbol{\beta} = 0$ and that $\gamma(x) \geq \gamma_0 > 0$ a.e. in Ω. This kind of problem occurs particularly in implicit time stepping for the dynamic problem

$$\frac{\partial u}{\partial t} - \text{div}(\alpha\nabla u) \;=\; f.$$

In each time step, one has to solve a problem of the form (2.97) with $\boldsymbol{\beta} = 0$ and $\gamma = 1/\Delta t$, where Δt is the time step.

Let W_i and W be the spaces from either the classical or the all-floating formulation and let $S_i : W_i \to W_i^*$ denote the corresponding (discrete) Steklov-Poincaré operators. On a FEM subdomain, S_i corresponds to the Schur complement of the matrix

$$\mathbf{K}_i + \mathbf{M}_i$$

where $\mathbf{K}_i$ is the stiffness matrix corresponding to $\int_\Omega \alpha \nabla u \cdot \nabla v \, dx$ and $\mathbf{M}_i$ the mass matrix corresponding to $\int_\Omega \gamma \, u \, v \, dx$. For a BEM subdomain, the operator S_i is given analogously to Sect. 1.3 but the fundamental solution in the boundary integral operators has to modified accordingly (see, e.g., [SS11]). Using $S := \operatorname{diag}(S_i)_{i=1}^s$, one derives the FETI/BETI saddle point formulation analogously to Sect. 2.2.1.1: find $(u, \lambda) \in W \times U$ such that

$$\begin{bmatrix} S & B^\top \\ B & 0 \end{bmatrix} \begin{bmatrix} u \\ \lambda \end{bmatrix} = \begin{bmatrix} g \\ 0 \end{bmatrix},$$

see also [FCM95, FM98, Tos01]. For convenience we split the local Steklov-Poincaré operators into the part corresponding to the stationary term $-\operatorname{div}(\alpha \nabla u)$ and a remainder,

$$S_i = S_i^K + S_i^M .$$

Observe that, even if S_i^K might have a non-trivial kernel, the operator S_i is always SPD, and so

$$\ker(S_i) = \{0\}.$$

That is, according to Definition 2.15, all subdomains are non-floating. Therefore, no projection is needed to eliminate the unknown u from the saddle point problem, and the resulting dual system simply takes the form

$$\text{find } \lambda \in U : \quad \underbrace{B S^{-1} B^\top}_{=:F} \lambda = \underbrace{B S^{-1} g}_{=:d},$$

where $S^{-1} = \operatorname{diag}(S_i^{-1})_{i=1}^s$. The solution λ is unique up to $\ker(B^\top)$. Hence, let $\widetilde{U} := U_{/\ker(B^\top)}$ and $\widetilde{U}^* = \operatorname{range}(B)$, then we can seek (formally) $\lambda \in \widetilde{U}$. Due to the lack of the projection, there is no coarse problem, and so using the (scaled) Dirichlet preconditioner only results in a method that is not scalable, i.e., it deteriorates when the number s of subdomains grows.

We now reinstall a coarse problem using an outer projection. This technique is also called deflation (cf. [VSMW01, KR12]) and was first used for a FETI method by Farhat, Chen, and Mandel [FCM95] (see also [FM98]) in the context of dynamic elasticity. Let R_i span the kernel of S_i^K and define $R = [R_i]_{i=1}^s$ and $G := Q B R$, where $Q : U^* \to U$ is a suitable SPD operator. The outer projector $P : \widetilde{U} \to \widetilde{U}$ is then given by

$$P := I - G(G^\top F G)^{-1} G^\top F,$$

Algorithm 8: FETI/BETI algorithm for a dynamic problem based on PCG

$\lambda^{(0)} = G(G^\top F\, G)^{-1}\, G^\top d$
$r^{(0)} = d - F\,\lambda^{(0)}$
$k = 0$
repeat
  $z^{(k)} = M_{\mathrm{sD}}^{-1}\, r^{(k)}$
  $s^{(k)} = P\, z^{(k)}$
  $p^{(k)} = s^{(k)} + \beta_{k-1}\, p^{k-1}$ where $\beta_{-1} = 0, \quad \beta_{k-1} = \frac{\langle r^{(k)}, s^{(k)}\rangle}{\langle r^{(k-1)}, s^{(k-1)}\rangle}$
  for $k > 0$
  $\lambda^{(k)} = \lambda^{(k)} + \alpha_k\, p^{(k)}$ where $\alpha_k = \frac{\langle r^{(k)}, s^{(k)}\rangle}{\langle \widehat{F}\, p^{(k)}, p^{(k)}\rangle}$
  $r^{(k+1)} = r^{(k)} - \alpha_k\, \widehat{F}\, p^{(k)}$
  $k = k + 1$
until *stopping criterion fulfilled for* $r^{(k)}$

cf. [FM98, Sect. 5] (where $Q = I$) and [Tos01, Sect. 3]. To assemble the (sparse) matrix corresponding to $(G^\top F\, G)$, one has to apply the operator to the kernel vectors. To do this efficiently in a parallel regime, one needs similar techniques as described in Remark 2.36. Since range(P) and range($I - P$) are F-orthogonal, we can split $\lambda = \lambda_0 + \widetilde{\lambda}$ where $\lambda_0 \in$ range($I - P$) and $\widetilde{\lambda} \in$ range(P). Due to the F-orthogonality,

$$\lambda_0 = G\,(G^\top F\, G)^{-1} G^\top d.$$

To solve for $\widetilde{\lambda}$, we apply a PCG to the (SPD) equation

$$P^\top F \widetilde{\lambda} = P^\top d$$

with initial value 0 and with the preconditioner $P\,M_{sD}^{-1}$, where

$$M_{\mathrm{sD}}^{-1} = B_D\, S\, B_D^\top, \qquad S = \mathrm{diag}(S_i)_{i=1}^s\,.$$

As for the algorithm in Sect. 2.3, one can rewrite this PCG algorithm as an iteration for the original variable λ, see Algorithm 8. Thanks to the choice of the initial value, one needs to perform only one projection step (cf. [FM98] and [Tos01, Lemma 3.1]).

Including Advection. A generalization of the above algorithm to the non-symmetric problem (2.97) with $\boldsymbol{\beta} \neq 0$ (but further assumptions on $\boldsymbol{\beta}$ and γ) can be found in [Tos01], where the PCG is replaced by a preconditioned GMRes method. For a BDDC method we refer to [TL08].

The Case $\boldsymbol{\beta} = 0, \gamma \geq 0$. Assume that on some subdomains $\gamma = 0$ and on the remaining ones $\gamma > 0$. This means that some of the Steklov-Poincaré operators already have kernels, and some do not. A method called *generalized FETI* has been proposed by Farhat and Mandel [FM97], see also [Mat08, Sect. 4.3]. It introduces a projection that includes both the existing kernels and the artificial ones.

Unification by Dual-Primal Methods. We note that using a dual-primal approach (see Chap. 5), no case distinction is necessary, at least for the symmetric problems, cf. Sect. 5.3.5.1.

Acceleration for Dynamic Problems. In [FCR94] it has been proposed to accelerate the iterative solves by recycling the Krylov space of previous time steps. This technique can be used for FETI, Neumann-Neumann, FETI-DP, and BDDC.

2.8.2.2 Continuum Mechanics

Linear Elasticity. We consider the primal formulation of linear elasticity in variational form, for simplicity with homogeneous Dirichlet boundary conditions: find the displacement $\boldsymbol{u} \in H^1(\Omega)^d$, $\boldsymbol{u}_{|\Gamma_D} = 0$ such that

$$\int_\Omega \boldsymbol{C}\, \boldsymbol{\varepsilon}(\boldsymbol{u}) : \boldsymbol{\varepsilon}(\boldsymbol{v})\, \mathrm{d}x = \int_\Omega \boldsymbol{f} \cdot \boldsymbol{v}\, \mathrm{d}x + \int_{\Gamma_N} \boldsymbol{t}_N \cdot \boldsymbol{v}\, \mathrm{d}s \qquad \forall \boldsymbol{v} \in H^1(\Omega)^d,\ \boldsymbol{v}_{|\Gamma_D} = 0,$$

where $\boldsymbol{\varepsilon}(\boldsymbol{v}) = \frac{1}{2}(\nabla \boldsymbol{v} + (\nabla \boldsymbol{v})^\top)$ is the linearized strain tensor and $\boldsymbol{C}$ the (linear) material tensor due to Hooke's law, such that $\boldsymbol{\sigma}(\boldsymbol{u}) = \boldsymbol{C}\, \boldsymbol{\varepsilon}(\boldsymbol{u})$ is the stress tensor. It depends on the Young modulus E and the Poisson ratio ν in the usual way, see e.g. [Bra01, SDH04] or [TW05, Sect. A.6.2]. Here, we assume that the Poisson ratio ν is bounded away from the incompressible limit $1/2$.

The classical FETI method was originally introduced for the above problem, cf. [FR91, FR94]. When using the continuous piecewise linear finite elements for the components of the displacement $\boldsymbol{u}$, the derivation of FETI is in large parts analogous to the presentation in Sect. 2.2. The crucial difference lies in the local kernels which can have dimension of 0 up to 6 in three dimensions. Here lies an advantage of all-floating (total) FETI (cf. [DHK06]), where the local kernel is always the space of rigid body modes,

$$\mathscr{RB} := \begin{cases} \left\{ \mathbf{a} + b \begin{bmatrix} x_2 \\ -x_1 \end{bmatrix} : \mathbf{a} \in \mathbb{R}^2,\ b \in \mathbb{R} \right\} & \text{for } d = 2, \\ \{ \mathbf{a} + \mathbf{b} \times \mathbf{x} : \mathbf{a},\ \mathbf{b} \in \mathbb{R}^3 \} & \text{for } d = 3. \end{cases}$$

For an analysis of FETI for linear elasticity see [KW00] and [TW05, Sect. 8.5]. Here, the key ingredient is a spectral equivalence between the stiffness matrix of linear elasticity and the stiffness matrix of the vector Laplacian, which holds at least for vectors orthogonal to the kernel of the elasticity matrix, cf. [KW00, Lemma 5]. Using that equivalence, the FETI analysis can in large parts be reduced to the scalar elliptic case of Sect. 2.6.

For the FETI-DP method (where the local kernels do not have to be known explicitly) we refer to [FLL+01, KR06, KW06, KR07b, KR10]. For the BETI method see [Of06, Of08, OS09]. For FETI(-DP) and BDDC for mortar discretizations of linear elasticity see [Kim08a, Kim08c, Kim08b]. Finally, for a novel approach using the Smith factorization, see [CDNQ12].

Plate and Shell Problems. The FETI method was generalized for plate and shell problems by Farhat, Mandel, and Tezaur, see [MTF99, FM98]. Without a special treatment, however, the convergence is not satisfying, and a further outer projection of the Lagrange multipliers is necessary, see [FM98, FCMR98]. Due to the introduction of this second level, the resulting method is now called FETI-2. Note also that this method was important for the development of the dual-primal methods (cf. Chap. 5).

FETI-DP and BDDC methods for plate problems can be found in [FLL+01, BCLP10]. For an analysis see also [Bre03b].

Almost Incompressible Elasticity. In the almost incompressible case of elasticity, the Poisson ratio is close to $1/2$, which needs special treatment. Here we refer to [KRW07, PWZ10, GKR12], see also the related papers [DW09, DW10].

Nonlinear Elasticity and Contact Problems. There are three major sources of nonlinearities in solid mechanics:

(i) *Geometric nonlinearities* (due to large deformations),
(ii) *Nonlinear material laws* (other than Hooke's linear law),
(iii) *Nonlinearities due to contact* (introducing active or inactive constraints).

FETI type methods for contact problems can, e.g., be found in [BDS08, DHK+05, DKV+10, HKD04, JKR12]. For FETI type methods for nonlinear material laws in biomechanics, we refer to [BKRS08, BBK+09, KNRV11].

The Stokes Problem. For FETI-DP and BDDC for Stokes we refer to [PW02, Li05, LW06a, KLP10a, KLP10b, KL10, ŠSB+11], see also [TW05, Sect. 9.4.2]. For a novel approach using the Smith factorization, see [DNR08, DNR09].

Porous Media Flow. For BDDC methods on mixed and hybrid discretizations of porous media flow problems, also called Darcy's problem (cf. [TW05, Sect. A.7.2]), we refer to [Tu05, Tu07a, Tu11, Sou12].

2.8.2.3 Acoustic Scattering

The acoustic scattering problem is governed by the (scalar) Helmholtz equation

$$-\Delta u - k^2 u = f,$$

where k is the wave number. Besides Dirichlet and Neumann conditions, one often considers non-reflecting boundary conditions modeling waves that are only outgoing. Besides the "standard" difficulties of this indefinite problem (including a proper discretization), there are more difficulties when considering non-overlapping domain decomposition: local Dirichlet and Neumann problems can be unsolvable, when k^2 hits an eigenvalue of the corresponding local Helmholtz problem. A remedy was proposed by Farhat, Macedo, and Tezaur [FMT99], see also [FML00a]. The local problems are consistently supplemented with Robin boundary conditions, which guarantee the solvability of the local problems. This method is now called FETI-H. The dual-primal generalization, FETI-DPH, was introduced later in [FLLA05, FATL05]. The generalization to BEM discretizations ("BETI-H") was investigated by Steinbach and Windisch, see [Win10, SW11a, SW11b]. For a related method called FETI-2LM method see [FML$^+$00b].

2.8.2.4 Electromagnetic Problems

Eddy current problems are governed by the equation

$$\mathbf{curl}(\alpha\, \mathbf{curl}\, \boldsymbol{u}) + \beta \boldsymbol{u} \;=\; \boldsymbol{f}\,,$$

with α, $\beta > 0$, where $\boldsymbol{u}$ is a vector field in $H(\mathbf{curl})$, usually discretized by Nédélec edge elements. Neumann-Neumann and FETI type methods have been first investigated by Toselli and coworkers [Tos99, Tos00, RT01, TK01, Tos06]. The results were further refined by Dohrmann and Widlund [DW12a, DW12b].

In electromagnetic scattering problems, the coefficient β above is negative. For a BETI approach on such kind of problems we refer to [Win10, SW12].

Chapter 3
Multiscale Problems

In the analysis in Chap. 2, we have assumed that the coefficient $\mathscr{A}$ is isotropic and constant in each subdomain, cf. Assumption 2.99 in Sect. 2.6. In this chapter, we consider *varying* (but still isotropic) coefficients that are not necessarily *resolved* by the subdomain partition. Such situations can easily occur when applying a mesh partitioner. A typical application with highly varying coefficients is heterogeneous porous media flow (see e.g. [GLS07, GE10a]). For particular coefficients, we are able to give benign condition number bounds of FETI that are robust with respect to the heterogeneities. The theory of this chapter has evolved from a series of papers by Rob Scheichl and the author [PS08, PS09, PS11c, PS11b, PS12b] (see also [LP06]). All the key ideas can already be found there, but here we have tried to give a coherent presentation, and also to refine the estimates. The only other known *theoretical* works on iterative substructuring methods for such non-resolved heterogeneous problems seem to be [DS11, GKR12], while suitable methods (and choices of scalings, etc.) were already defined earlier, see e.g. [RF98a, RF99, BDF$^+$00, KR07b]. Note that we consider pure FETI methods throughout the chapter, but the whole analysis generalizes to FETI/BETI straightforwardly (with constant coefficients in the BEM subdomains).

The chapter is organized as follows. In Sect. 3.1 we give an introduction to multiscale problems themselves. Furthermore, we discuss the coefficient model for our subsequent analysis: it is the superposition of a piecewise constant coefficient with possibly high variation (not necessarily resolved by the subdomains) and some noise on the element level of low variation. In Sect. 3.2, we give a rough condition number bound which is comparable to that of Theorem 2.106 (see Sect. 2.6) but it contains the maximal variation of the coefficient in each subdomain. For many coefficient distributions this bound is overly pessimistic. In Sect. 3.3.5, we begin to analyze the case where the coefficient heterogeneities (without noise) are not resolved by the subdomains. There, we need to make geometric assumptions on the coefficient, we need to discuss which scaling to use for the weighted jump operator, and finally we need to adapt and generalize the technical tools from

C. Pechstein, *Finite and Boundary Element Tearing and Interconnecting Solvers for Multiscale Problems*, Lecture Notes in Computational Science and Engineering 90, DOI 10.1007/978-3-642-23588-7_3, © Springer-Verlag Berlin Heidelberg 2013

Sect. 2.5. As the most important tool we introduce *weighted Poincaré inequalities* in Sect. 3.4. Refined condition number bounds are presented in Sect. 3.5. The chapter is concluded with a few numerical results and remarks in Sect. 3.6.

3.1 Introduction and Notation

In this section, we first give an introduction to multiscale problems, and introduce the special coefficient model that will be considered in the theory later on. Also, we need to fix some notation.

3.1.1 Multiscale Problems

The term *multiscale* refers to the different scales on which the solution varies. For example the solution of Laplace's problem usually varies on a single scale associated to the domain. When we impose a periodically oscillating diffusion coefficient, the solution additionally varies on the smaller periodic scale. When the coefficient variation is not periodic, other scales appear. The problem treated in Chap. 2, where the diffusion coefficient is constant in each subdomain, can be a multiscale problem if many subdomains are present. In this chapter, however, we allow that the coefficient varies from element to element, although we will have additional assumptions on the type of variation. This means that the solution can vary on *all* possible scales.

Multiscale *methods* usually aim to discretize the problem at a scale which is larger than the finest variation, but build in the fine variation, e.g., by using problem-adapted, so-called multiscale basis functions. The process of approximating the fine scale equation by such a special coarse scale equation is known as *upscaling*. If the coefficient pattern is periodic this can be done very effectively. For an overview on multiscale methods and homogenization see [EH09] and the references therein. For the variational multiscale method we refer to [HFMQ98, Bre00, HS07, CGH10], for mixed multiscale methods to [CH02, AB06], and for heterogeneous multiscale methods to [CH02, AB06]. Finally, for an extension to high order see [AB05].

If the coefficient pattern is not periodic and if no scale separation is possible, then upscaling typically involves the computation of basis functions on coarse cells via cell problems that are discretized at the fine scale and which are all different, see e.g. [CGH10]. If the cost of this precomputation is of the order of fine unknowns in each cell, and the size of coarse system is of the order of the number of cells, the total cost (for one solve) is at least $\mathcal{O}(n)$ where n is the number of global fine unknowns. We see that from the compuational cost, upscaling methods and (optimal) domain decomposition method are comparable, see also [PS09, GE10a, GE10b]. It seems also that upscaling a (fine) system with heterogeneities and preconditioning the fine system (by means of a good coarse solver) is very much related.

Of course, there are other types of solvers for such multiscale problems than FETI type methods [RF98a, LP06, PS08, PS11b, DS11, GKR12]. Here we mention multigrid methods [ABDP81, RS85, XZ08], multilevel methods [VMB96, Osw99, CFH+03, GKM08, SVZ12, Kra12] and additive (overlapping) Schwarz methods [GH99, GLS07, GS07, SV07, DKW08a, GE10a, GE10b, EGLW12, SVZ12, DNSS11, SDH+11]. See also the related papers [BDV97, Sar97, VSM99, CGSS00, CDS03, AGKS08, SVZ11] as well as the survey article [CM94] and the conference proceedings [GHLS12].

3.1.2 The Coefficient Model

In the following, we roughly describe which coefficient model we have in mind for the subsequent analysis. Detailed assumptions will be given later on in Sect. 3.3.1.

Definition 3.1. For any bounded Lipschitz domain $\Omega \subset \mathbb{R}^d$, the set of bounded and uniformly positive coefficients is given by

$$L^\infty_+(\Omega) := \{\alpha \in L^\infty(\Omega) : \operatorname*{ess.inf}_{x\in\Omega} \alpha(x) > 0\}.$$

Our coefficient model is as follows.

1. We take a function $\alpha_{\text{coarse}} \in L^\infty_+(\Omega)$ that is piecewise constant with respect to a quasi-uniform coarse partitioning of Ω and may have arbitrary large jumps across the interfaces between the subregions, i.e., α_{coarse} has *large variation*. Let η be the diameter of the subregions (with $\eta > h$), then we can say that the function α_{coarse} *varies at the coarse scale* η.
2. We take a second function $\alpha_{\text{noise}} \in L^\infty_+(\Omega)$ that is piecewise constant with respect to the finite element mesh $\mathcal{T}^h(\Omega)$, but which fulfills the property

$$\operatorname*{ess.sup}_{x,y\in\Omega} \frac{\alpha_{\text{noise}}(x)}{\alpha_{\text{noise}}(y)} \leq c_{\text{noise}},$$

   where $c_{\text{noise}} \geq 1$ is a given noise level which is typically small (e.g., 10). We can say that α_{noise} *varies at the fine scale* h and that it has *small variation*. However, this function may be highly oscillatory, which we will have to address in a certain part of our theory.
3. The coefficient α in our PDE is the superposition of α_{coarse} and α_{noise}:

$$\alpha(x) = \alpha_{\text{coarse}}(x)\,\alpha_{\text{noise}}(x) \qquad \text{for } x \in \Omega \text{ a.e.}$$

We can say that α has large variation on the coarse scale η, and *additional* noise on the fine scale h.

Remark 3.2. Note that by choosing either the noise level high enough or $\eta = h$, this model captures *any* coefficient resolved by $\mathcal{T}^h(\Omega)$.

We would like to emphasize that a large part of the theory in this chapter will allow the scale η to be *in between* the fine scale h and the scale H of the subdomain decomposition. In particular, we will look at situations where the subdomain decomposition does *not* resolve the heterogeneities in α_{coarse}.

For certain coefficient patterns α_{coarse} we can provide condition number bounds for FETI that are independent of the *values* of α_{coarse}, but instead depend on the scale η and on the noise level c_{noise}. For small noise level, one can say that our bound (and therefore the FETI method) is *robust in α and explicit in the scale η.*

Unfortunately, this robustness does not hold in general. We will provide sufficient conditions using *weighted Poincaré inequalities* (WPI) and the concept of *quasi-monotone coefficients* (Sect. 3.4). Our condition number bound will be robust when the coefficient is quasi-monotone in a suitable part of each subdomain Ω_i. For a fixed coefficient pattern α_{coarse}, this of course restricts the class of subdomain partitions that lead to robustness. However, we will never have to assume that α_{coarse} is fully *resolved* by the subdomain partitioning.

3.1.3 Notation

Definition 3.3. Let D be a bounded Lipschitz domain. A coefficient (weight function) $\alpha \in L^\infty_+(D)$ induces the weighted norm and seminorm

$$\|u\|_{L^2(D),\alpha} := \Big(\int_D \alpha\, |u|^2\, \mathrm{d}x\Big)^{1/2} \qquad \text{for } u \in L^2(D),$$

$$|u|_{H^1(D),\alpha} := \Big(\int_D \alpha\, |\nabla u|^2\, \mathrm{d}x\Big)^{1/2} \qquad \text{for } u \in H^1(D).$$

Definition 3.4. For $\alpha \in L^\infty_+(D)$, we set

$$\alpha_D^{\min} := \operatorname*{ess.inf}_{x\in D} \alpha(x), \qquad \alpha_D^{\max} := \operatorname*{ess.sup}_{x\in D} \alpha(x).$$

Apparently,

$$\begin{aligned} \alpha_D^{\min}\, \|u\|^2_{L^2(D)} &\le \|u\|^2_{L^2(D),\alpha} \le \alpha_D^{\max}\, \|u\|^2_{L^2(D)} \quad \forall v \in L^2(D),\\ \alpha_D^{\min}\, |u|^2_{H^1(D)} &\le |u|^2_{H^1(D),\alpha} \le \alpha_D^{\max}\, |u|^2_{H^1(D)} \quad \forall v \in H^1(D). \end{aligned} \tag{3.1}$$

Remark 3.5. For a fixed triangulation $\mathscr{T}^h(\Omega)$ and if $V^h(\Omega)$ is the space of continuous and piecewise *linear* functions, we can replace the coefficient $\alpha \in L^\infty_+(\Omega)$ in our bilinear form

$$a(v, w) = \int_\Omega \alpha\, \nabla v \cdot \nabla w\, \mathrm{d}x \qquad \text{for } v, w \in V^h(\Omega)$$

by the *effective* coefficient $\alpha_{\text{eff}} \in L^\infty_+(\Omega)$ given by

$$\alpha_{\text{eff}|\tau} = \frac{1}{\operatorname{meas}_d(\tau)} \int_\tau \alpha \,\mathrm{d}x \quad \text{for } \tau \in \mathscr{T}^h(\Omega)$$

because ∇v, ∇w are constant on each element τ. This may result in different values of $\alpha^{\min}_{\text{eff},D}$, $\alpha^{\max}_{\text{eff},D}$ for different discretization levels.

Recall that S_i denotes the operator corresponding to the FEM Schur complement of the local bilinear form

$$a_i(v, w) = \int_{\Omega_i} \alpha \,\nabla v \cdot \nabla w \,\mathrm{d}x,$$

cf. Sect. 1.2.6. Let $\mathscr{H}_i^{\alpha,h}$ denote the discrete PDE-harmonic extension with respect to $a_i(\cdot,\cdot)$, cf. Definition 1.55. Then we have

$$|v|^2_{S_i} = |\mathscr{H}_i^{\alpha,h} v|^2_{H^1(\Omega_i),\alpha} = \min_{\substack{\tilde{v} \in V^h(\Omega_i) \\ \tilde{v}_{|\partial\Omega_i} = v}} |\tilde{v}|^2_{H^1(\Omega_i),\alpha} \qquad \forall v \in V^h(\partial\Omega_i). \tag{3.2}$$

3.2 A Rough Condition Number Bound

In this section, we provide a condition number estimate for the case that we use the simple scaling $\rho_i(x^h) = \alpha^{\max}_{\Omega_i}$ as suggested in Sect. 2.6.4. As the following theorem (cf. [LP06, Proposition 2.2]) shows, the condition number then depends linearly on the maximal subdomain variation of α. In the special case where the coefficient has small variation in each subdomain (with the notation from Sect. 3.1.2, each subdomain resolves α_{coarse}), we get the same behavior (up to the noise level) as if it would be constant, cf. Theorem 2.106. At the end of this section, we briefly discuss alternative scalings that lead to similar bounds. The more interesting case where α_{coarse} is not resolved by the subdomains will be treated in the subsequent sections.

Theorem 3.6. *Let Assumptions 2.53–2.56 and 2.102 hold. If we choose the weights*

$$\rho_i(x^h) = \alpha^{\max}_{\Omega_i}, \tag{3.3}$$

then for both the classical and the all-floating FETI method, either with $Q = M_{\text{sD}}^{-1}$ or with $Q = Q_{\text{diag}}$ (see Definition 2.107),

$$\kappa(P\, M_{\text{sD}}^{-1} P^\top F_{|\widetilde{U}_{\text{ad}}}) \leq C \left(\max_{i=1}^{s} \frac{\alpha^{\max}_{\Omega_i}}{\alpha^{\min}_{\Omega_i}} \right) \max_{i=1}^{s} (1 + \log(H_i/h_i))^2,$$

where the constant C is independent of α.

Proof. By Lemmas 2.45 and 2.103 it suffices to show a bound of the form

$$|P_D(w+z_w)|_S^2 \;\le\; \omega\,|w|_S^2 \qquad \forall w \in W^\perp,$$

where $W^\perp$ is defined as in (2.86). Let $S_i^{\max} : W_i \to W_i^*$ be defined as the FEM Schur complement of the bilinear form

$$a_i^{\max}(v,\, w) \;=\; \alpha_{\Omega_i}^{\max} \int_{\Omega_i} \nabla v \cdot \nabla w \, dx,$$

see Definition 1.53, and let $|\cdot|_{S_i^{\max}}$ be the corresponding norm. Using that

$$|w_i|_{S_i}^2 \;=\; |\mathscr{H}_i^{\alpha,h} w_i|_{H^1(\Omega_i),\alpha}^2, \qquad |w_i|_{S_i^{\max}}^2 \;=\; \alpha_{\Omega_i}^{\max}\,|\mathscr{H}_i^h w_i|_{H^1(\Omega_i)}^2,$$

and (3.1), we can show that

$$|w_i|_{S_i}^2 \;\le\; |w_i|_{S_i^{\max}}^2 \;\le\; \frac{\alpha_{\Omega_i}^{\max}}{\alpha_{\Omega_i}^{\min}}\,|w_i|_{S_i}^2 \qquad \forall w_i \in W_i\,.$$

With $S^{\max} := \operatorname{diag}(S_i^{\max})_{i=1}^s$, it follows that

$$|w|_S^2 \;\le\; |w|_{S^{\max}}^2 \;\le\; \max_{i=1}^{s} \frac{\alpha_{\Omega_i}^{\max}}{\alpha_{\Omega_i}^{\min}}\,|w|_S^2 \qquad \forall w \in W.$$

Hence, we can conclude from Lemmas 2.104 and 2.105 (applied for $S \mapsto S^{\max}$) that for any $w \in W^\perp$,

$$\begin{aligned}
|P_D(w+z_w)|_S^2 \;&\le\; |P_D(w+z_w)|_{S^{\max}}^2 \;\le\; C\,\max_{i=1}^{s}(1+\log(H_i/h_i))^2\,|w|_{S^{\max}}^2\\
&\le\; C\,\Big(\max_{i=1}^{s}\frac{\alpha_{\Omega_i}^{\max}}{\alpha_{\Omega_i}^{\min}}\Big)\max_{i=1}^{s}(1+\log(H_i/h_i))^2\,|w|_S^2\,,
\end{aligned}$$

which implies the condition number bound. The case $Q = Q_{\text{diag}}$ is analogous. □

Remark 3.7. As in Chap. 2 and in the remainder of the current chapter, the above theorem assumes an *ideal* implementation with *exact* subdomain solves. Recall, however, from Remark 2.32 that the conditioning of the subdomain stiffness matrix $\mathbf{K}_i$ grows linearly with $\alpha_{\Omega_i}^{\max}/\alpha_{\Omega_i}^{\min}$. Therefore, even direct solvers (see Sect. 1.2.4.1) may have stability problems when this ratio gets very large.

Corollary 3.8. *The statement of Theorem 3.6 remains true for the (classical or all-floating) FETI method with the following modifications of the weights $\rho_i(x^h)$.*

(i) For $Q = M_{\text{sD}}^{-1}$ we choose $\rho_i(x^h)$ as the maximal diagonal entry of the subdomain stiffness matrix $\mathbf{K}_i$.

(ii) *For* $Q = Q_{\mathrm{diag}}$ *we choose* $\rho_i(x^h)$ *as the maximal diagonal entry of the subdomain stiffness matrix* $\mathbf{K}_i$ *scaled by* h_i^{2-d}.

Proof. Recall that $\mathscr{T}^h(\Omega_i)$ is quasi-uniform and so the maximal diagonal entry of $\mathbf{K}_i$ is equivalent to $h_i^{d-2}\alpha_{\Omega_i}^{\max}$. Since $h_i \eqsim h_j$ for neighboring subdomains, we can conclude that the function $\delta_j^\dagger$ is constant at the nodes of each glob $\mathscr{G}_i$, and $\delta_{j|\mathscr{G}_i}^\dagger \eqsim \alpha_{\Omega_j}^{\max} / \sum_{k\in\mathscr{N}_{\mathscr{G}_i}} \alpha_{\Omega_k}^{\max}$. For the choice $Q = Q_{\mathrm{diag}}$, we need to scale the diagonal entry because $\rho_i(x^h)$ is used in Q_{diag}. □

Note that the extra scaling in Corollary 3.8 (ii) poses no *additional* difficulties in the implementation, because Q_{diag} requires already the knowledge (or calculation) of h_i and H_i.

It is tempting to set $\rho_i(x^h)$ simply to the diagonal of $\mathbf{K}_i$ at x^h or to the maximum of α on the elements touching x^h. However, at high probability (unless the mesh is structured and the coefficient piecewise constant) this will lead to oscillatory functions $\delta_j^\dagger$. As we know from Sect. 2.6.4 (p. 142), this can result in very high condition numbers and very slow convergence.

Remark 3.9. Following the line of the proof of Theorem 3.6, one can show bounds for the choice

$$\rho_i(x^h) = (\alpha_{\Omega_i}^{\max})^{\beta_i}\, (\alpha_{\Omega_i}^{\min})^{1-\beta_i}$$

with exponents $\beta_i \in [0,\ 1]$, cf. [LP06, Proposition 2.2]. Then we have to replace the factor $\max_{i=1}^{s} \frac{\alpha_{\Omega_i}^{\max}}{\alpha_{\Omega_i}^{\min}}$ in the bound from Theorem 3.6 by

$$\max_{i=1}^{s} \left(\frac{\alpha_{\Omega_i}^{\max}}{\alpha_{\Omega_i}^{\min}}\right)^{\beta_i - 1} \max_{j=1}^{s} \left(\frac{\alpha_{\Omega_j}^{\max}}{\alpha_{\Omega_j}^{\min}}\right)^{\beta_j},$$

which, in the worst case, results in a quadratic dependence on the maximal local variation. If $\beta_i = \beta$ is uniform, we regain the linear dependence.

3.3 Towards a Refined Condition Number Bound

In case that the coefficient variation in each subdomain is large, the condition number bound from Theorem 3.6 is also large. However, it has been numerically observed by several authors [RF98a, RF99, LP06, KR07b] that the bound is often too pessimistic. This section provides a refined bound by generalizing the theory from Sect. 2.6 to the weighted norms introduced in Sect. 3.1. First, Sect. 3.3.1 introduces assumptions on the multiscale coefficient. Section 3.3.2 discusses the choice of the scalings $\rho_i(x^h)$. In Sect. 3.3.3 we provide technical tools that we need to prove the refined condition number bound in Sect. 3.3.5. Note that this bound requires a weighted Poincaré inequality, which will be the subject of Sect. 3.4.

3.3.1 Assumptions on the Coefficient

In the following, we make geometrical assumptions on the subregions underlying the function α_{coarse} from Sect. 3.1.2, at least in the vicinity of the interface Γ and the Dirichlet boundary Γ_D. In a somewhat different form, these assumptions can found in [PS08, PS11b], see also [GLS07, SV07].

Definition 3.10 (Skeleton Cover). An open domain $\Lambda \subset \Omega$ is called a *skeleton cover*[1] *compatible with* $\mathscr{T}^h(\Omega)$ if

- Λ is the union of elements from $\mathscr{T}^h(\Omega)$ and
- If Λ contains all the elements $\tau \in \mathscr{T}^h(\Omega)$ with $\overline{\tau} \cap (\Gamma \cup \Gamma_D) \neq \emptyset$.

The *local covers* are defined by

$$\Lambda_i := \Lambda \cap \Omega_i .$$

For an illustration see Fig. 3.1 (left, middle).

Let us fix a skeleton cover Λ that is compatible with $\mathscr{T}^h(\Omega)$. Each local cover Λ_i is decomposed into non-overlapping subregions $Y_i^{(\ell)}$ such that

$$\overline{\Lambda}_i = \bigcup_{\ell=1}^{n_i} \overline{Y}_i^{(\ell)} . \tag{3.4}$$

Later on in our theory, these subregions will play the role of the regions underlying the function α_{coarse}. Next, we need to make regularity assumptions on the subregions $Y_i^{(\ell)}$ (similar to the assumptions on the subdomains from Sect. 2.5.2).

Assumption 3.11. There exists a shape regular triangulation $\mathscr{T}^\eta(\Lambda)$ of the skeleton cover Λ such that each of the subregions $Y_i^{(\ell)}$ is an $\mathscr{T}^\eta$-agglomerate and at the same time a union of elements from $\mathscr{T}^h(\Omega)$. The number of elements from $\mathscr{T}^\eta(\Omega)$ per subregion $Y_i^{(\ell)}$ is uniformly bounded.

For a simple illustration of Assumption 3.11 see Fig. 3.1 (right). Note that Assumption 3.11 is not a local but global assumption because the triangulation $\mathscr{T}^\eta(\Lambda)$ is globally conforming.

Definition 3.12 (Boundary Layer). Let Assumption 3.11 hold. For each subdomain, the open *boundary layer* $\Omega_{i,\eta}$ is given by

$$\overline{\Omega}_{i,\eta} = \bigcup \left\{ \overline{Y}_i^{(\ell)} : \partial Y_i^{(\ell)} \cap \partial\Omega_i \neq \emptyset \right\}.$$

[1]The name skeleton cover is not completely adequate because it is a cover of $\Gamma \cup \Gamma_D$ only.

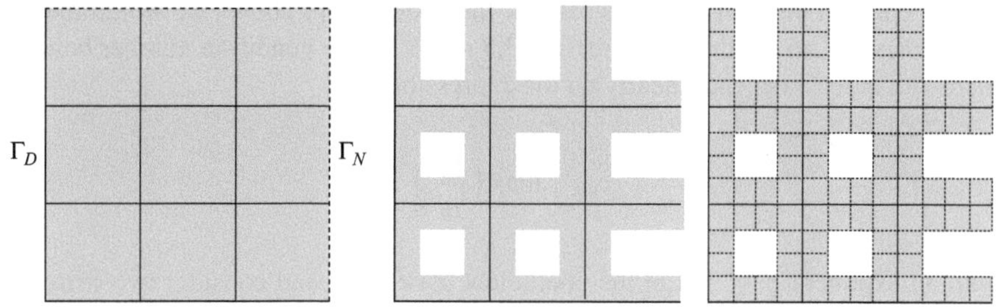

Fig. 3.1 Example of a skeleton cover. *Left*: subdomain decomposition. *Middle*: skeleton cover (*dark shaded region*). *Right*: subregions $Y_i^{(\ell)}$ indicated by *dashed lines*

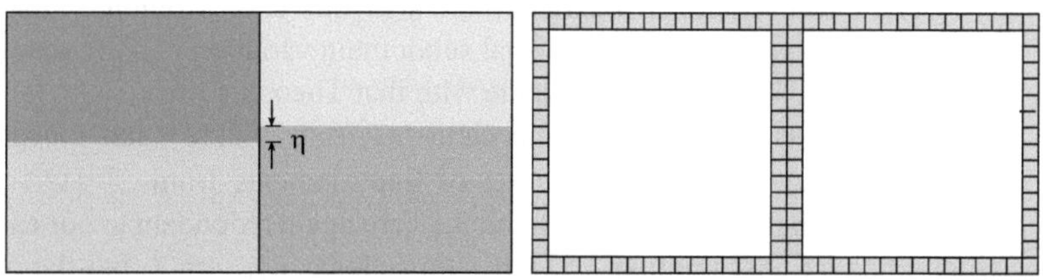

Fig. 3.2 Illustration of Assumption 3.11 for two adjacent subdomains. *Left*: coefficient distribution (different shadings indicate different values of α). *Right*: skeleton cover

Note that in many cases $\Lambda_i = \Omega_{i,\eta}$, e.g., in the example in Fig. 3.1. Nevertheless, we allow Λ_i to be larger than $\Omega_{i,\eta}$ because this can lead to better condition numbers for certain situations, see e.g. Theorem 3.70 and Corollary 3.72 below.

Assumption 3.13. For each $i = 1, \dots, s$, the restriction $\mathscr{T}^\eta(\Omega_{i,\eta})$ of $\mathscr{T}^\eta(\Lambda)$ to the boundary layer $\Omega_{i,\eta}$ is quasi-uniform with mesh parameter η_i. This implies that $\eta_i \eqsim \eta_j$ when Ω_i, Ω_j are neighboring.

Note that none of the definitions and assumptions above contains the coefficient α. The next assumption closes this gap via the noise level that we informally introduced in Sect. 3.1.2.

Assumption 3.14. There is a uniform constant c_{noise} such that

$$\max_{i=1}^{s} \max_{Y_i^{(\ell)} \subset \Omega_{i,\eta}} \frac{\alpha^{\max}_{Y_i^{(\ell)}}}{\alpha^{\min}_{Y_i^{(\ell)}}} \leq c_{\text{noise}} \,.$$

Note that Assumptions 3.11, 3.13, and 3.14 have to hold simultaneously. In most situations, Assumption 3.14 has influence on the other two assumptions. E.g., for the example shown in Fig. 3.2, the coefficient creates a staggered geometry, which forces the mesh $\mathscr{T}^\eta$ underlying the skeleton cover to be relatively fine compared to the subdomains if Assumption 3.14 should be fulfilled with $c_{\text{noise}} = 1$.

To explain a little bit more the roles of the assumptions above, we would like to give a preview on the theory in Sects. 3.3.5 and 3.5. The condition number bounds there will at least depend linearly on the expression

$$c_{\text{noise}} \max_{i=1}^{s} \left(\frac{H_i}{\eta_i} \right)^{\beta} \tag{3.5}$$

with an exponent $\beta \geq 1$. Let the coefficient α be fixed and consider two extreme choices of the skeleton cover Λ and the subregions from (3.4).

(a) We choose $\Lambda_i = \Omega_i = Y_i^{(1)}$ and $\mathscr{T}^{\eta}(\Omega) = \mathscr{T}^{H}(\Omega)$. Then Assumptions 3.11 and 3.13 are redundant to our earlier assumptions (cf. Assumption 2.54), and $\eta_i = H_i$. However, for a fixed coefficient α, this choice requires that the noise level coincides with the maximal subdomain variation of α. Indeed, the bounds in Sects. 3.3.5 and 3.5 coincide with that Theorem 3.6.

(b) We choose Λ as the union of fine elements from $\mathscr{T}^{h}(\Omega)$ that touch the skeleton Γ_S and $\{Y_i^{(\ell)}\}_{i=1}^{n_i}$ as the set of fine elements from $\mathscr{T}^{h}(\Omega_i)$ that touch $\partial\Omega_i$. Then Assumptions 3.11 and 3.13 are again redundant to our earlier assumptions (cf. Assumption 2.53). For this choice, $c_{\text{noise}} = 1$, but the other factor in (3.5) equals $(H_i/h_i)^{\beta}$. Even though, for extremely large coefficient variation and subdomain meshes of medium size, this might be acceptable.

The bounds in Sect. 3.5 also hold for cases in between (a) and (b). However, we would like to warn the reader that the full condition number bound does not only contain the expression (3.5) but also the constants in weighted Poincaré inequalities associated with a suitable part of each subdomain. These constants may depend on the heterogeneities in α, and we will clarify when there are benign bounds.

It will turn out that for the FETI methods considered in this chapter, we can drop (unless stated otherwise) all the assumptions from Sect. 2.5.2 (Assumptions 2.53–2.56) and replace them by Assumptions 3.11, 3.13, and 3.14 from above and the following two assumptions.

Assumption 3.15. The fine mesh $\mathscr{T}^{h}(\Omega)$ restricted to each boundary layer $\Omega_{i,\eta}$ is quasi-uniform with mesh parameter h_i.

Assumption 3.16. The Dirichlet boundary Γ_D is the union of η-facets (faces/edges) of the coarse triangulation $\mathscr{T}^{\eta}(\Lambda)$ from Assumption 3.11.

3.3.2 *Choice of Scalings*

According to the subsequent theory, the ideal scaling is

$$\rho_i(x^h) = \max_{x^h \in \overline{Y}_i^{(\ell)}} \alpha_{Y_i^{(\ell)}}^{\max} \qquad \text{for } x^h \in \partial\Omega_i \setminus \Gamma_N^h. \tag{3.6}$$

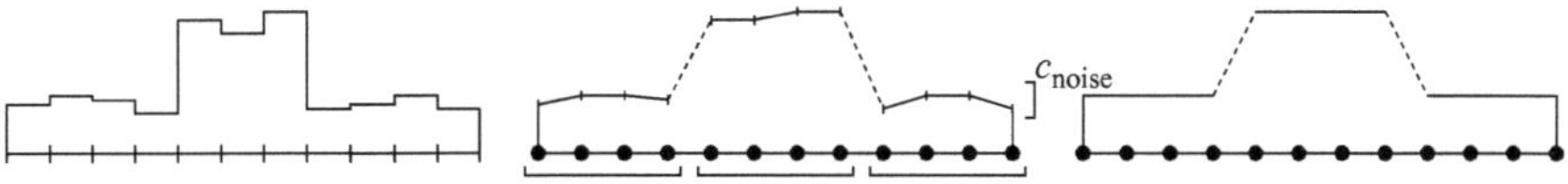

Fig. 3.3 Illustration of the heuristics described in Sect. 3.3.2. *Left*: edge with the associated coefficients (1D projection). *Middle*: values of $\widetilde{\rho}_i(x^h)$ and resulting groups. *Right*: final scaling $\rho_i(x^h)$

To mimic this choice in practice is a non-trivial task: typically the regions $Y_i^{(\ell)}$ are not known, but only the subdomain stiffness matrix $\mathbf{K}_i$ is at hand, and in some cases the element coefficient $\alpha_\tau^{\max}$. To give some ideas, let us consider the two-dimensional case. In a first step, we produce

$$\widetilde{\rho}_i(x^h) = \max_{\tau \in \mathcal{T}^h(\Omega_i):\, x^h \in \bar{\tau}} \alpha_\tau^{\max} \quad \text{for } x^h \in \partial\Omega_i^h . \tag{3.7}$$

If the mesh is quasi-uniform, we may alternatively set $\widetilde{\rho}_i(x^h)$ to the diagonal entry of the stiffness matrix $\mathbf{K}_i$ corresponding to x^h, see Sect. 2.6.4. In a second step, we consider a subdomain edge $\mathcal{E}_i \subset \Gamma$, collect all the nodes on $\mathcal{E}_i$ and sort them such that we can traverse the edge from one endpoint to the other. With a given noise level c_{noise}, we split the nodes on $\mathcal{E}_i$ into (connected) groups $\{e_\ell\}$ such that

$$\frac{\max_{x^h \in e_\ell} \widetilde{\rho}_i(x^h)}{\min_{y^h \in e_\ell} \widetilde{\rho}_i(y^h)} \le c_{\text{noise}} ,$$

where $\widetilde{\rho}_i$ is defined as in (3.7). This can be simply done by traversing the edge and keep adding nodes to a group until the ratio is beyond the tolerance c_{noise}. After that, the next group is started. For an illustration see Fig. 3.3. Having now the η-subglobs of $\mathcal{E}_i$, one can move inside the subdomain and collect more information in the same fashion. Note that here, only the graph information from the stiffness matrix is needed. Finally, we set $\rho_i(y^h)$ to the maximum of $\widetilde{\rho}_i(x^h)$ where x^h is in the same group as y^h. This covers all values $\rho_i(x^h)$ on the subdomain edges. The weight $\rho_i(\mathcal{V})$ at a subdomain vertex $\mathcal{V}$ is set to the maximal values of $\widetilde{\rho}_i(x^h)$ of all the groups touching $\mathcal{V}$.

A comparable algorithm in three dimensions would require a kind of edge detection (e.g. through total variation minimization) as used in image processing, see e.g. [Wei98] and the references therein.

3.3.3 *Technical Tools*

In this section, we work out technical tools that we need for the analysis in Sect. 3.3.5. Most of the tools are adaptions of Sect. 2.5 to the multiscale setting.

3.3.3.1 η-Globs and Cut-Off Functions

Recall that up to the noise, the coefficient varies with respect to the medium scale η. Recall also that the analysis in Sect. 2.6 was performed using a splitting into the contributions from subdomain globs. In order to address the heterogeneities, we use a *finer* splitting into so-called η-globs, cf. [PS08, PS11b].

Definition 3.17. Consider the decomposition

$$\overline{\Lambda} = \bigcup_{i=1}^{s} \bigcup_{\ell=1}^{n_i} \overline{Y_i^{(\ell)}}.$$

Analogously to Definition 2.48, we define globs of this decomposition that lie either on the interface Γ or on the Dirichlet boundary Γ_D (there would be more globs but we do not need them). Under Assumption 3.11 these globs are all $\mathscr{T}^\eta$-agglomerates of dimension 0 up to $(d-1)$. To distinguish them from the subdomain globs from Definition 2.48 we will refer to them as η-globs and denote them generically by g. We have

- η-*facets* that are globs of dimension $(d-1)$ and generically denoted by f,
- η-*edges* that are globs of dimension one and generically denoted by e,
- η-*vertices* that are globs of dimension zero and generically denoted by v.

For any η-glob g, we define $\mathscr{N}_\mathsf{g} = \{j = 1, \dots, s : \mathsf{g} \subset \partial\Omega_j\}$.

Notation. By writing $\sum_{\mathsf{g}\subset\partial\Omega_i} \dots$ we mean that we sum over all η-globs that lie on $\partial\Omega_i \cap (\Gamma \cup \Gamma_D)$.

Definition 3.18. For an η-glob $\mathsf{g} \subset \Gamma \cup \Gamma_D$, we define the open neighborhood

$$\overline{\mathsf{U}}_\mathsf{g} = \bigcup \{\overline{Y_j^{(\ell)}} : j = 1, \dots, s,\ \ell = 1, \dots, n_j,\ \mathsf{g} \subset \partial Y_j^{(\ell)}\}.$$

For an η-glob $\mathsf{g} \subset \partial\Omega_i$, the local neighborhood is given by

$$\mathsf{U}_{i,\mathsf{g}} = \mathsf{U}_\mathsf{g} \cap \Omega_i .$$

Definition 3.19. Similarly to Definitions 2.51 and 2.72, for an η-glob $\mathsf{g} \subset \Gamma \cup \Gamma_D$, let

$$\theta_\mathsf{g} \in V^h(\Gamma_S)$$

be the function that equals one at all nodes on g and zero elsewhere.

We have the partition of unity property

$$\sum_{\mathsf{g}_i \subset \partial\Omega_i} I^h(\theta_{\mathsf{g}_i} v) = v \qquad \forall v \in V^h(\partial\Omega_i),\ v_{|\partial\Omega_i \cap \Gamma_N^h} = 0. \tag{3.8}$$

Above, we may also replace $\partial\Omega_i$ by Γ_S.

Definition 3.20. For an η-facet $\mathsf{f} \subset \Gamma \cup \Gamma_D$, the function

$$\vartheta_{\mathsf{f}} \in V^h(\Omega)$$

equals one at all interior nodes of f, zero at all nodes in $\overline{\Omega} \setminus \mathsf{U}_{\mathsf{f}}$, and it is discrete harmonic in $\mathsf{U}_{i,\mathsf{f}}$ for each $i \in \mathcal{N}_{\mathsf{f}}$. For an η-glob g of dimension less that $(d-1)$, the function $\vartheta_{\mathsf{g}} \in V^h(\Omega)$ equals one at all interior nodes of g and zero at all other nodes. Note that

$$\mathrm{supp}(\vartheta_{\mathsf{g}}) \subset \overline{\mathsf{U}}_{\mathsf{g}}\,.$$

3.3.3.2 A Weighted Cut-Off Estimate

The following lemma (cf. [PS11b, Lemma 5.4]) estimates the energy norm in terms of contributions from the η-globs.

Lemma 3.21. *Let Assumption 3.11 hold. For $v \in V^h(\partial\Omega_i)$ with $v_{|\partial\Omega_i \cap \Gamma_N^h} = 0$, let $\tilde{v} \in V^h(\Omega_i)$ be an arbitrary extension of v. Then the function $\sum_{\mathsf{g}\subset\partial\Omega_i} I^h(\vartheta_{\mathsf{g}}\,\tilde{v}) \in V^h(\Omega_i)$ is an extension of v. Furthermore,*

$$|v|_{S_i}^2 \;\le\; C \sum_{\mathsf{g}\subset\partial\Omega_i} |I^h(\vartheta_{\mathsf{g}}\,\tilde{v})|^2_{H^1(\mathsf{U}_{\mathsf{g}}),\alpha}\,,$$

where the constant C only depends on the shape regularity of $\mathcal{T}^\eta(\Omega)$.

Proof. Thanks to Definition 3.20, the function $u := \sum_{\mathsf{g}\subset\partial\Omega_i} I^h(\vartheta_{\mathsf{g}}\,\tilde{v})$ is indeed in $V^h(\Omega_i)$. Since θ_{g} is the trace of ϑ_{g}, the partition of unity property (3.8) implies that u coincides with v on $\partial\Omega_i$. From (3.2) we obtain

$$|v|_{S_i}^2 \;\le\; |u|^2_{H^1(\Omega_i),\alpha} \;=\; |u|^2_{H^1(\Omega_{i,\eta}),\alpha}\,,$$

where in the last step we have used that $\mathrm{supp}(u) \subset \overline{\Omega}_{i,\eta}$. Since $\mathcal{T}^\eta(\Omega)$ is shape regular, the neighborhoods $\mathsf{U}_{i,\mathsf{g}}$ have finite overlap, i.e., each $\mathcal{T}^\eta$-agglomerate $Y_i^{(\ell)} \subset \Omega_{i,\eta}$ is only contained in C of the neighborhoods $\mathsf{U}_{i,\mathsf{g}}$ (where the functions $I^h(\vartheta_{\mathsf{g}}\,\tilde{v})$ are supported), where C depends on the shape regularity constant of $\mathcal{T}^\eta(\Omega)$. This implies the estimate in the lemma. □

3.3.3.3 An FE Transfer Operator on η-Glob Neighborhoods

Similarly to Sect. 2.5.7, the following lemma provides a discrete transfer operator between the η-agglomerates of different subdomains that are associated to an η-glob, cf. [PS11b, Lemma 5.5]. For an illustration see Fig. 3.4.

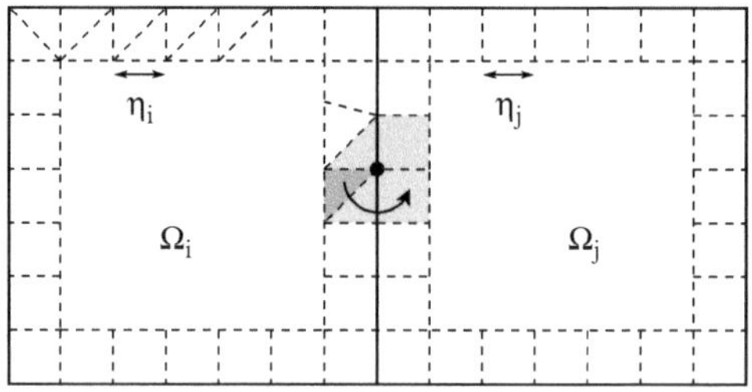

Fig. 3.4 Illustration of the transfer operator $E_{i,\mathsf{g}}^{h,(k)}$ from Lemma 3.22 for an η-vertex $\mathsf{g} = \mathsf{v}$. *Dashed lines* indicate η-agglomerates in the boundary layers $\Omega_{i,\eta}$, $\Omega_{j,\eta}$, • indicates g, *shaded*: neighborhood U_g, *dark shaded*: "source" agglomerate $Y_i^{(k)}$ from where the functions are extended

Lemma 3.22. *Let Assumption 3.11 hold. For a fixed η-glob $\mathsf{g} \subset \partial\Omega_i$, let $Y_i^{(k)} \subset \mathsf{U}_{i,\mathsf{g}}$ be one of the η-agglomerates from Assumption 3.11. Then there exists a discrete transfer operator*

$$E_{i,\mathsf{g}}^{h,(k)} : V^h(Y_i^{(k)}) \to V^h(\mathsf{U}_\mathsf{g})$$

such that for all $v \in V^h(Y_i^{(k)})$, $(E_{i,\mathsf{g}}^{h,(k)} v)_{|Y_i^{(k)}} = v$ and

$$|E_{i,\mathsf{g}}^{h,(k)} v|_{H^1(\mathsf{U}_\mathsf{g})} \leq C\, |v|_{H^1(Y_i^{(k)})}, \qquad \|E_{i,\mathsf{g}}^{h,(k)} v\|_{L^2(\mathsf{U}_\mathsf{g})} \leq C\, \|v\|_{L^2(Y_i^{(k)})}.$$

The constant C depends only on the uniform constants from Assumption 3.11.

Proof. The construction and proof follows completely the line of Lemma 2.87, see also Remark 2.87. □

3.3.4 *Weighted Poincaré Inequalities: Introduction*

In this section, we introduce two weighted Poincaré inequalities. The constants in these inequalities will appear in our condition number bound in Sect. 3.3.5. A detailed analysis of these constants will be the subject of Sect. 3.4.

Definition 3.23. For a bounded Lipschitz domain D and a weight $\alpha \in L^\infty_+(D)$, let $C_{P,\alpha}(D)$ be the smallest constant such that

$$\inf_{c\in\mathbb{R}} \|u - c\|_{L^2(D),\alpha} \leq C_{P,\alpha}(D)\,\mathrm{diam}(D)\,|u|_{H^1(D),\alpha} \qquad \forall u \in H^1(D). \qquad (3.9)$$

We call (3.9) a *weighted Poincaré inequality*. Like in Sect. 1.2.1.6, the parameter $C_{P,\alpha}(D)$ depends on the shape of the domain D, but not on its diameter.

Lemma 3.24. *The infimum in* (3.9) *is attained at*

$$c = \bar{u}^{D,\alpha} := \frac{\int_D \alpha\, u \,\mathrm{d}x}{\int_D \alpha \,\mathrm{d}x}, \tag{3.10}$$

i.e., at the α-weighted average of u over D.

Proof. The expression on the left hand side of (3.30) is convex with respect to c, so the infimum is attained if and only if $0 = \frac{\mathrm{d}}{\mathrm{d}c} \int_D \alpha\, |c-u|^2 \,\mathrm{d}x = 2 \int_D \alpha\, (c-u) \,\mathrm{d}x$. □

Remark 3.25. Let λ_2 be the second smallest eigenvalue of the generalized eigenvalue problem

$$-\nabla \cdot (\alpha\, \nabla u) = \lambda\, \alpha\, u \qquad \text{in } D, \tag{3.11}$$

$$\alpha\, \nabla u \cdot n = 0 \qquad \text{on } \partial D. \tag{3.12}$$

Then (cf. [GE10b]),

$$C_{P,\alpha}(D) = \frac{1}{\sqrt{\lambda_2}\, \mathrm{diam}(D)}.$$

In our theory, it will be enough to have a weighted Poincaré inequality for finite element functions only.

Definition 3.26. Let D be a bounded Lipschitz domain, $\mathscr{T}^h(D)$ a regular triangulation of D, and $V^h(D)$ the corresponding finite element space of continuous piecewise linear functions. For $\alpha \in L^\infty_+(D)$ let $C_{P,\alpha}(D;\, h)$ denote the smallest parameter such that

$$\inf_{c \in \mathbb{R}} \|u - c\|_{L^2(D),\alpha} \leq C_{P,\alpha}(D;\, h)\, \mathrm{diam}(D)\, |u|_{H^1(D),\alpha} \qquad \forall u \in V^h(D). \tag{3.13}$$

Again, the infimum is attained at the weighted average $c = \bar{u}^{D,\alpha}$.

3.3.5 Semi-abstract Condition Number Bounds

This section provides condition number bounds of all-floating (total) FETI using the technical tools exposed above. The bounds will contain

- A polylogarithmic term involving the discretization parameter h,
- The noise level c_{noise} from Assumption 3.14,
- The scale η,
- A weighted Poincaré parameter, which requires further investigations in Sect. 3.4.

3.3.5.1 A Condition Number Bound Involving Boundary Layers

The FETI condition number bound in Theorem 3.28 below involves the Poincaré parameters $C_{P,\underline{\alpha}_i}(\Omega_{i,\eta}; h)$ of the boundary layers $\Omega_{i,\eta}$ and will be independent of the coefficient values of α in the subdomain interiors $\Omega_i \setminus \Omega_{i,\eta}$. The central ingredient is the stability estimate of the next lemma, which goes back to [PS11b, Lemma 5.6].

Lemma 3.27. *Let Assumptions 3.11–3.16 (p. 164f) hold. Moreover, let the scalings $\rho_i(x^h)$ be chosen according to (3.6). For $j = 1, \ldots, s$, let the lower and upper bound functions $\underline{\alpha}_j, \overline{\alpha}_j \in L^\infty(\Omega_{j,\eta})$ be given by*

$$\underline{\alpha}_j(x) = \alpha^{\min}_{Y_j^{(\ell)}}, \qquad \overline{\alpha}_j(x) = \alpha^{\max}_{Y_j^{(\ell)}} \qquad \text{for } x \in Y_j^{(\ell)},\ \ell = 1, \ldots, n_j .$$

Then for any $w \in W$ and each $i = 1, \ldots, s$,

$$|(P_D w)_i|^2_{S_i} \leq C \max_{k \in \mathcal{N}_i} \left\| \frac{\overline{\alpha}_k}{\underline{\alpha}_k} \right\|_{L^\infty(\Omega_{k,\eta})} \sum_{j \in \mathcal{N}_i} \left[\left(1 + \log\left(\frac{\eta_i}{h_i}\right)\right)^2 |\mathcal{H}_j^{\alpha,h} w_j|^2_{H^1(\Omega_{j,\eta}),\underline{\alpha}_j} \right.$$
$$\left. + \frac{1}{\eta_i^2}\left(1 + \log\left(\frac{\eta_i}{h_i}\right)\right) \|\mathcal{H}_j^{\alpha,h} w_j\|^2_{L^2(\Omega_{j,\eta}),\underline{\alpha}_j} \right],$$

where $\mathcal{N}_i = \{j = 1, \ldots, s : \partial\Omega_i \cap \partial\Omega_j \neq \emptyset\}$, cf. Definition 2.12. In the estimate above, we may replace the weight $\underline{\alpha}_j$ in the norms by α. The constant C is independent of α, h_i, η_i, and H_i. It depends only on the uniform constants from the mentioned assumptions.

Proof. Let the function $w \in W$ and the index $i = 1, \ldots, s$ be fixed. For each η-glob $\mathsf{g} \subset \partial\Omega_i$ and each $j \in \mathcal{N}_\mathsf{g}$, we define the function $\tilde{w}_{ij}^\mathsf{g} \in V^h(\mathsf{U}_{i,\mathsf{g}})$ by

$$\tilde{w}_{ij}^\mathsf{g} := \left(E_{j,\mathsf{g}}^{h,(k_{j,\mathsf{g}})} \mathcal{H}_j^{\alpha,h} w_j\right)_{|\mathsf{U}_{i,\mathsf{g}}}, \tag{3.14}$$

where the index $k_{j,\mathsf{g}}$ is such that

$$\alpha^{\max}_{\mathsf{U}_{j,\mathsf{g}}} = \alpha^{\max}_{Y_j^{(k_{j,\mathsf{g}})}} . \tag{3.15}$$

In other words, we extend from the η-agglomerate where the locally maximal coefficient is attained. Then the function

$$v_i := \sum_{\mathsf{g} \subset \partial\Omega_i \cap \Gamma} \sum_{j \in \mathcal{N}_\mathsf{g}} \delta^\dagger_{j|\mathsf{g}} I^h(\vartheta_\mathsf{g}(\tilde{w}_{ii}^\mathsf{g} - \tilde{w}_{ij}^\mathsf{g})) + \sum_{\mathsf{g} \subset \Gamma_D} I^h(\vartheta_\mathsf{g} \tilde{w}_{ii}^\mathsf{g}) \in V^h(\Omega_i)$$

is an extension of $(P_D w)_i$, cf. (2.70), p. 108. Here, $\delta^\dagger_{j|\mathsf{g}}$ denotes the constant value that $\delta_j^\dagger$ attains at the nodes on g. To see that the value is constant, recall from (3.6) that $\rho_{i|\mathsf{g}} = \alpha^{\max}_{\mathsf{U}_{i,\mathsf{g}}}$ and compare with the definition (2.53) of $\delta_j^\dagger(x^h)$, p. 91. With Lemma 3.21 and using the fact that the number $\#\mathscr{N}_i$ of neighboring subdomains is uniformly bounded, we obtain

$$|(P_D w)_i|^2_{S_i} \tag{3.16}$$

$$\lesssim \sum_{\mathsf{g}\subset\partial\Omega_i\cap\Gamma} \underbrace{\sum_{j\in\mathscr{N}_\mathsf{g}} (\delta^\dagger_{j|\mathsf{g}})^2 \left| I^h\big(\vartheta_\mathsf{g}(\tilde{w}^\mathsf{g}_{ii} - \tilde{w}^\mathsf{g}_{ij})\big)\right|^2_{H^1(\mathsf{U}_{i,\mathsf{g}}),\alpha}}_{=:\,\Upsilon(\mathsf{g},\,j)} + \sum_{\mathsf{g}\subset\Gamma_D\cap\partial\Omega_i} \left| I^h(\vartheta_\mathsf{g}\tilde{w}^\mathsf{g}_{ii})\right|^2_{H^1(\mathsf{U}_{i,\mathsf{g}}),\alpha}.$$

Estimate (3.1) and the elementary inequality from Lemma 2.98 yield

$$\begin{aligned}\Upsilon(\mathsf{g},\,j) &\le \alpha^{\max}_{\mathsf{U}_{i,\mathsf{g}}}(\delta^\dagger_{j|\mathsf{g}})^2 \left| I^h\big(\vartheta_\mathsf{g}(\tilde{w}^\mathsf{g}_{ii} - \tilde{w}^\mathsf{g}_{ij})\big)\right|^2_{H^1(\mathsf{U}_{i,\mathsf{g}})}\\ &\le \min\left(\alpha^{\max}_{\mathsf{U}_{i,\mathsf{g}}},\, \alpha^{\max}_{\mathsf{U}_{j,\mathsf{g}}}\right) \left| I^h\big(\vartheta_\mathsf{g}(\tilde{w}^\mathsf{g}_{ii} - \tilde{w}^\mathsf{g}_{ij})\big)\right|^2_{H^1(\mathsf{U}_{i,\mathsf{g}})}.\end{aligned}$$

Inserting this estimate into (3.16) and using the triangle inequality, we get

$$|(P_D\, w)_i|^2_{S_i} \lesssim \sum_{\mathsf{g}\subset\partial\Omega_i} \sum_{j\in\mathscr{N}_\mathsf{g}} \alpha^{\max}_{\mathsf{U}_{j,\mathsf{g}}} |I^h(\vartheta_\mathsf{g}\tilde{w}^\mathsf{g}_{ij})|^2_{H^1(\mathsf{U}_{i,\mathsf{g}})}. \tag{3.17}$$

Using Corollary 2.78 (replacing $\vartheta_{\mathscr{G}_i}$ by ϑ_g and Ω_i by $\mathsf{U}_{i,\mathsf{g}}$ and using Assumptions 3.11 and 3.15), the definition (3.14) of $\tilde{w}^\mathsf{g}_{ij}$, and Lemma 3.22, as well as the identities $\alpha^{\max}_{\mathsf{U}_{j,\mathsf{g}}} = \alpha^{\max}_{Y_j^{(k_j,\mathsf{g})}}$ and $\alpha^{\min}_{Y_j^{(k_j,\mathsf{g})}} = \underline{\alpha}_{j|Y_j^{(k_j,\mathsf{g})}}$, we can conclude (with the short hand $\omega_i := 1 + \log(\eta_i/h_i))$ that

$$\begin{aligned}&\alpha^{\max}_{\mathsf{U}_{j,\mathsf{g}}} |I^h(\vartheta_\mathsf{g}\tilde{w}^\mathsf{g}_{ij})|^2_{H^1(\mathsf{U}_{i,\mathsf{g}})}\\ &\le \alpha^{\max}_{\mathsf{U}_{j,\mathsf{g}}} \left[\omega_i^2\, |\tilde{w}^\mathsf{g}_{ij}|^2_{H^1(\mathsf{U}_{i,\mathsf{g}})} + \eta_i^{-2}\,\omega_i\, \|\tilde{w}^\mathsf{g}_{ij}\|^2_{L^2(\mathsf{U}_{i,\mathsf{g}})}\right]\\ &\lesssim \alpha^{\max}_{Y_j^{(k_j,\mathsf{g})}} \left[\omega_i^2\, |\mathscr{H}^{\alpha,h}_j w_j|^2_{H^1(Y_j^{(k_j,\mathsf{g})})} + \eta_i^{-2}\,\omega_i\, \|\mathscr{H}^{\alpha,h}_j w_j\|^2_{L^2(Y_j^{(k_j,\mathsf{g})})}\right]\\ &\lesssim \frac{\alpha^{\max}_{Y_j^{(k_j,\mathsf{g})}}}{\alpha^{\min}_{Y_j^{(k_j,\mathsf{g})}}} \left[\omega_i^2\, |\mathscr{H}^{\alpha,h}_j w_j|^2_{H^1(Y_j^{(k_j,\mathsf{g})}),\underline{\alpha}_j} + \eta_i^{-2}\,\omega_i\, \|\mathscr{H}^{\alpha,h}_j w_j\|^2_{L^2(Y_j^{(k_j,\mathsf{g})}),\underline{\alpha}_j}\right].\end{aligned}$$

Inserting this local estimate into (3.17), rearranging the sums, and using a finite overlap argument yields the estimate in the lemma. □

The next theorem gives a refined condition number estimate by combining the P_D-estimate from Lemma 3.27 with weighted Poincaré inequalities in the boundary layers $\Omega_{i,\eta}$ (see [PS11b, Theorem 6.4] for an earlier version of this theorem).

Theorem 3.28. *Let Assumptions 3.11–3.16 (p. 164f) hold. Moreover, let the scalings $\rho_i(x^h)$ be chosen according to* (3.6). *For $j = 1, \dots, s$, let the lower bound function $\underline{\alpha}_j \in L^\infty(\Omega_{j,\eta})$ be defined as in Lemma 3.27. Then the all-floating FETI method with $Q = M_{\mathrm{sD}}^{-1}$ satisfies*

$$\kappa(P\, M_{\mathrm{sD}}^{-1}\, P^\top F_{|\tilde{U}_{\mathrm{ad}}}) \;\le\; C\, c_{\mathrm{noise}}\, \max_{i=1}^{s} \Big(\frac{H_i}{\eta_i}\Big)^2 \big(1 + \log\big(\frac{\eta_i}{h_i}\big)\big)^2 \Big(1 + C_{P,\underline{\alpha}_i}(\Omega_{i,\eta};\, h)^2\Big),$$

where c_{noise} is the constant from Assumption 3.14. The constant C depends only of the uniform constants from the remaining assumptions and is independent of α, h_i, η_i, and H_i.

Proof. We set

$$W_i^\perp := \Big\{ w_i \in W_i : \overline{\mathscr{H}_i^{\alpha,h} w_i}^{\,\Omega_{i,\eta},\underline{\alpha}_i} = 0 \Big\}, \tag{3.18}$$

and $W^\perp := \prod_{i=1}^s W_i^\perp$. From the discrete weighted Poincaré inequality (3.13) (see Definition 3.26) and from Lemma 3.24, we can conclude that for all $w \in W^\perp$ and $j = 1, \dots, s$,

$$\begin{aligned}\|\mathscr{H}_j^{\alpha,h} w_j\|_{L^2(\Omega_{j,\eta}),\underline{\alpha}_j} &\le C_{P,\underline{\alpha}_j}(\Omega_{j,\eta};\, h)\, H_j\, |\mathscr{H}_j^{\alpha,h} w_j|_{H^1(\Omega_{j,\eta_j}),\underline{\alpha}_j} \\ &\le C_{P,\underline{\alpha}_j}(\Omega_{j,\eta};\, h)\, H_j\, |\mathscr{H}_j^{\alpha,h} w_j|_{H^1(\Omega_j),\alpha}\,.\end{aligned}$$

Combining this estimate with Lemma 3.27, using that $|\mathscr{H}_j^{\alpha,h} w_j|_{H^1(\Omega_j),\alpha} = |w_j|_{S_j}$ and Assumption 3.14, we obtain

$$|(P_D\, w)_i|_{S_i}^2 \lesssim c_{\mathrm{noise}} \sum_{j \in \mathscr{N}_i} \Big(\omega_i^2\, |w_j|_{S_j}^2 + \eta_i^{-2}\, \omega_i\, H_j^2\, C_{P,\underline{\alpha}_j}(\Omega_{i,\eta};\, h)^2\, |w_j|_{S_j}^2\Big) \quad \forall w \in W^\perp,$$

where $\omega_i = 1 + \log(\eta_i / h_i)$. Since $Q = M_{\mathrm{sD}}^{-1}$, Lemma 2.105 implies that $|P_D\, z_w|_S^2 \le |P_D\, w|_S^2$. The estimate in the theorem now follows from Lemma 2.45 and the fact that $\eta_i \simeq \eta_j$, $h_i \simeq h_j$, and $\eta_j \le H_j$. □

3.3.5.2 Special Cases of Theorem 3.28

Here we would like to discuss special cases of the theorem above in order to make the statement more intelligible, and also to motivate the need of investigating the weighted Poincaré parameters.

The Case $\eta \eqsim H$. If $\Omega_{i,\eta} = \Omega_i$ (and $\eta_i \eqsim H_i$ since $\mathscr{T}^\eta(\Omega)$ contains just a few elements per subdomain), we reproduce the estimate from Theorem 3.6. This is because $\underline{\alpha}_i$ is constant and so $C_{P,\underline{\alpha}_i}(\Omega_i;\, h) = C_P(\Omega_i) \lesssim 1$ for all $i = 1, \dots, s$ and $c_{\mathrm{noise}} = \max_{i=1}^s \frac{\alpha_{\Omega_i}^{\max}}{\alpha_{\Omega_i}^{\min}}$.

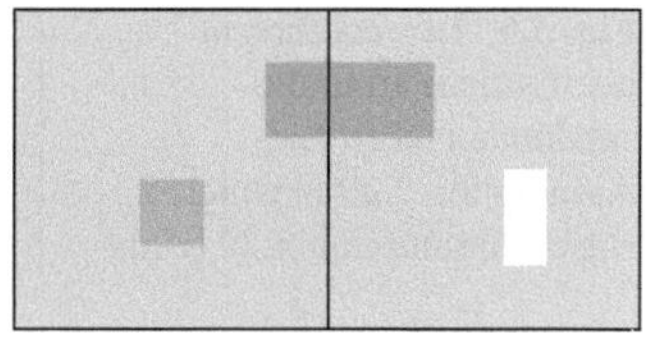

Fig. 3.5 Coefficient that does not jump *across* a subdomain interface

Constant Coefficients in the Boundary Layers. We will see in Sect. 3.5.1 that if additionally Assumption 2.54 (p. 115) holds, then $C_P(\Omega_{i,\eta})$ is uniformly bounded and independent of H_i and η_i. Hence, if α is constant on $\Omega_{i,\eta}$, then we get the condition number bound

$$\kappa(P\,M_{\mathrm{sD}}^{-1}P^\top F_{|\tilde{U}_{\mathrm{ad}}}) \;\le\; C\,\max_{i=1}^{s}\left(\frac{H_i}{\eta_i}\right)^2\left(1+\log(\tfrac{\eta_i}{h_i})\right)^2,$$

and the constant C is independent of α, h_i, η_i, and H_i. In particular, C is independent of the values of α in the subdomain interiors $\Omega_i \setminus \Omega_{i,\eta}$. We will see that the quadratic dependence on H_i/η_i can be reduced to a linear one under additional assumptions on α in the interior regions $\Omega_i \setminus \Omega_{i,\eta}$, see Theorem 3.64.

The Case $\eta \precsim h$. Let $\Omega_{i,h}$ denote the union of elements from $\mathcal{T}^h(\Omega_i)$ touching $\Gamma \cup \Gamma_D$. Setting $\Lambda_i = \Omega_{i,h}$ and choosing $\mathcal{T}^\eta(\Lambda) = \bigcup_{i=1}^N \mathcal{T}^h(\Omega_{i,h})$ yields

$$\kappa(P\,M_{\mathrm{sD}}^{-1}P^\top F_{|\tilde{U}_{\mathrm{ad}}}) \;\le\; C\,\max_{i=1}^{s}\left(\frac{H_i}{h_i}\right)^2\left(1+C_{P,\alpha}(\Omega_{i,h};\,h)\right).$$

Note that for certain coefficients, this bound might already be valuable, provided that $C_{P,\alpha}(\Omega_{i,h};\,h)$ can be robustly bounded and that the subdomain meshes are not too large.

3.3.5.3 A Result for the Multiplicity Scaling

Assume that there are no jumps across subdomain interfaces, i.e., for each glob $\mathcal{G} \subset \Gamma$, and for all η-globs $\mathsf{g} \subset \mathcal{G}$ of the same dimension as $\mathcal{G}$,

$$\alpha_{\mathsf{U}_{i,\mathsf{g}}}^{\max} \qquad \forall i,\,j \in \mathcal{N}_{\mathsf{g}} = \alpha_{\mathsf{U}_{j,\mathsf{g}}}^{\max}\,.$$

For an illustration, see Fig. 3.5. Then the statement of Theorem 3.28 remains true for the multiplicity scaling (i.e., $\rho_i(x^h) = 1$). This is seen by following the proof of Lemma 3.27 and using that $\delta_j^\dagger(x^h) \le 1$.

3.3.5.4 Diagonal Choice of Q

In this section, we discuss how to choose Q_{diag} in order to get the same (or at least a similar) condition number bound as in Theorem 3.28 (where $Q = M_{\mathrm{sD}}^{-1}$). This issue

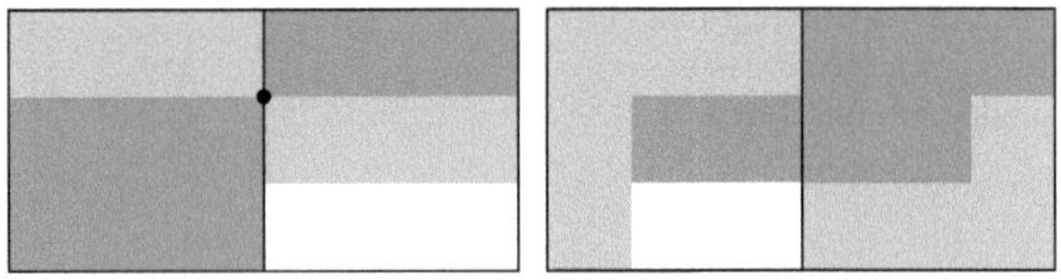

Fig. 3.6 Two coefficient distributions around a subdomain facet. *Left*: Assumption 3.29 is violated. *Right*: Assumption 3.29 holds

is rather technical, and so the reader may initially bypass this section and return to it at a later stage. The key ideas of this section can already be found in [PS11b].

Recall from (2.91) and (2.92) that $Q_{\text{diag}} : U^* \to U$ is defined by

$$(Q_{\text{diag}}\,\mu)_{ij}(x^h) := \min(\rho_i(x^h), \rho_j(x^h))\, q_{ij}(x^h)\, \mu_{ij}(x^h) \qquad \text{for } \mu \in U^*,$$

where $q_{ij}(x^h) = \min(q_i(x^h), q_j(x^h))$ and (in case of the all-floating formulation)

$$(Q_{\text{diag}}\,\mu)_{iD}(x^h) := \rho_i(x^h)\, q_i(x^h) \qquad \text{for } \mu \in U^*.$$

with the weights $q_i(x^h)$ given by

$$q_i(x^h)\mu_{iD}(x^h) := \begin{cases} (1 + \log(H_i/h_i))\,\frac{h_i^{d-1}}{H_i} & \text{if } x^h \in \mathscr{F}, \\ h_i^{d-2} & \text{else}, \end{cases} \tag{3.19}$$

Hence, Q_{diag} requires the scalings $\rho_i(x^h)$ and some mesh information. Unfortunately, by using $Q = Q_{\text{diag}}$, we cannot (theoretically) show the same bound as in Theorem 3.28 (cf. Theorem 3.31 below). However, the bound can be improved by using one of the two following alternative diagonal choices of Q.

- The operator $Q_{\text{diag}}^{(2)}$ is defined like Q_{diag} but each weight $q_i(x^h)$ is replaced by

$$q_i^{(2)}(x^h) := \begin{cases} (1 + \log(\eta_i/h_i))\,\frac{h_i^{d-1}}{\eta_i} & \text{if } x^h \in \mathscr{F}, \\ h_i^{d-2} & \text{else}. \end{cases} \tag{3.20}$$

  This choice requires the additional knowledge of the scale η_i.
- The operator $Q_{\text{diag}}^{(3)}$ is defined like Q_{diag} but each weight $q_i(x^h)$ is replaced by

$$q_i^{(3)}(x^h) := \begin{cases} (1 + \log(\eta_i/h_i))\,\frac{h_i^{d-1}}{\eta_i} & \text{if } x^h \in \mathsf{f}, \\ h_i^{d-2} & \text{else}. \end{cases} \tag{3.21}$$

  Note that this choice requires additional knowledge of both the scale η_i and of the η-globs. In practice, this is hardly ever possible.

For the choices $Q = Q_{\text{diag}}$ and $Q = Q_{\text{diag}}^{(2)}$ we will require an additional assumption on the coefficient α. Essentially, this assumptions rules out local checkerboard distributions in the interior of subdomain faces or edges, cf. Fig. 3.6.

Assumption 3.29. For all subdomain facets $\mathscr{F}_{ij}$ and for all η-vertices $\mathsf{v} \in \mathscr{F}_{ij}$ there exists an η-facet $\mathsf{f} \subset \mathscr{F}_{ij}$ with $\mathsf{v} \in \bar{\mathsf{f}}$ such that

$$\alpha_{\mathsf{U}_{i,\mathsf{v}}}^{\max} = \alpha_{\mathsf{U}_{i,\mathsf{f}}}^{\max} \quad \text{and} \quad \alpha_{\mathsf{U}_{j,\mathsf{v}}}^{\max} = \alpha_{\mathsf{U}_{j,\mathsf{f}}}^{\max} .$$

If $d = 3$, then for all η-edges $\mathsf{e} \in \mathscr{F}_{ij}$, there exists an η-facet $\mathsf{f} \subset \mathscr{F}_{ij}$ with $\mathsf{e} \in \bar{\mathsf{f}}$ such that

$$\alpha_{\mathsf{U}_{i,\mathsf{e}}}^{\max} = \alpha_{\mathsf{U}_{i,\mathsf{f}}}^{\max} \quad \text{and} \quad \alpha_{\mathsf{U}_{j,\mathsf{e}}}^{\max} = \alpha_{\mathsf{U}_{j,\mathsf{f}}}^{\max} .$$

Finally, for $d = 3$ and for all η-vertices v contained in a subdomain edge $\mathscr{E}$, there exists an η-edge $\mathsf{e} \subset \mathscr{E}$ with $\mathsf{v} \in \bar{\mathsf{e}}$ such that

$$\forall j \in \mathscr{N}_{\mathsf{e}} : \alpha_{\mathsf{U}_{j,\mathsf{v}}}^{\max} = \alpha_{\mathsf{U}_{j,\mathsf{e}}}^{\max} .$$

Lemma 3.30. *Let the assumptions of Lemma 3.27 hold and let z_w be the unique element from Lemma 2.44. Then the following statements hold for the both the classical and the all-floating formulation. For $Q = Q_{\mathrm{diag}}^{(3)}$ (see (3.21)),*

$$|P_D\, z_w|_S^2 \lesssim c_{\mathrm{noise}} \sum_{i=1}^{s} \Big(1 + \log\big(\frac{\eta_i}{h_i}\big)\Big) \Big[|\mathscr{H}_i^{\alpha,h} w_i|^2_{H^1(\Omega_{i,\eta}),\underline{\alpha}_i} + \frac{1}{\eta_i^2} \| \mathscr{H}_i^{\alpha,h} w_i \|^2_{L^2(\Omega_{i,\eta}),\underline{\alpha}_i} \Big]. \tag{3.22}$$

For $Q = Q_{\mathrm{diag}}^{(2)}$ (see (3.20)) with Assumption 3.29 fulfilled, bound (3.22) holds as well.
For $Q = Q_{\mathrm{diag}}$ (see (3.19)) with Assumption 3.29 fulfilled, bound (3.22) holds with an additional (global) factor of $\max_{k=1}^{s} H_k/\eta_k$.

Proof. As in the proof of Lemma 2.109, z_w is constant in each subdomain, and we denote the components by z_i. With the same arguments as in the proof of Lemma 3.27, the function

$$v_i := \sum_{\mathsf{g} \subset \partial\Omega_i \cap \Gamma} \sum_{j \in \mathscr{N}_{\mathsf{g}}} \delta_{j|\mathsf{g}}^{\dagger}\, I^h\big(\vartheta_{\mathsf{g}}(z_i - z_j)\big) + \sum_{\mathsf{g} \subset \Gamma_D} I^h(\vartheta_{\mathsf{g}} z_i) \in V^h(\Omega_i)$$

is an extension of $(P_D\, z_w)_i$. With Lemma 3.21 and using the fact that the number $\#\mathscr{N}_i$ of neighboring subdomains is uniformly bounded, we obtain

$$\begin{aligned} &|(P_D\, z_w)_i|_{S_i}^2 \qquad (3.23) \\ &\lesssim \sum_{\mathsf{g} \subset \partial\Omega_i \cap \Gamma} \sum_{j \in \mathscr{N}_{\mathsf{g}}} \big(\delta_{j|\mathsf{g}}^{\dagger}\big)^2 |z_i - z_j|^2 |\vartheta_{\mathsf{g}}|^2_{H^1(\mathsf{U}_{i,\mathsf{g}}),\alpha} + \sum_{\mathsf{g} \subset \Gamma_D \cap \partial\Omega_i} |z_i|^2 |\vartheta_{\mathsf{g}}|^2_{H^1(\mathsf{U}_{i,\mathsf{g}}),\alpha} \\ &\lesssim \sum_{\mathsf{g} \subset \partial\Omega_i \cap \Gamma} \sum_{j \in \mathscr{N}_{\mathsf{g}}} \min(\alpha_{\mathsf{U}_{i,\mathsf{g}}}^{\max}, \alpha_{\mathsf{U}_{j,\mathsf{g}}}^{\max}) |\vartheta_{\mathsf{g}}|^2_{H^1(\mathsf{U}_{i,\mathsf{g}})} |z_i - z_j|^2 + \sum_{\mathsf{g} \subset \Gamma_D \cap \partial\Omega_i} \alpha_{\mathsf{U}_{i,\mathsf{g}}}^{\max} |\vartheta_{\mathsf{g}}|^2_{H^1(\mathsf{U}_{i,\mathsf{g}})} |z_i|^2, \end{aligned}$$

where we have used that $\rho_{i|\mathsf{g}} = \alpha^{\max}_{\mathsf{U}_{i,\mathsf{g}}}$ and the elementary inequality from Lemma 2.98. Thanks to Assumptions 3.11 and 3.15, Lemmas 2.74 and 2.76 are applicable on $\mathsf{U}_{i,\mathsf{g}}$ and yield

$$\left.\begin{array}{l} \alpha^{\max}_{\mathsf{U}_{i,\mathsf{f}}} |\vartheta_\mathsf{f}|^2_{H^1(\mathsf{U}_{i,\mathsf{f}})} \lesssim p_{i,\mathsf{f}} := \alpha^{\max}_{\mathsf{U}_{i,\mathsf{f}}} (1 + \log(\eta_i / h_i))\, \eta_i^{d-2}, \\ \alpha^{\max}_{\mathsf{U}_{i,\mathsf{e}}} |\vartheta_\mathsf{e}|^2_{H^1(\mathsf{U}_{i,\mathsf{e}})} \lesssim p_{i,\mathsf{e}} := \alpha^{\max}_{\mathsf{U}_{i,\mathsf{e}}} \eta_i \qquad\qquad \text{if } d = 3, \\ \alpha^{\max}_{\mathsf{U}_{i,\mathsf{v}}} |\vartheta_\mathsf{v}|^2_{H^1(\mathsf{U}_{i,\mathsf{v}})} \lesssim p_{i,\mathsf{v}} := \alpha^{\max}_{\mathsf{U}_{i,\mathsf{v}}} h_i^{d-2}. \end{array}\right\} \tag{3.24}$$

Due to Assumptions 3.11 and 3.15, there are $\mathscr{O}((\eta_i / h_i)^{d-1})$ nodes contained in an η-facet $\mathsf{f} \subset \partial\Omega_i$ and $\mathscr{O}(\eta_i / h_i)$ in an η-edge (if $d = 3$). Hence, in (3.23), we get

- η-facet terms:

$$\min(\alpha^{\max}_{\mathsf{U}_{i,\mathsf{f}}}, \alpha^{\max}_{\mathsf{U}_{j,\mathsf{f}}})\, |\vartheta_\mathsf{f}|_{H^1(\mathsf{U}_{i,\mathsf{f}})}\, |z_i - z_j|^2$$
$$\lesssim \sum_{x^h \in \mathsf{f}^h} \min(\rho_i(x^h), \rho_j(x^h)) \underbrace{\left(\frac{h_i}{\eta_i}\right)^{d-1} \eta_i^{d-2}}_{= h_i^{d-1}/\eta_i} (1 + \log(\eta_i / h_i))\, |z_i - z_j|^2,$$

- η-edge terms (if $d = 3$):

$$\min(\alpha^{\max}_{\mathsf{U}_{i,\mathsf{e}}}, \alpha^{\max}_{\mathsf{U}_{j,\mathsf{e}}})\, |\vartheta_\mathsf{e}|_{H^1(\mathsf{U}_{i,\mathsf{e}})}\, |z_i - z_j|^2 \lesssim \sum_{x^h \in \mathsf{e}^h} \min(\rho_i(x^h), \rho_j(x^h)) \underbrace{\frac{h_i}{\eta_i}\, \eta_i}_{= h_i} |z_i - z_j|^2,$$

- η-vertex terms:

$$\min(\alpha^{\max}_{\mathsf{U}_{i,\mathsf{v}}}, \alpha^{\max}_{\mathsf{U}_{j,\mathsf{v}}})\, |\vartheta_\mathsf{e}|_{H^1(\mathsf{U}_{i,\mathsf{v}})}\, |z_i - z_j|^2 \lesssim \min(\rho_i(\mathsf{v}), \rho_j(\mathsf{v}))\, h_i^{d-2}\, |z_i - z_j|^2,$$

In the all-floating formulation, we get the analogous terms on the local Dirichlet boundary involving $|z_i|^2$. Inserting the estimates of the different terms into (3.23) and noticing that $\pm(z_i - z_j)$ are the components of $B\, z_w$, we see that under the condition

$$\left.\begin{array}{ll} q_i(x^h) \gtrsim \left(1 + \log(\frac{\eta_i}{h_i})\right) \frac{h_i^{d-1}}{\eta_i} & \text{for } x^h \in \mathsf{f}, \\ q_i(x^h) \gtrsim h_i & \text{for } x^h \in \mathsf{e} \text{ and } d = 3, \\ q_i(x^h) \gtrsim h_i^{d-2} & \text{for } x^h = \mathsf{v}, \end{array}\right\} \tag{3.25}$$

we get

$$|(P_D\, z_w)|^2_S \lesssim \|B\, z_w\|^2_{Q_{\text{diag}}} \le \|B\, w\|^2_{Q_{\text{diag}}},$$

where the second inequality follows from Lemma 2.44.

Recall that $q_{ij}(x^h) = \min(q_i(x^h), q_i(x^h))$. Splitting the components of $B\, w$ into η-globs yields

$$\|B\,w\|^2_{Q_{\mathrm{diag}}} \lesssim \sum_{\mathsf{g}\subset\Gamma}\sum_{i<j\in\mathcal{N}_{\mathsf{g}}} \min(\rho_{i|\mathsf{g}},\,\rho_{j|\mathsf{g}}) \sum_{x^h\in\mathsf{g}^h} q_{ij}(x^h)\,|w_i(x^h)-w_j(x^h)|^2$$

$$+\sum_{i=1}^{s}\sum_{\mathsf{g}\subset\partial\Omega_i\cap\Gamma_D} \rho_{i|\mathsf{g}} \sum_{x^h\in\mathsf{g}^h} q_i(x^h)\,|w_i(x^h)|^2$$

$$\lesssim \sum_{i=1}^{s}\sum_{\mathsf{g}\subset\partial\Omega_i\cap(\Gamma\cup\Gamma_D)} \alpha^{\max}_{\mathsf{U}_{i,\mathsf{g}}} \sum_{x^h\in\mathsf{g}^h} q_i(x^h)\,|w_i(x^h)|^2$$

$$\lesssim \sum_{i=1}^{s}\sum_{\mathsf{g}\subset\partial\Omega_i\cap(\Gamma\cup\Gamma_D)} \alpha^{\max}_{\mathsf{U}_{i,\mathsf{g}}}\, q_{i|\mathsf{g}}\, h_i^{-d_{\mathsf{g}}}\, \|w_i\|^2_{L^2(\mathsf{g})}\,.$$

Applying the trace inequality from Lemma 2.69 (for $\mathcal{G}$ replaced by g and Ω_i replaced by $Y_i^{(k_{i,\mathsf{g}})}$ where $k_{i,\mathsf{g}}$ is defined as in (3.15)), we obtain

$$\|B\,w\|^2_{Q_{\mathrm{diag}}}$$

$$\lesssim \sum_{i=1}^{s}\sum_{\mathsf{g}\subset\partial\Omega_i\cap(\Gamma\cup\Gamma_D)} \alpha^{\max}_{\mathsf{U}_{i,\mathsf{g}}}\, r_{\mathsf{g}}(\eta_i,\,h_i)\Big(|\mathcal{H}_i^{\alpha,h} w_i|^2_{H^1(Y_i^{(k_{i,\mathsf{g}})})} + \eta_i^{-2}\,\|\mathcal{H}_i^{\alpha,h} w_i\|^2_{L^2(Y_i^{(k_{i,\mathsf{g}})})}\Big),$$

where

$$r_{\mathsf{g}}(\eta_i,\,h_i) := q_{i|\mathsf{g}}\, h_i^{d_{\mathsf{g}}}\, \sigma^{d-d_{\mathsf{g}}}\big(\tfrac{\eta_i}{h_i}\big)\, \eta_i^{2+d_{\mathsf{g}}-d}\,. \tag{3.26}$$

A short computation shows that $r_{\mathsf{g}}(\eta_i,\,h_i) \lesssim (1+\log(\eta_i/h_i))$ if

$$\left.\begin{array}{ll} q_i(x^h) \lesssim (1+\log(\eta_i/h_i))\,h_i^{d-1}/\eta_i & \text{for } x^h\in\mathsf{f},\\ q_i(x^h) \lesssim h_i & \text{for } x^h\in\mathsf{e} \text{ if } d=3,\\ q_i(x^h) \lesssim (1+\log(\eta_i/h_i))^{d-2}\,h_i^{d-2} & \text{for } x^h=\mathsf{v}. \end{array}\right\} \tag{3.27}$$

Conditions (3.25) and (3.27) are fulfilled by the choice (3.21) ($Q^{(3)}_{\mathrm{diag}}$). With Assumption 3.14 we obtain from the above estimates that

$$|P_D\,z_w|^2_S \lesssim c_{\mathrm{noise}} \sum_{i=1}^{s}\big(1+\log\big(\tfrac{\eta_i}{h_i}\big)\big)\Big[|\mathcal{H}_i^{\alpha,h} w_i\|^2_{H^1(\Omega_{i,\eta}),\underline{\alpha}_i} + \tfrac{1}{\eta_i^2}\,\|\mathcal{H}_i^{\alpha,h} w_i\|^2_{L^2(\Omega_{i,\eta}),\underline{\alpha}_i}\Big].$$

We now turn to the estimate for the choice $Q^{(2)}_{\mathrm{diag}}$ (see (3.20)) and assume that Assumption 3.29 holds. Then for each η-vertex $\mathsf{v}\subset\mathcal{F}_{ij}$ there exists an η-facet $\mathsf{f}\subset\mathcal{F}_{ij}$ with $\mathsf{v}\in\bar{\mathsf{f}}$ and

$$p_{i,\mathsf{v}} \le p_{i,\mathsf{f}}\,, \qquad p_{j,\mathsf{v}} \le p_{j,\mathsf{f}}\,,$$

where $p_{i,\mathsf{v}}$ etc. are defined in (3.24). Similarly for $d = 3$ and $\mathsf{e} \subset \mathscr{F}_{ij}$ there exists an η-facet $\mathsf{f} \subset \mathscr{F}_{ij}$ with $\mathsf{e} \subset \bar{\mathsf{f}}$ and

$$p_{i,\mathsf{e}} \leq p_{i,\mathsf{f}}, \qquad p_{j,\mathsf{e}} \leq p_{j,\mathsf{f}}.$$

Finally, for $d = 3$ and $\mathsf{v} \subset \mathscr{E}$ there exists an η-edge $\mathsf{e} \subset \mathscr{E}$ with $\mathsf{v} \subset \bar{\mathsf{e}}$ and

$$\forall j \in \mathscr{N}_{\mathscr{E}} : p_{i,\mathsf{v}} \leq p_{i,\mathsf{e}}.$$

In case of the all-floating formulation we also need to consider η-globs on the Dirichlet boundary. However, for an η-vertex $\mathsf{v} \subset \mathscr{F} \subset \Gamma_D$, we can easily find an η-facet f containing v such that $p_{i,\mathsf{v}} \leq p_{i,\mathsf{f}}$. In three dimensions, for an η-edge $\mathsf{e} \subset \mathscr{F} \subset \Gamma_D$, there exists an η-facet f containing e with $p_{i,\mathsf{e}} \leq p_{i,\mathsf{f}}$, and for $\mathsf{v} \subset \mathscr{E} \subset \Gamma_D$, there exists an η-edge e containing v with $p_{i,\mathsf{v}} \leq p_{i,\mathsf{e}}$.

Therefore, we can skip the η-edge and η-vertex terms in the interior of subdomain facets and the η-vertex terms in the interior of subdomain edges because their upper bounds already pop up in the η-facet terms or the η-edge terms. Summarizing, we get

$$\begin{aligned}
|P_D\, z_w|_S^2 &\lesssim \sum_{i=1}^{s} \Bigg[\sum_{\mathscr{G} \subset \partial\Omega_i \cap \Gamma} \sum_{j \in \mathscr{N}_{\mathscr{G}}} \sum_{\substack{\mathsf{g} \subset \mathscr{G} \\ d_{\mathsf{g}} = d_{\mathscr{G}}}} \min(q^{(2)}_{i|\mathsf{g}}, q^{(2)}_{j|\mathsf{g}}) \sum_{x^h \in \mathsf{g}} \min(\rho_i(x^h), \rho_j(x^h)) |z_i - z_j|^2 \\
&\quad \sum_{\mathscr{G} \subset \partial\Omega_i \cap \Gamma_D} \sum_{\substack{\mathsf{g} \subset \mathscr{G} \\ d_{\mathsf{g}} = d_{\mathscr{G}}}} q^{(2)}_{i|\mathsf{g}} \sum_{x^h \in \mathsf{g}} \rho_i(x^h)\, |z_i|^2 \Bigg] \\
&\lesssim \|B z_w\|^2_{Q^{(2)}_{\text{diag}}} \lesssim \|B w\|^2_{Q^{(2)}_{\text{diag}}},
\end{aligned} \tag{3.28}$$

under the condition that

$$\left.\begin{aligned}
&q_i^{(2)}(x^h) \gtrsim \left(1 + \log(\tfrac{\eta_i}{h_i})\right) \frac{h_i^{d-1}}{\eta_i} && \text{for } x^h \in \mathsf{f} \subset \mathscr{F}, \\
&q_i^{(2)}(x^h) \gtrsim h_i && \text{for } x^h \in \mathsf{e} \subset \mathscr{E} \text{ and } d = 3, \\
&q_i^{(2)}(x^h) \gtrsim h_i^{d-2} && \text{for } x^h = \mathscr{V},
\end{aligned}\right\} \tag{3.29}$$

which is fulfilled (see (3.20)). Note that there are so far no restrictions on $q_i^{(2)}(x^h)$ on the remaining η-globs. For the particular choice (3.20) we have the following properties. For $\mathsf{g} \subset \mathscr{F}$,

$$q^{(2)}_{i|\mathsf{g}} = q^{(2)}_{i|\mathscr{F}} \quad \text{and} \quad \mathscr{N}_{\mathsf{g}} = \mathscr{N}_{\mathscr{F}}.$$

For $d = 3$ and $\mathsf{g} \subset \mathscr{E}$,

$$q^{(2)}_{i|\mathsf{g}} = q^{(2)}_{i|\mathscr{E}} \quad \text{and} \quad \mathscr{N}_{\mathsf{g}} = \mathscr{N}_{\mathscr{E}}.$$

Due to these properties and because of Assumption 3.29, the terms corresponding to η-vertices (and η-edges if $d = 3$) that are contained in the interior of a subdomain face can be dropped because they already occur in the terms corresponding to η-facets. The additional factor that we have to introduce is uniform because the number of neighboring η-globs of an η-glob is uniformly bounded. Similarly, for $d = 3$ the terms corresponding to η-vertices contained in the interior of a subdomain edge can be dropped. Summarizing, we get

$$\begin{aligned} \|B\,w\|^2_{Q^{(2)}_{\text{diag}}} &\lesssim \sum_{i=1}^{s} \sum_{\mathcal{G}\subset\partial\Omega_i\cap(\Gamma\cup\Gamma_D)} \sum_{\mathsf{g}\subset\mathcal{G},\,d_{\mathsf{g}}=d_{\mathcal{G}}} \alpha^{\max}_{\mathsf{U}_{i,\mathsf{g}}}\, q^{(2)}_{i|\mathsf{g}}\, h_i^{d_{\mathsf{g}}}\, \|w_i\|^2_{L^2(\mathsf{g})} \\ &\lesssim \sum_{i=1}^{s} \sum_{\mathcal{G}\subset\partial\Omega_i\cap(\Gamma\cup\Gamma_D)} \sum_{\mathsf{g}\subset\mathcal{G},\,d_{\mathsf{g}}=d_{\mathcal{G}}} \alpha^{\max}_{\mathsf{U}_{i,\mathsf{g}}}\, \underbrace{r^{(2)}_{i,\mathsf{g}}(\eta_i,\,h_i)}_{\lesssim(1+\log(\eta_i/h_i))} \Big(|\mathcal{H}_i^{\alpha,h} w_i|^2_{H^1(Y_i^{(k_i,\mathsf{g})})} \\ &\qquad + \eta_i^{-2}\, \|\mathcal{H}_i^{\alpha,h} w_i\|^2_{L^2(Y_i^{(k_i,\mathsf{g})})} \Big), \end{aligned}$$

with $r^{(2)}_{i,\mathsf{g}}(\eta_i,\,h_i)$ defined analogously to (3.26). The estimate above leads to the same bound as for the choice $Q = Q^{(3)}$.

The proof of choice $Q = Q_{\text{diag}}$ (see (3.19)) follows that of $Q = Q^{(2)}_{\text{diag}}$. Due to the mismatch of η_i and H_i in $q_i^{(2)}(x^h)$ and $q_i(x^h)$, we obtain (analogously to (3.28))

$$|P_D\, z_w|^2_S \;\lesssim\; \max_{k=1,\dots,s} \frac{H_k}{\eta_k}\, \|B\,w\|^2_{Q_{\text{diag}}}\,.$$

The rest of the proof is identical. □

The previous lemma allows for a bound of the condition number of all-floating FETI, cf. [PS11b, Theorems 4.1 and 6.4].

Theorem 3.31. *Let Assumptions 3.11–3.16 (p. 164f) hold. Moreover, let the scalings $\rho_i(x^h)$ be chosen according to (3.6). For $j = 1,\dots,s$, let the lower bound function $\underline{\alpha}_i \in L^\infty(\Omega_{i,\eta})$ be defined as in Lemma 3.27. Then for all-floating FETI method with $Q = Q^{(3)}_{\text{diag}}$ (see (3.21)),*

$$\kappa(P\,M_{\text{sD}}^{-1}\,P^\top F_{|\widetilde{U}_{\text{ad}}}) \;\le\; C\,c_{\text{noise}}\, \max_{i=1}^{s} \Big(\frac{H_i}{\eta_i}\Big)^{\beta} \big(1+\log\big(\frac{\eta_i}{h_i}\big)\big)^2 \Big(1 + C_{P,\underline{\alpha}_i}(\Omega_{i,\eta};\,h)^2\Big),$$

where $\beta = 2$ and c_{noise} is the constant from Assumption 3.14. The constant C depends only of the uniform constants from the remaining assumptions and is independent of α, h_i, η_i, and H_i. For $Q = Q^{(2)}_{\text{diag}}$ (see (3.20)) with Assumption 3.29 fulfilled, the bound holds with $\beta = 2$ as well. Finally, for $Q = Q_{\text{diag}}$ (see (3.19)) with Assumption 3.29 fulfilled, the bound holds with $\beta = 3$.

Proof. The proof is analogous to that of Theorem 3.28 but uses Lemma 3.30 instead of Lemma 2.105. □

3.4 Weighted Poincaré Inequalities (WPI)

The FETI condition number bound in Theorem 3.28 contains the constants $C_{P,\alpha}(\Omega_{i,\eta})$ in the weighted Poincaré inequalities of the boundary layer $\Omega_{i,\eta}$. In this section, which is taken from [PS11c, PS12b], we further investigate these constants and discuss when benign bounds for them can be given.

In the following, let $D \subset \mathbb{R}^d$ ($d = 2$ or 3) be a Lipschitz domain and $\alpha \in L^\infty_+(D)$. Recall the definition of $C_{P,\alpha}(D)$ from Definition 3.23:

$$\inf_{c\in\mathbb{R}} \|u - c\|_{L^2(D),\alpha} \le C_{P,\alpha}(D)\,\mathrm{diam}(D)\,|u|_{H^1(D),\alpha} \qquad \forall u \in H^1(D). \tag{3.30}$$

We are interested in finding bounds for the constant $C_{P,\alpha}(D)$ that are independent of the values that a piecewise constant weight function α takes on D. Of course, we can obtain a straightforward bound for $C_{P,\alpha}(D)$ in (3.30) from the usual Poincaré inequality (Lemma 1.27)

$$C_{P,\alpha}(D)^2 \le \operatorname*{ess.sup}_{x,y\in D} \frac{\alpha(x)}{\alpha(y)}\, C_P(D)^2.$$

However, this bound depends on the *global variation* $\mathrm{ess.sup}_{x,y\in D} \frac{\alpha(x)}{\alpha(y)}$, and if α is highly variable, this may be very large and in many cases very pessimistic.

In this section, we characterize when and in what sense robust bounds for $C_{P,\alpha}(D)$ can be achieved. Also, for a couple of situations we will be able to make explicit the dependence on geometrical parameters hidden in α, such as the scale η, see Sects. 3.1.2 and 3.3.1.

Remark 3.32. Although weighted Poincaré inequalities have been investigated a lot in the literature, estimates of the Poincaré constant $C_{P,\alpha}(D)$ that show certain robustness in α have been hardly known. Chua [Chu93] showed that the weighted Poincaré inequality holds for domains satisfying the Boman chain condition with weights α from a Muckenhoupt class (see [Muc72]). See also [IN85, FKS82, Maz85] for related work. The constant in the Poincaré inequality depends in general on the weight. A similar result can be found in [ZP08, Lemma 2.6] for weights $\alpha \in L^r$ with $\alpha^{-1} \in L^s$ where $2d^{-1} = r^{-1} + s^{-1}$. Also there, the Poincaré constant depends on α. In [CW06], Chua and Wheeden provide explicit estimates for the Poincaré constant for the class of convex domains Ω with weights α that are a positive power of a non-negative concave function. Note that concavity implies *continuity*. Recently, Veeser and Verfürth [VV11] refined these results to star shaped domains, where the weight function satisfies a certain concavity property with respect to the central point of the star (see Condition (2.3) in [VV11] for more details, and see [VV09] on how to use these inequalities in (explicit) a-posteriori error estimation). We also note that [Chu93, CW06, VV11] cover the general case of L^p, not only L^2. To the best of our knowledge, the first paper that deals with robust estimates of the weighted Poincaré constant for *discontinuous* weight functions is [GE10a]. There, Galvis and Efendiev show the following. If α is piecewise constant, if the largest value is attained in a connected region Ω_1 and if all the other regions of constant α are inclusions of (or at least bordering) Ω_1, then $C_{P,\alpha}(D)$ is independent of the *values* of α, in particular of possibly high *contrast*. The material contained in this section stems from [PS11c, PS12b], where these results were refined and extended to more general situations. Note finally, that weighted Poincaré inequalities were already sucessfully used in analyses of other iterative solvers [SVZ12].

Our strategy for bounding $C_{P,\alpha}(D)$ is not to use the weighted average $\bar{u}^{D,\alpha}$, but an unweighted one over a suitably chosen manifold.

Definition 3.33. Let $X \subset \overline{D}$ be a either a subset or a $(d-1)$-dimensional manifold with positive (surface) measure. Let $C_{P,\alpha}(Y, X)$ be the smallest constant such that

$$\|u - \bar{u}^X\|_{L^2(D),\alpha} \leq C_{P,\alpha}(D, X)\,\mathrm{diam}(D)\,|u|_{H^1(D),\alpha} \qquad \forall u \in H^1(D). \tag{3.31}$$

For a subset $W \subset D$ with positive (d-dimensional) measure, let $C_{P,\alpha}(Y, X, W)$ be the smallest constant such that

$$\|u - \bar{u}^X\|_{L^2(W),\alpha} \leq C_{P,\alpha}(D, X, W)\sqrt{\frac{\mathrm{meas}_d(W)}{\mathrm{meas}_d(D)}}\,\mathrm{diam}(D)\,|u|_{H^1(D),\alpha} \quad \forall u \in H^1(D). \tag{3.32}$$

Apparently,

$$C_{P,\alpha}(D) \leq C_{P,\alpha}(D, X).$$

Bounds for $C_{P,\alpha}(D, X)$ will be treated in Sect. 3.4.2. They are robust under a quasi-monotonicity condition on α. If X is a lower-dimensional manifold, we can still get inequalities for finite element functions (similarly as in Sect. 2.5.4), which will be the subject of Sect. 3.4.3. In Sect. 3.4.4, we discuss the dependence on geometrical parameters. Before we come to the bounds for $C_{P,\alpha}(D, X)$, we need to make some regularity assumptions on the weight function α in Sect. 3.4.1. Note that these assumptions are slightly related to those in Sect. 3.3.1.

3.4.1 Regularity Assumptions on the Weight Function

Let the weight function $\alpha \in L^\infty_+(D)$ be piecewise constant with respect to a non-overlapping partitioning $\mathscr{Y} := \{Y^{(\ell)} : \ell = 1, \ldots, n\}$ of D into open, connected Lipschitz polytopes, i.e.,

$$\overline{D} = \bigcup_{\ell=1}^{n} \overline{Y}^{(\ell)} \qquad \text{and} \qquad \alpha_{|Y^{(\ell)}} = \alpha_\ell = \text{const.} \tag{3.33}$$

Note that the lower bound function $\underline{\alpha}_i$ from Lemma 3.27 has exactly this property. Note also that in Definition 3.39 and Corollary 3.40 below, assumption (3.33) will be replaced by a weaker one, which allows to treat oscillating coefficients as well.

We assume further that there exists a regular triangulation $\mathcal{T}^\eta(D)$ such that the regions $Y^{(\ell)}$ from above are $\mathcal{T}^\eta$-agglomerates in the sense of Definition 2.58 (p. 117). Recall from Definition 2.65 (p. 119) that an m-dimensional $\mathcal{T}^\eta$-agglomerate is the union of m-facets of the triangulation $\mathcal{T}^\eta(D)$. Recall also the definition of the average $\bar{u}^X$ for agglomerates X of dimension $0 \le d_X \le d$ (see Definition 2.66, p. 119).

3.4.2 *Weighted Poincaré Inequalities in H^1*

The following definition will be essential to characterize when the WPI in Definition 3.23, (3.30), is robust with respect to the values α_ℓ. It was first formulated in [PS11c] (see also [PS12b]) and is a generalization of the notion of quasi-monotone coefficients introduced by Dryja, Sarkis, and Widlund [Sar94, DSW96] in the context of Schwarz solvers.

Definition 3.34 (quasi-monotone paths and weights). Let $\mathcal{Y} = \{Y^{(\ell)}\}_{\ell=1}^n$ be a non-overlapping partitioning of the Lipschitz polytope D into $\mathcal{T}^\eta$-agglomerates.

(a) The region $P_{\ell_1,\ell_s} := \text{interior}(\overline{Y}^{(\ell_1)} \cup \overline{Y}^{(\ell_2)} \cup \cdots \cup \overline{Y}^{(\ell_s)})$, $1 \le \ell_1, \ldots, \ell_s \le n$ is called *path* from $Y^{(\ell_1)}$ to $Y^{(\ell_s)}$ of *length* s if for each $i = 1, \ldots, s-1$, the regions $\overline{Y}^{(\ell_i)}$ and $\overline{Y}^{(\ell_i+1)}$ share a common $(d-1)$-dimensional $\mathcal{T}^\eta$-agglomerate X_i.

(b) Let $\alpha \in L_+^\infty(D)$ be piecewise constant with respect to $\mathcal{Y}$. We say that a path P_{ℓ_1,ℓ_s} is *quasi-monotone* with respect to α if

$$\alpha_{\ell_1} \le \alpha_{\ell_2} \le \cdots \le \alpha_{\ell_s}.$$

(c) Let ℓ^* be a fixed index such that

$$\alpha_{\ell^*} = \max_{\ell=1,\ldots,n} \alpha_\ell .$$

We say that α is *quasi-monotone on D*, if for any $k = 1, \ldots, n$ there exists a quasi-monotone path P_{k,ℓ^*} from $Y^{(k)}$ to $Y^{(\ell^*)}$ with respect to α.

In Fig. 3.7a–c we give some examples of weight functions that satisfy Definition 3.34. The weight function shown in Fig. 3.7d fails to be quasi-monotone.

Remark 3.35. Note that if the maximum of α is attained in two or more subregions (such that the index ℓ^* is not unique), then the coefficient α is either not quasi-monotone, or all subregions where the maximum of α is attained must be connected. Hence, Definition 3.34 (c) is independent of the actual choice of ℓ^*.

The next theorem (cf. [PS11c, Lemma 1], [PS12b, Theorem 2.2]) gives a bound of the constant $C_{P,\alpha}(D)$ in the weighted Poincaré inequality that is independent of

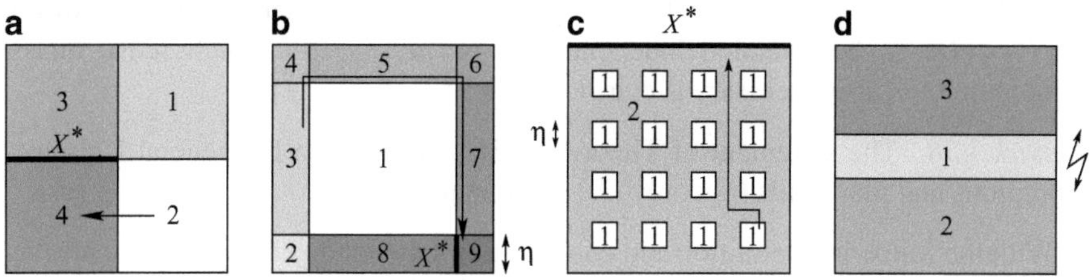

Fig. 3.7 Examples of piecewise constant weight functions. The numbering of the regions $Y^{(\ell)}$ is according to the relative sizes of the weights α_ℓ on each region, with the smallest weight in region $Y^{(1)}$. **(a)**–**(c)** Quasi-monotone, **(d)** not quasi-monotone

the values $\{\alpha_\ell\}_{\ell=1}^n$ if α is quasi-monotone. Recall the definitions of the Poincaré type constants $C_P(Y, X)$, $C_P(Y, X, W)$ from Definition 2.67, p. 120.

Theorem 3.36 (WPI in H^1). *Let $\alpha \in L^\infty_+(D)$ be quasi-monotone on D in the sense of Definition 3.34 such that the assumptions from Sect. 3.4.1 hold. Then for any $\mathscr{T}^\eta$-agglomerate $X^* \subset \overline{Y}^{(\ell^*)}$ of dimension d or $d-1$,*

$$C_{P,\alpha}(D)^2 \;\le\; C_{P,\alpha}(D, X^*)^2 \;\le\; \sum_{k=1}^{n} C_P(P_{k,\ell^*}, X^*, Y^{(k)})^2\, \frac{|Y^{(k)}|\,\mathrm{diam}(P_{k,\ell^*})^2}{|P_{k,\ell^*}|\,\mathrm{diam}(D)^2}. \tag{3.34}$$

In particular, $C_{P,\alpha}(D)$ and $C_{P,\alpha}(D, X^*)$ are independent of the values $\{\alpha_\ell\}_{\ell=1}^n$.

Proof. Let $k \in \{1,\dots,n\}$ be fixed. Then, due to assumption (3.33) and Definition 2.67,

$$\begin{aligned}\|u - \bar{u}^{X^*}\|^2_{L^2(Y^{(k)}),\alpha} &= \alpha_k\,\|u - \bar{u}^{X^*}\|^2_{L^2(Y^{(k)})}\\ &\le \alpha_k\, C_P(P_{k,\ell^*}, X^*, Y^{(k)})^2\, \frac{|Y^{(k)}|}{|P_{k,\ell^*}|}\,\mathrm{diam}(P_{k,\ell^*})^2\, |u|^2_{H^1(P_{k,\ell^*})}.\end{aligned}$$

Thanks to the quasi-monotonicity of α,

$$\alpha_k\,|u|^2_{H^1(P_{k,\ell^*})} \;\le\; |u|^2_{H^1(D),\alpha}.$$

Combining these two estimates and summing over $k = 1,\dots,n$ gives

$$\begin{aligned}\inf_{c\in\mathbb{R}}\|u - c\|^2_{L^2(D)} &\le \|u - \bar{u}^{X^*}\|^2_{L^2(D),\alpha}\\ &\le \Big(\sum_{k=1}^{n} C_P(P_{k,\ell^*}, X^*, Y^{(k)})^2\,\frac{|Y^{(k)}|}{|P_{k,\ell^*}|}\,\mathrm{diam}(P_{k,\ell^*})^2\Big)\,|u|^2_{H^1(D),\alpha}.\end{aligned}$$

A comparison with (3.31) concludes the proof. □

Remark 3.37. Theorem 3.36 states that quasi-monotonicity is a sufficient condition for $C_{P,\alpha}(D)$ to be bounded independently of $\{\alpha_\ell\}_{\ell=1}^s$. In which sense quasi-monotonicity is also necessary is clarified in [PS12b, Sect. 3.1].

Remark 3.38. The statement of Theorem 3.36 remains true for general Lipschitz subregions and manifolds (not only $\mathscr{T}^\eta$-agglomerates).

With the following definition and corollary, we extend to general coefficients that are not necessarily piecewise constant, cf. [PS12b, Sect. 2.2].

Definition 3.39. Let $\mathscr{T}^\eta(D)$ be a shape regular triangulation and let $\mathscr{Y} = \{Y^{(\ell)}\}_{\ell=1}^n$ be a non-overlapping partition of D into $\mathscr{T}^\eta$-agglomerates. For a general weight function $\alpha \in L^\infty_+(D)$, we define the lower and upper bound functions $\underline{\alpha}$, $\overline{\alpha} \in L^\infty_+(D)$ by

$$\underline{\alpha}(x) := \alpha^{\min}_{Y^{(\ell)}}, \qquad \overline{\alpha}(x) := \alpha^{\max}_{Y^{(\ell)}} \qquad \text{for } x \in Y^{(\ell)}.$$

Note that these functions depend on the partition $\mathscr{Y}$. We say that α is *macroscopically quasi-monotone* on D with respect to $\mathscr{Y}$ if the lower bound function $\underline{\alpha}$ is quasi-monotone on D.

Corollary 3.40. *Let α, $\mathscr{Y}$, and $\underline{\alpha}$ be as in Definition 3.39. If α is macroscopically quasi-monotone on D with respect to $\mathscr{Y}$, then*

$$C_{P,\alpha}(D)^2 \le \left(\max_{\ell=1}^{n} \frac{\alpha^{\max}_{Y^{(\ell)}}}{\alpha^{\min}_{Y^{(\ell)}}} \right) C_{P,\underline{\alpha}}(D)^2 ,$$

i.e., the Poincaré constant depends additionally on the local variation α *in each of the subregions $Y^{(\ell)}$, but is independent of the values of $\underline{\alpha}$.*

Proof. For any $u \in H^1(D)$,

$$\begin{aligned}
\inf_{c\in\mathbb{R}} \|u - c\|^2_{L^2(D),\alpha} &\le \|u - \bar{u}^{D,\underline{\alpha}}\|^2_{L^2(D),\alpha} \le \max_{\ell=1}^{n} \frac{\alpha^{\max}_{Y^{(\ell)}}}{\alpha^{\min}_{Y^{(\ell)}}} \|u - \bar{u}^{D,\underline{\alpha}}\|^2_{L^2(D),\underline{\alpha}} \\
&\le \max_{\ell=1}^{n} \frac{\alpha^{\max}_{Y^{(\ell)}}}{\alpha^{\min}_{Y^{(\ell)}}} C_{P,\underline{\alpha}}(D)^2 \operatorname{diam}(D)^2 \, |u|^2_{H^1(D),\underline{\alpha}} \,.
\end{aligned}$$

The fact that $|u|^2_{H^1(D),\underline{\alpha}} \le |u|^2_{H^1(D),\alpha}$ concludes the proof. □

Remark 3.41. Note that any coefficient $\alpha \in L^\infty_+(D)$ is macroscopically quasi-monotone on D with respect to the trivial partition $\mathscr{Y} = \{D\}$. Then the bound in Corollary 3.40 corresponds to the worst case scenario containing the full variation of α on D.

3.4.3 Weighted Poincaré Inequalities for FE Functions

Recall from Sect. 3.3.5 that for our FETI analysis, it would be sufficient to have a WPI just for FE functions. Indeed there are coefficient patterns, where the constant in the discrete WPI (Definition 3.26) is robust, but not in the continuous one (Definition 3.23), cf. [PS11c, Lemma 2] and [PS12b, Sect. 3].

For simplicity, let $\{\mathcal{T}^h(D)\}_h$ be a family of regular triangulations that are refinements of the "coarse" triangulation $\mathcal{T}^\eta(D)$. See Remark 3.46 on how this assumption can be weakened. Let $V^h(D)$ denote the finite element space of continuous and piecewise linear functions with respect to $\mathcal{T}^h(D)$.

Definition 3.42. For $\alpha \in L^\infty_+(D)$ and a $\mathcal{T}^\eta$-agglomerate X of dimension $0 \le d_X \le d$, let $C_{P,\alpha}(D, X; h)$ denote the smallest parameter such that

$$\|u - \bar{u}^X\|_{L^2(D),\alpha} \le C_{P,\alpha}(D, X; h)\,\mathrm{diam}(D)\,|u|_{H^1(D)} \qquad \forall u \in V^h(D). \quad (3.35)$$

For a subset $W \subset D$ with positive (d-dimensional) measure, let $C_{P,\alpha}(Y, X, W; h)$ be the smallest parameter such that

$$\|u - \bar{u}^X\|_{L^2(W),\alpha} \le C_{P,\alpha}(D, X, W; h)\sqrt{\frac{\mathrm{meas}_d(W)}{\mathrm{meas}_d(D)}}\,\mathrm{diam}(D)\,|u|_{H^1(D),\alpha} \quad \forall u \in V^h(D). \quad (3.36)$$

Definition 3.43. Let $\mathcal{Y} = \{Y^{(\ell)}\}_{\ell=1}^n$ be a non-overlapping partitioning of the Lipschitz polytope D into $\mathcal{T}^\eta$-agglomerates and let $0 \le m \le d-1$.

(a) We call the region $P_{\ell_1,\ell_s} := \mathrm{interior}(\overline{Y}^{(\ell_1)} \cup \overline{Y}^{(\ell_2)} \cup \cdots \cup \overline{Y}^{(\ell_s)})$, $1 \le \ell_1, \ldots, \ell_s \le n$, a *type-$m$ path* from $Y^{(\ell_1)}$ to $Y^{(\ell_s)}$ of *length* s if for each $i = 1, \ldots, s-1$, the regions $\overline{Y}^{(\ell_i)}$ and $\overline{Y}^{(\ell_{i+1})}$ share a common m-dimensional $\mathcal{T}^\eta$-agglomerate X_i.

(b) Let $\alpha \in L^\infty_+(D)$ be piecewise constant with respect to $\mathcal{Y}$. We say that a type-m path P_{ℓ_1,ℓ_s} is *quasi-monotone* with respect to α if

$$\alpha_{\ell_1} \le \alpha_{\ell_2} \le \cdots \le \alpha_{\ell_s}.$$

(c) Let ℓ^* be a fixed index such that

$$\alpha_{\ell^*} = \max_{\ell=1,\ldots,n} \alpha_\ell .$$

We say that α is *type-m quasi-monotone on* D, if for any $k = 1, \ldots, n$ there exists a quasi-monotone type-m path P_{k,ℓ^*} from $Y^{(k)}$ to $Y^{(\ell^*)}$ (with respect to α).

(d) For a general coefficient $\alpha \in L^\infty_+(D)$ and a fixed partition $\{Y^{(\ell)}\}_{\ell=1}^n$ of D, we call α *macroscopically type-m quasi-monotone* if the lower bound function $\underline{\alpha}$ as defined in Definition 3.39 is type-m quasi-monotone.

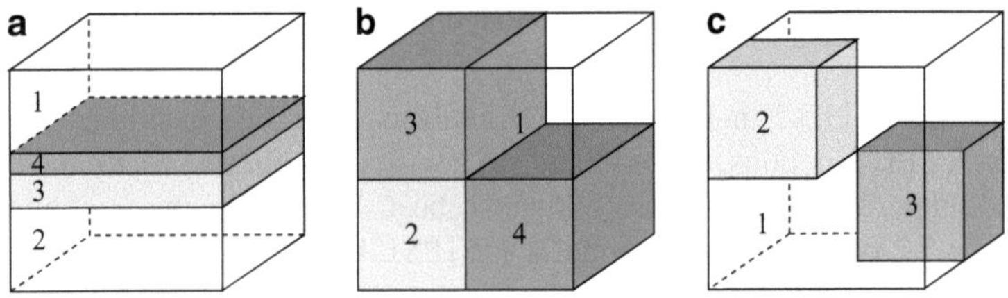

Fig. 3.8 Examples of type-m quasi-monotone weight functions for $d = 3$. The numbering of the regions $Y^{(\ell)}$ is according to the relative sizes of the weights α_ℓ on each region, with the smallest weight in region $Y^{(1)}$. (**a**) $m = 2$, (**b**) $m = 1$, and (**c**) $m = 0$

For an illustration in three dimensions see Fig. 3.8. Obviously, for $m > 0$, any type-m path is also a type-$(m-1)$ path, and any type-m quasi-monotone coefficient is also type-$(m-1)$ quasi-monotone.

Remark 3.44. Note that if the index ℓ^* is not unique, then the coefficient α is either not type-m quasi-monotone, or there exists a type-m quasi-monotone path connecting all the subregions where the maximum of α is attained. Hence, Definition 3.43 (c) is independent of the actual choice of ℓ^*.

Theorem 3.45 (discrete WPI). *For $0 \le m \le d-1$, let $\alpha \in L^\infty_+(D)$ be type-m quasi-monotone on D in the sense of Definition 3.43 and let the assumptions from Sect. 3.4.1 hold. Furthermore, let the triangulations $\{\mathcal{T}^h(D)\}_h$ be refinements of $\mathcal{T}^\eta(D)$ and let $\{V^h(D)\}_h$ be the corresponding FE spaces of continuous and piecewise linear functions. Then for any $\mathcal{T}^\eta$-agglomerate $X^* \subset \overline{Y}^{(\ell^*)}$ of dimension $0 \le d_X \le d$,*

$$C_{P,\alpha}(D;h)^2 \le C_{P,\alpha}(D,X^*;h)^2 \le \sum_{k=1}^{n} C_P(P_{k,\ell^*},X^*,Y^{(k)};h)^2 \, \frac{|Y^{(k)}|\,\mathrm{diam}(P_{k,\ell^*})^2}{|P_{k,\ell^*}|\,\mathrm{diam}(D)^2}.$$

In particular, the parameters $C_{P,\alpha}(D;h)$ and $C_{P,\alpha}(D,X^*;h)$ are independent of the values $\{\alpha_\ell\}_{\ell=1}^n$.

Proof. The proof is identical to that of Theorem 3.36; see also [PS11c, Lemma 2] and [PS12b, Lemma 3.1]. □

Remark 3.46. Almost all results in this subsection can be transferred to the more general case that the fine mesh $\mathcal{T}^h(D)$ resolves the subregions $Y^{(\ell)}$ but not necessarily the "coarse" elements from $\mathcal{T}^\eta(D)$, cf. [PS12b, Lemma 4.9]. In that case, we can find a virtual triangulation $\mathcal{T}^{h,2}(D)$ with comparable shape regularity constants that is a refinement of $\mathcal{T}^\eta(D)$ and which coincides with $\mathcal{T}^h(D)$ at the boundaries $\partial Y^{(\ell)}$, $\ell = 1,\dots,n$. Let $V^{h,2}(D)$ the finite element space corresponding to $\mathcal{T}^{h,2}(D)$. Furthermore, for each $\ell = 1,\dots,n$, let $\Pi_\ell^{h,2} : H^1(Y^{(\ell)}) \to V^{h,2}(Y^{(\ell)})$ be a Scott-Zhang quasi-interpolant (see Lemma 1.45) such that the values $(\Pi_\ell^{h,2} v)_{|\partial Y^{(\ell)}}$ depend only on $v_{|\partial Y^{(\ell)}}$ and such that piecewise linear boundary values on $\partial Y^{(\ell)}$ are preserved. The global operator $\Pi^{h,2} : V^h(D) \to V^{h,2}(D)$ given by $(\Pi^{h,2}v)_{|Y^{(\ell)}} := \Pi_\ell^{h,2} v_{|Y^{(\ell)}}$ is then well-defined and fulfills the estimates

$$\left.\begin{aligned} \|v - \Pi^{h,2}v\|_{L^2(D),\alpha} &\leq C_{SZ}\, h\, |v|_{H^1(D),\alpha} \\ |\Pi^{h,2}v|_{H^1(D),\alpha} &\leq C_{SZ}\, |v|_{H^1(D),\alpha} \end{aligned}\right\} \quad \forall v \in V^h(D),$$

where the constant C_{SZ} depends only on the shape regularity constant of $\mathcal{T}^{h,2}(D)$. (Note, however, that the global operator $\Pi^{h,2}$ would fail to produce a continuous FE function when interpolating a general H^1-function.) Let X^* be a $(d-1)$-dimensional $\mathcal{T}^h$-agglomerate contained in $\partial Y^{(\ell^*)}$. Then for any $u \in V^h(\Omega)$, $\bar{u}^{X^*} = \overline{\Pi^{h,2}u}^{X^*}$. Let $C_{P,\alpha}(D, X^*; h)$ be the weighted Poincaré parameter with respect to $V^h(\Omega)$ and $C_{P,\alpha}(D, X^*; h_2)$ that with respect to $V^{h,2}(\Omega)$. Using the estimates above, we get that

$$\begin{aligned} \|u - u^{X^*}\|_{L^2(D),\alpha} &\leq \|u - \Pi^{h,2}u\|_{L^2(D),\alpha} + \left\|\Pi^{h,2}u - \overline{\Pi^{h,2}u}^{X^*}\right\|_{L^2(D),\alpha} \\ &\leq C_{SZ}\, h\, |u|_{H^1(D),\alpha} + C_{P,\alpha}(D, X^*; h)\,\mathrm{diam}(D)\,|\Pi^{h,2}u|_{H^1(D),\alpha} \\ &\leq C_{SZ}\left(h/\mathrm{diam}(D) + C_{P,\alpha}(D, X^*; h)\right)\mathrm{diam}(D)\,|u|_{H^1(D),\alpha}\,. \end{aligned}$$

Since $h \leq \mathrm{diam}(D)$, this shows that $C_{P,\alpha}(D, X^*; h_2) \leq C_{SZ}(1+C_{P,\alpha}(D, X^*; h))$.

Remark 3.47. The union of $\mathcal{T}^\eta$-agglomerates that are connected via a type-m path is called a *type-m connected* region. Without any additional work, Theorem 3.45 and the theory below can in fact be extended to subregions $\{Y_\ell\}_{\ell=1}^n$ that are only type-m connected rather than connected. See also the example in Sect. 3.4.4.5.

3.4.4 Dependence on Geometrical Parameters

In this section, following [PS12b, Sect. 4], we investigate the dependence of $C_{P,\alpha}(D)$ and $C_{P,\alpha}(D; h)$ on the geometrical parameters. First, we would like to warn the reader that for the path constant $C_P(P_{k,\ell^*}, X^*, Y^{(k)})$ occurring in Theorem 3.36,

$$C_P(P_{k,\ell^*}, X^*, Y^{(k)}) \not\lesssim C_P(P_{k,\ell^*}, X^*).$$

This is because of the ratio of measures in Definition 2.67. One can easily show from $\|u - \bar{u}^{X^*}\|_{L^2(Y^{(k)})} \leq \|u - \bar{u}^{X^*}\|_{L^2(P_{k,\ell^*})}$ that

$$C_P(P_{k,\ell^*}, X^*, Y^{(k)}) \leq \sqrt{\frac{|P_{k,\ell^*}|}{|Y^{(k)}|}}\, C_P(P_{k,\ell^*}, X^*),$$

but this estimate is in general too pessimistic. The next lemma (cf. [PS11c, Lemma 4], [PS12b, Lemma 3.6]) bounds the Poincaré constant of a path in terms of Poincaré constants of the individual subregions. An alternative will be provided in Lemma 3.55.

Lemma 3.48. *Let* $P_{\ell_1,\ell_s} := \text{interior}(\overline{Y}^{(\ell_1)} \cup \overline{Y}^{(\ell_2)} \cup \cdots \cup \overline{Y}^{(\ell_s)})$, $1 \le \ell_1, \ldots, \ell_s \le n$ *be a type-m path with* $0 \le m \le d-1$ *and let* X_i *be the m-dimensional* $\mathcal{T}^\eta$*-agglomerates shared by* $Y^{(\ell_i)}$ *and* $Y^{(\ell_{i+1})}$*. For convenience, let* $X_0 = X_1$*. Furthermore, let* $X_s \subset \overline{Y}^{(\ell_s)}$ *be a* $\mathcal{T}^\eta$*-agglomerate of dimension* $0 \le d_{X_s} \le d$*. Then*

$$C_P(P_{\ell_1,\ell_s}, X_s, Y^{(\ell_1)}; h)^2 \le 2\sum_{i=1}^{s} \frac{|P_{\ell_1,\ell_s}|\,\text{diam}(Y^{(\ell_i)})^2}{|Y^{(\ell_i)}|\,\text{diam}(P_{\ell_1,\ell_s})^2}\left(C_P(Y^{(\ell_i)}, X_{i-1}; h)^2 + C_P(Y^{(\ell_i)}, X_i; h)^2\right).$$

If $m \ge d-1$ *and* $d_{X_s} \ge d-1$*, the parameters h above can be dropped and hence the constants refer to inequalities in* H^1*.*

Proof. Let $u \in V^h(D)$ be arbitrary but fixed. If $m, d_{X_s} \ge d-1$ we fix $u \in H^1(D)$. By a telescoping argument we have

$$\|u - \bar{u}^{X_s}\|_{L^2(Y^{(\ell_1)})} \le \|u - \bar{u}^{X_1}\|_{L^2(Y^{(\ell_1)})} + \sum_{i=2}^{s} \sqrt{|Y^{(\ell_1)}|}\,\left|\bar{u}^{X_{i-1}} - \bar{u}^{X_i}\right|.$$

With Definition 2.67,

$$\|u - \bar{u}^{X_1}\|_{L^2(Y^{(\ell_1)})} \le C_P(Y^{(k)}, X_1; h)\,\text{diam}(Y^{(\ell_1)})\,|u|_{H^1(Y^{(\ell_1)})}\,.$$

For a fixed index $i \in \{2, \ldots, s\}$, the same definition implies that

$$\begin{aligned}\left|\bar{u}^{X_{i-1}} - \bar{u}^{X_i}\right|^2 &\le \frac{2}{|Y^{(\ell_i)}|}\left(\|\bar{u}^{X_{i-1}} - u\|^2_{L^2(Y^{(\ell_i)})} + \|u - \bar{u}^{X_i}\|^2_{L^2(Y^{(\ell_i)})}\right)\\ &\le \frac{2}{|Y^{(\ell_i)}|}\left(C_P(Y^{(\ell_i)}, X_{i-1}; h)^2 + C_P(Y^{(\ell_i)}, X_i; h)^2\right)\text{diam}(Y^{(\ell_i)})^2\,|u|^2_{H^1(Y^{(\ell_i)})}\,.\end{aligned}$$

Combining the three estimates above and applying Cauchy's inequality (in $\mathbb{R}^s$) yields

$$\begin{aligned}&\|u - \bar{u}^{X^*}\|^2_{L^2(Y^{(\ell_1)})}\\ &\le \Big[\sum_{i=1}^{s} \frac{2|Y^{(\ell_1)}|\text{diam}(Y^{(\ell_i)})^2}{|Y^{(\ell_i)}|}\left(C_P(Y^{(\ell_i)}, X_{i-1}; h)^2 + C_P(Y^{(\ell_i)}, X_i; h)^2\right)\Big]|u|^2_{H^1(P_{\ell_1,\ell_s})}.\end{aligned}$$

A comparison with Definition 2.67 (p. 120) completes the proof. □

Recall the bound $C_{P,\alpha}(D)^2 \le \sum_{k=1}^{n} C_P(P_{k,\ell^*}, X^*, Y^{(k)})^2 \frac{|Y^{(k)}|\mathrm{diam}(P_{k,\ell^*})^2}{|P_{k,\ell^*}|\mathrm{diam}(D)^2}$ from Theorem 3.36 and the analogous bound from Theorem 3.45. Although these bound might suggest that the constant $C_{P,\alpha}(D)$ grows with the number n of subregions, this is typically *not* the case. Lemma 3.48 implies that for any quasi-monotone coefficient α,

$$C_{P,\alpha}(D)^2 \le 4 \qquad \text{if } d = 1,$$

see also [PS11c, Corollary 2.5]. The situation in higher dimensions is more complicated and will be subject of the following sections.

3.4.4.1 Shape Regular Partitions

Theorems 3.36 and 3.45, and Lemma 3.48 provide an abstract framework to reduce the estimation of $C_{P,\alpha}(D, X)$ or $C_{P,\alpha}(D, X; h)$ to individual Poincaré constants on the η-agglomerates $Y^{(\ell)}$. So far we have not made any assumptions on the triangulation $\mathcal{T}^\eta(D)$. In this section, we provide further results under the assumption that $\mathcal{T}^\eta(D)$ is *shape regular*. Throughout this subsection, let

$$\eta_{\min} := \min_{\ell=1}^{n} \mathrm{diam}(Y^{(\ell)}), \qquad \eta_{\max} := \max_{\ell=1}^{n} \mathrm{diam}(Y^{(\ell)}). \tag{3.37}$$

The next lemma (cf. [PS12b, Lemma 4.1]) bounds the weighted Poincaré constant in terms of $\eta_{\min}$ and $\eta_{\max}$.

Lemma 3.49. *For $d = 2$ or 3, let α be piecewise constant with respect to $\mathcal{Y}$ and type-m quasi-monotone on D. Assume that the underlying triangulation $\mathcal{T}^\eta(D)$ (see Definition 3.43) is shape regular and that the number of elements per agglomerate $Y^{(\ell)}$ is uniformly bounded. Moreover, let $X^* \subset \overline{Y}^{(\ell^*)}$ be a $\mathcal{T}^\eta$-agglomerate of dimension $0 \le d_{X^*} \le d$. If $m < d - 1$ or $d_{X^*} < d - 1$, assume additionally that the fine mesh $\mathcal{T}^h(D)$ is quasi-uniform. Then*

$$C_{P,\alpha}(D, X^*; h)^2 \le C \sum_{k=1}^{n} \frac{s_k\, |Y^{(k)}|}{\mathrm{diam}(D)^2\, \eta_{\min}^{d-2}}\, \sigma^{d-\min(m, d_{X^*})}\Big(\frac{\eta_{\max}}{h}\Big),$$

where s_k is the length of the path P_{k,ℓ^*}. The constant C depends only on the shape regularity constant of $\mathcal{T}^\eta(D)$, on the shape regularity and quasi-uniformity constants of $\mathcal{T}^h(D)$, and on the number of elements per agglomerate. If $m = d - 1$ and $d_{X^*} \ge d - 1$ then

$$C_{P,\alpha}(D, X^*)^2 \le C \sum_{k=1}^{n} \frac{s_k\, |Y^{(k)}|}{\mathrm{diam}(D)^2\, \eta_{\min}^{d-2}}.$$

Proof. Let $\eta_\ell := \operatorname{diam}(Y^{(\ell)})$. The assumptions on $Y^{(\ell)}$ and $\mathscr{T}^\eta(D)$ imply that

$$\frac{\operatorname{diam}(Y^{(\ell)})^2}{|Y^{(\ell)}|} \lesssim \eta_\ell^{2-d} \tag{3.38}$$

Let k be fixed and let $P_{k,\ell^*} = \operatorname{interior}(\overline{Y}^{(\ell_1)} \cup \ldots \cup \overline{Y}^{(\ell_{s_k})})$ denote the path from $Y^{(k)}$ to $Y^{(\ell^*)}$ of length s_k with the connecting m-dimensional $\mathscr{T}^\eta$-agglomerates X_i. For convenience we set $X_0 = X_1$ and $X_{s_k} := X^*$. Using Lemma 3.48, the local estimates (3.38), and Corollary 2.71 we obtain

$$\begin{aligned} C_P(P_{k,\ell^*}, X^*, Y^{(k)}; h)^2 &\\ \lesssim \frac{|P_{k,\ell^*}|}{\operatorname{diam}(P_{k,\ell^*})^2} \sum_{i=1}^{s_k} \eta_{\ell_i}^{2-d} \left(C_P(Y^{(\ell_i)}, X_{i-1}; h)^2 + C_P(Y^{(\ell_i)}, X_i; h)^2 \right) &\\ \lesssim \frac{|P_{k,\ell^*}|}{\operatorname{diam}(P_{k,\ell^*})^2}\, s_k\, \eta_{\min}^{2-d}\, \sigma^{d-\min(m,d_{X^*})} \left(\tfrac{\eta_{\max}}{h}\right). & \end{aligned} \tag{3.39}$$

Substituting into the estimate from Theorem 3.45 yields

$$C_{P,\alpha}(D, X^*; h)^2 \lesssim \sum_{k=1}^{n} \frac{|Y^{(k)}|}{\operatorname{diam}(D)^2}\, s_k\, \eta_{\min}^{2-d}\, \sigma^{d-\min(m,d_{X^*})} \left(\tfrac{\eta_{\max}}{h}\right),$$

which completes the proof. □

The following corollary gives the worst case scenario.

Corollary 3.50. *With the assumptions of Lemma 3.49,*

$$C_{P,\alpha}(D, X^*; h)^2 \lesssim \left(\frac{\operatorname{diam}(D)}{\eta_{\min}}\right)^{2(d-1)} \sigma^{d-\min(m,d_{X^*})} \left(\tfrac{\eta_{\max}}{h}\right).$$

If we assume in addition that $s_k \lesssim \operatorname{diam}(D)/\eta_{\min}$ *for all* $k = 1, \ldots, n$ *(which is satisfied, e.g., if the sum of the diameters of the subregions in each path does not exceed* $\operatorname{diam}(D)$*) then*

$$C_{P,\alpha}(D, X^*; h)^2 \lesssim \left(\frac{\operatorname{diam}(D)}{\eta_{\min}}\right)^{d-1} \sigma^{d-\min(m,d_{X^*})} \left(\tfrac{\eta_{\max}}{h}\right).$$

If $\min(m, d_{X^*}) = d - 1$*, the analogous statements hold for* $C_{P,\alpha}(D, X^*)$.

Proof. Due to the shape regularity, $s_k \le n \lesssim (\operatorname{diam}(D)/\eta_{\min})^d$ in the most general case. Furthermore, $\sum_{k=1}^{n} \operatorname{meas}_d(Y^{(k)}) = \operatorname{meas}_d(D) \le \operatorname{diam}(D)^d$. □

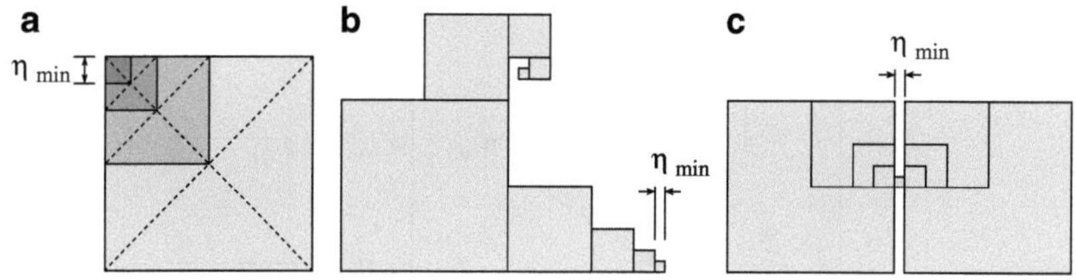

Fig. 3.9 Two-dimensional examples with shape regular partitions. In each case a corresponding family of partitions is defined by continuing the fractal structure and therefore halving $\eta_{\min}$. In (**a**), different colors mean different subregions, *dashed lines* indicate the triangulation $\mathscr{T}^{\eta}(D)$

To illustrate the result, let us go through a couple of examples, which can already be found in [PS11c, PS12b].

Example 3.51. Let $d = 2$ and consider the three domains shown in Fig. 3.9. In each of the three cases,

- The assumptions of Lemma 3.49 are fulfilled with $m \geq d - 1$,
- The underlying triangulation $\mathscr{T}^{\eta}(D)$ (only shown for (a)) is shape regular,
- $\text{meas}_2(D) \eqsim H^2$ where $H = \text{diam}(D)$, and
- $\eta_{\min} \eqsim 2^{-n} H$.

Since $\max_{k=1}^{n} s_k \leq n \lesssim \log_2(H/\eta_{\min})$ in each of these cases, it follows from Lemma 3.49 that if $X^* \subset \overline{Y}^{(\ell^*)}$ is an appropriate $(d-1)$-dimensional $\mathscr{T}^{\eta}$-agglomerate, then

$$C_{P,\alpha}(D)^2 \leq C_{P,\alpha}(D, X^*)^2 \lesssim 1 + \log\left(\frac{H}{\eta_{\min}}\right).$$

Remark 3.52. Example 3.51 shows that the (standard) Poincaré constant $C_P(D)$ of the two-dimensional "dumbbell" domain in Fig. 3.9c is $\mathscr{O}(1 + \log(H/\eta_{\min}))$. Note that the isoperimetric constant (often used to bound $C_P(D)$, see Remark 1.30, p. 15 and [DKW08a, FF60, Maz60, Maz85, Pec12]) is $\mathscr{O}(H/\eta_{\min})$ and thus yields a pessimistic bound for $C_P(D)$. See also [DW06] for a similar logarithmic behavior of the constant in the Bramble-Hilbert lemma.

Example 3.53. Let now $d = 3$ and consider the domain D in Fig. 3.10 (left) with $Y^{(1)}$ being the small cube in the top corner and the remaining subregions numbered away from $Y^{(1)}$, such that $\eta_k \eqsim 2^k \eta_{\min}$. Let α be piecewise constant with respect to $\{Y^{(\ell)}\}_{\ell=1}^{n}$ and quasi-monotone on D.

Let us first consider the case that $\ell^* = 1$, i.e. the largest coefficient is in the small cube, and let X^* be a suitable face on $\partial Y^{(\ell^*)}$. Let k be fixed, then $s_k = k$ and $\ell_i = k + 1 - i$. It follows from inequality (3.39) in the proof of Lemma 3.49 that

$$C_P(P_{k,\ell*}, X^*, Y^{(k)})^2 \lesssim \frac{|P_{k,\ell*}|}{\text{diam}(P_{k,\ell*})^2} \sum_{i=1}^{k} (\eta_{k+1-i})^{-1} \lesssim \frac{|P_{k,\ell*}|}{\text{diam}(P_{k,\ell*})^2}\, \eta_{\min}^{-1} \sum_{i=1}^{k} 2^{i-k-1}.$$

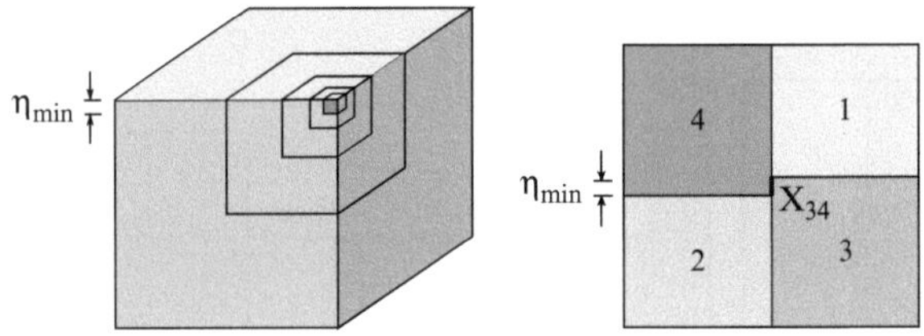

Fig. 3.10 *Left*: Example with shape regular polyhedral partition, consisting of a small cube and nested Fichiera corners. *Right*: Coefficient distribution with staggered structure. (The largest coefficient is in region $Y^{(4)}$)

Since the last sum is $\eqsim 1$, Theorem 3.36 yields

$$C_{P,\alpha}(D, X^*)^2 \lesssim \sum_{k=1}^{n} \eta_{\min}^{-1} \frac{\mathrm{meas}_3(Y^{(k)})}{\mathrm{diam}(D)^2}$$

Using that $\mathrm{meas}_3(Y^{(k)}) \eqsim 8^k\, \eta_{\min}^3$ and $\mathrm{diam}(D) \eqsim 2^n\, \eta_{\min}$, we obtain

$$C_{P,\alpha}(D)^2 \le C_{P,\alpha}(D, X^*)^2 \lesssim \sum_{k=1}^{n} \frac{8^k}{4^n} \lesssim 2^n \eqsim \frac{\mathrm{diam}(D)}{\eta_{\min}}.$$

For $\alpha \equiv 1$ and $\eta_{\min} = h$, this result reflects in a way the discrete Poincaré type inequality from Lemma 2.70, where the "average" is taken at a subdomain vertex.

If, on the other hand, the largest coefficient value is attained in the largest domain, i.e. $\ell^* = n$, then for fixed k, we have $s_k = n - k + 1$ and $\ell_i = k - 1 + i$. A short computation shows that this leads to the optimal bound

$$C_{P,\alpha}(D) \le C_{P,\alpha}(D, X^*) \lesssim 1.$$

In the same way, we can also show that $C_{P,\alpha}(D) \lesssim 1$ for the domains in Fig. 3.9a, b, if the largest coefficient is attained in the largest subregion.

Note that the examples in this section are not artificial. They arise naturally when interfaces between perfectly well-shaped coefficient regions are *small* compared to the size of the regions, see e.g. Fig. 3.10 (right).

Example 3.54. Consider the scenario in Fig. 3.10 (right). The quasi-monotone path $P_{3,4}$ from $Y^{(3)}$ to $Y^{(4)}$ contains the interface $X_{3,4}$ which has $\mathrm{diam}(X_{3,4}) = \eta_{\min} \ll \mathrm{diam}(D)$. Subdividing both $Y^{(3)}$ and $Y^{(4)}$ further as in Fig. 3.9a, we get (as in Example 3.51)

$$C_{P,\alpha}(D)^2 \le C_{P,\alpha}(D, X_{34})^2 \lesssim 1 + \log\left(\frac{\mathrm{diam}(D)}{\eta_{\min}}\right).$$

3.4.4.2 Estimates with Poincaré Constants of Paths

Before we move to other constellations, such as anisotropic subregions, inclusions, and checkerboard distributions, we need more tools than just Theorems 3.36 and 3.45 and Lemma 3.48.

Recall that $C_P(P_{k,\ell^*}, X^*, Y^{(k)}) \not\lesssim C_P(P_{k,\ell^*}, X^*)$ and that the straightforward estimate $C_P(P_{k,\ell^*}, X^*, Y^{(k)}) \leq \sqrt{\frac{|P_{k,\ell^*}|}{|Y^{(k)}|}}\, C_P(P_{k,\ell^*}, X^*)$ is too pessimistic in general. However, sometimes there may be good estimates for $C_P(P_{k,\ell^*}, X^*)$ or $C_P(P_{k,\ell^*}, X^*; h)$ at hand. In such a situation, the next lemma (cf. [PS12b, Lemma 4.13]) is a good alternative to Theorems 3.36 and 3.45.

Lemma 3.55. *Let the assumptions of Theorem 3.45 be fulfilled (see Remark 3.46 on how these can be weakened). Furthermore, for each $k = 1, \dots, n$, let $X_k \subset \overline{Y}^{(k)}$ be an $\mathscr{T}^\eta$-agglomerate of dimension 0 up to d. Then*

$$C_{P,\alpha}(D, X^*; h)^2 \leq 2 \max_{k=1}^{n} \Big(\frac{\operatorname{diam}(Y^{(k)})^2}{\operatorname{diam}(D)^2}\, C_P(Y^{(k)}, X_k; h)^2 \Big)$$
$$+ 4 \sum_{k=1}^{n} \frac{\operatorname{meas}_d(Y^{(k)}) \operatorname{diam}(P_{k,\ell^*})^2}{\operatorname{meas}_d(P_{k,\ell^*}) \operatorname{diam}(D)^2} \big(C_P(P_{k,\ell^*}, X_k; h)^2 + C_P(P_{k,\ell^*}, X^*; h)^2\big).$$

If the coefficient is type-$(d-1)$ quasi-monotone and if all the dimensions of X^* and X_k are at least $d-1$, then the analogous statement holds for H^1 (with the dependence on h removed).

Proof. Let $k = 1, \dots, n$ be fixed. Then

$$\begin{aligned}
&\tfrac{1}{2}\, \|u - \bar{u}^{X^*}\|^2_{L^2(Y^{(k)}),\alpha} \\
&\leq \alpha_k\, \|u - u^{X_k}\|^2_{L^2(Y^{(k)})} + \alpha_k \operatorname{meas}_d(Y^{(k)})\, |\bar{u}^{X_k} - \bar{u}^{X^*}|^2 \\
&\leq \alpha_k\, \|u - \bar{u}^{X_k}\|^2_{L^2(Y^{(k)})} + 2\alpha_k \frac{\operatorname{meas}_d(Y^{(k)})}{\operatorname{meas}_d(P_{k,\ell^*})} \Big(\|u - \bar{u}^{X_k}\|^2_{L^2(P_{k,\ell^*})} + \|u - \bar{u}^{X^*}\|^2_{L^2(P_{k,\ell^*})}\Big).
\end{aligned}$$

The first term is bounded from above by $C_P(Y^{(k)}, X_k; h)^2 \operatorname{diam}(Y^{(k)})^2\, |u|^2_{H^1(Y^{(k)}),\alpha}$. The second term can be bounded using $C_P(P_{k,\ell^*}, X_k; h)$ and $C_P(P_{k,\ell^*}, X^*; h)$, as well as the type-$m$ quasi-monotonicity of α. □

The following two lemmas (cf. [PS12b, Lemmas 4.15 and 4.16]) are interesting and useful on their own, but they can also be used to get good bounds for the Poincaré parameters $C_P(P_{k,\ell^*}, X_k; h)$ and $C_P(P_{k,\ell^*}, X^*; h)$ occurring in the bound of Lemma 3.55. See also [PS08, Lemmas A.2 and A.3], [PS11b, Lemma 3.3] for earlier results.

Lemma 3.56. *Assume that $\mathscr{T}^\eta(D)$ is quasi-uniform with mesh parameter η. Let $\mathscr{Y} := \{Y^{(\ell)}\}_{\ell=1}^n$ be a subdivision of D into $\mathscr{T}^\eta$-agglomerates such that the number of elements per subregion $Y^{(\ell)}$ is uniformly bounded. Let the triangulations $\{\mathscr{T}^h(D)\}_h$ be refinements of $\mathscr{T}^\eta(D)$ and let $\{V^h(D)\}_h$ be the corresponding*

FE spaces. Moreover, let $G_j \subset \overline{Y}^{(j)}$ *be a* d_{G_j}*-dimensional* $\mathcal{T}^\eta$*-agglomerate. Then for any type-m path* $P_{k,j}$ *from* $Y^{(k)}$ *to* $Y^{(j)}$,

$$\int_{G_j}\int_{Y^{(k)}} |u(x)-u(y)|^2 \mathrm{d}y\, \mathrm{d}s_x \lesssim s_{k,j}\, \eta^{d_{G_j}+2} \sigma^{d-\min(m,d_{G_j})}\left(\tfrac{\eta}{h}\right)|u|^2_{H^1(P_{k,j})} \quad \forall u \in V^h(P_{k,j}),$$

where $s_{k,j}$ *is the length of* $P_{k,j}$*. If* $\min(m, d_{G_j}) = d-1$*, the inequality also holds for all* $u \in H^1(P_{k,j})$.

Proof. Let $u \in V^h(D)$ be fixed. Note first that

$$\begin{aligned}\int_{G_j}\int_{Y^{(k)}} \left|u(x)-u(y)\right|^2 \mathrm{d}y\, \mathrm{d}s_x &\lesssim \int_{G_j}\int_{Y^{(k)}} \left|u(x)-\bar{u}^{G_j}\right|^2 + \left|\bar{u}^{G_j}-u(y)\right|^2 \mathrm{d}y\, \mathrm{d}s_x \\ &\lesssim \operatorname{meas}_{d_{G_j}}(G_j)\, \|u-\bar{u}^{G_j}\|^2_{L^2(Y^{(k)})} + \operatorname{meas}_d(Y^{(k)})\, \|u-\bar{u}^{G_j}\|^2_{L^2(G_j)}\,. \end{aligned} \tag{3.40}$$

It follows from Definition 2.67 and Lemma 3.48 (with $D = P_{k,j}$ and $X^* = G_j$) that

$$\|u-\bar{u}^{G_j}\|^2_{L^2(Y^{(k)})} \lesssim s_{k,j}\, \eta^2\, \sigma^{d-\min(m,d_{G_j})}\left(\tfrac{\eta}{h}\right)|u|^2_{H^1(P_{k,j})}\,, \tag{3.41}$$

where we have used that $\operatorname{meas}_d(Y^{(\ell)}) \eqsim \eta^d$ and $\operatorname{diam}(Y^{(\ell)}) \eqsim \eta$ for all $\ell = 1,\ldots,n$. Secondly, by Lemma 2.69,

$$\|u-\bar{u}^{G_j}\|^2_{L^2(G_j)} \lesssim \eta^{d_{G_j}-d+2}\, \sigma^{d-d_{G_j}}\left(\tfrac{\eta}{h}\right)|u|^2_{H^1(Y^{(j)})}\,. \tag{3.42}$$

Substituting these last two bounds into (3.40) and using that $\operatorname{meas}_d(Y^{(k)}) \eqsim \eta^d$ and $\operatorname{meas}_m(G_j) \eqsim \eta^{d_{G_j}}$ yields the final result. □

The following lemma uses Lemma 3.56 as well as a counting argument and delivers a bound for the parameter $C_{P,\alpha}(D, X^*; h)$, which can be also used to estimate the parameters $C_P(P_{k,\ell^*}, X_k; h)$ and $C_P(P_{k,\ell^*}, X^*; h)$ in Lemma 3.55.

Lemma 3.57. *Assume that* $\mathcal{T}^\eta(D)$ *is quasi-uniform with mesh parameter* η*. Let* $\mathcal{Y} := \{Y^{(\ell)}\}_{\ell=1}^n$ *be a subdivision of* D *into* $\mathcal{T}^\eta$*-agglomerates such that the number of elements per subregion* $Y^{(\ell)}$ *is uniformly bounded. Let the triangulations* $\{\mathcal{T}^h(D)\}_h$ *be refinements of* $\mathcal{T}^\eta(D)$ *and let* $\{V^h(D)\}_h$ *be the corresponding FE spaces. Moreover, let* $J \subset \{1,\ldots,n\}$ *be an index set such that* $\overline{X}^* = \bigcup_{j\in J}\overline{G}_j$*, where each* $G_j \subset \overline{Y}^{(j)}$ *is a* d_{X^*}*-dimensional* $\mathcal{T}^\eta$*-agglomerate. Let* $\alpha \in L^\infty_+(D)$ *be type-m quasi-monotone on* D *(with respect to* $\mathcal{Y}$*) such that* X^* *lies in the region(s) where the maximal coefficient is attained. For each* $k \in I := \{1,\ldots,n\}$ *and* $j \in J$*, let* $P_{k,j}$ *be a type-m quasi-monotone path from* $Y^{(k)}$ *to* $Y^{(j)}$ *with respect to* α *and define*

$$r_{\max} := \max_{\ell=1,\ldots,n} \#\{(k,j) \in I \times J : Y^{(\ell)} \in P_{k,j}\},$$

i.e., the maximum number of times any of the agglomerates $Y^{(\ell)}$ is contained in the paths. Then

$$C_{P,\alpha}(D, X^*; h)^2 \lesssim \frac{s_{\max}\, r_{\max}\, \eta^{2+d_{X^*}}}{\mathrm{meas}_{d_{X^*}}(X^*)\,\mathrm{diam}(D)^2}\, \sigma^{d-\min(m,d_{X^*})}\big(\tfrac{\eta}{h}\big),$$

where $s_{\max} := \max\{s_{k,j} : (k, j) \in I \times J\}$ denotes the maximal path length. If $m \geq d-1$ and $d_{X^*} \geq d-1$, the analogous inequality holds for $C_{P,\alpha}(D, X^*)$.

Let W be a subset of D with $\overline{W} = \bigcup_{\ell\in I} \overline{Y}^{(\ell)}$, where $I \subset \{1,\ldots,n\}$ is a non-empty index set. Then, with the analogous definitions of $s_{\max}$ and $r_{\max}$,

$$C_{P,\alpha}(D, X^*, W; h)^2\, \frac{\mathrm{meas}_d(W)}{\mathrm{meas}_d(D)} \lesssim \frac{s_{\max}\, r_{\max}\, \eta^{2+d_{X^*}}}{\mathrm{meas}_{d_{X^*}}(X^*)\,\mathrm{diam}(D)^2}\, \sigma^{d-\min(m,d_{X^*})}\big(\tfrac{\eta}{h}\big).$$

Proof. We give the proof for the second statement. The first one then follows immediately by setting $W = D$. To this end, let $W \subset D$ be a suitable subset as above. Without loss of generality assume that $u \in V^h(D)$ with $\bar{u}^{X^*} = 0$. We now integrate the identity $u(x)^2 - 2u(x)u(y) + u(y)^2 = (u(x) - u(y))^2$ over X^* with respect to x, multiply it by $\alpha(y)$, and finally integrate over W with respect to y:

$$\begin{aligned}\int_W \alpha(y)\,\mathrm{d}y\, \|u\|^2_{L^2(X^*)} - 2\int_{X^*} u(x)\,\mathrm{d}s_x \int_W \alpha(y)\, u(y)\,\mathrm{d}y\, + \\ +\, \mathrm{meas}_{d_{X^*}}(X^*)\, \|u\|^2_{L^2(W),\alpha} = \int_{X^*}\int_W \alpha(y)\, |u(x) - u(y)|^2\,\mathrm{d}y\,\mathrm{d}s_x\,.\end{aligned}$$

The first term on the left hand side is positive, and the middle term vanishes since $\bar{u}^{X^*} = 0$. Thus,

$$\begin{aligned}\mathrm{meas}_{d_{X^*}}(X^*)\, \|u\|^2_{L^2(W),\alpha} &\leq \int_{X^*}\int_W \alpha(y)\, |u(x) - u(y)|^2\,\mathrm{d}y\,\mathrm{d}s_x \\ &= \sum_{j\in J}\sum_{k\in I} \alpha_k \int_{G_j}\int_{Y^{(k)}} |u(x) - u(y)|^2\,\mathrm{d}y\,\mathrm{d}s_x\,.\end{aligned}$$

Using Lemma 3.56, type-m quasi-monotonicity, and the definitions of $s_{\max}$ and $r_{\max}$, we obtain

$$\begin{aligned}\mathrm{meas}_{d_{X^*}}(X^*)\, \|u\|^2_{L^2(W),\alpha} &\lesssim \sum_{j\in J}\sum_{k\in I} s_{k,j}\, \eta^{2+d_{X^*}} \sigma^{d-\min(m,d_{X^*})}\big(\tfrac{\eta}{h}\big)\, |u|^2_{H^1(P_{k,j}),\alpha} \\ &\leq s_{\max}\, \eta^{2+d_{X^*}} \sigma^{d-\min(m,d_{X^*})}\big(\tfrac{\eta}{h}\big) \sum_{\ell\in I} \#\{(k,j) \in I \times J : Y^{(\ell)} \subset P_{k,j}\}\, |u|^2_{H^1(Y^{(\ell)}),\alpha} \\ &\leq s_{\max}\, r_{\max}\, \eta^{2+d_{X^*}} \sigma^{d-\min(m,d_{X^*})}\big(\tfrac{\eta}{h}\big)\, |u|^2_{H^1(D),\alpha}\,.\end{aligned}$$

Comparing with Definition 3.42 concludes the proof. □

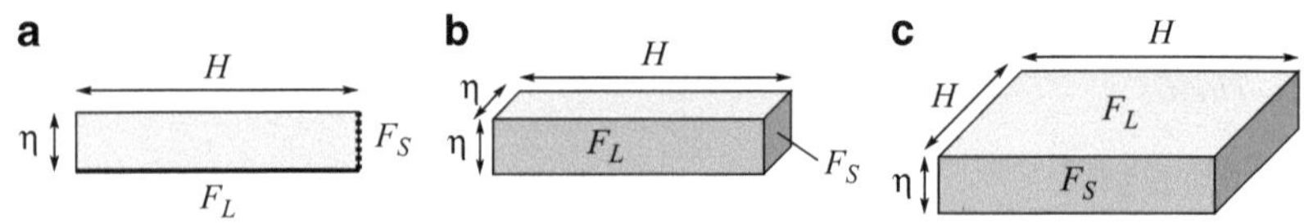

Fig. 3.11 Three model cases of anisotropic domains in two (Case (**a**)) and three dimensions (Cases (**b**) and (**c**))

3.4.4.3 Anisotropic Subregions

In this subsection, we treat cases where the partition $\mathscr{Y}$ contains anisotropic subregions. We will see that it is often advantageous *not* to further subdivide this into a shape regular partition. We start by showing an elementary result for the Poincaré constant of a parallelepiped.

Lemma 3.58. *Let* $\{\mathbf{e}_i\}_{i=1}^d$ *be a (normalized) basis of* $\mathbb{R}^d$ *and let* Y *be the parallelogram/parallelepiped* $\left\{\sum_{i=1}^d \beta_i\, \mathbf{e}_i : \beta_i \in (0, L_i)\right\}$. *If* X *is one of the facets (edges/faces) of* Y, *then*

$$C_P(Y, X) \lesssim 1,$$

and the hidden constant is independent of the lengths L_i, *the aspect ratios* L_i/L_j, *and the angles between* $\mathbf{e}_i$ *and* $\mathbf{e}_j$, *for any* $1 \le i, j \le d$.

Proof. The result can easily be shown by transforming Y to the (isotropic) reference cube $Q = (0,1)^d$ using the linear transformation $F(x) = J^{-1}x$ where $J = (L_1\mathbf{e}_1|\cdots|L_d\mathbf{e}_d)$. See also [VV11, Sect. 2.1] and [PS12b, Lemma A.1] for an explicit estimate ($C_P(Y, X) \le \sqrt{7/5}$). □

Example 3.59. For any of the regions Y in Fig. 3.11a–c and for any $(d-1)$–facet X of Y, Lemma 3.58 implies

$$C_P(Y, X) \lesssim 1,$$

independent of the aspect ratio H/η.

The following examples (cf. [PS12b, Sect. 4.3]) all involve anisotropic subregions and make use of the theoretical results above.

Example 3.60. Let Y be one of the two annular subregions shown in Fig. 3.12, and let X be an edge of length H (left figure) or a face of area H^2 (right figure). Then

$$C_P(Y, X) \lesssim 1.$$

This can be shown by further subdividing the subregions into a few anisotropic rectangles/cuboids and using Lemma 3.48 (with $D = Y$ and $X^* = X$) together with the estimates in Example 3.59, see also [PS08].

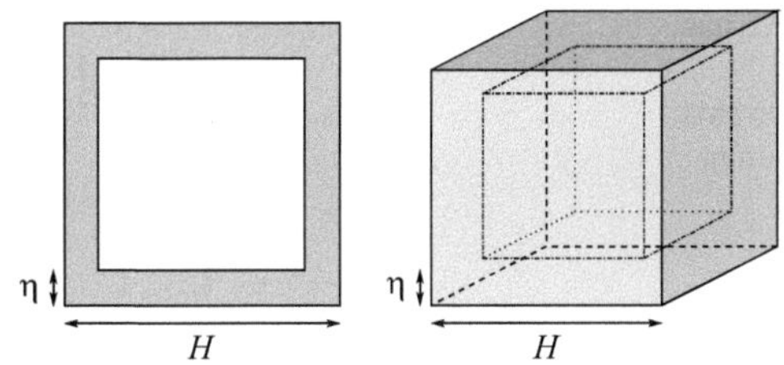

Fig. 3.12 "Annular" subregions in Example 3.60 in two and three dimensions. The smaller cube sketched inside is cut out from the larger cube

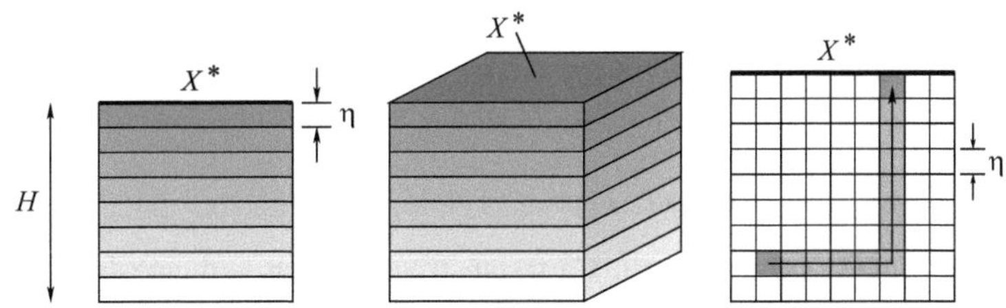

Fig. 3.13 *Left/middle*: Layered coefficient distributions in two and three dimensions. *Right*: Partitioning and quasi-monotone paths for Example 3.61

Example 3.61. For the two scenarios in Fig. 3.13 (left/middle) where the coefficient grows towards the top facet X^*, we have

$$C_{P,\alpha}(D) \lesssim 1.$$

First, let $d = 2$. We subdivide each anisotropic region in Fig. 3.13 (left), such that the resulting partition $\mathscr{Y}$ consists of $(H/\eta)^2$ square regions $Y^{(k)}$, as shown in Fig. 3.13 (right). The manifold X^* (on the top of ∂D) with $\mathrm{meas}_{d-1}(X^*) = H$ is the union of H/η edges G_j, $j \in J$. By using generic "L"–shaped paths $P_{k,j}$ from $Y^{(k)}$ to G_j as depicted in Fig. 3.13 (right), for any pair $(k, j) \in I \times J$ (with the notation as in Lemma 3.57), it is easy to see that

(i) Each of the paths is quasi-monotone with respect to the given coefficient distribution in Fig. 3.13 (left),
(ii) $s_{\max} \eqsim H/\eta$ and
(iii) $r_{\max} \eqsim (H/\eta)^2$.

Therefore it follows from Lemma 3.57 that $C_{P,\alpha}(D, X^*) \lesssim \frac{H}{\eta}\left(\frac{H}{\eta}\right)^2 \frac{\eta^3}{H\,H^2} = 1$.

For $d = 3$, the partition $\mathscr{Y}$ consists of $\mathscr{O}((H/\eta)^3)$ cubic regions and the manifold X^* is the union of $\mathscr{O}((H/\eta)^2)$ faces. Extending the idea of L-shaped paths to three dimensions, one can choose the quasi-monotone paths in such a way that $s_{\max} \lesssim H/\eta$ and $r_{\max} \lesssim (H/\eta)^3$. Lemma 3.57 then yields $C_{P,\alpha}(D, X^*) \lesssim \frac{H}{\eta}\left(\frac{H}{\eta}\right)^3 \frac{\eta^4}{H^2\,H^2} = 1$.

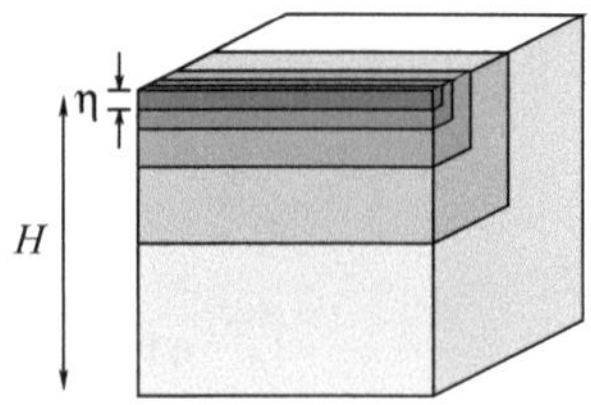

Fig. 3.14 Piecewise constant coefficient distribution increasing gradually towards an edge in three dimensions

Next we consider the example shown in Fig. 3.14, where a piecewise constant coefficient increases gradually towards an edge of a cube in three dimensions. To get an optimal bound in this case is surprisingly difficult.

Lemma 3.62. *Let Q be a cube with sidelengths H and let F_η be an anisotropic quadrilateral on ∂Q with sidelengths H and η such that ∂F_η contains an edge E of the cube Q. Then*

$$C_P(Q, F_\eta)^2 \lesssim 1 + \log(H/\eta).$$

Proof. Let $\mathcal{T}^\eta(Q)$ be a uniform Cartesian triangulation with mesh size η such that F_η is resolved by just one layer of element faces, and let $V^\eta(Q)$ denote the corresponding continuous piecewise linear finite element space. Let $\Pi^\eta : H^1(Q) \to V^\eta(Q)$ be a Scott-Zhang operator (see Lemma 1.45) such that for all $u \in H^1(Q)$,

$$\begin{aligned} \|v - \Pi^\eta v\|_{L^2(Q)} &\le C_{SZ}\, \eta\, |v|_{H^1(Q)}, \\ |\Pi^\eta v|_{H^1(Q)} &\le C_{SZ}\, |v|_{H^1(Q)}, \\ \overline{\Pi^\eta v}^E &= \bar{v}^{F_\eta}. \end{aligned}$$

The last property is achieved by choosing the averaging manifolds suitably: for every node x on E, we choose the associated manifold f_x as shown in the sketch below (for more details we refer to [PS12b, Example 4.14]).

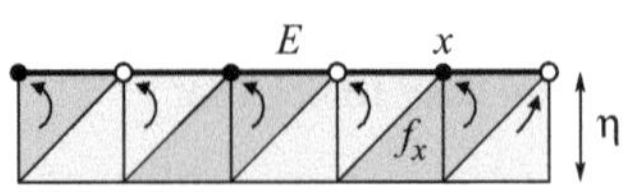

The constant C_{SZ} above depends only on the shape regularity of the auxiliary Cartesian grid $\mathcal{T}^\eta(Q)$ and so $C_{SZ} \lesssim 1$. Using the properties of Π^η above and Lemma 2.70 (for $h = \eta$), we can conclude that

$$\begin{aligned} \|u - \bar{u}^{F_\eta}\|^2_{L^2(Q)} &\lesssim \|u - \Pi^\eta u\|^2_{L^2(Q)} + \|\Pi^\eta u - \overline{\Pi^\eta u}^E\|^2_{L^2(Q)} \\ &\lesssim \eta^2\, |u|^2_{H^1(Q)} + H^2\,(1 + \log(\tfrac{H}{\eta}))\, |\Pi^\eta u|^2_{H^1(Q)} \\ &\lesssim H^2\,(1 + \log(\tfrac{H}{\eta}))\, |u|^2_{H^1(Q)}, \end{aligned}$$

since $\eta \le H$. □

Example 3.63. For the scenario depicted in Fig. 3.14, one can show that

$$C_{P,\alpha}(D)^2 \lesssim (1 + \log(H/\eta))^2 .$$

The underlying subdivision $\overline{D} = \bigcup_{\ell=1}^{n} \overline{Y}^{(\ell)}$ consists of $n \eqsim 1 + \log(H/\eta)$ subdomains, where $\eta \eqsim 2^{-n}H$. Suppose that the subdomains are numbered sequentially from the domain with the smallest coefficient ($Y^{(1)}$) to that with the largest one ($Y^{(n)}$).

Let $X_n = X^* \subset Y^{(n)}$ be a long rectangular face with side lengths H and η, and for each $k = 1, \dots, n-1$, let $X_k \subset \partial Y^{(k)}$ be a face of $P_{k,n}$ with side lengths H and $2^{1-k}H$. Using Lemma 3.55, it suffices to estimate $C_P(Y^{(k)}, X_k)$, $C_P(P_{k,n}, X_k)$ and $C_P(P_{k,n}, X_n)$.

Note that for each $k = 1, \dots, n$, the path $P_{k,n}$ is a parallelepiped with $\operatorname{diam}(Y^{(k)}) \eqsim \operatorname{diam}(P_{k,n}) \eqsim H$ and $\operatorname{meas}_3(Y^{(k)}) \eqsim \operatorname{meas}_3(P_{k,n}) = 4^{1-k}H^3$. By Lemma 3.58, it follows that $C_P(P_{k,n}, X_k) \lesssim 1$. It is also easy to show that $C_P(Y^{(k)}, X_k) \lesssim 1$.

In order to estimate $C_P(P_{k,n}, X_n)$ for each k, we transform $P_{k,n}$ to the unit cube $\widehat{Q}$ (as in Lemma 3.58) by multiplying two of the coordinates by $2^{k-1}H^{-1}$ and the remaining one by H^{-1}. Let $\widehat{F}_k$ be the corresponding image of X_n under this mapping and note that $\widehat{F}_k$ is a face of side length 1 and 2^{k-n}. With Lemma 3.62 and $\operatorname{meas}_3(\widehat{Q}) = 1$,

$$\begin{aligned} \|u - \bar{u}^{X_n}\|^2_{L^2(P_{k,n})} &= \operatorname{meas}_3(P_{k,n})\, \|\hat{u} - \bar{\hat{u}}^{\widehat{F}_k}\|^2_{L^2(\widehat{Q})} \\ &\lesssim \operatorname{meas}_3(P_{k,n})\,(1 + n - k)\,|\hat{u}|^2_{H^1(\widehat{Q})} \le (1 + n - k)\, H^2\, |u|^2_{H^1(P_{k,n})}, \end{aligned}$$

where in the last step we have used that the spectral norm of the Jacobian of the transformation mapping $\widehat{Q}$ to $P_{k,n}$ is $\le H$. Hence, $C_P(P_{k,n}, X_n)^2 \lesssim 1 + n - k$. The estimate from Lemma 3.55 thus yields

$$C_{P,\alpha}(D)^2 \lesssim 1 + \sum_{k=1}^{n} (1 + (1 + n - k)) \lesssim n^2 \eqsim (1 + \log(H/\eta))^2 .$$

From the numerical computations in [PS12b, Sect. 5.4], one sees that $C_{P,\alpha}(D)^2$ behaves like $(1 + \log(H/\eta))$.

3.4.4.4 Inclusions

Consider the region depicted in Fig. 3.15 with a large number of square inclusions with coefficient α_1 inside a background medium with coefficient $\alpha_2 > \alpha_1$. We choose X^* as a boundary edge of D of length $H \eqsim \operatorname{diam}(D)$. To bound the weighted Poincaré constant $C_{P,\alpha}(D)$ for this case, we treat all the inclusions as

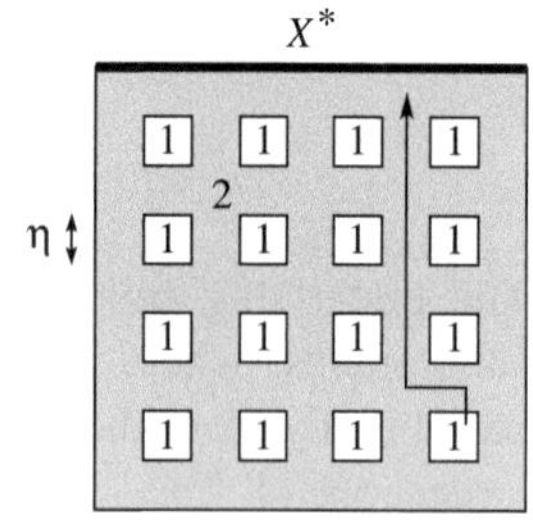

Fig. 3.15 Coefficient distribution based on inclusions

one subregion $Y^{(1)}$ and the remainder as $Y^{(2)}$, use Theorem 3.36, and bound the constants therein directly. Whereas $Y^{(1)}$ is disconnected, the *path* $P_{12} = D$ is connected and

$$C_P(P_{12}, X^*, Y^{(1)})^2 \frac{|Y^{(1)}|}{|P_{12}|} \leq C_P(D, X^*)^2 \lesssim 1.$$

To get a bound for the Poincaré constant $C_P(Y^{(2)}, X^*, Y^{(2)}) = C_P(Y^{(2)}, X^*)$ of the *perforated* domain $Y^{(2)}$ without the inclusions, we use Lemma 3.57. It is straightforward to find a quasi-uniform (square) partition $\{\widetilde{Y}_i\}_{i=1}^n$ of $Y^{(2)}$ with mesh size equal to the diameter η of the holes (see Fig. 3.7c). We construct a (quasi-monotone) path from each region $\widetilde{Y}_i$ to one of the faces $F_j \subset X^*$ by following (essentially) the same construction as in Example 3.61 (with some small modifications at the start and at the end of the path). It is easy to see that again $s_{\max} \lesssim H/\eta$ and $r_{\max} \lesssim (H/\eta)^2$. Hence, $C_P(Y^{(2)}; X^*) \lesssim 1$ and so with Theorem 3.36,

$$C_{P,\alpha}(D) \leq C_{P,\alpha}(D, X^*) \lesssim 1.$$

If there are p distinct values in the inclusions, which are all smaller than α_2, following the same technique we see that

$$C_{P,\alpha}(D)^2 \lesssim p.$$

On first glance this would suggest, that in the worst case $C_{P,\alpha}(D)^2 \lesssim n$, where n is the number of inclusions, but this is not quite true. Using the concept of macroscopically quasi-monotone coefficients (cf. Definition 3.39), we may combine subregions with weights of similar size, even if they are not connected. Assume, for example, that the values of α range from $\alpha_1 \ll 1$ to $\alpha_n = 1$, where $Y^{(n)}$ is now the perforated (background) region. Let us fix $b > 1$ (e.g., $b = 2$) and combine all subregions with values in $[b^{-i}, b^{-i+1}]$ into one subregion. Then we have $\log_b(\alpha_1^{-1})$ such compound subregions with a local variation of b. Therefore,

$$C_{P,\alpha}(D)^2 \lesssim b \log_b(\alpha_1^{-1})$$

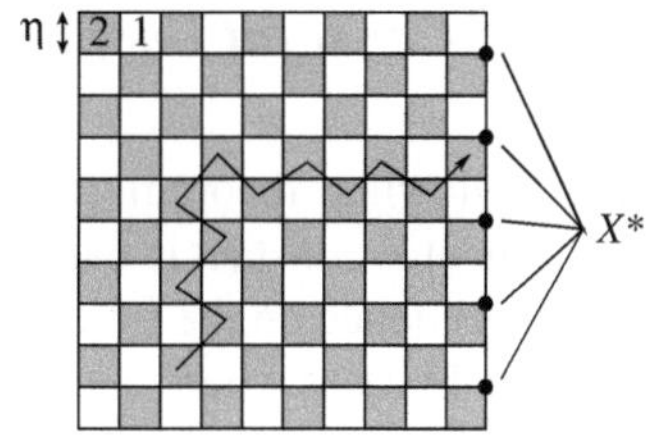

Fig. 3.16 A checkerboard distribution with a type-0 quasi-monotone path

uniformly, even for $n \to \infty$. We note that estimates for $C_{P,\alpha}(D)$ for this example have been shown in [GE10a, Lemma 4], but they depend on the number n of inclusions and are not explicit in the geometric parameters.

3.4.4.5 A Checkerboard Distribution

Our last type of example is that of the checkerboard distribution depicted in Fig. 3.16, cf. [PS12b, Sect. 4.5]. There, we assume that the coefficient takes two values α_1 and $\alpha_2 \gg \alpha_1$, see also [PS12b, Example 4.19]. We will show that

$$C_{P,\alpha}(D;\, h)^2 \;\lesssim\; 1 + \log\left(\frac{\eta}{h}\right).$$

To get this bound, we choose X^* as the union of $\mathcal{O}(H/\eta)$ vertices on the boundary of D, as shown in the figure. Note that $\operatorname{meas}_0(X^*) \eqsim H/\eta$, cf. Definition 2.66, p. 119. We construct type-0 quasi-monotone paths $P_{k,j}$ from every square $Y^{(k)} \in \mathscr{Y}$ to every vertex $V_j \in X^*$, as shown in Fig. 3.16. As in Example 3.61 and in Sect. 3.4.4.4, it is easy to see that these paths satisfy

$$s_{\max} \lesssim \frac{H}{\eta} \quad \text{and} \quad r_{\max} \lesssim \left(\frac{H}{\eta}\right)^2.$$

Finally, from $\operatorname{meas}_0(X^*) \eqsim H/\eta$ and Lemma 3.57, we obtain

$$C_{P,\alpha}(D;\, h)^2 \;\le\; C_{P,\alpha}(D,\, X^*;\, h)^2 \;\lesssim\; \frac{H}{\eta}\left(\frac{H}{\eta}\right)^2 \frac{\eta^{2+0}}{(H/\eta)\, H^2}\sigma^{2-0}\left(\frac{\eta}{h}\right) \;\lesssim\; 1 + \log\left(\frac{\eta}{h}\right).$$

3.5 Robust FETI Condition Number Bounds Explicit in η

In this section, we combine the bounds from Theorems 3.28, 3.31, and 3.70 using the estimates for the parameter in the weighted Poincaré inequality from Sect. 3.4.

3.5.1 Constant Coefficients in the Boundary Layers

The following theorem (cf. [PS08, Theorem 3.3]) treats the case where the coefficient is constant in each boundary layer (up to noise). Unlike in Theorem 3.28, the dependence on the ratio H_i/η_i is made explicit and turns out to be quadratic in general. In cases where the coefficient in each subdomain interior $\Omega_i \setminus \Omega_{i,\eta}$ is *larger* than in the boundary layer $\Omega_{i,\eta}$ we get a linear behavior.

Theorem 3.64. *Let the assumptions of Theorem 3.28 hold and let Assumption 2.54 (p. 115) be fulfilled additionally. Furthermore, for each $i = 1, \dots, s$, let the lower bound function $\underline{\alpha}_i$ be constant on the boundary layer $\Omega_{i,\eta}$. Then for the all-floating FETI method with $Q = M_{\mathrm{sD}}^{-1}$, $Q = Q_{\mathrm{diag}}^{(2)}$ (with Assumption 3.29 fulfilled), or $Q = Q_{\mathrm{diag}}^{(3)}$,*

$$\kappa(M_{\mathrm{sD}}^{-1} F) \le C\, c_{\mathrm{noise}} \max_{i=1}^{s} \left(\frac{H_i}{\eta_i}\right)^{\beta} \left(1 + \log\left(\tfrac{\eta_i}{h_i}\right)\right)^2,$$

with $\beta = 2$, where the constant C is independent of α, H_i, η_i, and h_i. In particular, C does not depend on the values of α in the subdomain interiors $\Omega_i \setminus \Omega_{i,\eta}$. If for each $i = 1, \dots, s$,

$$\alpha(x) \ge \underline{\alpha}_{i|\Omega_{i,\eta}} \qquad \forall x \in \Omega_i\,, \tag{3.43}$$

then the bound holds with $\beta = 1$. If Assumption 3.29 holds and $Q = Q_{\mathrm{diag}}$, then the above bound holds with $\beta = 3$, if (3.43) is fulfilled with $\beta = 2$.

Proof. Assumption 2.54 states that each subdomain Ω_i is composed of a few coarse elements. Let $\mathscr{F}$ be a fixed subdomain facet of Ω_i. Then $\mathscr{F}$ is the union of $\mathscr{O}((H_i/\eta_i)^{d-1})$ η-facets and $\Omega_{i,\eta}$ is the union of $\mathscr{O}((H_i/\eta_i)^{d-1})$ η-agglomerates. An application of Lemma 3.57 yields

$$C_{P,\underline{\alpha}_i}(\Omega_{i,\eta})^2 = C_P(\Omega_{i,\eta})^2 \le C_P(\Omega_{i,\eta}, \mathscr{F})^2 \lesssim \frac{s_{\max}\, r_{\max}\, \eta_i^{2+d-1}}{H_i^{d-1}\, H_i^2}$$

We can choose the paths such that $s_{\max} \lesssim H_i/\eta_i$ and $r_{\max} \lesssim (H_i/\eta_i)^d$, cf. Example 3.61. Hence,

$$C_{P,\underline{\alpha}_i}(\Omega_{i,\eta})^2 \lesssim \left(\frac{H_i}{\eta_i}\right)^{d+1} \left(\frac{\eta_i}{H_i}\right)^{d+1} \lesssim 1.$$

The bounds without assumption (3.43) follow now directly from Theorem 3.28.

The improved bounds under assumption (3.43) will follow as a special case of Theorem 3.73 in Sect. 3.5.3 below, and the proof of Theorem 3.73 will not rely on the current theorem. □

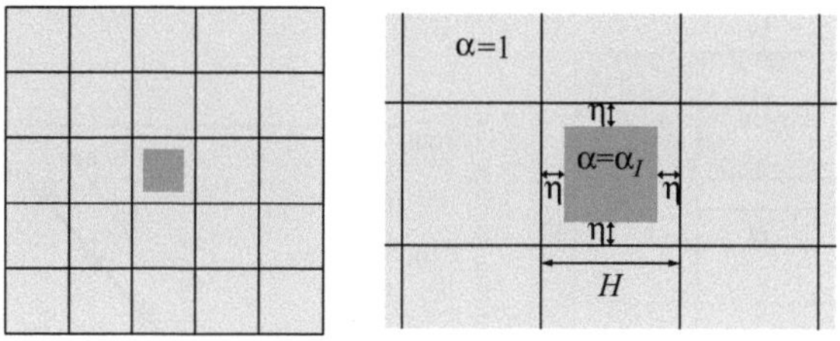

Fig. 3.17 Subdomain decomposition and coefficient distribution for the study in Fig. 3.18

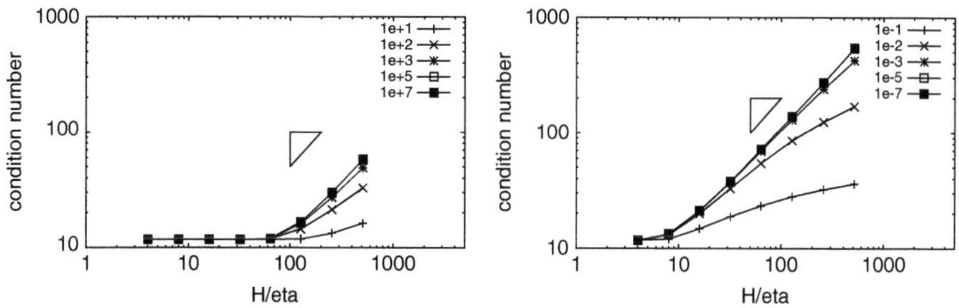

Fig. 3.18 Estimated condition number $\kappa(M_{\mathrm{sD}}^{-1}F)$ for all-floating FETI (with $Q = M_{\mathrm{sD}}^{-1}$) for the problem shown in Fig. 3.17. *Triangle* indicates linear growth in H/η. *Left*: $\alpha_I > 1$. *Right*: $\alpha_I < 1$

The bound in Theorem 3.64 is sharp in general. First, we show an example for the linear bound ($\beta = 1$). The subdomain decomposition and the coefficient distribution are shown in Fig. 3.17. The coefficient α is varied from 10^{-7} to 10^{+7} within the shaded square, and set to one elsewhere; the Dirichlet boundary is chosen as $\Gamma_D = \partial\Omega$. We use a fixed mesh with $H/h = 512$ and vary η between h and $H/4$. The condition numbers (estimated by the Lanczos method) are shown in Fig. 3.18. For the case $\alpha_I > 1$, Theorem 3.64 predicts a condition number of $\mathcal{O}\big(\frac{H}{\eta}\,(1+\log(\eta/h))^2\big)$ and the linear behavior on H/η is clearly seen. The condition numbers for the case $\alpha_I < 1$ are higher but do not grow quadratically in practice. For a theoretical support of this phenomenon see [DS11]. The quadratic bound is seen for the scenario in Fig. 3.19 (left), which is an example due to Marcus Sarkis (cf. [PS11b, Sect. 7.5]). The estimated condition number bounds are shown in Fig. 3.19 (right) and clearly behave quadratically, which shows the sharpness of the bound in Theorem 3.64.

3.5.2 *Quasi-Monotone Coefficients in the Boundary Layers*

In this section, we consider the case where the coefficient α is quasi-monotone in the boundary layers (up to some noise). Results of this type for particular coefficients have been treated in [PS11b].

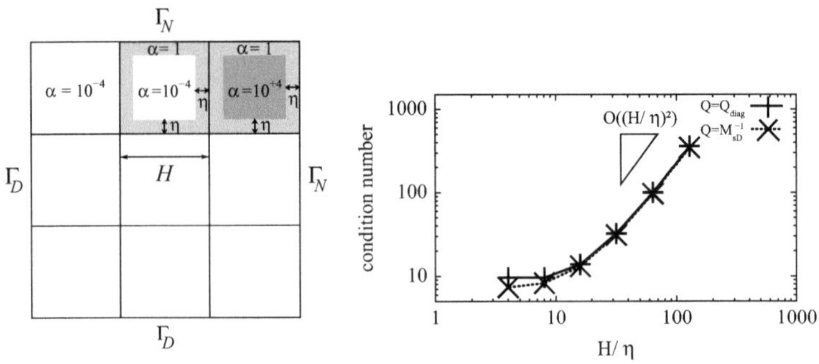

Fig. 3.19 *Left*: Subdomain decomposition and coefficient distribution. *Right*: Estimated condition numbers for all-floating FETI with $Q = M_{\mathrm{sD}}^{-1}$ and $Q = Q_{\mathrm{diag}}$ (see (3.19))

Lemma 3.65. *Let the assumptions of Theorem 3.28 hold and let Assumption 2.54 (p. 115) be fulfilled additionally. Furthermore, suppose that $\underline{\alpha}_i$ is quasi-monotone on $\Omega_{i,\eta}$. Then*

$$C_{P,\underline{\alpha}_i}(\Omega_{i,\eta})^2 \lesssim \begin{cases} 1 & \text{for } d = 2, \\ \left(\frac{H_i}{\eta_i}\right)^2 & \text{for } d = 3, \end{cases}$$

in general. If the length of all quasi-monotone paths is $\mathcal{O}(H_i/\eta_i)$, then the improved estimate $C_{P,\underline{\alpha}_i}(\Omega_{i,\eta})^2 \lesssim H_i/\eta_i$ holds for $d = 3$. See also Example 3.68 for an optimal bound.

Proof. Quasi-monotonicity implies that there is an η-facet X^* contained in the agglomerate with the largest coefficient in $\Omega_{i,\eta}$ such that Lemma 3.57 yields

$$C_{P,\underline{\alpha}_i}(\Omega_{i,\eta})^2 \le C_{P,\underline{\alpha}_i}(\Omega_{i,\eta}, X^*)^2 \lesssim \frac{s_{\max}\, r_{\max}\, \eta_i^{2+d-1}}{\eta_i^{d-1}\, H_i^2} = s_{\max}\, r_{\max} \left(\frac{\eta_i}{H_i}\right)^2.$$

For $d = 2$, we have $s_{\max} \lesssim H_i/\eta_i$ and $r_{\max} \lesssim H_i/\eta_i$, which leads to $C_{P,\underline{\alpha}_i}(\Omega_{i,\eta}) \lesssim 1$. For $d = 3$, in the worst case, $s_{\max} \lesssim (H_i/\eta_i)^2$ and $r_{\max}\ (H_i/\eta_i)^2$ since there are $\mathcal{O}((H_i/\eta_i)^2)$ many η-agglomerates in $\Omega_{i,\eta}$, and so $C_{P,\underline{\alpha}_i}(\Omega_{i,\eta})^2 \lesssim (H_i/\eta_i)^2$. If $s_{\max} \lesssim H_i/\eta_i$ then the bound is linear. □

Combining Theorem 3.28 and Lemma 3.65 leads to bounds for the total condition number of all-floating FETI. In the following, we work out the explicit behavior on η_i for a series of examples.

Example 3.66. In both scenarios in Fig. 3.20, assume that the coefficient $\underline{\alpha}_i$ takes two values in the boundary layer, where the larger one is attained in the dark shaded region. According to Lemma 3.65, we get

$$C_{P,\underline{\alpha}_i}(\Omega_{i,\eta}) \lesssim 1.$$

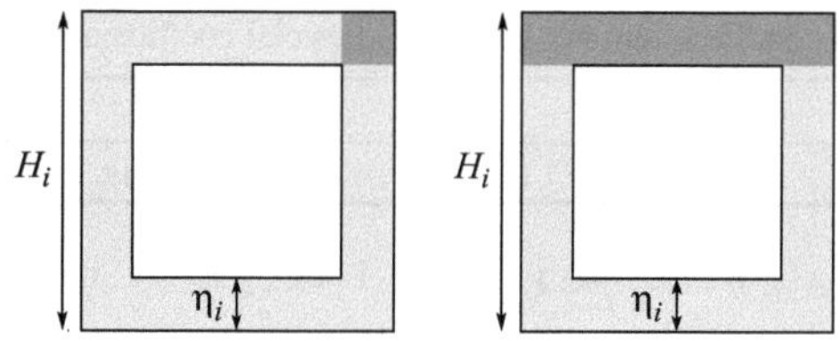

Fig. 3.20 Examples of coefficients with two values in the boundary layer (indicated by *dashed lines*) of a square ($d = 2$). Largest coefficient value attained in a small square (*left*) or a long thin rectangle (*right*)

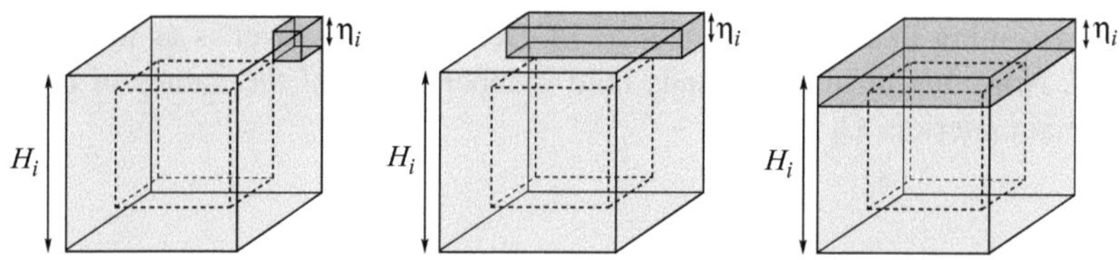

Fig. 3.21 Examples of coefficients with two values in the boundary layer (indicated by *dashed lines*) of a cube ($d = 3$). Largest coefficient value attained in a small cube (*left*), a beam (*middle*), or a plate (*right*)

Example 3.67. In the scenario in Fig. 3.21 (left), assume that the coefficient $\underline{\alpha}_i$ takes two values in the boundary layer, where the larger one is attained in the small dark shaded cube. According to Lemma 3.65, we get

$$C_{P,\underline{\alpha}_i}(\Omega_{i,\eta})^2 \lesssim \frac{H_i}{\eta_i}.$$

As the next example shows, the optimal bound $C_{P,\underline{\alpha}_i}(\Omega_{i,\eta}) \lesssim 1$ can also be achieved for special cases in three dimensions.

Example 3.68. In the scenario in Fig. 3.21 (middle), assume that the coefficient $\underline{\alpha}_i$ takes two values in the boundary layer, where the larger one is attained in the dark shaded beam. Choosing X^* as a long and thin rectangle within the beam and following the structure of the proof of Lemma 3.65, the application of Lemma 3.57 yields

$$C_{P,\underline{\alpha}_i}(\Omega_{i,\eta})^2 \lesssim \frac{s_{\max}\, r_{\max}\, \eta_i^4}{H_i\, \eta_i\, H_i^2} = s_{\max}\, r_{\max} \left(\frac{\eta_i}{H_i}\right)^3.$$

Note that there are $\mathcal{O}((H_i/\eta_i)^2)$ η-agglomerates contained in $\Omega_{i,\eta}$ and $\mathcal{O}(H_i/\eta_i)$ of them touch the manifold X^*. Similarly to the two-dimensional case in Example 3.61, we can choose the paths such that $s_{\max} \lesssim H_i/\eta_i$ and $r_{\max} \lesssim (H_i/\eta_i)^2$. Finally,

$$C_{P,\underline{\alpha}_i}(\Omega_{i,\eta}) \lesssim 1.$$

The same bound can be shown straightforwardly for the scenario in Fig. 3.21 (right).

Table 3.1 Theoretical condition number bounds for special coefficient patterns

$d = 2$	$d = 3$	
Fig. 3.20 (left/right)	Fig. 3.21 (left)	Fig. 3.21 (middle/right)
$C \max_{i=1} \left(\frac{H_i}{\eta_i}\right)^2 (1 + \log(\frac{\eta_i}{h_i}))^2$	$C \max_{i=1} \left(\frac{H_i}{\eta_i}\right)^3 (1 + \log(\frac{\eta_i}{h_i}))^2$	$C \max_{i=1} \left(\frac{H_i}{\eta_i}\right)^2 (1 + \log(\frac{\eta_i}{h_i}))^2$

Table 3.1 displays the total condition number bounds for the all-floating FETI method with $Q = M_{\mathrm{sD}}^{-1}$, $Q = Q_{\mathrm{diag}}^{(2)}$ (with Assumption 3.29 fulfilled), or $Q = Q_{\mathrm{diag}}^{(3)}$, assuming that the coefficient $\underline{\alpha}_i$ in each boundary layer is as in the stated figures. Note that all these bounds hold independently of the values of α in the subdomain interiors $\Omega_i \setminus \Omega_{i,\eta}$.

3.5.3 Condition Number Bounds Using Artificial Coefficients

In many situations, the coefficient $\underline{\alpha}_i$ is not quasi-monotone in the boundary layer $\Omega_{i,\eta}$, but the Poincaré parameters of regions *larger* than the boundary layers might be robust. Even if they are not, there are still situations where a WPI can be employed using a so-called *artificial* coefficient, cf. [PS11b, Sect. 6].

Definition 3.69. We call $\underline{\alpha}_i^{\mathrm{art}} \in L_+^\infty(\Lambda_i)$ an *artificial coefficient* (with respect to α, $\Omega_{i,\eta}$, and Λ_i) if

$$\begin{aligned} \underline{\alpha}_i^{\mathrm{art}}(x) &= \underline{\alpha}_i(x) && \forall x \in \Omega_{i,\eta} \text{ a.e.,} \\ \underline{\alpha}_i^{\mathrm{art}}(x) &\leq \alpha(x) && \forall x \in \Lambda_i \setminus \Omega_{i,\eta} \text{ a.e.} \end{aligned}$$

Furthermore, let $C_i^*(\underline{\alpha}_i^{\mathrm{art}}; h)$ be the smallest parameter such that

$$\inf_{c \in \mathbb{R}} \|u - c\|_{L^2(\Omega_{i,\eta}), \underline{\alpha}_i^{\mathrm{art}}} \leq C_i^*(\underline{\alpha}_i^{\mathrm{art}}; h)\, H_i\, |u|_{H^1(\Lambda_i), \underline{\alpha}_i^{\mathrm{art}}} \qquad \forall u \in V^h(\Lambda_i).$$

Note that above, any local skeleton cover Λ_i "between" $\Omega_{i,\eta}$ and Ω_i is allowed. Note also that (due to Definition 3.42) for any $X^* \subset \overline{\Lambda_i}$,

$$C_i^*(\underline{\alpha}_i^{\mathrm{art}}; h) \leq C_{P, \underline{\alpha}_i^{\mathrm{art}}}(\Lambda_i, X^*, \Omega_{i,\eta}; h) \sqrt{\frac{|\Omega_{i,\eta}|}{|\Lambda_i|}}. \tag{3.44}$$

The following theorem (cf. [PS11b, Sect. 6]) is an alternative to Theorem 3.28, but instead of $C_{P,\underline{\alpha}_i}(\Omega_{i,\eta}; h)$ it involves the parameter $C_i^*(\underline{\alpha}_i^{\mathrm{art}}; h)$. Robust bounds for this parameter will be discussed after the theorem.

Theorem 3.70. *Let the assumptions from Theorem 3.28 be fulfilled and let $\underline{\alpha}_i^{\mathrm{art}} \in L^\infty_+(\Lambda_i)$, $i = 1, \ldots, s$ be artificial coefficients in the sense of Definition 3.69. Then for the all-floating FETI method with $Q = M_{\mathrm{sD}}^{-1}$, $Q = Q_{\mathrm{diag}}^{(2)}$ (with Assumption 3.29 fulfilled), or $Q = Q_{\mathrm{diag}}^{(3)}$,*

$$\kappa(P\, M_{\mathrm{sD}}^{-1} P^\top F_{|\widetilde{U}_{\mathrm{ad}}}) \le C\, c_{\mathrm{noise}} \max_{i=1}^{s} \left(1 + \log(\tfrac{\eta_i}{h_i})\right)^2 \left[1 + \left(\frac{H_i}{\eta_i}\right)^2 C_i^*(\underline{\alpha}_i^{\mathrm{art}};\, h)^2\right],$$

with $C_i^*(\underline{\alpha}_i^{\mathrm{art}};\, h)$ as in Definition 3.69. If $Q = Q_{\mathrm{diag}}$ and Assumption 3.29 is fulfilled, then the bound holds with an additional (global) factor of $\max_{k=1}^s H_k/\eta_k$.

Proof. We follow the proof of Theorem 3.28 but we set

$$W_i^\perp := \left\{ w_i \in W_i : \overline{\mathcal{H}_i^{\alpha,h} w_i}^{\Omega_{i,\eta},\underline{\alpha}_i^{\mathrm{art}}} = 0 \right\}.$$

Using Lemmas 3.27 and 2.105 or 3.30, we obtain for $Q = M_{\mathrm{sD}}^{-1}$, $Q = Q_{\mathrm{diag}}^{(2)}$, or $Q = Q_{\mathrm{diag}}^{(3)}$,

$$\begin{aligned}
&|P_D(w + z_w)|_S^2 \\
&\lesssim \sum_{i=1}^{s} \left(1 + \log\left(\tfrac{\eta_i}{h_i}\right)\right)^2 \left[|\mathcal{H}_i^{\alpha,h} w_i|^2_{H^1(\Omega_{i,\eta_i}),\underline{\alpha}_i} + \eta_i^{-2}\, \|\mathcal{H}_i^{\alpha,h} w_i\|^2_{L^2(\Omega_{i,\eta_i}),\underline{\alpha}_i} \right] \\
&\lesssim \sum_{i=1}^{s} \left(1 + \log\left(\tfrac{\eta_i}{h_i}\right)\right)^2 \left[|\mathcal{H}_i^{\alpha,h} w_i|^2_{H^1(\Omega_i),\alpha} + \left(\tfrac{H_i}{\eta_i}\right)^2 C_i^*(\alpha_i^{\mathrm{art}};\, h)^2 \underbrace{|\mathcal{H}_i^{\alpha,h} w_i|^2_{H^1(\Lambda_i),\underline{\alpha}_i^{\mathrm{art}}}}_{\le |\mathcal{H}_i^{\alpha,h} w_i|^2_{H^1(\Omega_i),\alpha}} \right].
\end{aligned}$$

From here on the proof is again identical to that of Theorem 3.28. The extra factor of $\max_{k=1}^s H_k/\eta_k$ for the choice $Q = Q_{\mathrm{diag}}$ originates from Lemma 3.30. □

From (3.44) it is clear that $C_i^*(\underline{\alpha}_i^{\mathrm{art}};\, h)$ can be bounded robustly if the artificial coefficient $\underline{\alpha}_i^{\mathrm{art}}$ is type-m quasi-monotone on Λ_i. For an illustration see Fig. 3.22. The geometric dependence on H_i/η_i can be analyzed using the tools from Sect. 3.4. For simplicity, we restrict ourselves to the case that $\Lambda_i = \Omega_i$.

Lemma 3.71. *Let the assumptions of Theorem 3.28 hold and let Assumption 2.54 (p. 115) be fulfilled. Furthermore, for each $i = 1, \ldots, s$, let $\underline{\alpha}_i^{\mathrm{art}}$ be a type-m quasi-monotone artificial coefficient in the sense of Definition 3.69 with $\Lambda_i = \Omega_i$. Then (for $d = 2$ or $d = 3$), the parameter $C_i^*(\underline{\alpha}_i^{\mathrm{art}};\, h)$ from Theorem 3.70 fulfills in general the estimate*

$$C_i^*(\underline{\alpha}_i^{\mathrm{art}};\, h)^2 \lesssim \left(\frac{H_i}{\eta_i}\right)^{2d-2} \sigma^{d-m}\left(\tfrac{\eta_i}{h_i}\right).$$

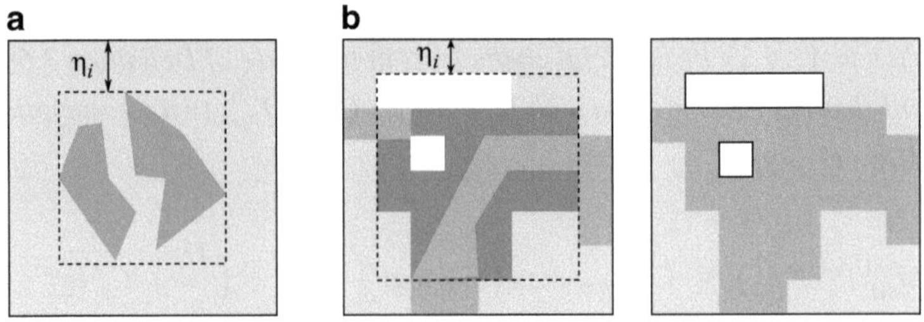

Fig. 3.22 Examples of quasi-monotone artificial coefficients. *Dashed line* indicates the boundary layer $\Omega_{i,\eta}$. (**a**) Coefficient distribution of α in Ω_i. Here, Λ_i can be set to Ω_i and $\underline{\alpha}_i^{\text{art}}$ can be chosen as the constant $\alpha_{|\Omega_{i,\eta}}$. (**b**) *Left*: Coefficient distribution of α in Ω_i. *Right*: Local cover Λ_i (domain with two holes) with artificial coefficient $\underline{\alpha}_i^{\text{art}}$. Alternatively, here α_i^{art} can also be defined on the whole of Ω_i with the original values of α in the "holes". Note that for both cases (**a**) and (**b**), α fails to be quasi-monotone on the whole of Ω_i

If all the path lengths are $\mathcal{O}(H_i/\eta_i)$, then we have the improved bound

$$C_i^*(\underline{\alpha}_i^{\text{art}};\, h)^2 \;\lesssim\; \left(\frac{H_i}{\eta_i}\right)^{d-1} \sigma^{d-m}\left(\tfrac{\eta_i}{h_i}\right).$$

Proof. Let X^* be an m-dimensional η-facet contained in the subregion where the coefficient $\underline{\alpha}_i^{\text{art}}$ attains its maximum. From (3.44) and an application of Lemma 3.57 (with $W = \Omega_{i,\eta}$), we obtain

$$C_i^*(\underline{\alpha}_i^{\text{art}};\, h)^2 \;\le\; C_{P,\underline{\alpha}_i^{\text{art}}}(\Omega_i,\, X^*,\, \Omega_{i,\eta};\, h)^2\, \frac{|\Omega_{i,\eta}|}{|\Omega_i|} \;\lesssim\; \frac{s_{\max}\, r_{\max}\, \eta_i^{2+m}}{\eta_i^m\, H_i^2}\, \sigma^{d-m}\left(\tfrac{\eta_i}{h_i}\right),$$

where in the worst case, $s_{\max} \lesssim (H_i/\eta_i)^d$ and $r_{\max} \lesssim (H_i/\eta_i)^d$ because Ω_i contains $\mathcal{O}((H_i/\eta_i)^d)$ many η-agglomerates. Combining these estimates yields

$$C_i^*(\underline{\alpha}_i^{\text{art}};\, h)^2 \;\lesssim\; \left(\frac{H_i}{\eta_i}\right)^{2d} \left(\frac{\eta_i}{H_i}\right)^2 \sigma^{d-m}\left(\tfrac{\eta_i}{h_i}\right) \;\lesssim\; \left(\frac{H_i}{\eta_i}\right)^{2d-2} \sigma^{d-m}\left(\tfrac{\eta_i}{h_i}\right).$$

If $s_{\max} \lesssim H_i/\eta_i$ then the exponent $2d - 2$ can be replaced by $d - 1$. □

Corollary 3.72. *Let the assumptions of Theorem 3.28 hold and let Assumption 2.54 (p. 115) be fulfilled additionally. Furthermore, for each $i = 1, \ldots, s$, let $\underline{\alpha}_i^{\text{art}}$ be a type-m quasi-monotone artificial coefficient in the sense of Definition 3.69 with $\Lambda_i = \Omega_i$. Then the total condition number bound of all-floating FETI with $Q = M_{\text{sD}}^{-1}$, $Q = Q_{\text{diag}}^{(2)}$ (with Assumption 3.29 fulfilled), or $Q = Q_{\text{diag}}^{(3)}$ fulfills*

$$\kappa(M_{\text{sD}}^{-1} F) \;\lesssim\; \left(\frac{H_i}{\eta_i}\right)^{2d} \sigma^{d-m}\left(\tfrac{\eta_i}{h_i}\right) (1 + \log(\eta_i/h_i))^2\,.$$

If the path lengths are $\lesssim H_i/\eta_i$, then

$$\kappa(M_{\text{sD}}^{-1}F) \lesssim \left(\frac{H_i}{\eta_i}\right)^{d+1} \sigma^{d-m}\left(\tfrac{\eta_i}{h_i}\right)(1+\log(\eta_i/h_i))^2 .$$

The analogous bounds hold for $Q = Q_{\text{diag}}$ with Assumption 3.29 fulfilled, but with an additional (global) factor of $\max_{i=1,\ldots,s} H_k/\eta_k$.

Proof. The proof follows directly from Theorem 3.70 and Lemma 3.71. □

The estimates in Corollary 3.72 may appear to be weaker than those derived after Lemma 3.65, but recall that they also hold when the coefficient $\underline{\alpha}_i$ is *not* quasi-monotone in the boundary layer $\Omega_{i,\eta}$, but only the *artificial* coefficient $\underline{\alpha}_i^{\text{art}}$ is quasi-monotone in the subdomains, such as in Fig. 3.22 (right) on p. 210.

Note also that for specific artificial coefficients, the bounds in Lemma 3.71 and Corollary 3.72 can be further improved. If the manifold X^* in the proof of Lemma 3.71 can be chosen as the union of enough η-agglomerates and if the paths can be suitably chosen, the multiplicity $r_{\max}$ reduces. Also, if the mesh $\mathscr{T}^\eta$ is geometrically coarsened towards the subdomain center and the largest coefficient is attained there, one can apply the theory from Sect. 3.4.4.1 and may replace one of the factors H_i/η_i by $(1+\log(H_i/\eta_i))$. Of course, all this can be rather technical. To give statements for simpler configurations, the theorem below assumes that the artificial coefficient is resolved by the coarse mesh $\mathscr{T}^H(\Omega_i)$. The following result is a generalization of the bounds in [PS11b, Lemma 3.6, Theorem 4.1].

Theorem 3.73. *Let the assumptions of Theorem 3.28 hold and let Assumption 2.54 (p. 115) be fulfilled. For each $i = 1,\ldots,s$, let $\underline{\alpha}_i^{\text{art}} \in L_+^\infty(\Omega_i)$ be an artificial coefficient in the sense of Definition 3.69 with $\Lambda_i = \Omega_i$. Assume further that $\underline{\alpha}_i^{\text{art}}$ is piecewise constant with respect to the coarse mesh $\mathscr{T}^H(\Omega_i)$ and type-m quasi-monotone on Ω_i. Then for the all-floating FETI method with $Q = M_{\text{sD}}^{-1}$,*

$$\kappa(P\, M_{\text{sD}}^{-1} P^\top F_{|\tilde{U}_{\text{ad}}}) \le C\, c_{\text{noise}} \max_{i=1}^{s} \frac{H_i}{\eta_i}\, \sigma^{d-m}\left(\tfrac{H_i}{h_i}\right)\left(1+\log\left(\tfrac{\eta_i}{h_i}\right)\right)^2 .$$

The same bound holds for $Q = Q_{\text{diag}}^{(3)}$ or $Q = Q_{\text{diag}}^{(2)}$ if in the latter case Assumption 3.29 is fulfilled. For $Q = Q_{\text{diag}}$ with Assumption 3.29 being fulfilled, the bound has to be multiplied by a factor of $\max_{k=1}^{s} H_k/\eta_k$.

Proof. To prove the statement, we show a direct bound for $C_i^*(\underline{\alpha}_i^{\text{art}}; h)$ by borrowing an argument from the classical overlapping Schwarz theory (cf. [TW05, Lemma 3.10]). The statement then simply follows from Theorem 3.70.

Let $\{Y_i^{(k)}\}_k$ be the η-agglomerates in $\Omega_{i,\eta}$ and assume without loss of generality that $\partial Y_i^{(k)} \cap \partial\Omega_i$ is always an η-facet. Otherwise, we have to combine a few agglomerates into one. For each of the subregions $Y_i^{(k)}$, a local Poincaré type inequality (combined with Cauchy's inequality) yields

$$\|v\|^2_{L^2(Y_i^{(k)})} \lesssim \eta_i^2 \, |v|^2_{H^1(Y_i^{(k)})} + \eta_i \, \|v\|^2_{L^2(\partial Y_i^{(k)} \cap \Omega_i)} \qquad \forall v \in H^1(Y_i^{(k)}). \tag{3.45}$$

Recall that $\underline{\alpha}_i^{\text{art}}$ is type-m quasi-monotone on Ω_i and resolved by $\mathscr{T}^H(\Omega_i)$. Let X^* be an m-dimensional $\mathscr{T}^H$-facet of a coarse element where the maximum of $\underline{\alpha}_i^{\text{art}}$ is attained. By summing (3.45) over all $Y_i^{(k)} \subset T$ with a fixed coarse element $T \subset \mathscr{T}^H(\Omega_i)$ we get

$$\begin{aligned} \|v\|^2_{L^2(\Omega_{i,\eta} \cap T)} &\lesssim \eta_i^2 \, |v|^2_{H^1(\Omega_{i,\eta} \cap T)} + \eta_i \, \|v\|^2_{L^2(\partial T \cap \Omega)} \\ &\lesssim \eta_i^2 \, |v|^2_{H^1(\Omega_{i,\eta} \cap T)} + \frac{\eta_i}{H_i} \, \|v\|^2_{L^2(T)} + \eta_i \, H_i \, |v|^2_{H^1(T)} \\ &\lesssim \frac{\eta_i}{H_i} \Big[H_i^2 \, |v|^2_{H^1(T)} + \|v\|^2_{L^2(T)} \Big], \end{aligned} \tag{3.46}$$

where the second step we have used Lemma 2.69 (p. 120).

Setting $v = u - \bar{u}^{X^*}$, multiplying (3.46) by the constant value $\underline{\alpha}_{i|T}^{\text{art}}$ and summing over the coarse elements $T \in \mathscr{T}^H(\Omega_i)$, we get

$$\begin{aligned} \|u - \bar{u}^{X^*}\|^2_{L^2(\Omega_{i,\eta}), \underline{\alpha}_i} &\lesssim \frac{\eta_i}{H_i} \Big[H_i^2 \, |u|^2_{H^1(\Omega_i), \underline{\alpha}_i^{\text{art}}} + \|u - \bar{u}^{X^*}\|^2_{L^2(\Omega_i), \underline{\alpha}_i^{\text{art}}} \Big] \\ &\lesssim \frac{\eta_i}{H_i} \, H_i^2 \left(1 + C_{P, \underline{\alpha}_i^{\text{art}}}(\Omega_i; h)^2\right) |u|^2_{H^1(\Omega_i), \underline{\alpha_i}^{\text{art}}} . \end{aligned} \tag{3.47}$$

Thanks to the tools from Sect. 3.4 (e.g., with Corollary 3.50), it follows that

$$C_{P, \underline{\alpha}_i^{\text{art}}}(\Omega_i; h)^2 \lesssim \sigma^{d-m}(\tfrac{H_i}{h_i}).$$

This shows that

$$C_i^*(\underline{\alpha}_i^{\text{art}}; h)^2 \lesssim \frac{\eta_i}{H_i} \, \sigma^{d-m}(\tfrac{H_i}{h_i}).$$

As pointed out before, an application of Theorem 3.70 concludes the proof. □

Note that in the special case where $\underline{\alpha}_i^{\text{art}}$ is constant, the improved bound of Theorem 3.64 follows from the above theorem.

3.6 Numerical Examples and Conclusions

In this section, we show two-dimensional numerical experiments for the "cross-point" and the "edge" island in Fig. 3.23. From the estimated condition numbers in Figs. 3.24 and 3.25 one can observe that the behavior in H/η is linear. In fact, this can be also rigorously shown following the same ideas as in the proof of Theorem 3.73 (by choosing $\alpha_i^{\text{art}} = \alpha_{|\Omega_i}$) and constructing paths with $s_{\max} \lesssim H/\eta$ and $r_{\max} \lesssim (H/\eta)^2$.

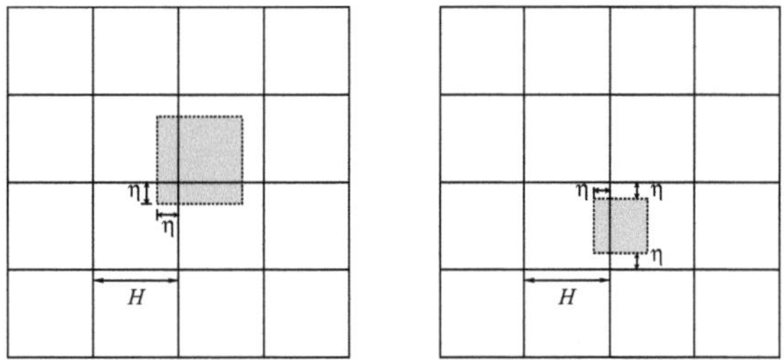

Fig. 3.23 Setup of a "crosspoint" island (*left*) and an "edge" island (*right*)

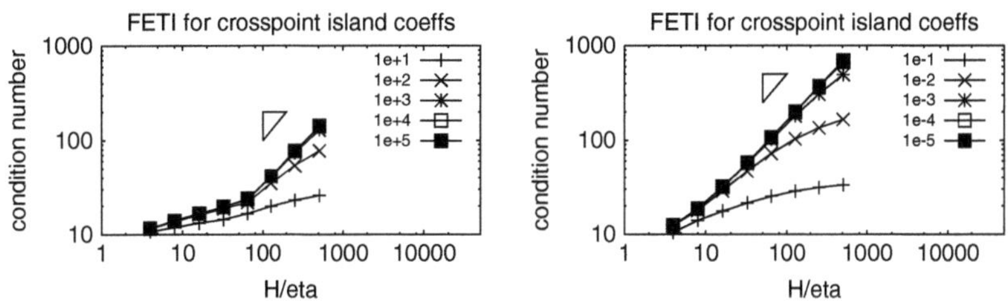

Fig. 3.24 Estimated condition numbers for the "crosspoint" island in Fig. 3.23 (*left*). *Left*: $\alpha_I > 1$. *Right*: $\alpha_I < 1$

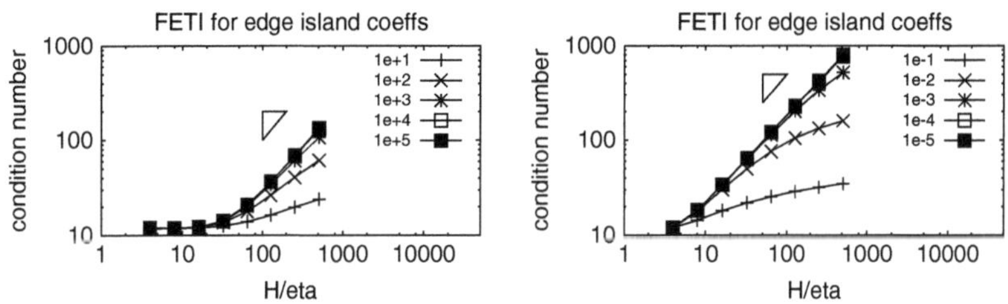

Fig. 3.25 Estimated condition numbers for the "edge" island in Fig. 3.23 (*right*). *Left*: $\alpha_I > 1$. *Right*: $\alpha_I < 1$

For more numerical experiments and for applications in two-dimensional magnetostatic problems we refer to [RF98a, RF99, LP06, PS08, PS10, Pec08b, PS11b], see also [KR07c, KR07b, DS11, GKR12] for dual-primal methods.
We conclude the chapter with further remarks.

(i) For an optimal estimate for the choice $Q = Q_{\text{diag}}$ see also [PS11b, Theorem 6.3].
(ii) All the theorems above treat the all-floating formulation. Bounds for the classical formulation can be achieved if the largest (artificial) coefficient touches the Dirichlet boundary, cf. [PS11b, Sect. 7.2].
(iii) We have seen that for all-floating FETI methods, the existence of boundary layers and suitable artificial coefficients which are type-m quasi-monotone on Λ_i is sufficient for the robustness. However, these conditions are not necessary. Further research will be needed to characterize when FETI methods are robust.

Chapter 4
Unbounded Domains

In this chapter, we extend the methods and theory from Chap. 2 to the case where we add an unbounded exterior subdomain (similar to Ω^{ext} from Sect. 1.3) to our potential equation and prescribe a radiation condition for $u(x)$ as $|x|$ goes to infinity. In Sect. 4.1, we introduce a precise description of our model problem and the corresponding variational skeleton formulation. In particular, Sect. 4.1.3 discusses a generalization of classical as well as all-floating FETI/BETI methods for this model problem, cf. [LP07, Pec09]. A precise analysis including condition number bounds is provided in Sect. 4.2. We conclude with numerical results in Sect. 4.3.

It might seem that the generalizations sketched above are straightforward since the exterior part can simply be seen as an additional subdomain, and the corresponding operator is simply the exterior Steklov-Poincaré operator from Sect. 1.3.4. However, some of the properties of the bounded case apparently change in the unbounded case, at least in general. In the bounded case, a shape regular subdomain partition (cf. Assumption 2.54) implies that each subdomain has a finite number of neighboring subdomains, and that the diameters of two neighboring subdomains are comparable. Adding an unbounded exterior domain to such a partition, however, easily creates situations where the exterior subdomain has arbitrarily many neighbors, and the diameter of its boundary can be arbitrarily large as well. This fundamental difference complicates the theory: a straightforward analysis of the general case usually leads to pessimistic condition number bounds.

Of course, we could just *require* that the number of neighbors is bounded, which would lead to guaranteed benign bounds. But such an assumption is in general very limiting (although satisfactory for many applications, in particular when FETI/BETI is used merely as a coupling method between a few subdomains). Our improved analysis covers the general case and delivers two types of condition number estimates (see Sects. 4.2.3 and 4.2.4 below) that are explicit in the number of subdomain neighbors and sizes of the subdomain boundaries. In special cases the bounds even turn out to be quasi-optimal.

C. Pechstein, *Finite and Boundary Element Tearing and Interconnecting Solvers for Multiscale Problems*, Lecture Notes in Computational Science and Engineering 90, DOI 10.1007/978-3-642-23588-7_4, © Springer-Verlag Berlin Heidelberg 2013

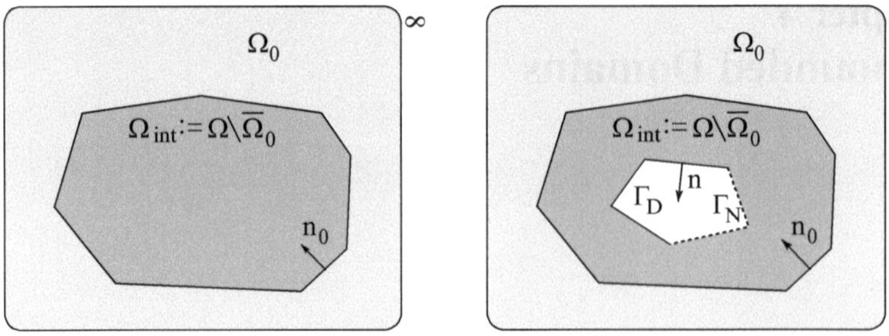

Fig. 4.1 Illustrations of the unbounded computational domain Ω with exterior part Ω_0 *Left*: $\partial\Omega = \emptyset$. *Right*: $\partial\Omega = \Gamma_D \cup \Gamma_N \neq \emptyset$

4.1 Model Problem and Skeleton Formulation

4.1.1 Model Problem

Let $\Omega \subset \mathbb{R}^d$ (with $d = 2$ or 3) be an unbounded domain with the following property. There exists a domain $\Omega_0 \subset \mathbb{R}^d$ such that the sets $\mathbb{R}^d \setminus \overline{\Omega_0}$ and $\Omega \setminus \overline{\Omega_0}$ are bounded Lipschitz domains. We call Ω_0 the *exterior part*, and define further

- The *interior part* $\Omega_{\text{int}} := \Omega \setminus \overline{\Omega_0}$, and
- The *complement* $\Omega_0^c := \mathbb{R}^d \setminus \overline{\Omega_0}$ of the exterior part.

To avoid complications in the theory, we assume that the complement Ω_0^c is simply connected and that its boundary $\partial\Omega_0$ has only one (simply) connected component. We restrict ourselves to coefficients $\mathscr{A} \in L^\infty(\Omega)$ with

$$\mathscr{A}_{|\Omega_0} = \alpha_0 I \qquad \text{with} \qquad \alpha_0 = \text{const} > 0. \tag{4.1}$$

Furthermore, let $f_\Omega \in L^2(\Omega)$ be a source function with

$$\text{supp}(f_\Omega) \subset \overline{\Omega}_{\text{int}}, \tag{4.2}$$

i.e., f_Ω vanishes in the exterior part Ω_0. Moreover, let $\partial\Omega = \Gamma_D \cup \Gamma_N$ with disjoint sets $\Gamma_D = \overline{\Gamma_D}$, Γ_N which are either empty or have positive surface measure, see Fig. 4.1 (right). Note that the boundary $\partial\Omega$ may be empty (if $\Omega = \mathbb{R}^d$), see Fig. 4.1 (left), but $\partial\Omega$ may also have several connected components, see Fig. 4.2 (left). Furthermore, it may happen that $\partial\Omega$ touches $\partial\Omega_0$, cf. Fig. 4.2 (right).

Recall the definition of $H^{1,\star}(\Omega)$ from Definition 1.62 and Lemma 1.63 (p. 40). For given functions $g_D \in H^{1/2}(\Gamma_D)$ and $g_N \in H_{00}^{-1/2}(\Gamma_N)$, we seek $u \in H^{1,\star}(\Omega)$ with $u_{|\Gamma_D} = g_D$ such that

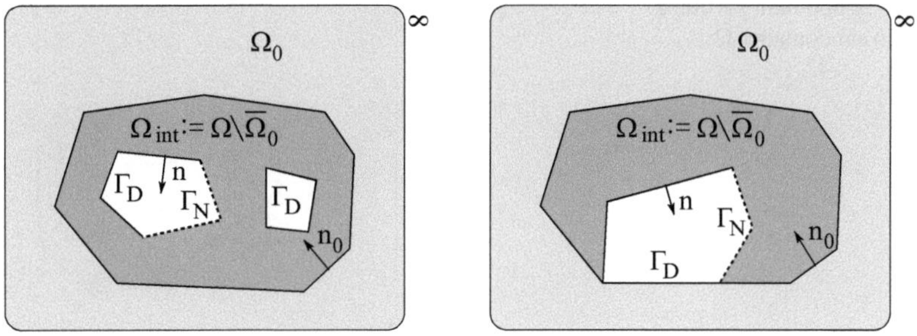

Fig. 4.2 Special cases of the unbounded computational domain Ω. *Left*: The boundary $\partial\Omega$ has two connected components. *Right*: The boundary $\partial\Omega$ touches $\partial\Omega_0$

$$\underbrace{\int_\Omega \mathscr{A}\,\nabla u\cdot\nabla v\,\mathrm{d}x}_{=:a(u,v)} = \underbrace{\int_\Omega f_\Omega\, v\,\mathrm{d}x + \langle g_N, v\rangle_{\Gamma_N}}_{:=\langle f,v\rangle} \qquad \forall v\in H^{1,\star}(\Omega),\; v_{|\Gamma_D}=0. \tag{4.3}$$

This is the weak formulation of the classical problem

$$\begin{aligned} -\mathrm{div}(\mathscr{A}\nabla u) &= f_\Omega && \text{in } \Omega,\\ u &= g_D && \text{on } \Gamma_D,\\ \mathscr{A}\nabla u\cdot n &= g_N && \text{on } \Gamma_N,\\ u(x) &= b+\mathscr{O}(|x|^{-1}) && \text{as } |x|\to\infty, \end{aligned}$$

where n is the outward unit normal to $\partial\Omega$, and $b=0$ for $d=3$ (cf. Sects. 1.2.7 and 1.3.3). In the following, we assume that if $d=2$, then $\Gamma_D\neq\emptyset$ (for $d=3$ no such restriction is required). This guarantees that problem (4.3) is well-posed and admits a unique solution. In the case $d=2$ and $\Gamma_D=\emptyset$, which will be briefly discussed in Sect. 4.2.6 below, we additionally need the compatibility condition

$$\int_\Omega f_\Omega\,\mathrm{d}x + \langle g_N, 1\rangle_{\Gamma_N} = 0.$$

Recall that in this special case, the solution of (4.3) is only unique up to a constant.

4.1.2 Skeleton Formulation

We consider a non-overlapping partition of the interior part Ω_{int} into open subdomains Ω_i, $i=1,\ldots,s$ such that

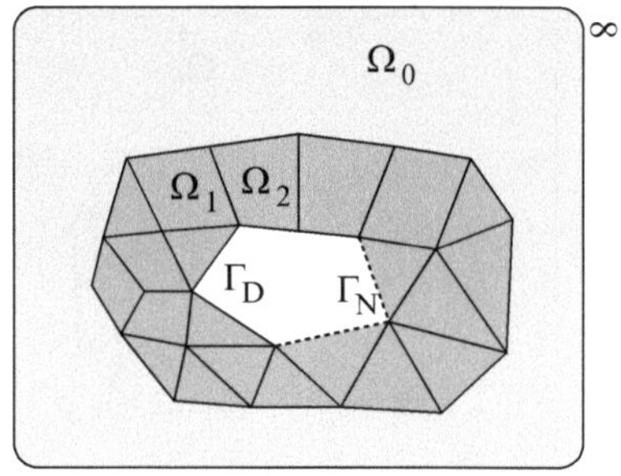

Fig. 4.3 Subdomain partition of Ω into subdomains Ω_i, $i = 0, 1, \ldots, \Omega_s$

$$\overline{\Omega} = \bigcup_{i=0}^{s} \overline{\Omega}_i , \qquad \Omega_i \cap \Omega_j = \emptyset \quad \text{for } i \neq j ,$$

where Ω_0 is the exterior part from above. For each $i = 0, \ldots, s$, let n_i denote the outward unit normal to $\partial\Omega_i$. In particular, n_0 points to the inside of the complement Ω_0^c. As in Table 2.1 (p. 65), we can define the interface Γ, the skeleton Γ_S, and the subdomain diameters $H_i := \text{diam}(\Omega_i)$ for $i = 1, \ldots, s$. An exception is the exterior part Ω_0 for which we set

$$H_0 := \text{diam}(\Omega_0^c) = \text{diam}(\partial\Omega_0). \tag{4.4}$$

Note that $H_i \leq H_0$ for all $i = 1, \ldots, s$. If the number of interior subdomains is large, then typically also $H_i \ll H_0$.

In the following, we assume additionally that

$$g_N \in L^2(\Gamma_N).$$

Under this condition, Problem (4.3) obeys the splitting property

$$a(u, v) = \sum_{i=0}^{s} \underbrace{\int_{\Omega_i} \mathscr{A} \nabla u \cdot \nabla v \, \mathrm{d}x}_{=:a_i(u_{|\Omega_i}, v_{|\Omega_i})}, \qquad \langle f, v \rangle = \sum_{i=0}^{s} \underbrace{\int_{\Omega_i} f_\Omega \, v \, \mathrm{d}x + \int_{\partial\Omega_i \cap \Gamma_N} g_N \, v \, \mathrm{d}s}_{=:\langle f_i, v_{|\Omega_i} \rangle}$$

Let S_i, $i = 1, \ldots, s$ be the Steklov-Poincaré operator corresponding to $a_i(\cdot, \cdot)$, and let S_0 be the exterior Steklov-Poincaré operator corresponding to $a_0(\cdot, \cdot)$. With the notations in Definition 1.67 (p. 42) and with Ω^{ext} replaced by Ω_0, we have $S_0 = \alpha_0 \, S^{\text{ext}}$. Furthermore, let N_i, $i = 1, \ldots, s$ be the Newton potentials corresponding to $a_i(\cdot, \cdot)$ (see Definition 1.41) and let us formally define $N_0 f_0$ by

$$\langle N_0 f_0, v \rangle_{\partial\Omega_0} = \int_{\partial\Omega_0 \cap \Gamma_N} g_N \, v \, \mathrm{d}s \qquad \text{for } v \in H^{1/2}(\partial\Omega_0).$$

Following the derivation in Sect. 2.1, we obtain the continuous skeleton formulation.

Find $u \in H^{1/2}(\Gamma_S)$ with $u_{|\Gamma_D} = g_D$ such that

$$\sum_{i=0}^{s} \langle S_i\, u_{|\partial\Omega_i},\, v_{|\partial\Omega_i}\rangle \;=\; \sum_{i=0}^{s} \langle N_i\, f_i,\, v_{|\partial\Omega_i}\rangle \qquad \forall v \in H^{1/2}(\Gamma_S),\; v_{|\Gamma_D} = 0. \quad (4.5)$$

Similarly to Sect. 2.1.2, we fix an index set $\mathcal{I}_{\text{BEM}} \subset \{0, 1, \ldots, s\}$ with $0 \in \mathcal{I}_{\text{BEM}}$ and its complement $\mathcal{I}_{\text{FEM}} = \{0, 1, \ldots, s\} \backslash \mathcal{I}_{\text{BEM}}$. The discrete skeleton formulation is obtained by choosing the approximations

$$S_{i,h} := \begin{Bmatrix} S_{i,\text{BEM}} & \text{if } i \in \mathcal{I}_{\text{BEM}} \\ S_{i,\text{FEM}} & \text{if } i \in \mathcal{I}_{\text{FEM}} \end{Bmatrix}, \qquad N_{i,h} := \begin{Bmatrix} N_{i,\text{BEM}} & \text{if } i \in \mathcal{I}_{\text{BEM}} \\ N_{i,\text{FEM}} & \text{if } i \in \mathcal{I}_{\text{FEM}} \end{Bmatrix}.$$

With the notations from Sect. 1.3.8.1 and with Ω^{ext} replaced by Ω_0,

$$S_{0,\text{BEM}} \;=\; \alpha_0\, S_{\text{BEM}}^{\text{ext}}\,.$$

4.1.3 *FETI/BETI Formulation*

Based on the skeleton formulation from the previous section, the derivation of classical and all-floating FETI/BETI for the unbounded case is analogous to Sects. 2.2.1 and 2.2.2. The unbounded part Ω_0 takes simply the role of an additional subdomain. Only the following issues have to be taken into account.

(i) We mention explicitly that for the all-floating formulation, there might be Lagrange multipliers acting on $\partial\Omega_0 \cap \Gamma_D$, should this set be non-empty. Also,

$$W_0 \;:=\; \begin{cases} \{w \in V^h(\partial\Omega_0) : w_{|\Gamma_D} = 0\} & \text{for the classical formulation,} \\ V^h(\partial\Omega_0) & \text{for the all-floating formulation.} \end{cases}$$

(ii) Recall that Ω_0 is a floating subdomain in the sense of Definition 2.15 if $S_{0,\text{BEM}}$ is singular. Hence, for $d = 3$, Ω_0 is always non-floating. For $d = 2$, Ω_0 is floating, unless we use the classical formulation and $\partial\Omega_0 \cap \Gamma_0 \neq \emptyset$.

(iii) For $d = 2$, we need the condition $\operatorname{diam}(\Omega_{\text{int}}) < 1$ in order to ensure the invertibility of the discrete single layer potential operators occurring in $S_{0,\text{BEM}}$ and $S_{i,\text{BEM}}$ for all $i \in \mathcal{I}_{\text{BEM}}$.

4.1.3.1 Preconditioning

As in Chap. 2, from now on we drop the discretization parameter h in the Steklov-Poincaré and Newton potential approximants:

$$S_i : W_i \to W_i^*, \qquad N_i : H^1(\Omega_i)^* \to W_i^* \qquad \forall i = 0, \dots, s.$$

Moreover, we write $W = \prod_{i=0}^{s} W_i$ and $S := \operatorname{diag}(S_i)_{i=0}^{s}$.

Since the coefficient α_0 in Ω_0 is scalar and constant, the notions of multiplicity and coefficient scaling, see Sect. 2.2.4.2 are straightforwardly generalized. This allows to define the weighted jump operator $B_D : W^* \to U$ analogously to Sect. 2.2.4.2, and we obtain the following formula for the operator $P_D := B_D^\top B$:

$$(P_D\, w)_i(x^h) = \begin{cases} \sum_{j \in \mathcal{N}_{x^h}} \delta^\dagger(x^h) \left(w_i(x^h) - w_j(x^h) \right) & \text{for } x^h \in \Omega_i^h \cap \Gamma^h, \\ w_i(x^h) & \text{for } x^h \in \Omega_i^h \cap \Gamma_D, \\ 0 & \text{otherwise.} \end{cases} \tag{4.6}$$

Furthermore, $BP_D = B$, and so the statement of Corollary 2.40 holds true. Finally, the scaled Dirichlet preconditioner reads

$$M_{\text{sD}}^{-1} := B_D\, S B_D^\top = B_D \operatorname{diag}(S_i)_{i=0}^{s}\, B_D^\top\,. \tag{4.7}$$

Remark 4.1. Similarly as outlined in Remark 2.29, the block S_0 appearing in M_{sD}^{-1} may be replaced by a spectrally equivalent operator. From the estimates in Sect. 1.3.8.3 it turns out that (in three dimensions) a proper choice is the discretization of the operator $\alpha_0 \widetilde{D}_0$, where $\widetilde{D}_0$ is a suitable regularization of the hypersingular operator D_0 on $\partial\Omega_0$, e.g., given by

$$\langle \widetilde{D}_0\, v, w \rangle = \langle D_0\, v, w \rangle + \frac{1}{H_0^d} \int_{\partial\Omega_0} v\, ds \int_{\partial\Omega_0} w\, ds\,.$$

4.1.3.2 Positivity of the Preconditioner

Recall from Sect. 2.2.1.4, that in the projection $P : U \to U_{\text{ad}}$ we have to choose the operator $Q : U^* \to U$. As in Chap. 2, we consider the choices $Q = M_{\text{sD}}^{-1}$ or $Q = Q_{\text{diag}}$ (see the next section). Recall, that for either of the two choices, we have to make sure that Q is SPD on range(G), cf. Lemma 2.43 (p. 108).

Lemma 4.2. *The scaled Dirichlet preconditioner M_{sD}^{-1} defined in* (4.7) *is SPD on* range(G). *Consequently, if $Q = M_{\text{sD}}^{-1}$, then the projections P and $P^\top$ are well-defined and the operator $P\, M_{\text{sD}}^{-1}$ is SPD on $\widetilde{U}_{\text{ad}}^*$.*

Proof. Since the positive semi-definiteness of M_{sD}^{-1} on the whole of U^* is immediate, we only need to check the definiteness on range$(G) = B(\ker(S))$. Assume that $\langle B\, z, M_{\text{sD}}^{-1} B\, z \rangle = 0$ for $z \in \ker(S)$. By the same arguments as in Lemma 2.42 (p. 108), it follows that $(I - P_D)\, z = 0$, which implies that the piecewise constant function z is continuous across all subdomain interfaces and vanishes at the Dirichlet boundary. Since there is at least one subdomain Ω_i with $z_i = 0$ (for $d = 3$, $z_0 = 0$), this continuity implies that $z = 0$, which shows the definiteness. The rest follows from Lemma 2.43. □

4.1.3.3 Diagonal Choice of Q

As in Chap. 2, there is an alternative diagonal choice Q_{diag} of Q based on the decomposition of $\Gamma \cup \Gamma_D$ into *globs*. This decomposition follows Definition 2.48 (p. 113) in an analogous way, by treating Ω_0 as an additional subdomain. Finally, we define $Q_{\text{diag}} : U^* \to U$ as follows. For $\mu \in U^*$, we set

$$(Q_{\text{diag}}\,\mu)_{ij}(x^h) := \min(\rho_i(x^h),\, \rho_j(x^h))\, q_{ij}(x^h)\, \mu_{ij}(x^h) \quad \text{for } x^h \in \partial\Omega_i^h \cap \partial\Omega_j^h\,, \tag{4.8}$$

where

$$\begin{aligned} q_{0i}(x^h) &:= q_{i0}(x^h) := q_i(x^h)\,, \\ q_{ij}(x^h) &:= \min(q_i(x^h),\, q_j(x^h)) \qquad \text{for } i,\, j \neq 0\,, \end{aligned}$$

and $q_i(x^h)$, $i = 1, \ldots, s$ is as in Definition 2.107 (p. 137). In the all-floating formulation, we additionally set

$$(Q_{\text{diag}}\,\mu)_{iD}(x^h) := \rho_i(x^h)\, q_i(x^h)\, \mu_{iD}(x^h) \qquad \text{for } x^h \in \partial\Omega_i^h \cap \Gamma_D\,, \tag{4.9}$$

where

$$q_0(x^h) = 0 \qquad \text{for } x^h \in \partial\Omega_0^h \cap \Gamma_D\,. \tag{4.10}$$

Note that by this construction, the possibly large diameter H_0 never occurs in the entries of Q_{diag}. See also Sect. 4.2.6 for the case of two dimensions.

Lemma 4.3. *For the unbounded setting in three dimensions, the operator Q_{diag} is SPD on* range(G). *Consequently, if $Q = Q_{\text{diag}}$, then the projections P and $P^\top$ are well-defined and the operator $P\, M_{\text{sD}}^{-1}$ is SPD on $\widetilde{U}^*_{\text{ad}}$.*

Proof. In the classical formulation, Q_{diag} is SPD on the whole of $\widetilde{U}^*$ by definition. In the all-floating formulation, Q_{diag} is at least symmetric and positive semi-definite, and it suffices to show the definiteness on range$(G) = B(\ker(S))$. Hence, assume that $\langle Bz,\, Q_{\text{diag}} Bz\rangle = 0$ for some $z \in \ker(S)$. We find that $(B\,z)_{ij}(x^h) = 0$ for all interface nodes $x^h \in \Gamma_{ij}^h$ where $i,\, j = 0, \ldots, s$, and $(B\,z)_{iD}(x^h) = 0$ for all Dirichlet nodes $x^h \in \partial\Omega_i^h \cap \Gamma_D$, where $i = 1, \ldots, s$. However, since $z_0 = 0$, we conclude that $(B\,z)_{0D}(x^h) = 0$ for all $x^h \in \partial\Omega_0^h \cap \Gamma_D$. Summarizing, it follows that $B\,z = 0$, which shows the definiteness. The rest follows from Lemma 2.43. □

4.2 Analysis

In this section, we analyze the scaled Dirichlet preconditioner for both the classical and all-floating BETI/FETI method including the unbounded subdomain Ω_0. We will restrict ourselves to the case of three dimensions, which is practically more relevant. See Sect. 4.2.6 for some notes on the two-dimensional case. The main line of the proof follows the abstract framework in Sect. 2.4.2, and the exterior part Ω_0 will simply play the role of an additional subdomain. However, Ω_0 has to be treated differently when estimating the energy of the P_D-operator.

As outlined in the beginning of this chapter, we have to face two crucial difficulties that are not present in the bounded case (Chap. 2).

(i) The number of neighbors of Ω_0 can be arbitrarily large.
(ii) The diameter $H_0 = \mathrm{diam}(\partial\Omega_0)$ can be very large compared to the diameter H_i of a neighboring interior subdomain.

Obviously, these two issues go hand in hand. We will therefore draw special attention on how the condition number $\kappa(P\, M_{\mathrm{sD}}^{-1}\, P^\top F_{|U_{\mathrm{ad}}})$ depends on the number of neighbors of Ω_0 and on a ratio of the form

$$\max_{\mathscr{F}_0 \subset \partial\Omega_0} \frac{H_0}{\mathrm{diam}(\mathscr{F}_0)}, \tag{4.11}$$

where the maximum is taken over the subdomain faces on $\partial\Omega_0$ (including possible Dirichlet faces). It will turn out that the behavior is not robust in general. After stating some assumptions in Sect. 4.2.1 and providing some tools in Sect. 4.2.2, we will provide two types of condition number bounds. The first bound, presented in Sect. 4.2.3, depends linearly on the ratio (4.11). The second bound, presented in Sect. 4.2.4, depends on a geometrical factor that is related to the interior Dirichlet conditions, and on an additional factor of $\max_{i=0,\ldots,s} \dfrac{\alpha_i}{\alpha_0}$. Complete robustness can only be achieved in special cases, which can also be seen in the numerical results in Sect. 4.3.

4.2.1 *Assumptions on the Subdomains and on the Data*

For the subsequent analysis, we need to adapt the assumptions from Sect. 2.5.2 (Assumptions 2.53–2.56) to the new setting that includes the unbounded part Ω_0. The following two assumptions are simple counterparts to Assumptions 2.53 and 2.54.

Assumption 4.4. The subdomain triangulations $\mathscr{T}^h(\Omega_i)$, $i \in \mathscr{I}_{\mathrm{FEM}}$ and $\mathscr{T}^h(\partial\Omega_i)$, $i \in \mathscr{I}_{\mathrm{BEM}}$ (including the case $i = 0$) are quasi-uniform with mesh parameter h_i.

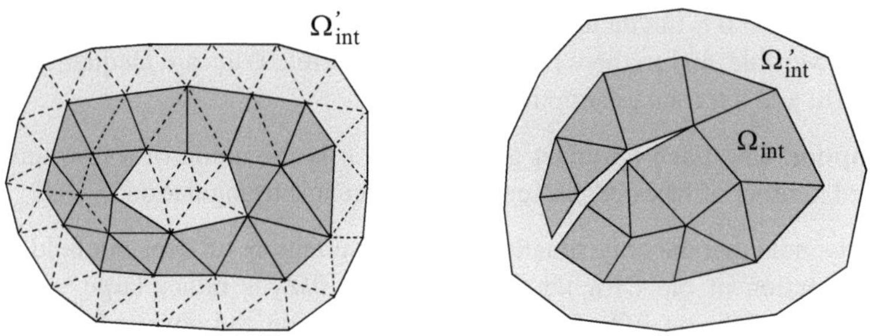

Fig. 4.4 Illustration of Assumption 4.6. *Left*: Sketch of Ω'_{int} and the triangulation $\mathscr{T}^H(\Omega'_{\text{int}})$ for the example from Fig. 4.3 (*dark shaded*: Ω_{int}). *Right*: Setting where Assumption 4.6 is only fulfilled with a very bad shape regularity constant

Assumption 4.5. There is a shape regular simplicial coarse triangulation $\mathscr{T}^H(\Omega_{\text{int}})$ of the interior part $\Omega_{\text{int}} = \Omega \setminus \Omega_0$, such that each interior subdomain $\Omega_i \subset \Omega_{\text{int}}$ is the union of coarse elements and the number of coarse elements per subdomain is uniformly bounded.

As in Sect. 2.5.2, the previous assumption implies that the subdomain diameter H_i is equivalent to the local mesh parameter of $\mathscr{T}^H(\Omega_i)$ for $i = 1, \ldots, s$. The next assumption can be seen as an analogue to Assumptions 2.55 and 2.56.

Assumption 4.6. There exists a bounded Lipschitz domain Ω'_{int} such that

(i) $\overline{\Omega_0^c} \subset \Omega'_{\text{int}}$,
(ii) The coarse triangulation $\mathscr{T}^H(\Omega_{\text{int}})$ from Assumption 4.5 can be extended to a shape regular simplicial coarse triangulation $\mathscr{T}^H(\Omega'_{\text{int}})$,
(iii) The Dirichlet boundary Γ_D is the union of facets (faces/edges) from $\mathscr{T}^H(\Omega'_{\text{int}})$.

This assumption ensures that the exterior angles of Ω_{int} do not degenerate. This will be crucial for the exterior angles on $\partial\Omega_0$. In case that the domain Ω_{int} has interior holes separated from $\partial\Omega_0$, one may relax Assumption 4.6 and allow degenerate angles there (while the interior subdomains themselves still should satisfy Assumption 2.54). However, this would complicate the following proofs, and so for the sake of a simple presentation, we have chosen to use the slightly more restrictive conditions in Assumption 4.6. For an illustration see Fig. 4.4.

Furthermore, we have to add a few more assumptions, which are analogous to Assumptions 2.99–2.102.

Assumption 4.7. (i) In each subdomain Ω_i, for $i = 0, \ldots, s$, there exists a constant $\alpha_i > 0$ such that

$$\mathscr{A}_{|\Omega_i} = \alpha_i \, I.$$

(ii) The *coefficient scaling* is used, i.e., $\rho_i(x^h) = \alpha_i$, cf. Sect. 2.2.4.2.
(iii) For the classical formulation of FETI/BETI in three dimensions, we assume that for all $i = 1, \ldots, s$, the set $\Gamma_D \cap \partial\Omega_i$ is either empty or contains at least a subdomain edge (i.e., it should not collapse to a subdomain vertex).

The case $i = 0$ is not included in Assumption 4.7 (iii) because the operator S_0 is a priori invertible, and so there is no need for a discrete Poincaré inequality. Finally, we have to add a formal assumption.

Assumption 4.8. Each Dirichlet glob $\mathcal{G}_0 \subset \Gamma_D$ is the union of a few, uniformly bounded number of faces/edges/vertices of the coarse triangulation $\mathcal{T}^H(\Omega'_{\text{int}})$.

Firstly, note that the only situation where Assumption 4.8 does not hold is when the intersection of the Dirichlet boundary with $\partial\Omega_0$ is rather large, and so are the Dirichlet globs on $\partial\Omega_0$. However, the assumption can always be fulfilled by further subdividing these globs. Secondly, note that for the diagonal operator Q_{diag}, this further subdivision never needs to be performed in practice, since the entries of Q_{diag} corresponding to the constraints on $\Gamma_D \cap \partial\Omega_0$ are set to zero. Actually, the only purpose of Assumption 4.8 is to simplify the presentation of our theory. For example, the assumption ensures that for any coarse element $T \in \mathcal{T}^H(\Omega'_{\text{int}})$ touching a Dirichlet facet $\mathcal{F}_0$ (or edge $\mathcal{E}_0$), we have that $\text{diam}(\mathcal{F}_0) \gtrsim \text{diam}(T)$ (or $\text{diam}(\mathcal{E}_0) \gtrsim \text{diam}(T)$).

4.2.2 *Auxiliary Meshes, Globs, and Cut-Off Functions*

If we let Ω_0 take the role of a subdomain and apply Definition 2.48, then the interface Γ and the Dirichlet boundary Γ_D are the union of globs of the decomposition $\overline{\Omega} = \bigcup_{i=0}^{s} \Omega_i$. As in Sect. 2.5, we denote by $\mathcal{G}_i$ generically a glob that is shared by subdomain Ω_i. In particular, when writing $\mathcal{G}_0$, we refer to a glob that is either at the interface between the exterior part Ω_0 and an interior subdomain, or that is part of $\Gamma_D \cap \partial\Omega_0$.

For any glob $\mathcal{G}_i$ of an interior subdomain Ω_i, $i \neq 0$, the subdomain itself serves as a local neighborhood of that glob. For the globs of the exterior part Ω_0, we need to define such local neighborhoods "manually" (for an illustration see Fig. 4.5).

Definition 4.9. For each glob $\mathcal{G}_0 \subset \partial\Omega_0$, we define the open neighborhood $\mathsf{U}_{0,\mathcal{G}_0} \subset \Omega_0$ by the relation

$$\overline{\mathsf{U}_{0,\mathcal{G}_0}} = \bigcup \{\overline{T} : T \in \mathcal{T}^H(\Omega'_{\text{int}}),\ T \subset \Omega_0,\ \mathcal{G}_0 \cap \overline{T} \neq \emptyset\}.$$

In the following, we refer to the region $\Omega'_{\text{int}} \cap \Omega_0$ as the *annulus*. Note that the neighborhoods in Definition 4.9 all lie in the annulus. Next, we need auxiliary meshes and finite element spaces on all local neighborhoods of globs. We construct these based on a single triangulation of the auxiliary domain Ω'_{int} from Assumption 4.6.

Definition 4.10. Let $\mathcal{T}^h(\Omega'_{\text{int}})$ be an auxiliary shape regular triangulation with the following properties.

(i) The triangulation $\mathcal{T}^h(\Omega'_{\text{int}})$ matches with the existing triangulations $\mathcal{T}^h(\Omega_i)$, $i \in \mathcal{I}_{\text{FEM}}$ and $\mathcal{T}^h(\partial\Omega_i)$, $i \in \text{BEM}$, in particular with $\mathcal{T}^h(\partial\Omega_0)$.

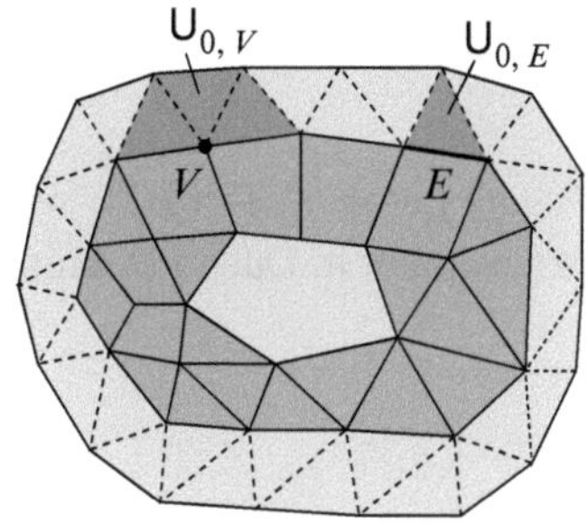

Fig. 4.5 Illustration of the neighborhoods $\mathsf{U}_{0,\mathscr{V}}$, $\mathsf{U}_{0,\mathscr{E}}$ from Definition 4.9

(ii) The triangulation $\mathscr{T}^h(\Omega'_{\text{int}})$ resolves the coarse elements of $\mathscr{T}^H(\Omega'_{\text{int}} \cap \Omega_0)$, i.e., those in the annulus.
(iii) For each index $i \in \mathscr{I}_{\text{BEM}} \setminus \{0\}$, the restriction of $\mathscr{T}^h(\Omega'_{\text{int}})$ to Ω_i is quasi-uniform with mesh parameter h_i. Furthermore, the restriction of $\mathscr{T}^h(\Omega'_{\text{int}})$ to $\Omega'_{\text{int}} \cap \Omega_0$ is quasi-uniform with mesh parameter h_0.
(iv) For any coarse element $T \in \mathscr{T}^H(\Omega'_{\text{int}})$ not contained in Ω but touching $\partial\Omega_0$, the restriction of $\mathscr{T}^h(\Omega'_{\text{int}})$ to T is quasi-uniform with mesh parameter h_0.

Let $V^h(\Omega'_{\text{int}})$ denote the corresponding FE space of continuous piecewise linear functions.

The existence of $\mathscr{T}^h(\Omega'_{\text{int}})$ is guaranteed from Assumptions 4.4–4.6. Note also that property (iv) in Definition 4.10 is only relevant if Ω has holes touching $\partial\Omega_0$. Finally, we define cut-off functions $\theta_{\mathscr{G}_i}$ and $\vartheta_{\mathscr{G}_i}$, very similar to those in Sect. 2.5, which are supported only in the local neighborhoods of the glob $\mathscr{G}_i$.

Definition 4.11. For each glob $\mathscr{G}_i$, let $\theta_{\mathscr{G}_i} \in V^h(\Gamma_S)$ be the function which takes value one at all the nodes contained in $\mathscr{G}_i$ and zero at all other nodes. Furthermore, the function $\vartheta_{\mathscr{G}_i} \in V^h(\Omega'_{\text{int}} \cap \Omega)$ is uniquely defined by the following properties.

- $\vartheta_{\mathscr{G}_i|\Gamma_S} = \theta_{\mathscr{G}_i}$,
- $\vartheta_{\mathscr{G}_i}$ is discrete harmonic in all interior subdomains Ω_j with $\mathscr{G}_i \subset \partial\Omega_j$, $j \neq 0$, and vanishes entirely in all other interior subdomains.
- If $\mathscr{G}_i \not\subset \partial\Omega_0$, then $\vartheta_{\mathscr{G}_i}$ vanishes entirely in the annulus $\Omega_0 \cap \Omega'_{\text{int}}$.
- If $\mathscr{G}_i \subset \partial\Omega_0$, then
  - $\vartheta_{\mathscr{G}_i}$ vanishes entirely in $(\Omega_0 \cap \Omega'_{\text{int}}) \setminus \mathsf{U}_{0,\mathscr{G}_i}$ and at all nodes on $\partial\mathsf{U}_{0,\mathscr{G}_i} \setminus \mathscr{G}_i$,
  - $\vartheta_{\mathscr{G}_i}$ is discrete harmonic in $\mathsf{U}_{0,\mathscr{G}_i}$.

The following lemma helps to split functions on $\partial\Omega_0$ and is actually similar to Lemma 3.21 (p. 169).

Lemma 4.12. *Let the assumptions of Sect. 4.2.1 hold. Then there exists a positive constant C depending only on the shape regularity of the coarse triangulation $\mathscr{T}^H(\Omega'_{\text{int}})$ such that for any $w \in V^h(\Omega_0)$ with $w_{|\Gamma_N^h} = 0$ and for any extension $\tilde{w} \in V^h(\Omega'_{\text{int}})$ of w,*

$$|w|_{S_0}^2 \le C\,\alpha_0 \sum_{\mathcal{G}_0} |I^h(\vartheta_{\mathcal{G}_0}\tilde{w})|_{H^1(\mathsf{U}_{0,\mathcal{G}_0})}^2 .$$

Proof. Let $w \in V^h(\partial\Omega_0)$ with $w_{|\Gamma_N^h} = 0$ be fixed and let $\tilde{w} \in V^h(\Omega'_{\text{int}})$ be an extension of it. Due to Definition 4.11, the function

$$\psi := \sum_{\mathcal{G}_0} I^h(\vartheta_{\mathcal{G}_0}\,\tilde{w})$$

is an extension of w from $\partial\Omega_0$ to $H^{1,\star}(\Omega_0)$ (see Definition 1.62), because each of the functions $\vartheta_{\mathcal{G}_0}$ can be extended to $\Omega_0 \setminus \Omega'_{\text{int}}$ by zero. Thus, with the spectral equivalence relation "$S_{\text{BEM}}^{\text{ext}} \le S^{\text{ext}}$" from Lemma 1.93 (ii) (p. 58) and the minimizing property of S^{ext} (Definition 1.67 (ii), p. 42),

$$\begin{aligned}
|w|_{S_0}^2 &\le \alpha_0 \min_{\substack{u \in H^{1,\star}(\Omega_0)\\ u_{|\partial\Omega_0} = w}} \int_{\Omega_0} |\nabla u|^2 \,\mathrm{d}x \\
&\le \alpha_0 \int_{\Omega_0 \cap \Omega'_{\text{int}}} |\nabla\psi|^2 \,\mathrm{d}x \;\lesssim\; \alpha_0 \sum_{\mathcal{G}_0} |I^h(\vartheta_{\mathcal{G}_0}\,\tilde{w})|_{H^1(\mathsf{U}_{0,\mathcal{G}_0})}^2 ,
\end{aligned}$$

where in the last step, we have used that the local neighborhoods $\mathsf{U}_{0,\mathcal{G}_0}$, where the functions $\vartheta_{\mathcal{G}_0}$ are supported, have finite overlap. □

Using Lemma 4.12, we can give a first estimate for the P_D operator. Recall our convention (2.79) from p. 122, which explains the expression $\mathcal{H}_i^h(\theta_{\mathcal{G}_i}\,v)$ for any function v with well-defined values at the nodes contained in $\mathcal{G}_i$. The next definition provides a similar notion related to the annulus $\Omega'_{\text{int}} \cap \Omega_0$.

Definition 4.13. For a glob $\mathcal{G}_0$ and any function v that has well-defined values at the nodes contained in $\mathcal{G}_0$, let $\mathcal{H}_{0,\mathsf{U}_{0,\mathcal{G}_0}}^h(\theta_{\mathcal{G}_0}v) \in V^h(\mathsf{U}_{0,\mathcal{G}_0})$ be the unique function that

- Coincides with v at all the nodes contained in $\mathcal{G}_0$,
- Vanishes at all the nodes contained in $\partial\mathsf{U}_{0,\mathcal{G}_0} \setminus \mathcal{G}_0$, and
- Is discrete harmonic inside of $\mathsf{U}_{0,\mathcal{G}_0}$.

Lemma 4.14. *Let Assumptions 4.4–4.7 hold. Then for all $w \in W$,*

$$\begin{aligned}
|(P_D w)_i|_{S_i}^2 \;\lesssim\; &\sum_{\mathcal{G}_i \subset \Gamma} \sum_{j \in \mathcal{N}_{\mathcal{G}_i}} \min(\alpha_i, \alpha_j)\, |\mathcal{H}_i^h(\theta_{\mathcal{G}_i}(w_i - w_j))|_{H^1(\Omega_i)}^2 \\
&+ \sum_{\mathcal{G}_i \subset \Gamma_D} \alpha_i\, |\mathcal{H}_i^h(\theta_{\mathcal{G}_i} w_i)|_{H^1(\Omega_i)}^2 \qquad \forall i = 1, \ldots, s
\end{aligned} \tag{4.12}$$

and

$$|(P_D w)_0|^2_{S_0} \lesssim \sum_{\mathcal{G}_0 \subset \Gamma} \sum_{j \in \mathcal{N}_{\mathcal{G}_0}} \min(\alpha_0, \alpha_j) \, |\mathcal{H}^h_{0,\mathsf{U}_{0,\mathcal{G}_0}}(\theta_{\mathcal{G}_0}(w_0 - w_j))|^2_{H^1(\mathsf{U}_{0,\mathcal{G}_0})}$$
$$+ \sum_{\mathcal{G}_0 \subset \Gamma_D} \alpha_0 \, |\mathcal{H}^h_{0,\mathsf{U}_{0,\mathcal{G}_0}}(\theta_{\mathcal{G}_0} w_0)|^2_{H^1(\mathsf{U}_{0,\mathcal{G}_0})} \,. \tag{4.13}$$

The hidden constants only depend on the shape regularity of the coarse mesh $\mathcal{T}^H(\Omega'_{\text{int}})$ from Assumption 4.6.

Proof. Recall formula (4.6). Thanks to Assumption 4.7, the factor $\delta^\dagger_j(x^h)$ is constant per glob and can be bounded by $\min(\alpha_i, \alpha_j)$. The result finally follows from the fact that interior subdomains have a bounded number of globs, and from the estimate in Lemma 4.12. □

The next two sections follow different strategies to estimate further the expressions in Lemma 4.14, resulting in two different condition number estimates of the FETI/BETI method. Before, we state a lemma that will be used quite frequently and that helps to "join" the contributions $\theta_{\mathcal{G}_0} w_0$ appearing in Lemma 4.14 and bound them in terms of $|w_0|_{S_0}$.

Lemma 4.15. *If $d = 3$, then for each $w_0 \in V^h(\partial\Omega_0)$ there exists an extension $\hat{w} \in V^h(\Omega_0^c)$ of w_0 such that*

$$\alpha_0 \left[|\hat{w}|^2_{H^1(\Omega_0^c)} + \frac{1}{H_0} \|w_0\|^2_{L^2(\partial\Omega_0)} \right] \lesssim |w_0|^2_{S_0} \,.$$

Proof. For a fixed function $w_0 \in V^h(\partial\Omega_0)$, we choose $\hat{w}$ as the discrete harmonic extension of w such that

$$|\hat{w}|_{H^1(\Omega_0^c)} = |w|_{S^{\text{int}}_{0,\text{FEM}}} ,$$

where $S^{\text{int}}_{0,\text{FEM}}$ is the interior FEM Steklov-Poincaré approximant defined via $V^h(\Omega_0^c)$. The spectral estimates from Corollary 1.57 (p.38) and Lemma 1.93 (iii) (p. 61) imply that

$$\alpha_0 \left[|\hat{w}|^2_{H^1(\Omega_0^c)} + \frac{1}{H_0} \|w_0\|^2_{L^2(\partial\Omega_0)} \right] = \alpha_0 \left[|w_0|^2_{S^{\text{int}}_{0,\text{FEM}}} + \frac{1}{H_0} \|w_0\|^2_{L^2(\partial\Omega_0)} \right] \lesssim |w_0|^2_{S_0},$$

which concludes the proof. □

Remark 4.16. We note that in order to deal with the contributions $\theta_{\mathcal{G}_0} w_0$ appearing in Lemma 4.14, one could also work in the $H^{1/2}(\partial\Omega_0)$ norm and use (related) techniques from the theory of iterative substructuring and additive Schwarz methods for the BEM, see, e.g., [AG02, Heu01, HS98, HS01, ST98, Pet89]. However, to keep the involved constants as explicit as possible, we have decided to work with harmonic extensions to Ω_0^c and the annulus $\Omega'_{\text{int}} \cap \Omega_0$.

4.2.3 A Condition Number Estimate Involving H_0

We define the local subspaces

$$W_i^\perp := \begin{cases} \left\{ w \in W_i : \overline{\mathscr{H}_i^h w}^{\Omega_i} = 0 \right\} & \text{if } i \in \mathscr{I}_{\text{float}}, \\ W_i & \text{otherwise,} \end{cases}$$

and $W^\perp := \prod_{i=0}^s W_i^\perp$. Recall that we restricted ourselves to the case $d = 3$. Hence the exterior is non-floating and $W_0^\perp = W_0$.

For any $w \in W$, we define the projections $\Pi_0 w \in W$ and $\Pi_{\text{int}} w \in W$ by

$$\begin{aligned} (\Pi_0 w)_0 &:= w_0, & (\Pi_0 w)_i &:= 0 & \forall i = 1, \ldots, s, \\ (\Pi_{\text{int}} w)_0 &:= 0, & (\Pi_{\text{int}} w)_i &:= w_i & \forall i = 1, \ldots, s, \end{aligned} \tag{4.14}$$

i.e., Π_0 selects the component associated to $\partial\Omega_0$ and $\Pi_{\text{int}} = I - \Pi_0$.

Recall the linear operator $w \mapsto z_w$ from Lemma 2.44 (p. 109). Thanks to the linearity, we obtain that for all $w \in W$,

$$\tfrac{1}{2} \, |P_D(w + z_w)|_S^2 \;\le\; |P_D(\Pi_0 w + z_{(\Pi_0 w)})|_S^2 + |P_D(\Pi_{\text{int}} w + z_{(\Pi_{\text{int}} w)})|_S^2 \,. \tag{4.15}$$

We now draw our attention to the first term in (4.15). In the sequel, we use the abbreviation

$$H_{\mathscr{F}_0} := \operatorname{diam}(\mathscr{F}_0), \tag{4.16}$$

for any subdomain face $\mathscr{F}_0$ of $\partial\Omega_0$. Note that for $\mathscr{F}_{0j} \subset \Gamma$, we have $H_{\mathscr{F}_{0j}} \eqsim H_j$, and for a Dirichlet face $\mathscr{F}_0 \subset \Gamma_D$, Assumption 4.8 ensures that $H_{\mathscr{F}_0}$ is proportional to the diameters of the coarse elements in $\mathscr{T}^H(\Omega'_{\text{int}})$ attached to $\mathscr{F}_0$.

Lemma 4.17. *Let $d = 3$, let Assumptions 4.4–4.8 hold, and let Π_0 be defined as in (4.14). Then for any $w \in W$,*

$$|P_D(\Pi_0 w)|_S^2 \;\lesssim\; \max_{\mathscr{F}_0 \subset \partial\Omega_0} (1 + \log(H_{\mathscr{F}_0}/h_0))^2 \, \frac{H_0}{H_{\mathscr{F}_0}} \, |w_0|_{S_0}^2 .$$

For the classical formulation, the maximum is taken only over faces $\mathscr{F}_0 \subset \Gamma$ (i.e., excluding the Dirichlet faces).

Proof. An application of Lemma 4.14 (with $w \mapsto \Pi_0 w$) yields

$$\begin{aligned} |P_D(\Pi_0 w)|_S^2 \;\lesssim\; & \sum_{\mathscr{G}_{0j} \subset \Gamma} \alpha_0 \, |\mathscr{H}^h_{0,\mathsf{U}_{0,\mathscr{G}_{0j}}}(\theta_{\mathscr{G}_{0j}} w_0)|^2_{H^1(\mathsf{U}_{0,\mathscr{G}_{0j}})} + \sum_{\mathscr{G}_{0j} \subset \Gamma} \alpha_0 \, |\mathscr{H}_i^h(\theta_{\mathscr{G}_{0j}} w_0)|^2_{H^1(\Omega_j)} \\ & + \sum_{\mathscr{G}_0 \subset \Gamma_D} \alpha_0 \, |\mathscr{H}^h_{0,\mathsf{U}_{0,\mathscr{G}_0}}(\theta_{\mathscr{G}_0} w_0)|^2_{H^1(\mathsf{U}_{\mathscr{G}_0})} , \end{aligned} \tag{4.17}$$

where we have used that $\min(\alpha_0, \alpha_j) \le \alpha_0$. Using the techniques from Sect. 2.5.7, the second sum can be estimated in terms of the first one. Hence, we concentrate on the terms in the first and the third sum in (4.17). To this end, let $\mathscr{G}_0 \subset \Gamma \cup \underline{\Gamma_D}$ be a fixed glob. Furthermore, let $\mathscr{F}_0$ be a subdomain face on $\partial\Omega_0$ with $\mathscr{G}_0 \subset \overline{\mathscr{F}_0}$. Without loss of generality, we may assume that $\mathscr{F}_0 \subset \partial \mathsf{U}_{0,\mathscr{G}_0}$ (otherwise, we can enlarge the neighborhood $\mathsf{U}_{0,\mathscr{G}_0}$ by a few coarse elements). From Lemmas 2.75, 2.76, and Assumption 4.8 we obtain that

$$\begin{aligned} |\mathscr{H}^h_{0,\mathsf{U}_{0,\mathscr{G}_0}}(\theta_{\mathscr{G}_0} w_0)|^2_{H^1(\mathsf{U}_{0,\mathscr{G}_0})} &\lesssim (1+\log(H_{\mathscr{F}_0}/h_0))^2 \Big[|\hat{w}|^2_{H^1(\mathsf{U}_{0,\mathscr{G}_0})} + \frac{1}{H^2_{\mathscr{F}_0}} \|\hat{w}\|^2_{L^2(\mathsf{U}_{0,\mathscr{G}_0})} \Big] \\ &\lesssim (1+\log(H_{\mathscr{F}_0}/h_0))^2 \Big[|\hat{w}|^2_{H^1(\mathsf{U}_{0,\mathscr{G}_0})} + \frac{1}{H_{\mathscr{F}_0}} \|w_0\|^2_{L^2(\mathscr{F}_0)} \Big], \end{aligned} \tag{4.18}$$

where $\hat{w}$ is the extension of w_0 from Lemma 4.15, and where in the last step we have used Lemma 2.57 (p. 116). Combining (4.17) with (4.18), we obtain with a finite overlap argument that

$$\begin{aligned} |P_D(\Pi_0 w)|^2_S \lesssim\ & \max_{\mathscr{F}_0 \subset \partial\Omega_0} (1+\log(H_{\mathscr{F}_0}/h_0))^2 \alpha_0 \, |\hat{w}|^2_{H^1(\Omega_0 \cap \Omega'_{\text{int}})} \\ & + \sum_{\mathscr{F}_0 \subset \partial\Omega_0} (1+\log(H_{\mathscr{F}_0}/h_0))^2 \frac{\alpha_0}{H_{\mathscr{F}_0}} \|w_0\|^2_{L^2(\mathscr{F}_0)} . \end{aligned}$$

In the classical formulation, the third sum in (4.17) vanishes, and so the maximum and sum need only to be taken over faces $\mathscr{F}_0 \subset \Gamma$. An application of Lemma 4.15 concludes the proof. □

The next lemma bounds the energy of $P_D(\Pi_0 w + z_{(\Pi_0 w)})$ for the case $Q = M_{\text{sD}}^{-1}$.

Lemma 4.18. *Let $d = 3$, let Assumptions 4.4–4.8 hold, and let Π_0 be defined as in (4.14). Then for $Q = M_{\text{sD}}^{-1}$,*

$$|P_D(\Pi_0 w + z_{(\Pi_0 w)})|^2_S \le C \max_{\mathscr{F}_0 \subset \partial\Omega_0} (1+\log(H_{\mathscr{F}_0}/h_0))^2 \frac{H_0}{H_{\mathscr{F}_0}} |w_0|^2_{S_0} \qquad \forall w \in W.$$

The constant C depends only on the shape regularity and quasi-uniformity constants from the stated assumptions, as well as on the shape of Ω_0^c.

Proof. The identity $P_D^\top S P_D = B^\top M_{\text{sD}}^{-1} B = B^\top Q B$ (cf. (4.7)) and Lemma 2.44 (p. 109) imply

$$\begin{aligned} |P_D(\Pi_0 w + z_{(\Pi_0 w)})|^2_S &= \|B(\Pi_0 w + z_{(\Pi_0 w)})\|^2_Q \\ &= \min_{z \in \ker(S)} \|B(\Pi_0 w + z)\|^2_Q \le \|B \Pi_0 w\|^2_Q = |P_D(\Pi_0 w)|^2_S . \end{aligned} \tag{4.19}$$

To conclude the proof, we simply apply Lemma 4.17. □

Next, we treat the case $Q = Q_{\text{diag}}$. We need two lemmas.

Lemma 4.19. *Let $d = 3$ and let Assumptions 4.4–4.8 hold, then*

$$|P_D\, z|_S^2 \;\lesssim\; \|Bz\|_{Q_{\text{diag}}}^2 \qquad \forall z \in \ker(S). \tag{4.20}$$

Proof. Let z_i denote the (constant) components of $z \in \ker(S)$. Since Ω_0 is a non-floating subdomain, $z_0 = 0$. Following the proof of Lemma 2.109 (p. 139), we obtain from formula (4.6) that

$$\begin{aligned}
&|P_D\, z|_S^2 \\
&\lesssim \sum_{\mathcal{G}_i \subset \Gamma, i \neq 0} \; \sum_{j \in \mathcal{N}_{\mathcal{G}_i}} \min(\alpha_i, \alpha_j) |\vartheta_{\mathcal{G}_i}|_{H^1(\Omega_i)}^2 \, |z_i - z_j|^2 + \sum_{\mathcal{G}_i \subset \Gamma_D, i \neq 0} \alpha_i \, |\vartheta_{\mathcal{G}_i}|_{H^1(\Omega_i)}^2 \, |z_i|^2 \\
&\quad + \sum_{\mathcal{G}_0 \subset \Gamma} \; \sum_{j \in \mathcal{N}_{\mathcal{G}_0}} \min(\alpha_0, \alpha_j) |\vartheta_{\mathcal{G}_0}|_{H^1(\mathsf{U}_{0,\mathcal{G}_0})}^2 \, |z_j|^2 .
\end{aligned}$$

Using the estimates from Lemmas 2.74 and 2.76 (p. 123ff), counting the number of nodes per subdomain glob, and comparing with the entries of Q_{diag} yields (4.20). Note that in the all-floating formulation, (4.20) holds for any non-negative values of $q_0(x^h)$ at Dirichlet nodes $x^h \in \partial\Omega_0^h \cap \Gamma_D$. □

Lemma 4.20. *Let $d = 3$, let Assumptions 4.4–4.8 hold, and let Π_0 be defined as in* (4.14)*. Then for $Q = Q_{\text{diag}}$,*

$$|P_D\, z_{(\Pi_0 w)}|_S^2 \;\le\; C \max_{\mathcal{F}_0 \subset \partial\Omega_0} (1 + \log(H_{\mathcal{F}_0}/h_0))^2 \, \frac{H_0}{H_{\mathcal{F}_0}} \, |w_0|_{S_0}^2 \qquad \forall w \in W.$$

The constant C depends only on the shape regularity and quasi-uniformity constants from the stated assumptions, as well as on the shape of Ω_0^c.

Proof. Let $w \in W$. Firstly, we use the elementary estimate

$$\tfrac{1}{2}\, |P_D(\Pi_0 w + z_{(\Pi_0 w)})|_S^2 \;\le\; |P_D(\Pi_0 w)|_S^2 + |P_D\, z_{(\Pi_0 w)}|_S^2 .$$

An upper bound for the first term has already been shown in Lemma 4.17. For the second term, we can apply Lemma 4.19. Together with Lemma 2.44 (p. 109), we obtain

$$\|P_D z_{(\Pi_0 w)}\|_S^2 \;\lesssim\; \|B\, z_{(\Pi_0 w)}\|_{Q_{\text{diag}}}^2 \;\le\; \|B\, \Pi_0 w\|_{Q_{\text{diag}}}^2 . \tag{4.21}$$

Applying the same technique as in the proof of Lemma 2.109 (p. 139) and using the defining property of Π_0, we get that

$$\|B\Pi_0 w\|^2_{Q_{\text{diag}}} \lesssim \sum_{\mathcal{G}_{0j}\subset\Gamma} \alpha_0\, q_{j|\mathcal{G}^h_{0j}}\, h_0^{-d_{\mathcal{G}_{0j}}}\, \|w_0\|^2_{L^2(\mathcal{G}_{0j})}, \tag{4.22}$$

since $\min(\alpha_0, \alpha_j) \le \alpha_0$. Note that in the classical formulation, there are no Lagrange multipliers acting on $\partial\Omega_0 \cap \Gamma_D$. In the all-floating formulation, we get no contributions from Dirichlet globs $\mathcal{G}_0 \subset \Gamma_D$ either, because $q_{0|\mathcal{G}^h_0} = 0$. Using the definition of $q_{j|\mathcal{G}^h_{0j}}(x^h)$, inequality (4.22) implies

$$\|B\Pi_0 w\|^2_{Q_{\text{diag}}}$$
$$\lesssim \sum_{\mathcal{F}_{0j}\subset\Gamma} \alpha_0\,(1+\log(H_j/h_j))\,\frac{1}{H_j}\,\|w_0\|^2_{L^2(\mathcal{F}_{0j})} + \sum_{\mathcal{E}_{0j}\subset\Gamma} \alpha_0\,\|w_0\|^2_{L^2(\mathcal{E}_{0j})}, \tag{4.23}$$

where the vertex contributions have been subsumed in the second sum. Using Lemma 2.69 (p. 120), it can be shown that

$$\|w_0\|^2_{L^2(\mathcal{E}_{0j})} \lesssim (1+\log(H_j/h_j))\Big[|\mathcal{H}^h_{0,\mathrm{U}_{0,\mathcal{E}_{0j}}} w_0|^2_{H^1(\mathrm{U}_{0,\mathcal{E}_{0j}})} + \frac{1}{H_j}\,\|w_0\|^2_{L^2(\mathcal{F}_{0k})}\Big], \tag{4.24}$$

where $\mathcal{F}_{0k}$ is a suitable face with $\mathcal{E}_{0j} \subset \partial\mathcal{F}_{0k}$ (as in the proof of Lemma 4.17 we may assume that $\mathcal{F}_{0k} \subset \partial\mathrm{U}_{0,\mathcal{E}_{0j}}$). Finally, combining (4.21)–(4.24) and applying a finite overlap argument, we get

$$|P_D\, z_{(\Pi_0 w)}|^2_S \lesssim \max_{j\in\mathcal{N}_0}(1+\log(H_j/h_j))\,\alpha_0\Big[|\hat{w}|^2_{H^1(\Omega_0\cap\Omega'_{\text{int}})} + \max_{j\in\mathcal{N}_0}\frac{1}{H_j}\sum_{\mathcal{F}_0}\|w_0\|^2_{L^2(\mathcal{F}_0)}\Big]$$
$$\lesssim \max_{j\in\mathcal{N}_0}(1+\log(H_j/h_j))\,\frac{H_0}{H_j}\,|w_0|^2_{S_0},$$

where $\hat{w}$ is as in Lemma 4.15. The proof is concluded by noting that $H_j \simeq H_{\mathcal{F}_0}$ if $\mathcal{F}_0$ touches Ω_j, and that $h_0 \eqsim h_j$ for all $j \in \mathcal{N}_0$. □

The previous lemmas can now be used to estimate the first term in (4.15). A bound for the second term (including the interior contributions) does not require new ideas and is achieved by combining the results from Lemmas 2.104 (p. 135) and 2.105 (p. 137) with Lemma 4.14. For the case $Q = M^{-1}_{\text{sD}}$, this yields

$$|P_D(\Pi_{\text{int}} w + z_{(\Pi_{\text{int}} w)})|^2_S \lesssim \max_{i=1,\dots,s}(1+\log(H_i/h_i))^2\,|\Pi_{\text{int}} w|^2_S \qquad \forall w \in W^\perp. \tag{4.25}$$

The same bound can be shown for the case $Q = Q_{\text{diag}}$, by following the proof of Lemma 2.109 (p. 139).

Theorem 4.21. *Let $\Omega \subset \mathbb{R}^3$ be unbounded, fulfilling the assumptions stated in Sect. 4.1, and subdivided as described in Sect. 4.1.2. Furthermore, let Assumptions 4.4–4.8 hold. Then for the classical and the all-floating FETI/BETI method in the unbounded setting of this chapter with either $Q = M_{\mathrm{sD}}^{-1}$ or $Q = Q_{\mathrm{diag}}$,*

$$\kappa(P\, M_{\mathrm{sD}}^{-1}\, P^\top F_{|\widetilde{U}_{\mathrm{ad}}}) \leq C \max\Big[\max_{j=1,\ldots,s} \big(1+\log\big(\tfrac{H_j}{h_j}\big)\big)^2,\ \max_{\mathscr{F}_0 \subset \partial\Omega_0} \big(1+\log\big(\tfrac{H_{\mathscr{F}_0}}{h_0}\big)\big)^2 \frac{H_0}{H_{\mathscr{F}_0}} \Big],$$

where the constant C depends only on the uniform constants from the stated assumptions and on the shape of Ω_0^c, but it is independent of H_i, h_i, α_i, the number of subdomains, and the number of neighbors of Ω_0.

Proof. The proof follows from (4.15), (4.25), Lemmas 4.17, 4.18, and 4.20, as well as from the abstract framework in Sect. 2.4.2. □

Remark 4.22. In the special case, where the faces $\mathscr{F}_0$ on $\partial\Omega_0$ have a diameter comparable to $H_0 = \mathrm{diam}(\Omega_0^c)$ (which means that the number of such faces is small), then the condition number behaves as in the bounded case. In general, of course, the factor $H_0/H_{\mathscr{F}_0}$ can be large. For example, if Ω_0^c is a cube and if the coarse triangulation $\mathscr{T}^H(\Omega'_{\mathrm{int}})$ is quasi-uniform, then the factor $H_0/H_{\mathscr{F}_0}$ is proportional to the square root of the number of faces contained in $\partial\Omega_0$.

4.2.4 An Alternative Condition Number Bound for Classical FETI/BETI

In this subsection, we will try to obtain robustness with respect to the possibly large ratio $\max_{\mathscr{F}_0 \subset \partial\Omega_0} H_0/H_{\mathscr{F}_0}$ that is present in the estimate of Theorem 4.21. Here, we restrict ourselves to the classical formulation (see Sect. 4.2.5 for the all-floating formulation). Our alternative estimate will depend on the coefficients α_i and on a geometrical parameter η that is linked to the interior Dirichlet boundary conditions. To model this parameter, we need the following rather mild assumption, on which we comment in detail below.

Assumption 4.23. If $\Gamma_D \cap \Omega_0^c \neq \emptyset$ then there exists a set $\widetilde{\Gamma}_D \subset \Omega_0^c$ of positive surface measure such that

(i) There exists a constant η with $\mathrm{dist}(\widetilde{\Gamma}_D, \partial\Omega_0) \geq \eta > 0$,
(ii) The triangulation $\mathscr{T}^H(\Omega'_{\mathrm{int}})$ resolves $\widetilde{\Gamma}_D$,
(iii) For all $i = 1,\ldots,s$ with $\partial\Omega_i \cap \Gamma_D \neq \emptyset$, either the set $\partial\Omega_i \cap \widetilde{\Gamma}_D$ contains at least a subdomain edge of Ω_i, or there is a neighboring subdomain Ω_j, $j \in \mathscr{N}_i \setminus \{0\}$ such that $\partial\Omega_j \cap \widetilde{\Gamma}_D$ contains at least a subdomain edge of Ω_j.

Note firstly, that Assumption 4.23 is not relevant if Γ_D is empty or fully contained in $\partial\Omega_0$ (cf. Fig. 4.1). Secondly, if the interior Dirichlet boundary conditions are separated from the boundary $\partial\Omega_0$ (cf. Fig. 4.2, left), i.e., $\Gamma_D \cap \Omega_0^c$ has positive

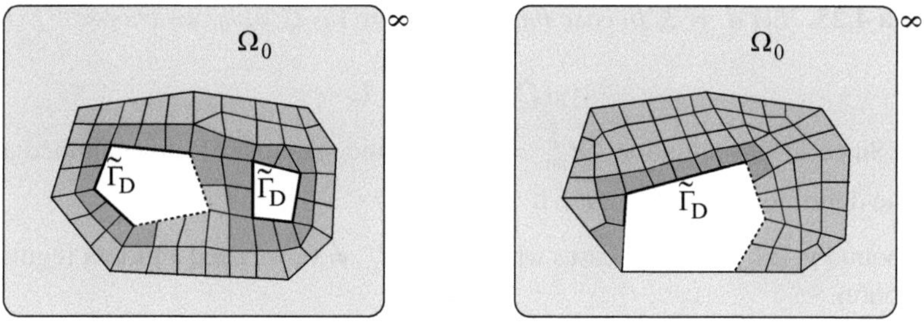

Fig. 4.6 Possible choice of $\widetilde{\Gamma}_D$ for two sample partitions. *Left*: corresponding to Fig. 4.2 (*left*), $\widetilde{\Gamma}_D = \Gamma_D$. *Right*: corresponding to Fig. 4.2 (*right*), $\widetilde{\Gamma}_D \subsetneq \Gamma_D$. *Dark-shared*: non-floating interior subdomains

distance η from $\partial\Omega_0$, then one can simply set $\widetilde{\Gamma}_D := \Gamma_D \cap \Omega_0^c$. In the crucial case where Γ_D touches $\partial\Omega_0$ (cf. Fig. 4.2, right), we choose set $\widetilde{\Gamma}_D$ as a genuine subset of $\Gamma_D \cap \Omega_0^c$ in order to ensure the positive distance to $\partial\Omega_0$, cf. Fig. 4.6 (however it is not necessary to have $\widetilde{\Gamma}_D \subset \Gamma_D$). Finding a set $\widetilde{\Gamma}_D$ fulfilling Assumption 4.23 is not too difficult: even in the case where Γ_D consists only of a single coarse face touching $\partial\Omega_0$, we can perform one step of refinement of the coarse triangulation $\mathscr{T}^H(\Omega_{\text{int}}')$ and choose $\widetilde{\Gamma}_D$ as a refined face that has positive distance from $\partial\Omega_0$, such that $\eta \gtrsim H$ if H is the coarse mesh parameter near Γ_D. This shows that Assumption 4.23 is not restrictive at all, because $\widetilde{\Gamma}_D$ can always be chosen such that $\eta \gtrsim \min_{\mathscr{F}_0 \subset \partial\Omega_0} H_{\mathscr{F}_0}$. Nevertheless, as we will see below, the larger the parameter η, the smaller the condition number bound, and in many situations this is possible (at least in cases like Fig. 4.6, left).

For the definition below, recall the triangulation $\mathscr{T}^h(\Omega_{\text{int}}')$ from Definition 4.10 as well as the corresponding FE space $V^h(\Omega_{\text{int}}')$, and recall that we denote its restriction to the interior part Ω_{int} by $V^h(\Omega_{\text{int}})$.

Definition 4.24. For $\mathscr{T}^h(\Omega_{\text{int}}')$, $V^h(\Omega_{\text{int}}')$, and $\widetilde{\Gamma}_D$ given as above, we define

(i) The discrete harmonic extension operator $\mathscr{H}_{\text{int}}^h : V^h(\partial\Omega_0) \to V^h(\Omega_{\text{int}})$,

$$\mathscr{H}_{\text{int}}^h w_0 := \operatorname{argmin}\left\{|v|_{H^1(\Omega^{\text{int}})} : v \in V^h(\Omega_{\text{int}})\right\},$$

(ii) The *restricted* extension operator $\mathscr{H}_{\text{int},\widetilde{\Gamma}_D}^h : V^h(\partial\Omega_0) \to V^h(\Omega_{\text{int}})$,

$$\mathscr{H}_{\text{int},\widetilde{\Gamma}_D}^h w_0 := \operatorname{argmin}\left\{|v|_{H^1(\Omega^{\text{int}})} : v \in V^h(\Omega_{\text{int}}),\ v_{|\widetilde{\Gamma}_D} = 0\right\},$$

(iii) The *extension indicator*

$$\gamma_h(\Omega_{\text{int}}, \widetilde{\Gamma}_D) := \sup_{w_0 \in V^h(\partial\Omega_0)} \frac{|\mathscr{H}_{\text{int},\widetilde{\Gamma}_D}^h w_0|_{H^1(\Omega_{\text{int}})}^2}{|\mathscr{H}_{\text{int}}^h w_0|_{H^1(\Omega_{\text{int}})}^2 + \frac{1}{H_0}\|w_0\|_{L^2(\partial\Omega_0)}^2}.$$

Lemma 4.25. *Let $d = 3$. In case that $\Gamma_D = \emptyset$ or $\Gamma_D \subset \partial\Omega_0$, we choose $\widetilde{\Gamma}_D = \emptyset$. Then*

$$\gamma_h(\Omega_{\text{int}}, \widetilde{\Gamma}_D) \leq 1.$$

Proof. Since in this situation, $\mathscr{H}^h_{\text{int}} = \mathscr{H}^h_{\text{int},\tilde{\Gamma}_D}$, the statement follows immediately from the definition of $\gamma_h(\Omega_{\text{int}}, \widetilde{\Gamma}_D)$. □

To bound $\gamma_h(\Omega_{\text{int}}, \widetilde{\Gamma}_D)$ in cases where $\Gamma_D \cap \Omega_0^c \neq \emptyset$, we need a further regularity assumption.

Assumption 4.26. For $\Gamma_D \cap \Omega_0^c \neq \emptyset$, let Assumption 4.23 hold with the associated set $\widetilde{\Gamma}_D$ and the parameter $\eta > 0$. Assume further that there exists a boundary layer $\Omega_{\text{int},\eta} \subset \Omega_{\text{int}}$ such that

(i) $\Omega_{\text{int},\eta}$ contains all points $x \in \Omega_{\text{int}}$ with $\text{dist}(x, \partial\Omega_0) < \eta$,
(ii) For all $x \in \Omega_{\text{int},\eta}$: $\text{dist}(x, \partial\Omega_0) \leq 2\,\eta$,
(iii) There exists a quasi-uniform triangulation $\mathscr{T}^\eta$ with mesh parameter η which resolves $\Omega_{\text{int},\eta}$.

With this assumption, the boundary layer consists of one or a few layers of "coarse" elements from $\mathscr{T}^\eta$. Note that Assumptions 4.23 and 4.26 can always be fulfilled by choosing

$$\eta \eqsim \min_{\mathscr{F}_0 \subset \partial\Omega_0} H_{\mathscr{F}_0}\,.$$

However, to get an enhanced bound compared to Theorem 4.21, we will need that

$$\eta > \max_{\mathscr{F}_0 \subset \partial\Omega_0} H_{\mathscr{F}_0}\,,$$

or in the ideal case $\eta \eqsim H_0$. Of course, it depends on the geometric setup of Ω_{int} and Γ_D whether this is possible.

Lemma 4.27. *Let $d = 3$, suppose that $\Gamma_D \cap \Omega_0^c \neq \emptyset$, and let Assumption 4.23 as well as Assumption 4.26 be fulfilled. Then,*

$$\gamma_h(\Omega_{\text{int}}, \widetilde{\Gamma}_D) \leq C\,\frac{H_0}{\eta},$$

where C depends only on the shape regularity constant of $\mathscr{T}^h(\Omega_{\text{int}})$ and on the quasi-uniformity constant of $\mathscr{T}^\eta$.

Proof. Let $\chi \in V^h(\Omega_{\text{int}})$ be a discrete cut-off function satisfying

- $\chi(x) \in [0, 1] \quad \forall x \in \Omega_{\text{int}}$,
- $\chi(x) = 1 \quad \forall x \in \partial\Omega_0$,
- $\chi(x) = 0 \quad \forall x : \text{dist}(x, \partial\Omega_0) \geq \eta$,
- $\|\nabla\chi\|_{L^\infty(\Omega_{\text{int}})} \lesssim \eta^{-1}$.

This cut-off function can, e.g., be chosen as

$$\chi(x^h) := \begin{cases} 1 - \frac{\operatorname{dist}(x^h, \partial\Omega_0)}{\eta} & \text{if } \operatorname{dist}(x^h, \partial\Omega_0) \le \eta, \\ 0 & \text{otherwise.} \end{cases}$$

for all nodes $x^h \in \Omega^h_{\text{int}}$. See also [TW05, Lemma 3.4] for a similar construction. For a fixed function $w_0 \in V^h(\partial\Omega_0)$, we set $\hat{w} := I^h(\chi\, \mathscr{H}^h_{\text{int}} w_0)$, where I^h is the nodal interpolator. The following arguments are borrowed from [TW05, Lemma 3.10]. Since $\hat{w}$ is supported in $\Omega_{\text{int},\eta}$, we can conclude from the properties of the cut-off function χ that

$$\begin{aligned} \int_{\Omega_{\text{int}}} |\hat{w}|^2 \,\mathrm{d}x &\lesssim \int_{\Omega_{\text{int},\eta}} |\nabla(\chi\, \mathscr{H}^h_{\text{int}} w_0)|^2 \,\mathrm{d}x \\ &\lesssim \int_{\Omega_{\text{int},\eta}} \underbrace{|\nabla\chi|^2}_{\lesssim \eta^{-2}} \,|\mathscr{H}^h_{\text{int}} w_0|^2 + \underbrace{|\chi|^2}_{\le 1} \,|\nabla \mathscr{H}^h_{\text{int}} w_0|^2 \,\mathrm{d}x \\ &\lesssim \eta^{-2} \,\|\mathscr{H}^h_{\text{int}} w_0\|^2_{L^2(\Omega_{\text{int},\eta})} + |\mathscr{H}^h_{\text{int}} w_0|^2_{H^1(\Omega_{\text{int},\eta})} \,. \end{aligned}$$

We now choose non-overlapping $\mathscr{T}^\eta$-agglomerates ω_j, $j \in J$ each of which consists of a few elements from $\mathscr{T}^\eta$ such that

$$\overline{\Omega}_{\text{int},\eta} = \bigcup_{j\in J} \overline{\omega}_j \,,$$

and such that $\partial\omega_j \cap \Omega_0$ contains a face of $\mathscr{T}^\eta$ for all $j \in J$. Assumption 4.26 makes such a choice possible. Continuing the above estimate, and using Lemma 2.57, we obtain that

$$\begin{aligned} |\hat{w}|^2_{H^1(\Omega_{\text{int}})} &\lesssim \sum_{j\in J} \Big[\eta^{-2} \,\|\mathscr{H}^h_{\text{int}} w_0\|^2_{L^2(\omega_j)} + |\mathscr{H}^h_{\text{int}} w_0|^2_{H^1(\omega_j)} \Big] \\ &\lesssim \sum_{j\in J} \Big[\eta^{-1} \,\|\mathscr{H}^h_{\text{int}} w_0\|^2_{L^2(\partial\omega_j \cap \partial\Omega_0)} + |\mathscr{H}^h_{\text{int}} w_0|^2_{H^1(\omega_j)} \Big] \\ &\lesssim \eta^{-1} \,\|w_0\|^2_{L^2(\partial\Omega_0)} + |\mathscr{H}^h_{\text{int}} w_0|^2_{H^1(\Omega_{\text{int}})} \,. \end{aligned}$$

Since $\hat{w}$ is an extension of w_0 with $\hat{w}_{|\tilde{\Gamma}_D} = 0$ and since $\eta \le H_0$, it follows that

$$|\mathscr{H}^h_{\text{int},\tilde{\Gamma}_D} w_0|^2_{H^1(\Omega_{\text{int}})} \le |\hat{w}|^2_{H^1(\Omega_{\text{int}})} \lesssim \frac{H_0}{\eta} \Big[|\mathscr{H}^h_{\text{int}} w_0|^2_{H^1(\Omega_{\text{int}})} + \frac{1}{H_0} \,\|w_0\|^2_{L^2(\partial\Omega_0)} \Big].$$

Comparing this estimate with the definition of $\gamma_h(\Omega_{\text{int}}, \widetilde{\Gamma}_D)$ concludes the proof.

□

Lemma 4.28. *Let $d = 3$ and let Assumptions 4.4–4.8 and 4.23 be fulfilled. Then for the classical FETI/BETI method with $Q = M_{\mathrm{sD}}^{-1}$, and for all $w \in W$,*

$$|P_D(\Pi_0 w + z_{(\Pi_0 w)})|_S^2 \le C\, \gamma_h(\Omega_{\mathrm{int}}, \widetilde{\Gamma}_D) \max_{j=0,\dots,s} \frac{\alpha_j}{\alpha_0} \max_{i=1,\dots,s} (1 + \log(H_i/h_i))^2\, |w_0|_{S_0}^2,$$

where the constant C depends only on the uniform constants from the stated assumptions and on the shape of Ω_0^c.

Proof. Let $w \in W$ be arbitrary but fixed. From (4.19), we see that

$$|P_D(\Pi_0 w + z_{(\Pi_0 w)})|_S^2 \le \min_{z \in \ker(S)} |P_D(\Pi_0 w + z)|_S^2 .$$

Firstly, we choose the extension

$$\tilde{w} := \mathscr{H}^h_{\mathrm{int},\tilde{\Gamma}_D} w_0$$

of w_0. Secondly, we now construct the special element $\tilde{z} \in \ker(S)$ given by

$$\tilde{z}_i := \begin{cases} \overline{\tilde{w}}^{\Omega_i} & \text{for } i \in \mathscr{I}_{\mathrm{float}}, \\ 0 & \text{otherwise.} \end{cases}$$

From Lemma 4.14, we can now conclude that

$$|P_D(\Pi_0 w + \tilde{z})|_S^2$$

$$\lesssim \sum_{\mathscr{G}_{0j} \subset \Gamma} \min(\alpha_0, \alpha_j)\, |\mathscr{H}^h_{0,\mathsf{U}_{0,\mathscr{G}_{0j}}} (\theta_{\mathscr{G}_{0j}} (w_0 - \tilde{z}_j))|^2_{H^1(\mathsf{U}_{0,\mathscr{G}_{0j}})} \tag{4.26}$$

$$+ \sum_{\mathscr{G}_0 \subset \Gamma_D} \alpha_0\, |\mathscr{H}^h_{0,\mathsf{U}_{0,\mathscr{G}_0}} (\theta_{\mathscr{G}_0} w_0)|^2_{H^1(\mathsf{U}_{0,\mathscr{G}_0})} \tag{4.27}$$

$$+ \sum_{\mathscr{G}_{ij} \subset \Gamma \setminus \partial\Omega_0} \min(\alpha_i, \alpha_j)\, |I^h(\vartheta_{\mathscr{G}_{ij}} (\tilde{z}_i - \tilde{z}_j))|^2_{H^1(\Omega_i)} \tag{4.28}$$

$$+ \sum_{\mathscr{G}_i \cap \Gamma_D,\, i \neq 0} \alpha_i\, |I^h(\vartheta_{\mathscr{G}_i} \tilde{z}_i)|^2_{H^1(\Omega_i)} \tag{4.29}$$

Since we are considering the classical formulation only, the terms (4.27) and (4.29) vanish entirely. We begin with the term (4.26). Let $E^h_{j,\mathscr{G}_{0j}}$ be a suitable transfer operator from Ω_j to $\mathsf{U}_{0,\mathscr{G}_{0j}}$ as constructed in Sect. 2.5.7 such that the function $E^h_{j,\mathscr{G}_{0j}} I^h(\vartheta_{\mathscr{G}_{0j}} (\tilde{w} - \tilde{z}_j))$ is an extension of $I^h(\theta_{\mathscr{G}_{0j}} (w_0 - \tilde{z}_j))$ from $\mathscr{G}_{0j}$ to $\mathsf{U}_{0,\mathscr{G}_{0j}}$ which has larger energy than $\mathscr{H}^h_{0,\mathsf{U}_{0,\mathscr{G}_{0j}}} (\theta_{\mathscr{G}_{0j}} (w_0 - \tilde{z}_j))$. An application of Corollary 2.78 yields

$$|\mathscr{H}^h_{0,\mathsf{U}_{0,\mathscr{G}_{0j}}}(\theta_{\mathscr{G}_{0j}}(w_0 - \tilde{z}_j))|^2_{H^1(\mathsf{U}_{0,\mathscr{G}_{0j}})}$$
$$\lesssim (1+\log(H_j/h_j))^2\, |\tilde{w}|^2_{H^1(\Omega_j)} + (1+\log(H_j/h_j))\, \frac{1}{H_j^2}\, \|\tilde{w} - \tilde{z}_j\|^2_{L^2(\Omega_j)}\,.$$

If Ω_j is a floating subdomain, the definition of $\tilde{z}_j$ and Poincaré's inequality imply that

$$\frac{1}{H_j^2}\, \|\tilde{w} - \tilde{z}_j\|^2_{L^2(\Omega_j)} \lesssim |\tilde{w}|^2_{H^1(\Omega_j)}\,. \tag{4.30}$$

If Ω_j is non-floating, then $\tilde{z}_j = 0$. However, Assumption 4.7 ensures that $\tilde{w}$ vanishes on a coarse edge of Ω_j or at least on a coarse edge of a neighboring subdomain. Let $\mathscr{U}_j \supset \Omega_j$ be a connected neighborhood of Ω_j that contains this coarse edge and that is the union of a few coarse elements from $\mathscr{T}^H(\Omega_{\text{int}})$. Then the discrete Poincaré inequality (Lemma 2.70) implies

$$\frac{1}{H_j^2}\, \|\tilde{w} - \underbrace{\tilde{z}_j}_{=0}\|^2_{L^2(\Omega_j)} \lesssim (1+\log(H_j/h_j))\, |\tilde{w}|^2_{H^1(\mathscr{U}_j)}\,. \tag{4.31}$$

Summarizing, since the neighborhoods $\mathscr{U}_j$ have finite overlap, we obtain for the term (4.26) that

$$\sum_{\mathscr{G}_{0j}\subset\Gamma} \min(\alpha_0, \alpha_j)\, |\mathscr{H}^h_{0,\mathsf{U}_{0,\mathscr{G}_{0j}}}(\theta_{\mathscr{G}_{0j}}(w_0 - \tilde{z}_j))|^2_{H^1(\mathsf{U}_{0,\mathscr{G}_{0j}})}$$
$$\lesssim \alpha_0 \max_{j\in\mathscr{N}_0}(1+\log(H_j/h_j))^2\, |\tilde{w}|^2_{H^1(\Omega_{\text{int}})}\,. \tag{4.32}$$

Next, we investigate the term (4.28). Let $\mathscr{G}_{ij} \subset \Gamma$ be fixed. If Ω_i and Ω_j are both non-floating, then $\tilde{z}_0 = \tilde{z}_j = 0$ and there is nothing to estimate. Hence, assume without loss of generality that Ω_i is floating. From Corollary 2.78, the fact that $H_i \eqsim H_j$, $h_i \eqsim h_j$, and from Cauchy's inequality we get

$$|I^h(\vartheta_i\,(\tilde{z}_i - \tilde{z}_j))|^2_{H^1(\Omega_i)} \lesssim (1+\log(H_j/h_j))\, \frac{|\Omega_i|}{H_i^2}\, |\tilde{z}_i - \tilde{z}_j|^2$$
$$\lesssim (1+\log(H_j/h_j))\, \frac{1}{H_i^2}\, \|\tilde{w} - \tilde{z}_j\|^2_{L^2(\Omega_i\cup\Omega_j)}\,.$$

If Ω_j is floating as well, the L^2-term can be bounded by corresponding H^1-norm using Poincaré's inequality. If Ω_j is non-floating, we can conclude by the same arguments as in (4.30) and (4.31) that

$$|I^h(\vartheta_i\,(\tilde{z}_i - \tilde{z}_j))|^2_{H^1(\Omega_i)} \lesssim (1+\log(H_j/h_j))^2\, |\tilde{w}|^2_{L^2(\mathscr{U}_{ij})}\,,$$

where $\mathscr{U}_{ij}$ is a suitable neighborhood of $\Omega_i \cup \Omega_j$ which contains at least a coarse edge where $\tilde{w}$ vanishes. By summing these terms and using that the neighborhoods have finite overlap, we obtain the following bound for the term (4.28),

$$\sum_{\mathscr{G}_{ij} \subset \Gamma \setminus \partial\Omega_0} \min(\alpha_i, \alpha_j) \, |I^h(\vartheta_i \, (\tilde{z}_i - \tilde{z}_j))|^2_{H^1(\Omega_i)} \lesssim \max_{i=1,\ldots,s} \alpha_i \, (1 + \log(H_i/h_i))^2 \, |\tilde{w}|^2_{H^1(\Omega_{\mathrm{int}})}. \tag{4.33}$$

Finally, by combining (4.26)–(4.29) with the bounds (4.32), (4.33) we get that

$$|P_D(\Pi_0 w + z_{(\Pi_0 w)})|^2_S \lesssim \max_{j=0,\ldots,s} \frac{\alpha_j}{\alpha_0} \max_{i=1,\ldots,s} (1 + \log(H_i/h_i))^2 \, \alpha_0 \, |\tilde{w}|^2_{H^1(\Omega_{\mathrm{int}})}.$$

Since $\tilde{w} = \mathscr{H}^h_{\mathrm{int},\tilde{\Gamma}_D}$, we see from Definition 4.24 and Lemma 4.15 that

$$\begin{aligned}
\alpha_0 \, |\tilde{w}|^2_{H^1(\Omega_{\mathrm{int}})} &\leq \gamma_h(\Omega_{\mathrm{int}}, \widetilde{\Gamma}_D) \, \alpha_0 \left[|\mathscr{H}^h_{\mathrm{int}} w_0|^2_{H^1(\Omega_{\mathrm{int}})} + \frac{1}{H_0} \|w_0\|^2_{L^2(\partial\Omega_0)} \right] \\
&\lesssim \gamma_h(\Omega_{\mathrm{int}}, \widetilde{\Gamma}_D) \, |w_0|^2_{S_0},
\end{aligned}$$

which concludes the proof. □

Lemma 4.29. *Let $d = 3$ and let Assumptions 4.4–4.8 and 4.23 be fulfilled. Then for the classical FETI/BETI method with $Q = Q_{\mathrm{diag}}$, and for all $w \in W$,*

$$|P_D(\Pi_0 w + z_{(\Pi_0 w)})|^2_S \leq C \, \gamma_h(\Omega_{\mathrm{int}}, \widetilde{\Gamma}_D) \max_{j=0,\ldots,s} \frac{\alpha_j}{\alpha_0} \max_{i=1,\ldots,s} (1 + \log(H_i/h_i))^2 \, |w_0|^2_{S_0},$$

where the constant C depends only on the uniform constants from the stated assumptions and on the shape of Ω_0^c.

Proof. Let the functions $\tilde{w} \in V^h(\Omega_{\mathrm{int}})$ and $\tilde{z} \in \ker(S)$ be defined as in the proof of Lemma 4.28. Then

$$\tfrac{1}{2} |P_D(\Pi_0 \, w + z_{(\Pi_0 w)})|^2_S \leq |P_D(\Pi_0 \, w + \tilde{z})|^2_S + |P_D(\tilde{z} - z_{(\Pi_0 w)})|^2_S. \tag{4.34}$$

The first term can be bounded as in the proof of Lemma 4.28. For the second term, note that $\tilde{z} - z_{(\Pi_0 w)} \in \ker(S)$. Hence, Lemma 4.19 implies that

$$\begin{aligned}
|P_D(\tilde{z} - z_{(\Pi_0 w)})|^2_S &\lesssim \|B \, (\tilde{z} - z_{(\Pi_0 w)})\|^2_{Q_{\mathrm{diag}}} \\
&\leq 2 \, \|B(\tilde{z} - \Pi_0 \, w)\|^2_{Q_{\mathrm{diag}}} + 2 \, \|B(\Pi_0 \, w - z_{(\Pi_0 w)})\|^2_{Q_{\mathrm{diag}}} \\
&= 2 \, \|B(\tilde{z} - \Pi_0 \, w)\|^2_{Q_{\mathrm{diag}}} + 2 \min_{z \in \ker(S)} \|B(\Pi_0 \, w - z)\|^2_{Q_{\mathrm{diag}}} \\
&\leq 4 \, \|B(\tilde{z} - \Pi_0 \, w)\|^2_{Q_{\mathrm{diag}}},
\end{aligned} \tag{4.35}$$

where in an intermediate step we have used Lemma 2.44 (p. 109). We can now easily adapt an argument from the proof of Lemma 2.109 and split the contributions from $\| B(\tilde{z} - \Pi_0 w)\|^2_{Q_{\text{diag}}}$ into globs. Following estimate (2.94), p. 140, and using that we are using the classical formulation, we obtain

$$\| B\,(\tilde{z} - \Pi_0\, w)\|^2_{Q_{\text{diag}}} \tag{4.36}$$

$$\lesssim \sum_{\mathscr{G}_{ij} \subset \Gamma, i<j} \min(\alpha_i, \alpha_j)\, q_{ij|\mathscr{G}_{ij}{}^h}\, h_j^{-d_{\mathscr{G}_{ij}}} \underbrace{\|(\tilde{z} - \Pi_0\, w)_i - (\tilde{z} - \Pi_0\, w)_j\|^2_{L^2(\mathscr{G}_{ij})}}_{=:\Upsilon_{ij}}.$$

We treat two cases.

- If $i = 0$, i.e., we consider the contribution from a glob $\mathscr{G}_{0j} \subset \Gamma \cap \partial\Omega_0$, we can use that $\tilde{w} - \tilde{z}_j$ is an extension of $w_0 - \tilde{z}_j$. Setting $\mathscr{U} = \Omega_j$ if Ω_j is floating, and choosing $\mathscr{U} \supset \Omega_j$ as a neighborhood where $\tilde{w}$ vanishes at least at a subdomain edge (cf. Assumption 4.23), we can conclude from Lemma 2.70 (p. 121) that

$$\Upsilon_{ij} = \|w_0 - \tilde{z}_j\|^2_{L^2(\mathscr{G}_{ij})} \lesssim \sigma^{d-\min(d_{\mathscr{G}_{ij}},1)}\left(\tfrac{H_j}{h_j}\right) H_j^{2+d_{\mathscr{G}_{ij}}-d}\, |\tilde{w}|^2_{H^1(\mathscr{U})}.$$

  Using the definitions of q_{ij} (Sect. 4.1.3.3) and σ^j (Definition 2.68, p. 120) reveals that the contribution of $\mathscr{G}_{0j}$ in (4.36) is bounded by

$$\min(\alpha_0, \alpha_j)\,(1 + \log(H_j/h_j))^2\, |\tilde{w}|^2_{H^1(\Omega_j)}.$$

- If $i \neq 0$, i.e., we consider the contribution from a glob $\mathscr{G}_{ij} \subset \Gamma \setminus \partial\Omega_0$, we set $\mathscr{U} = \Omega_i \cup \Omega_j$ if both Ω_i and Ω_j are floating, and otherwise choose $\mathscr{U} \subset \Omega_i \cup \Omega_j$ as neighborhood containing a coarse edge where $\tilde{w}$ vanishes (see also Lemma 4.28). By Cauchy's inequality and using again Lemma 2.70, we get that

$$\Upsilon_{ij} = \|\tilde{z}_i - \tilde{z}_j\|^2_{L^2(\mathscr{G}_{ij})} \leq H_j^{d_{\mathscr{G}_{ij}}-d+2}\,(1 + \log(\tfrac{H_j}{h_j}))\, |\tilde{w}|^2_{H^1(\mathscr{U})}.$$

  Comparing again with the definition of q_{ij}, this implies that the contribution of $\mathscr{G}_{0j}$ in (4.36) is bounded by $\min(\alpha_i, \alpha_j)\,(1 + \log(H_j/h_j))^2\, |\tilde{w}|^2_{L^2(\mathscr{U})}$.

Summarizing, from (4.34)–(4.36), the auxiliary estimates for Υ_{ij}, and a finite overlap argument we conclude that

$$|P_D(\tilde{z} - z_{(\Pi_0 w)})|^2_S \lesssim \max_{j=1,\ldots,s} \alpha_j\,(1 + \log(H_j/h_j))^2\, |\tilde{w}|^2_{H^1(\Omega_{\text{int}})}.$$

The rest of the proof follows the line of Lemma 4.28. □

With these results, we obtain an alternative bound for the classical formulation.

Theorem 4.30. *Let $d = 3$ and let Assumptions 4.4–4.8, 4.23, and 4.26 be fulfilled. Then for the classical FETI/BETI method with $Q = M_{\text{sD}}^{-1}$ or $Q = Q_{\text{diag}}$,*

$$\kappa(P\, M_{\text{sD}}^{-1} P^\top F_{|\tilde{U}_{\text{ad}}}) \le \begin{cases} C \max_{j=0,\dots,s} \dfrac{\alpha_j}{\alpha_0} \max_{i=1,\dots,s} (1+\log(H_i/h_i))^2 & \text{if } \Gamma_D \cap \Omega_0^c = \emptyset, \\ C \max_{j=0,\dots,s} \dfrac{\alpha_j}{\alpha_0} \dfrac{H_0}{\eta} \max_{i=1,\dots,s} (1+\log(H_i/h_i))^2 & \text{otherwise.} \end{cases}$$

The constant C depends only on the uniform constants from the stated assumptions and on the shape of Ω_0^c, but is independent of α_i, H_i, h_i, η, the number of subdomains, and the number of neighbors of Ω_0.

Proof. The proof follows from (4.15), (4.25), Lemmas 4.17, 4.25, 4.27–4.29, as well as from the abstract framework in Sect. 2.4.2. □

Remark 4.31. In the special case where $\Gamma_D \cap \Omega_0^c = \emptyset$ or $\eta \gtrsim H_0$, and if additionally $\alpha_0 \ge \alpha_i$ for all $i = 1, \dots, s$, the condition number bound for the classical FETI/BETI method simplifies to

$$\kappa(P\, M_{\text{sD}}^{-1} P^\top F_{|\tilde{U}_{\text{ad}}}) \le C \max_{i=1,\dots,s} (1+\log(H_i/h_i))^2 ,$$

where the constant C is in particular independent of H_0. Hence, in this case, the condition number bound in Theorem 4.30 is smaller than that from Theorem 4.21.

4.2.5 Further Remarks on All-Floating FETI/BETI

In this section, we explain how much of the previous section can be carried over to the all-floating formulation. Thus, let the notations of the previous section hold and let Assumptions 4.23 and 4.26 be fulfilled. First, we treat the case $Q = M_{\text{sD}}^{-1}$. In this situation, we can follow the proof of Lemma 4.28, but now we have to bound the terms from (4.27) and (4.29). Using similar ideas as in that proof, one shows for the terms in (4.29) that for a glob $\mathcal{G}_i \subset \Gamma_D$, $i \ne 0$,

$$|I^h(\vartheta_{\mathcal{G}_i})\tilde{z}_i|^2_{H^1(\Omega_i)} \lesssim (1+\log(H_i/h_i)) \frac{1}{H_i^2} \|\tilde{w}\|^2_{L^2(\Omega_i)} \lesssim (1+\log(H_i/h_i))^2 |\tilde{w}|^2_{H^1(\mathcal{U}_i)}, \tag{4.37}$$

where $\mathcal{U}_i$ is a suitable neighborhood of Ω_i. This implies the bound

$$\sum_{\mathcal{G}_i \subset \Gamma_D, i \ne 0} \alpha_i, |I^h(\vartheta_{\mathcal{G}_i})\tilde{z}_i|^2_{H^1(\Omega_i)} \lesssim \max_{i=1,\dots,s} \alpha_i\, (1+\log(H_i/h_i))^2 |\tilde{w}|^2_{H^1(\Omega_{\text{int}})}$$

for (4.29), which can be subsumed in the existing bounds for (4.26) and (4.28). However, for the contributions in (4.27), new ideas are required. In the following, we give a sketch on how a suitable estimate can be achieved.

For a Dirichlet glob $\mathcal{G}_0 \subset \Gamma_D$, let $\mathcal{U}_{\mathcal{G}_0} \subset \Omega_{\text{int}}$ a suitable neighborhood consisting of the coarse elements attached to $\mathcal{G}_0$. Then

$$|\mathcal{H}^h_{0,\mathrm{U}_{0,\mathcal{G}_0}}(\theta_{\mathcal{G}_0} w_0)|^2_{H^1(\mathrm{U}_{0,\mathcal{G}_0})} \lesssim (1 + \log(\operatorname{diam}(\mathcal{U}_{\mathcal{G}_0})/h_0)) \frac{1}{\operatorname{diam}(\mathcal{U}_{\mathcal{G}_0})^2} \|\tilde{w}\|^2_{L^2(\mathcal{U}_{\mathcal{G}_0})},$$

where above $\tilde{w} \in V^h(\Omega_0^c)$ denotes the minimal extension of w_0 from $\partial\Omega_0$ to the whole of Ω_0^c that vanishes on $\widetilde{\Gamma}_D$.

However, note that it is not guaranteed that $\tilde{w}$ vanishes on a coarse edge of $\mathcal{U}_{\mathcal{G}_0}$. It is not even ensured that $\tilde{w}$ vanishes *anywhere* in Ω_0^c because $\widetilde{\Gamma}_D$ could be empty. To tackle this problem we need two steps.

(i) We choose $\widetilde{\Gamma}_D$ in Assumption 4.23 not necessarily as a subset of Γ_D but larger, in order to guarantee that $\tilde{w}$ vanishes at least somewhere in the center of Ω_0^c, and we adapt Assumption 4.26 accordingly.
(ii) We assume that for each Dirichlet glob $\mathcal{G}_0 \subset \Gamma_D$ there exists a path of coarse elements from $\mathcal{T}^H(\Omega_0^c)$ joining $\mathcal{G}_0$ and a coarse edge of $\widetilde{\Gamma}_D$.

This way, one can show the bound

$$\sum_{\mathcal{G}_0 \subset \Gamma_D} \alpha_0 \, |\mathcal{H}^h_{0,\mathrm{U}_{0,\mathcal{G}_0}}(\theta_{\mathcal{G}_0} w_0)|^2_{H^1(\mathrm{U}_{0,\mathcal{G}_0})} \lesssim C^\star \alpha_0 \, |\tilde{w}|^2_{H^1(\Omega_0^c)}.$$

By slightly adapting Definition 4.24 and formally replacing Ω_{int} by Ω_0^c there, the term $\alpha_0 \, |\tilde{w}|^2_{H^1(\Omega_0^c)}$ can be bounded by $\frac{H_0}{\eta} \, |w_0|^2_{S_0}$. In general, however, the constant $C^\star$ may depend on the number of Dirichlet globs, the nature of the coarse triangulation $\mathcal{T}^H(\Omega_0^c)$, and the shape of Ω_0^c. Using the techniques from Sect. 3.4.4 (p. 189ff), this dependence can be made explicit, and in various situations, benign bounds for $C^\star$ will be found.

For the case $Q = Q_{\text{diag}}$, the same ideas will lead to identical bounds.

Remark 4.32. Let us note that the difficulties described above are only present in the all-floating formulation and only in cases where $\Gamma_D \cap \partial\Omega_0 \neq \emptyset$. The difficulties can be circumvented by using a mixture of classical and all-floating formulation, i.e., by enforcing the Dirichlet conditions on $\Gamma_D \cap \partial\Omega_0$ directly, and the remaining ones by Lagrange multipliers, see also [Pec09].

4.2.6 *The Case of Two Dimensions*

In this section, we briefly sketch what is expected for the case of two dimensions. First, recall that for $d = 2$, the Neumann problem on the unbounded part Ω_0 is not uniquely solvable. Hence, Ω_0 is floating if $\Gamma_D \cap \partial\Omega_0 \neq \emptyset$ and non-floating otherwise. This leads to serious complications.

- If Ω_0 is floating and if Ω_0 has a large number of neighboring interior subdomains, then the coarse matrix $(\mathbf{G}^\top \mathbf{Q}\,\mathbf{G})$ is not sparse anymore.
- If Ω_0 is non-floating, but if $\Gamma_D \cap \partial\Omega_0$ is small compared to $\partial\Omega_0$, we might get a large Friedrichs constant that depends on the ratio $|\partial\Omega_0|/|\Gamma_D \cap \partial\Omega_0|$ and that enters the total condition number bound. However, we expect from the results in Sect. 3.4.4, that if Ω_0^c is a square, and $\Gamma_D \cap \partial\Omega_0$ a straight edge of length η, then the (squared) Friedrichs constant will behave as $\mathcal{O}(1 + \log(H_0/\eta))$.
- In the special case $\Gamma_D = \emptyset$, the global problem needs to be regularized. Recall from Sect. 1.3.8.4 that the matrix $\mathbf{S}_0$ corresponding to S_0 is a rank-one correction of the matrix

$$\widetilde{\mathbf{S}}_0 := \alpha_0\big(\mathbf{D}_0 + (\tfrac{1}{2}\mathbf{M}_0^\top - \mathbf{K}_0^\top)\mathbf{V}_0^{-1}(\tfrac{1}{2}\mathbf{M}_0 - \mathbf{K}_0)\big), \tag{4.38}$$

which is always regular (since we assumed that $H_0 \leq 1$). Hence, to regularize the global problem, it is sufficient to use the matrix $\widetilde{\mathbf{S}}_0$ instead of $\mathbf{S}_0$. From the proof of Lemma 1.93 (p. 58), by formally setting the rank-one correction $S^{\text{corr}} = S^{\text{corr}}_{\text{BEM}} = 0$, we see that when fixing H_0 at, e.g., $H_0 = 1/2$, we get the spectral equivalence

$$\langle \widetilde{S}_0\, v, v\rangle \eqsim \alpha_0\big[\langle S^{\text{int}}\, v, v\rangle + \|v\|^2_{L^2(\partial\Omega_0)}\big] \qquad \forall v \in V^h(\partial\Omega_0),$$

where S^{int} is the Steklov-Poincaré operator corresponding to Laplace's equation on Ω_0^c. Therefore, the properties of the whole problem become essentially those of the three-dimensional one treated in the previous sections.

4.3 Numerical Results

In the following, we give some computational results on one-level BETI methods for unbounded domains in order to confirm the preceding analysis. Although Sect. 4.2 was restricted to the case of three dimensions, the actual difference to two dimensions are the properties of the operator S_0. To simplify the computations, we use a two-dimensional domain, but replace the matrix $\mathbf{S}_0$ by $\widetilde{\mathbf{S}}_0$ from (4.38), thus preserving the properties of the corresponding three-dimensional operator.

We consider three different geometric configurations, which are depicted in Fig. 4.7. In all cases, the unbounded part Ω_0 is the complement of the unit square. In the first case, Ω_{int} is simply the unit square and so $\Omega = \mathbb{R}^2$. In the second case, Ω_{int} is an annulus-shaped domain with Dirichlet conditions on the interior part of its boundary. Finally, in the third case, Ω_{int} is a C-shaped domain, and the intersection $\Gamma_D \cap \partial\Omega_0$ is non-empty. In all cases, we use conceptually the same coefficient distribution, as indicated in Fig. 4.7. Furthermore, we partition the interior part Ω_{int}

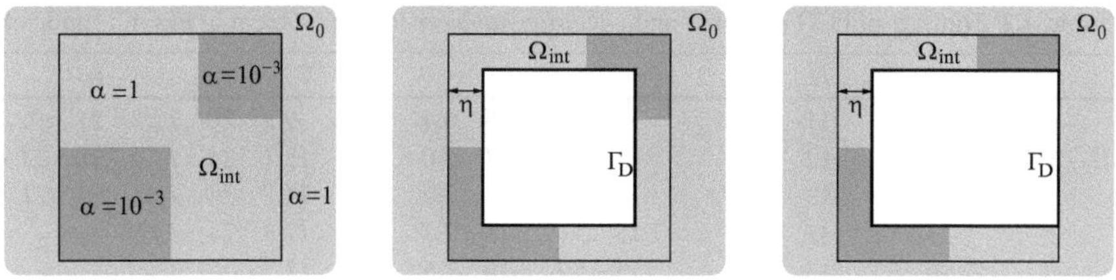

Fig. 4.7 Sketch of the problem geometries in this section. *Shaded* areas indicate coefficient values

Table 4.1 Number of PCG iterations and *condition numbers* for the problem in Fig. 4.7 (left)

H_0/H_{int}	$H_{\text{int}}/h = 2$		4		8		16		32		64	
6	9	*1.6*	10	*2.0*	12	*2.6*	14	*3.5*	16	*4.4*	18	*5.6*
12	9	*1.8*	11	*2.5*	14	*3.3*	16	*4.3*	18	*5.5*	20	*6.8*
24	10	*2.1*	12	*2.9*	14	*3.9*	17	*5.2*	18	*6.5*	20	*7.9*
48	10	*2.4*	12	*3.4*	15	*4.6*	18	*6.0*	20	*7.5*	–	–
96	11	*2.6*	13	*3.8*	16	*5.1*	18	*6.6*	–	–	–	–

into equally-sized square subdomains such that all coefficient jumps are resolved. In all interior subdomains, we use the same uniform discretization. To simplify the presentation, let H_0 denote the sidelength of Ω_0^c, let H_{int} denote the sidelength of the interior subdomains, and let $h_i = h_0 = h$ denote the mesh size of the FE/BE discretization.

Tables 4.1–4.3 display the number of PCG iterations and the condition numbers (estimated by the Lanczos method) for the classical BETI/FETI method applied to the three test cases. In our implementation, which builds upon OSTBEM [Ste00], we have used $Q = Q_{\text{diag}}$, and the PCG iteration was stopped when the relative error went below $\varepsilon = 10^{-8}$. As in our previous numerical experiments, the PARDISO solver [PAR05] was used for the coarse problem. The dense BEM-problems were solved using LAPACK (mainly interested in verifying our theoretical results we have not used any data-sparse approximation).

(i) In the first test case, corresponding to Fig. 4.7 (left) and Table 4.1, the expected behavior of the condition number is

$$\kappa(P\, M_{\text{sD}}^{-1} P^\top F_{|\tilde{U}_{\text{ad}}}) = \mathcal{O}((1 + \log(H_{\text{int}}/h))^2),$$

independently of H_0/H_{int}, see Theorem 4.30. We see that for fixed ratio H_0/H_{int} the logarithmic behavior is clearly seen. When doubling the ratio H_0/H_{int}, the condition numbers increases slightly as well; however, due to preasymptotic effects. Note that the case $H_0/H_{\text{int}} = 96$ corresponds to $96^2 = 9{,}216$ interior subdomains.

(ii) In the second and third test case, corresponding to Fig. 4.7 (middle/right) and Tables 4.2 and 4.3, the expected behavior of the condition number is

Table 4.2 Number of PCG iterations and *condition numbers* for the problem in Fig. 4.7 (middle)

	H_0/H_{int}	$H_{\mathrm{int}}/h=2$		4		8		16		32		64	
	6	11	*2.3*	13	*4.1*	15	*6.4*	18	*9.3*	20	*12.7*	21	*16.7*
$H_0/\eta=6$	12	12	*3.3*	15	*5.7*	17	*9.0*	21	*13.1*	24	*18.0*	28	*23.9*
	24	13	*3.1*	15	*5.0*	17	*7.9*	21	*11.7*	24	*16.3*	27	*21.7*
	48	12	*3.3*	15	*5.0*	18	*7.7*	21	*11.4*	24	*15.8*	–	–
	96	13	*3.5*	15	*5.1*	18	*7.6*	21	*11.2*	–	–	–	–
	H_0/H_{int}	$H_{\mathrm{int}}/h=2$		4		8		16		32			
	24	13	*4.6*	16	*7.2*	20	*11.8*	23	*15.1*	25	*18.4*		
$H_0/\eta=24$	48	15	*7.0*	20	*13.1*	22	*18.1*	25	*23.4*	27	*28.8*		
	96	17	*10.6*	20	*16.0*	24	*21.7*	28	*27.7*	–	–		
	192	17	*12.1*	21	*17.8*	24	*23.8*	–	–	–	–		
	H_0/H_{int}	$H_{\mathrm{int}}/h=2$		4		8		16					
	96	18	*19.1*	23	*31.5*	28	*43.8*	33	*56.1*				
$H_0/\eta=96$	192	23	*31.6*	30	*49.6*	35	*68.7*	–	–				
	384	27	*40.7*	33	*61.9*	–	–	–	—				

Table 4.3 Number of PCG iterations and *condition numbers* for the problem in Fig. 4.7 (right)

H_0/H_{int}		$H_{\mathrm{int}}/h=2$		4		8		16		32		64	
6		11	*2.3*	13	*4.1*	16	*6.4*	18	*9.3*	21	*12.7*	23	*16.7*
$H_0/\eta=6$	12	12	*3.3*	15	*5.7*	18	*9.0*	21	*13.1*	24	*18.0*	26	*23.9*
24		13	*3.1*	15	*5.1*	17	*8.0*	21	*11.7*	25	*16.3*	27	*21.7*
48		13	*3.3*	15	*5.1*	18	*7.8*	21	*11.4*	24	*15.8*	–	–
96		13	*3.5*	16	*5.2*	19	*7.6*	22	*11.2*	–	–	–	–
H_0/H_{int}		$H_{\mathrm{int}}/h=2$		4		8		16		32			
24		13	*5.0*	17	*7.8*	19	*10.7*	22	*13.7*	25	*16.8*		
$H_0/\eta=24$	48	16	*7.6*	19	*11.8*	22	*16.4*	24	*21.2*	28	*26.1*		
96		16	*9.6*	21	*14.5*	24	*19.7*	28	*25.1*	–	–		
192		18	*10.9*	21	*16.2*	25	*21.6*	–	–	–	–		
H_0/H_{int}		$H_{\mathrm{int}}/h=2$		4		8		16					
96		20	*17.1*	25	*28.1*	28	*39.1*	32	*50.0*				
$H_0/\eta=96$	192	24	*28.1*	30	*44.2*	35	*61.2*	–	–				
384		27	*36.3*	33	*55.2*	–	–	–	–				

$$\kappa(P\,M_{\mathrm{sD}}^{-1}P^{\top}F_{|\tilde{U}_{\mathrm{ad}}}) = \mathcal{O}\Big(\frac{H_0}{\eta}\,(1+\log(H_{\mathrm{int}}/h))^2\Big),$$

independently of the coefficient jumps, because $\alpha_0 \ge \alpha_i$, cf. Theorem 4.30. Indeed, when decreasing the parameter η (note that this changes the domain Ω), the condition numbers grow.

In all the experiments, it *seems* that the condition number *does depend* on the ratio H_0/H_{int}. We believe that this is due to preasymptotic effects which will vanish with larger problem sizes.

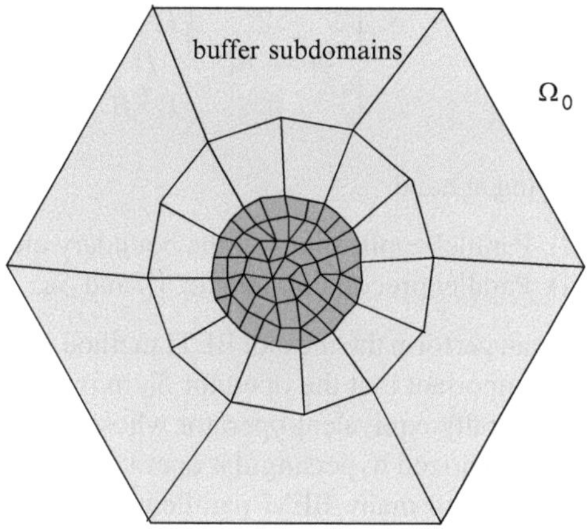

Fig. 4.8 Illustration of the buffering strategy. Original interior subdomains in *dark gray*, buffering subdomains in *light gray*

4.4 Implementation Issues

In this section, we restrict ourselves to the three-dimensional case. As outlined before, the principal differences to FETI/BETI methods for the bounded case are

(i) The possibly large number of neighbors to $\partial\Omega_0$, and
(ii) The possibly large number of DOFs in $V^h(\partial\Omega_0)$ compared to the other subdomains.

Issue (i) means that in practice, the processor which handles the exterior subdomain has to communicate with a lot of other processors. Issue (ii) implies that we might have no *load balancing*. The following strategies may help to overcome these difficulties.

Buffering. We introduce several layers of "buffer" subdomains in between the original interior subdomains and the exterior subdomain Ω_0, cf. Fig. 4.8. This way we can reduce the number of neighbors to Ω_0. If the problem setting permits, we can choose a coarser mesh on $\partial\Omega_0$ such that we regain load balancing. However, this strategy is contrary to the principle of using a boundary element approximation for the exterior subdomain instead of introducing a large layer (as the buffering layer) with Dirichlet boundary conditions on the outer boundary.

Sub-parallelization. We briefly discuss a possible procedure of further subdividing the computational work corresponding to the exterior subdomain Ω_0. The two bottlenecks in the method above are the application of S_0 in the preconditioner M^{-1} and the application of S_0^{-1} in the operator F, where the latter one is certainly the more critical one.

In order to sub-parallelize the action of S_0^{-1} we can move to the inexact method, cf. Sect. 2.2.6. Assume for the time being that we are in the pure BETI case. The system to be solved has the form

$$\begin{bmatrix} -V & \frac{1}{2}I + K & 0 \\ \frac{1}{2}I + K^\top & D & B^\top P \\ 0 & P^\top B & 0 \end{bmatrix} \begin{bmatrix} t \\ u \\ \lambda \end{bmatrix} = \begin{bmatrix} 0 \\ f \\ 0 \end{bmatrix}.$$

Having at hand

(i) Parallel applications of the boundary element matrices V_0, K_0, M_0, and D_0,
(ii) Parallel preconditioners for V_0 and S_0,

we can perform the inexact BETI method with a quasi-optimal load balancing. Here, it is important that the operator S_0 in the BETI-preconditioner M_{sD}^{-1} is replaced by a spectrally equivalent operator whose application can be done efficiently, such as the regularized hypersingular operator, cf. Remark 4.1.

Although many BEM parallelization techniques can be found in the literature (see, e.g., [BB08, BK05, CCC+03]) and parallel Schur complement preconditioners have been investigated extensively (see, e.g., [TW05, Sect. 5] and see also [FS97, Ste03a] for multilevel preconditioners), there is some space left for future research when combining such components as described above. Finally, we remark that if $\partial\Omega_0$ is a circle or a sphere, the BEM operators on $\partial\Omega_0$ can be realized very efficiently using the fast Fourier transform, see, e.g., [Rja90].

Chapter 5
Dual-Primal Methods

Dual-primal FETI (FETI-DP) methods were first introduced by Farhat, Lesoinne, Le Tallec, Pierson, and Rixen in [FLL$^+$01] (as a further development of a FETI method suitable for fourth order problems, [FLP00]). The main idea is to keep unknowns at the subdomain vertices (corners, crosspoints) *primal*, i.e., do not break the continuity and not introduce *dual* Lagrange multipliers there. This way, after an elimination of these primal unknowns (which can be seen as a coarse problem), the resulting subdomain operators are always invertible, because of vanishing values at the subdomain vertices. As a great advantage in the implementation of dual-primal methods (in contrast to one-level FETI methods), the local kernels do not have to be known explicitly.

The motivation to use corner constraints originated from the fact that the FETI method did not converge well for plate an shell problems. Farhat, Mandel, and Tezaur (see [MTF99, FM98]) proposed to project the Lagrange multipliers in each iteration to a subspace. Choosing this subspace such that the corresponding solutions are continuous at the subdomain vertices, the polylogarithmic bound was reinstalled for plate problems [FM98], and fast convergence observed for shell problems [FCMR98]. The resulting method is now called FETI-2.

A first analysis for the two-dimensional case of FETI-DP with corner constraints was performed by Mandel and Tezaur [MT01], showing that the method of this type equipped with a Dirichlet preconditioner leads to the condition number bound $C\,(1 + \log(H/h))^2$. Algorithms for the three-dimensional case were contributed by Farhat, Lesoinne, and Pierson [FLP00] (see also Pierson's doctoral dissertation [Pie00]), and finally by Klawonn, Widlund, and Dryja [KWD02a] where a rigorous analysis of the three-dimensional case can be found, including the case of heterogeneous coefficients. It turned out that in three dimensions, primal vertex constraints result in a poor condition number. A polylogarithmic bound can be obtained by additionally imposing edge and/or face average constraints. A comprehensive description of various FETI-DP algorithms and their analysis can be found in the monograph by Toselli and Widlund, see [TW05, Sect. 6.4].

C. Pechstein, *Finite and Boundary Element Tearing and Interconnecting Solvers for Multiscale Problems*, Lecture Notes in Computational Science and Engineering 90,
DOI 10.1007/978-3-642-23588-7_5, © Springer-Verlag Berlin Heidelberg 2013

A method very interrelated with FETI-DP was introduced independently by Dohrmann [Doh03], Fragakis and Papadrakakis [FP03], and Cros [Cro03], and is now called BDDC (balancing domain decomposition by constraints). It can be seen as the primal counterpart of FETI-DP and was further analyzed in [LW06b, LW07, MD03, MDT05]. Like for the Neumann-Neumann and FETI methods, the FETI-DP and BDDC methods have evolved with large synergies. For a historical overview and connections see [SM08]. The dual-primal BETI (BETI-DP) methods were introduced in [LPS05], an analysis for the unbounded case can be found in [Pec08a]. Extensions of FETI-DP to multiscale problems were done in [PS09, PSS12] and [KR07c, KR07b, GKR12].

The remainder of this chapter is organized as follows. In Sect. 5.1, we derive the formulation of dual-primal FETI/BETI methods, including the case of boundary element subdomains, and also that of unbounded domains. An analysis of FETI/BETI-DP (and also of BDDC) is contained in Sect. 5.2. Implementation issues are discussed in Sect. 5.3. This section also compares the one-level FETI/BETI and FETI/BETI-DP method in the case of unbounded domains.

5.1 Formulation of FETI/BETI-DP Methods

As in the previous sections, we consider the weak formulation of the potential equation

$$-\operatorname{div}(\mathscr{A}\,\nabla u) \;=\; f_\Omega \qquad \text{in } \Omega, \tag{5.1}$$

where the computational domain $\Omega \subset \mathbb{R}^d$ ($d = 2$ or 3) is either bounded (cf. Chap. 2) or unbounded (cf. Chap. 4). However, as in Chap. 4, the two-dimensional unbounded case is excluded from the subsequent considerations, cf. Sects. 1.3.8.4 and 4.2.6. Recall that in the unbounded case we have the radiation condition $|u(x)| = \mathscr{O}(|x|^{-1})$ as $|x| \to \infty$. This leads to the energy space $H^{1,\star}(\Omega)$ from Definition 1.62, whereas we use $H^1(\Omega)$ in the bounded case. The coefficient $\mathscr{A}$ is assumed to be uniformly elliptic, i.e., fulfilling Condition (1.22) from Lemma 1.39 (p. 20). Furthermore, we consider the mixed boundary conditions

$$\begin{aligned} u \;&=\; g_D \qquad \text{on } \Gamma_D,\\ \mathscr{A}\nabla u \cdot n \;&=\; g_N \qquad \text{on } \Gamma_N, \end{aligned}$$

where $\Gamma_D \subset \partial\Omega$ and $\Gamma_N = \partial\Omega \setminus \Gamma_D$. In the unbounded case, Γ_D may be empty. For the variational equations see (2.1), p. 64 and (4.3), p. 217.

In the following, for completeness, we summarize the assumptions on the data and the notation from Chaps. 2 and 4. Let $\mathscr{I}$ denote the set of subdomain indices, i.e., $\mathscr{I} = \{1,\dots,s\}$ in the bounded case, and $\mathscr{I} = \{0,1,\dots,s\}$ in the unbounded case, such that $\overline{\Omega} = \bigcup_{i\in\mathscr{I}} \overline{\Omega_i}$ and $\Omega_i \cap \Omega_j = \emptyset$ for $i \neq j$. Under the assumptions

$$f_\Omega \in L^2(\Omega), \qquad g_N \in L^2(\Gamma_N)$$

on the data, we can split the bilinear form and linear form into contributions $a_i(\cdot,\cdot)$ and f_i from the individual subdomains. In the unbounded case, we assume that $\mathrm{supp}(f_\Omega) \subset \Omega \setminus \Omega_0$. Let $\mathcal{I}_{\mathrm{FEM}} \subset \{1,\dots,s\}$ and $\mathcal{I}_{\mathrm{BEM}} = \mathcal{I} \setminus \mathcal{I}_{\mathrm{FEM}}$ be the index sets corresponding to the FEM and BEM subdomains, respectively. We assume that the coefficient is isotropic and constant in each BEM subdomain, i.e., $\mathcal{A}_{|\Omega_i} = \alpha_i I$ for all $i \in \mathcal{I}_{\mathrm{BEM}}$. Let $S_i : V^h(\partial\Omega_i) \to V^h(\partial\Omega_i)^*$, $i \in \mathcal{I}$, and $N_i : H^1(\Omega_i)^* \to V^h(\partial\Omega_i)^*$, $i \in \mathcal{I} \setminus \{0\}$ denote the corresponding Steklov-Poincaré and Newton potential approximants. As in previous sections, we have dropped the subscript h. For $i \in \mathcal{I}_{\mathrm{BEM}}$, we formally define $N_i f_i := g_{N|\partial\Omega_i \cap \Gamma_N}$. Finally, we set

$$g_i := N_i f_i - S_i \tilde{g}_{D|\partial\Omega_i},$$

where $\tilde{g}_D \in V^h(\Gamma_S)$ is a fixed extension of the given piecewise linear Dirichlet function $g_D \in V^h(\Gamma_D)$.

Recall from (2.14) (p. 68) and Sect. 4.1.2 that the discrete skeleton formulation is equivalent to the minimization problem

$$\tilde{u} = \operatorname*{argmin}_{v \in V_D^h(\Gamma_S)} \sum_{i \in \mathcal{I}} \Big(\tfrac{1}{2} \langle S_i v_{|\partial\Omega_i}, v_{|\partial\Omega_i} \rangle - \langle g_i, v_{|\partial\Omega_i} \rangle \Big), \tag{5.2}$$

where $V_D^h(\Gamma_S) = \{v \in V^h(\Gamma_S) : v_{|\Gamma_D} = 0\}$. As in the classical formulation of one-level methods, we use the definitions

$$W_i := V_D^h(\partial\Omega_i) \quad \text{and} \quad W := \prod_{i \in \mathcal{I}} W_i, \tag{5.3}$$

and we consider the Steklov-Poincaré operators S_i as mappings from W_i to W_i^*. Similarly, we set $S = \mathrm{diag}(S_i) : W \to W^*$ and $g = [g_i]_{i \in \mathcal{I}} \in W^*$. Let

$$B : W \to U^*$$

be the jump operator, defined as in Definition 2.9 and (2.23) (p. 73), and let

$$\widehat{W} = \ker(B) \equiv V_D^h(\Gamma_S)$$

be the subspace of functions that are continuous across the interface. Recall, that thanks to the assumptions above, $\ker(S) \cap \widehat{W} = \{0\}$.

5.1.1 Dual-Primal Spaces

Dual-primal methods work with an intermediate subspace $\widetilde{W}$ with $\widehat{W} \subset \widetilde{W} \subset W$, for which sufficiently many constraints are enforced such that the block operator S is SPD when restricted to $\widetilde{W}$, see [TW05, Sect. 6.4] as well as the original papers [FLL$^+$01, MT01, KWD02a]. The same technique is used in BDDC (balancing domain decomposition by constraints), cf. [Doh03, LW06b, LW07, MD03, MDT05], and it was shown that FETI-DP and BDDC are very interrelated and their preconditioned systems have identical spectra. Our presentation here will thus follow [MDT05] rather than [TW05, Sect. 6.4], in order to show the bridge to the BDDC methods, which will be briefly discussed in Sect. 5.1.4.

5.1.1.1 Primal Nodal Variables: Constraints

Recall from Definition 2.48 (p. 113) that the interface $\Gamma = \bigcup_{i \neq j \in \mathscr{I}} (\partial\Omega_i \cap \partial\Omega_j) \setminus \Gamma_D$ of the partition $\{\Omega_i\}_{i \in \mathscr{I}}$ subdivides into *globs*, i.e., subdomain facets, edges, and vertices. For any interface glob $\mathscr{G} \subset \Gamma$, let $V^h(\mathscr{G})$ denote the restriction of $V^h(\Gamma_S)$ to $\overline{\mathscr{G}}$.

Definition 5.1. (i) Let $\mathscr{G} \subset \Gamma$ be an interface glob. A linear functional $\psi \in V^h(\mathscr{G})^* \setminus \{0\}$ is *associated* to $\mathscr{G}$ if there exists no other glob $\mathscr{G}' \subset \overline{\mathscr{G}}$ of lower dimension than $\mathscr{G}$ such that $\psi(v)$ depends only on $v_{|\bar{\mathscr{G}}'}$, i.e., $\mathscr{G}$ is the unique minimal glob with $\psi \in V^h(\mathscr{G})^*$.

(ii) A *primal nodal variable* is a linear functional $\psi \in V^h(\mathscr{G})^* \setminus \{0\}$ that is associated to an interface glob $\mathscr{G} \subset \Gamma$.

(iii) For a primal nodal variable ψ, associated to glob $\mathscr{G}$, we set

$$\mathscr{N}_\psi := \mathscr{N}_\mathscr{G},$$

where $\mathscr{N}_\mathscr{G}$ is the index set of the subdomains sharing $\mathscr{G}$, cf. Definition 2.48 (ii), p. 113. If $i \in \mathscr{N}_\psi$, we say that ψ is associated to subdomain Ω_i.

(iv) A set Ψ of primal nodal variables is called *linearly independent* if for each glob $\mathscr{G} \subset \Gamma$, the subset $\{\psi \in \Psi : \psi \text{ is associated to } \mathscr{G}\} \subset V^h(\mathscr{G})^*$ is linearly independent. Note that in this case it follows from (i) and (ii) that the set Ψ, interpreted as a subset of $V^h(\Gamma_S)^*$, is linearly independent.

Example 5.2. Here we give five examples for primal nodal variables that will be important later on.

(a) Evaluation at a subdomain vertex $\mathscr{V}$:

$$\psi(v) = v(\mathscr{V}).$$

(b) Average over a subdomain glob $\mathscr{G}$:

$$\psi(v) = \bar{v}^\mathscr{G} := \frac{1}{|\mathscr{G}|} \int_\mathscr{G} v \, ds,$$

where $|\mathscr{G}|$ is the edge length or face area (or 1 for a subdomain vertex, cf. Definition 2.66, p. 119).

(c) Algebraic average over the "interior" nodes of a subdomain glob $\mathcal{G}$:

$$\psi(v) = \bar{v}^{\mathcal{G}^h} := \frac{1}{\#(\mathcal{G}^h)} \sum_{x^h \in \mathcal{G}^h} v(x^h).$$

Recall from Definition 2.50 that if $\mathcal{G}$ is a subdomain face, then the set $\mathcal{G}^h$ contains only its interior nodes (excluding those on $\partial\mathcal{G}$), and that if $\mathcal{G}$ is a subdomain edge, then $\mathcal{G}^h$ does not include the two endpoints.

(d) Algebraic average over the nodes on a "closed" subdomain glob $\overline{\mathcal{G}}$:

$$\psi(v) = \bar{v}^{\bar{\mathcal{G}}^h} := \frac{1}{\#(\overline{\mathcal{G}}^h)} \sum_{x^h \in \bar{\mathcal{G}}^h} v(x^h),$$

where $\overline{\mathcal{F}}^h$, $\overline{\mathcal{E}}^h$ denote the sets of nodes contained in $\overline{\mathcal{F}}$, $\overline{\mathcal{E}}$, respectively (i.e., *including* their boundary nodes), and $\overline{\mathcal{V}}^h = \mathcal{V}^h = \{\mathcal{V}\}$.

(e) The integral counterpart of (c) is the "interior" glob average over a function:

$$\psi(v) = \frac{\int_{\mathcal{G}} I^h(\theta_{\mathcal{G}}\, v)\, \mathrm{d}s}{\int_{\mathcal{G}} \theta_{\mathcal{G}}\, \mathrm{d}s},$$

where $\theta_{\mathcal{G}}$ are the cut-off functions from Definition 2.51, p. 114.

Apparently, if $\mathcal{G}$ is a vertex, the "averages" in (b)–(e) are identical to the evaluation in (a). In all the five examples, the nodal variables ψ reproduce constants, i.e., $\psi(1) = 1$. This feature is certainly not necessary for the definition of the method, but will simplify the analysis below.

5.1.1.2 Spaces with Constraints

For the following definition, note that for a primal nodal variable ψ and a function $w \in \widehat{W}$, the expression $\psi(w)$ is well-defined, because w has a well-defined restriction to the glob that ψ is associated to. Also for a function $w_i \in W_i$ with $i \in \mathcal{N}_\psi$, the expression $\psi(w_i)$ is well-defined.

Definition 5.3. Let Ψ be a fixed, linearly independent set of primal nodal variables.

(i) The subspaces $\widetilde{W} \subset W$ and $\widetilde{W}_\Delta \subset \widetilde{W}$ are defined as

$$\widetilde{W} := \{w \in W : \forall \psi \in \Psi\ \forall i, j \in \mathcal{N}_\psi : \psi(w_i) = \psi(w_j)\}, \tag{5.4}$$

$$\widetilde{W}_\Delta := \prod_{i \in \mathcal{I}} W_{\Delta,i} := \prod_{i \in \mathcal{I}} \{w_i \in W_i : \forall \psi \in \Psi,\ i \in \mathcal{N}_\psi : \psi(w_i) = 0\}, \tag{5.5}$$

where $\widetilde{W}_\Delta$ is called the *dual interface space*. By construction, we have that

$$\widehat{W} \subset \widetilde{W}. \tag{5.6}$$

(ii) We say that the set Ψ *controls* $\ker(S)$ if

$$\widetilde{W} \cap \ker(S) = \{0\}.$$

(iii) The operator $\widetilde{S} : \widetilde{W} \to \widetilde{W}^*$ shall denote the restriction of S to $\widetilde{W}$, i.e.,

$$\langle \widetilde{S}\, v,\, w\rangle = \langle S\, v,\, w\rangle \qquad \forall v,\, w \in \widetilde{W}.$$

If Ψ controls $\ker(S)$, then $\widetilde{S}$ is obviously invertible. The next lemma provides a sufficient condition for Ψ to control $\ker(S)$ for the case of the potential equation (5.1). For the case of linear elasticity, see [TW05, Sect. 8.5.3] and [KW05, KW06].

Lemma 5.4. *Let Ψ be a linearly independent set of primal nodal variables. Assume further that for each floating subdomain Ω_i, there is a path of subdomains from Ω_i to a non-floating subdomain, such that each pair of subsequent subdomains is connected through a glob with an associated primal nodal variable. Then Ψ controls* $\ker(S)$, *and so $\widetilde{S}$ is invertible.*

Proof. Let $w \in \widetilde{W} \cap \ker(S)$ be fixed. Recalling the properties of W and S we find that the components w_i are constant. For all non-floating subdomains Ω_j, it follows that $w_j = 0$. Let Ω_i be an arbitrary non-floating subdomain, with a path to a non-floating subdomain Ω_j. Without loss of generality, assume that $j < i$ and that the path is given by $\{\Omega_j,\, \Omega_{j+1},\, \ldots,\, \Omega_i\}$ and that for any $k \in \{j+1, \ldots, i\}$, ψ_k is a primal nodal variable associated to a glob $\mathcal{G}_{k,k+1}$. Because w_i is a constant and the functionals ψ_k are non-trivial, it follows from $\psi_k(w_k) = \psi_k(w_{k+1})$ and by induction that $w_k = 0$ for all $k = j, \ldots, i$. Since the subdomain Ω_i was arbitrary, this shows that $w = 0$, and so $\widetilde{W} \cap \ker(S) = \{0\}$. □

In the original construction [FLL+01] and in some analyses of FETI-DP [KWD02a, TW05], there is also a *primal space* $\widehat{W}_\Pi \subset \widehat{W}$ involved, where

$$\widetilde{W} = \widehat{W}_\Pi \oplus \widetilde{W}_\Delta$$

is a direct sum. To define and analyze the method, we will not need this particular space, but come back to it when discussing the implementation in Sect. 5.3.

5.1.1.3 Suitable Choices of Primal Nodal Variables

In the following, we display three possible choices for primal nodal variables, cf. [TW05, Algorithms 6.25–6.32] and [KWD02a]. The first one, Algorithm A, is

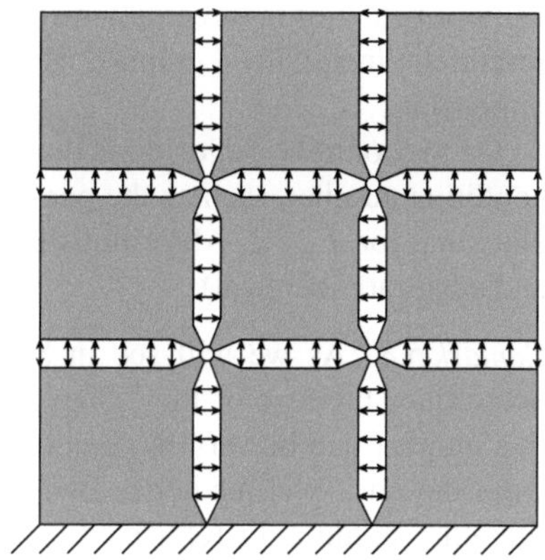

Fig. 5.1 Illustration of the dual-primal space in Algorithm A; "o" indicates primal constraints

illustrated in Fig. 5.1. For all choices, the assumptions of Lemma 5.4 are fulfilled, and so $\ker(S)$ is always controlled (in case of the potential equation).

Algorithm 5.5 (Algorithm A). The primal nodal variables are the evaluations at all subdomain vertices, i.e.,

$$\Psi := \{v \mapsto v(\mathcal{V})\}_{\mathcal{V}\subset\Gamma}\,.$$

Thus, $\widetilde{W}$ is the subspace of functions in W that are continuous at all subdomain vertices.

Algorithm 5.6 (Algorithm B). The primal nodal variables are the evaluations at all subdomain vertices, and the averages over all subdomian edges and facets, i.e.,

$$\Psi := \{v \mapsto v(\mathcal{V})\}_{\mathcal{V}\subset\Gamma} \;\cup\; \{v \mapsto \bar{v}^{\mathcal{E}}\}_{\mathcal{E}\subset\Gamma} \;\cup\; \{v \mapsto \bar{v}^{\mathcal{F}}\}_{\mathcal{F}\subset\Gamma}\,.$$

Thus, $\widetilde{W}$ is the subspace of functions in W that are continuous at the subdomain vertices and that have continuous edge and facet averages. Alternatively, the above averages may be replaced by the those from Example 5.2 (c)–(e).

Algorithm 5.7 (Algorithm C). The primal nodal variables are the evaluations at all subdomain vertices, and the averages over all subdomain edges, i.e.,

$$\Psi := \{v \mapsto v(\mathcal{V})\}_{\mathcal{V}\subset\Gamma} \;\cup\; \{v \mapsto \bar{v}^{\mathcal{E}}\}_{\mathcal{E}\subset\Gamma}.$$

Thus, $\widetilde{W}$ is the subspace of functions in W that are continuous at the subdomain vertices and that have continuous edge averages. Alternatively, the above averages may be replaced by the those from Example 5.2 (c)–(e).

Note that Algorithm A performs only well in two dimensions, cf. Lemma 5.24, [TW05, Remark 6.39], and see [FLL$^+$01, MT01] for some numerical experiments.

Its poor behavior in three dimensions relates to the vertex-based discrete Poincaré-Friedrichs inequality Lemma 2.70, whose parameter grows with H_i/h_i in three dimension.

On the contrary, Algorithms B and C will lead to a polylogarithmic bound for the condition number, such as the one-level FETI/BETI methods from Chaps. 2 and 4. Note that for $d = 2$, Algorithms B and C are identical (because subdomain facets and edges are identical).

Remark 5.8. As we will see in Sect. 5.3, the number of primal nodal variables determines the size of the "coarse" problem. However, for Algorithms B and C, this number can be very large in comparison to the dimension of the coarse space from the one-level methods. Two (reduced) choices of primal constraints (in the literature commonly referred to as Algorithms D and E) addressing this issue were proposed by Klawonn, Widlund, and Dryja. For their definition and analysis we refer to [KWD02a, KWD02b] and [TW05, Algorithms 6.28 and 6.29]. Note that these algorithms in the stated form require that the coefficient is constant in each subdomain. Related algorithms as well as computational results for linearized elasticity (including subdomain-unresolved material heterogeneities) can be found in [Rhe06, KW06, KR07b]. On the other hand, it has been observed numerically [ŠNM+10] that a larger coarse space can also have advantages for the total complexity. See [MS07a, Sou10, MSŠ12] on how primal nodal variables can be selected *adaptively* (based on local eigencomputations).

5.1.2 Saddle Point Formulation and Dual-Primal Method

In the following, we assume that Ψ is a fixed set of linearly independent primal nodal variables that controls $\ker(S)$. Recall the definition of the subspace $\widetilde{W} \subset W$ and let $\widetilde{I} : \widetilde{W} \to W$ be the natural embedding. Furthermore, let

$$\widetilde{B} := B\widetilde{I} : \widetilde{W} \to U^* \tag{5.7}$$

be the jump operator restricted to $\widetilde{W}$. Note that the adjoint operator $\widetilde{I}^\top : W^* \to \widetilde{W}$ is actually a partial assembling operator, see Sect. 5.3, and that $\widetilde{B}^\top = \widetilde{I}^\top B^\top$.

Since the solution $\tilde{u}$ of the minimization problem (5.2) lies in $\widetilde{W}$ as well, we can formally perform the derivation of classical FETI/BETI from Sect. 2.2.1.1 with W replaced by $\widetilde{W}$. The outcome is the saddle point problem: find $(\tilde{u}, \lambda) \in \widetilde{W} \times U$ such that

$$\begin{bmatrix} \widetilde{S} & \widetilde{B}^\top \\ \widetilde{B} & 0 \end{bmatrix} \begin{bmatrix} \tilde{u} \\ \lambda \end{bmatrix} = \begin{bmatrix} \tilde{g} \\ 0 \end{bmatrix}, \tag{5.8}$$

where $\tilde{g} := \widetilde{I}^\top g$. After eliminating the variable $\tilde{u}$, we are left with the dual equation:

$$\text{find } \lambda \in U : \quad F\lambda = \widetilde{B}\widetilde{S}^{-1}\tilde{g}, \tag{5.9}$$

where

$$F := \widetilde{B}\,\widetilde{S}^{-1}\,\widetilde{B}^\top. \tag{5.10}$$

The practical realization of $\widetilde{B}$ and $\widetilde{S}^{-1}$ will be discussed in Sect. 5.3.

Lemma 5.9. *Problems* (5.8) *and* (5.9) *are uniquely solvable up to adding elements from* $\ker(\widetilde{B})^\top$ *to* λ. *The solution* $\tilde{u} = \widetilde{S}^{-1}(\tilde{g} - \widetilde{B}^\top\lambda)$ *satisfies* $\tilde{u} \in \widehat{W} \equiv V_D^h(\Gamma_S)$ *and is the unique solution of* (5.2).

Proof. The proof is straightforward. □

Finally, we define the subspaces

$$\widetilde{U} := U_{/\ker(\tilde{B}^\top)}, \qquad \widetilde{U}^* := \operatorname{range}(\widetilde{B}). \tag{5.11}$$

According to Corollary 1.10, $\widetilde{U}^*$ is indeed the dual of $\widetilde{U}$. From the above, it follows that the operator $F_{|\widetilde{U}} : \widetilde{U} \to \widetilde{U}^*$ is SPD. Therefore, problem (5.9) can be solved with a (P)CG method.

Remark 5.10. In cases where a primal nodal variable ψ_k is the evaluation at a subdomain vertex $\mathscr{V}$, the corresponding coupling constraints in B (which enforce continuity at $\mathscr{V}$) can be omitted, and this will reduce the number of redundancies in $\widetilde{B}$. Note however, that regardless whether this is done or not, the spaces $\widetilde{U}$ and $\widetilde{U}^*$ remain invariant, and so does the operator $F_{|\widetilde{U}}$. The layout of the jump operator B should therefore be decided upon implementation aspects. On the one hand, in many situations, it might be easier to code a jump operator acting on all interface nodes. On the other hand, leaving out primal subdomain vertices might increase the parallel efficiency.

Under suitable assumptions, the condition number $\kappa(F_{|\widetilde{U}})$ behaves as that of the classical FETI method, cf. Theorem 2.38. Without further delay, we move to the preconditioned case.

5.1.3 Preconditioning

As a matter of fact, the scaled Dirichlet preconditioner from Sect. 2.2.4.2 can be reused verbatim for the dual-primal method. Following (2.52), (2.55), and (2.57) (see p. 91ff), let $\rho_i(x^h)$ be fixed weights, let $B_D : W^* \to U$ be the weighted jump operator from the classical formulation, and let

$$M_{\mathrm{sD}}^{-1} := B_D\, S\, B_D^\top : U^* \to U$$

denote the scaled Dirichlet preconditioner. In Lemma 5.14 below, we will show that (under appropriate conditions) M_{sD}^{-1} is SPD on $\widetilde{U}^* = \operatorname{range}(\widetilde{B})$. Since U is naturally embedded in the factor space $\widetilde{U}$, and because the duality pairing

Algorithm 9: FETI/BETI-DP method with scaled Dirichlet preconditioner

$g = [N_i f_i - S_i \tilde{g}_{D|\partial\Omega_i}]_{i\in\mathcal{I}}$
$\tilde{g} = \widetilde{I}^\top g$
$d = \widetilde{B}\widetilde{S}^{-1}\tilde{g}$
solve $F\lambda = d$ with PCG, with preconditioner M_{sD}^{-1} and any initial value $\lambda^{(0)} \in U$,
and stop after k iterations
$\tilde{u}^{(k)} = \widetilde{S}^{-1}(\tilde{g} - \widetilde{B}^\top\lambda^{(k)})$
$u^{(k)} = [\tilde{g}_{D|\partial\Omega_i} + \tilde{u}_i^{(k)}]_{i\in\mathcal{I}}$

$\langle\cdot,\cdot\rangle_{\widetilde{U}^*\times\widetilde{U}}$ is still realized by the Euclidean inner product in $\mathbb{R}^M$, the operator $M_{\mathrm{sD}}^{-1} : \widetilde{U}^* \to \widetilde{U}$ can indeed be used as a proper preconditioner within a PCG method for problem (5.9), cf. Corollary 1.50, p. 33. Algorithm 9 summarizes this procedure.

Remark 5.11. Analogously to Remark 2.20, one can show that the residual

$$d - F\lambda^{(k)} = \widetilde{B}\,\tilde{u}^{(k)} = B\,u^{(k)}$$

in the k-th step of PCG controls the jump of the approximate solution $u^{(k)}$.

In order to prove that M_{sD}^{-1} is SPD on $\widetilde{U}^*$, we need the two auxiliary results. Recall the projection operator $P_D : W \to W$ defined by

$$P_D := B_D^\top B,$$

with the properties $B\,P_D = B$ (cf. Lemma 2.39, p. 107) and $B^\top M_{\mathrm{sD}}^{-1} B = P_D^\top S\,P_D$. Recall also that the averaging operator $E_D := I - P_D$ projects into the continuous space, i.e., $E_D : W \to \widehat{W}$.

Lemma 5.12. *For any intermediate space $\widetilde{W}$ with $\widehat{W} \subset \widetilde{W} \subset W$,*

$$P_D\,\widetilde{I} : \widetilde{W} \to \widetilde{W} \quad \text{and} \quad \widetilde{B}\,P_D\,\widetilde{I} = \widetilde{B}.$$

In particular, this property holds for $\widetilde{W}$ constructed from a set Ψ of linearly independent nodal variables, according to Definition 5.3.

Proof. For any $w \in \widetilde{W}$, we have $P_D\,\widetilde{I}\,w = \tilde{w} - E_D\,\widetilde{I}\,w \in \widetilde{W}$ since $E_D : W \to \widehat{W} \subset \widetilde{W}$. The other formula follows then from $B\,P_D = B$ and the definition of $\widetilde{B}$. □

Lemma 5.13. *Let Ψ be a fixed set of linearly independent nodal variables with the corresponding space $\widetilde{W}$ according to Definition 5.3. Then, for all $\mu \in \widetilde{U}^*$ there exists an element $v \in \widetilde{W} \cap \mathrm{range}(P_D)$ such that $\mu = \widetilde{B}\,v$.*

Table 5.1 Overview on the spaces involved in FETI/BETI-DP

Spaces	
$U = \mathbb{R}^M$	Space of Lagrange multipliers
$\widetilde{U} = U_{/\ker(\widetilde{B}^\top)}$	Factor space modulo redundancies
$V_D^h(\Gamma_S)$	Skeletal FE space with homogeneous Dirichlet conditions
$W_i = V_D^h(\partial\Omega_i)$	Local spaces
$W = \prod_{i=1}^s W_i$	Product space ("discontinuous" functions)
$\widehat{W} = \ker(B) \subset \widetilde{W} \subset W$	Subspace of continuous functions in W, identifiable with $V_D^h(\Gamma_S)$
$\widetilde{W} \subset W$	Partially continuous space defined by Ψ
$\widetilde{W}_\Delta \subset \widetilde{W}$	(discontinuous) Subspace where $\psi(w_i) = 0$ for all $\psi \in \Psi$
$\widehat{W}_\Pi \subset \widehat{W}$	Continuous primal subspace, complement of $\widetilde{W}_\Delta$

Proof. Since $\widetilde{U}^* = \operatorname{range}(\widetilde{B})$, there exists an element $w \in \widetilde{W}$ such that $\mu = \widetilde{B}\, w$. We choose $v := P_D \widetilde{I}\, w$. Lemma 5.12 guarantees that $v \in \widetilde{W}$, and by construction $v \in \operatorname{range}(P_D)$. Finally, the previous lemma implies $\widetilde{B}\, v = \widetilde{B}\, P_D \widetilde{I}\, w = \widetilde{B}\, w = \mu$. □

We can now prove that the scaled Dirichlet preconditioner is invertible on $\widetilde{U}^*$.

Lemma 5.14. *Let Ψ be a fixed set of linearly independent nodal variables and let $\widetilde{W}$ and $\widetilde{U}$ be defined as above. Then the operator M_{sD}^{-1} is SPD on $\widetilde{U}^*$.*

Proof. It is immediate that M_{sD}^{-1} is symmetric positive semidefinite and so it suffices to show the definiteness. Let $\mu \in \widetilde{U}^*$ be fixed with $\langle \mu, M_{\text{sD}}^{-1}\, \mu \rangle = 0$. Now, let v be the element from Lemma 5.13 with $\mu = \widetilde{B}\, v$. Then

$$0 = \langle \widetilde{B}\, v, M_{\text{sD}}^{-1}\, \widetilde{B}\, v \rangle = |P_D \widetilde{I} v|_S^2 = |\widetilde{I}\, v|_S^2 = |v|_{\widetilde{S}}^2,$$

where we have used that $v \in \widetilde{W} \cap \operatorname{range}(P_D)$ and so $P_D \widetilde{I}\, v = \widetilde{I}\, v$, cf. Lemma 5.12. Since $\widetilde{S}$ is SPD, it follows that $v = 0$ and thus $\mu = 0$, which shows the definiteness. □

Remark 5.15. As the attentive reader will have noticed, the previous lemmas only use that $\widehat{W} \subset \widetilde{W} \subset W$ and that $\widetilde{S}$ is SPD on $\widetilde{W}$. For an abstract framework based on these assumptions see [MDT05, MS07b]. For a matrix-based approach see [LW06b].

Note also that for *any* space $\widetilde{W} \subset W$ with $\widetilde{W} \supset \widehat{W} = \ker(B)$ there exists (by standard linear algebra) a full-rank matrix $D \in \mathbb{R}^{m \times M}$ such that $\widetilde{W} = \ker(D\, B)$. This structure is exactly seen in the definition (5.4) of $\widetilde{W}$ using primal nodal variables.

Tables 5.1 and 5.2 summarize the spaces and operators introduced so far.

Table 5.2 Overview on the operators involved in FETI/BETI-DP

Operators	
$B : W \to U^*$	Jump operator
$\widetilde{B} : \widetilde{W} \to U^*$	Restricted jump operator, $\widetilde{B} = B\,\widetilde{I}$
$B_D : W^* \to U$	Weighted jump operator
$\widetilde{I} : \widetilde{W} \to W$	Embedding operator
$E_D : W \to \widehat{W}$	Averaging operator, $E_D = I - P_D$
$F : U \to U^*$	$F = \widetilde{B}\,\widetilde{S}^{-1}\widetilde{B}^\top$
$P_D : W \to W$	Projection operator, $P_D = B_D^\top B$, property $B\,P_D = B$
$S : W \to W^*$	Block operator of local approximate Steklov-Poincaré operators
$\widetilde{S} : \widetilde{W} \to \widetilde{W}^*$	$\widetilde{S} = \widetilde{I}^\top S\,\widetilde{I}$
Ψ	Set of linearly independent primal nodal variables

5.1.4 The BDDC Method

Balancing domain decomposition by constraints (BDDC) is a technique for preconditioning the Schur complement system. It was introduced independently by Dohrmann [Doh03], Fragakis and Papadrakakis [FP03], and Cros [Cro03], and further developed by Mandel and Dohrmann [MD03]. It can be seen as a generalization of the balancing Neumann-Neumann methods, see Sect. 2.3.

With the previous notations, let $\widehat{I} : \widehat{W} \equiv V_D^h(\Gamma_S) \to W$ be the natural embedding and let $\widehat{S} : \widehat{W} \to \widehat{W}^*$ denote the restriction of S to $\widehat{W}$, i.e., $\widehat{S} = \widehat{I}^\top S\,\widehat{I}$. We can think of $\widehat{W} \equiv V_D^h(\Gamma_S)$ as the global FE space and of $\widehat{I}^\top$ as the assembling operator that assembles the contributions from the subdomain boundaries. Problem (5.2) is then equivalent to

$$\text{find } \hat{u} \in \widehat{W} : \quad \widehat{S}\,\hat{u} = \hat{g}, \tag{5.12}$$

where $\hat{g} := \widehat{I}\,g$. Apparently, the operator $\widehat{S}$ is SPD, and so (5.12) can be solved by (P)CG. In the following, assume that Ψ is a fixed set of linearly independent nodal variables (cf. Definition 5.1) with the corresponding space $\widetilde{W}$ (cf. Definition 5.3). Moreover, assume that Ψ controls $\ker(S)$.

Let again $\widetilde{I} : \widetilde{W} \to W$ be the natural embedding and set $\widetilde{S} := \widetilde{I}^\top S\,\widetilde{I}$. Let $\rho_i(x^h)$ be fixed weights and let the weighted averaging operator $E_D : W \to \widehat{W}$ be defined as above (cf. Lemma 2.39, p. 107). Furthermore, we define the operator

$$\widetilde{E}_D := E_D\,\widetilde{I} : \widetilde{W} \to \widehat{W} \tag{5.13}$$

and its adjoint $\widetilde{E}_D^\top : \widehat{W}^* \to \widetilde{W}^*$. The latter operator distributes residuals to subdomain residuals. The BDDC preconditioner is finally given by

$$M_{\mathrm{BDDC}}^{-1} := \widetilde{E}_D \, \widetilde{S}^{-1} \widetilde{E}_D^\top : \widehat{W}^* \to \widehat{W}. \tag{5.14}$$

Remark 5.16. Note that the definition (5.14) uses the notation of the previous FETI/BETI-DP method, but it coincides with the original definition that can be found in [Doh03], see also [MDT05, MS07b, BS07]. With a FETI-DP code at hand (see Sect. 5.3), the application of $\widetilde{E}_D$ and $\widetilde{E}_D^\top$ can be realized via the existing jump operators (and thus the Lagrange multipliers). However, one can also realize them by just using local-global mappings of the degrees of freedom. Note also, that a method introduced as P-FETI-DP [FP03] has turned out to be equivalent to BDDC, cf. [MS07b, Sect. 3.4].

Lemma 5.17. *Let* $\Psi = \{\psi_k\}_{k=1}^{n_\Pi}$ *be a linearly independent set of primal nodal variables that controls* $\ker(S)$. *Then the BDDC preconditioner* M_{BDDC}^{-1} *is SPD.*

Proof. Since $\widetilde{S}^{-1}$ is SPD (cf. Definition 5.3), it suffices to show the definiteness. Let $r \in \widehat{W}^*$ with $0 = \langle r, M_{\mathrm{BDDC}}^{-1} r\rangle = \langle \widetilde{E}_D^\top r, \widetilde{S}^{-1} \widetilde{E}_D^\top r\rangle$. Then $\widetilde{E}_D^\top r = 0$, which implies that for any $w \in \widetilde{W}$, $\langle r, \widetilde{E}_D \tilde{w}\rangle = 0$. Since $\mathrm{range}(\widetilde{E}_D) = \widehat{W}$, this yields $r = 0$. □

The BDDC preconditioner will be briefly analyzed in Sect. 5.2.5.

5.2 Analysis of FETI/BETI-DP and BDDC

In this section, we perform an analysis of FETI/BETI-DP. Firstly, in Sect. 5.2.1 we show that an estimate for the P_D operator implies an estimate for the condition number. Secondly, we provide with such bounds for P_D in Sect. 5.2.2 under assumptions that are similar to those in Chaps. 2 and 4. The main results are finally summarized in Sect. 5.2.3. Extensions to multiscale coefficients are discussed in Sect. 5.2.4, and extensions to the BDDC method in Sect. 5.2.5.

5.2.1 An Abstract Condition Number Bound

The following lemma is an analogue of Lemma 2.45 (p. 109) and states that an upper bound for P_D implies a condition number estimate for FETI/BETI-DP. Its proof follows [MDT05, MS07b].

Lemma 5.18. *Let* Ψ *be a linearly independent set of primal nodal variables that controls* $\ker(S)$. *Assume further that there exists a parameter* ω *such that*

$$|P_D \widetilde{I}\, w|_{\widetilde{S}}^2 \le \omega\, |\widetilde{I}\, w|_{\widetilde{S}}^2 \qquad \forall w \in \widetilde{W}. \tag{5.15}$$

Then

$$\kappa(M_{\mathrm{sD}}^{-1} F_{|\widetilde{U}}) \le \omega.$$

Proof. Let $M_{\mathrm{sD}} : \widetilde{U} \to \widetilde{U}^*$ be the inverse of $(M_{\mathrm{sD}}^{-1})_{|\tilde{U}^*}$ (cf. Lemma 5.14). From Corollary 1.50, we see that

$$\kappa(M_{\mathrm{sD}}^{-1} F_{|\tilde{U}}) \;=\; \frac{\displaystyle\sup_{\lambda\in\tilde{U}} \frac{\langle F\lambda, \lambda\rangle}{\langle M_{\mathrm{sD}}\lambda, \lambda\rangle}}{\displaystyle\inf_{\lambda\in\tilde{U}} \frac{\langle F\lambda, \lambda\rangle}{\langle M_{\mathrm{sD}}\lambda, \lambda\rangle}}. \tag{5.16}$$

In order to estimate the numerator in (5.16) from above, we bound F in terms of M_{sD}. Let $\lambda \in \widetilde{U}$ be arbitrary but fixed. From the definitions of F and $\widetilde{S}$, a standard duality formula (cf. Lemma 1.14), from (5.15), and from the identity $B\, M_{\mathrm{sD}}^{-1}\, B^\top = P_D^\top\, S\, P_D$, it follows that

$$\langle F\lambda, \lambda\rangle \;=\; \sup_{w\in\tilde{W}} \frac{\langle \widetilde{B}\, w, \lambda\rangle^2}{|\widetilde{I}\, w|_S^2} \;\le\; \omega \sup_{w\in\tilde{W}} \frac{\langle \widetilde{B}\, w, \lambda\rangle^2}{|P_D\, \widetilde{I}\, w|_S^2} \;=\; \omega \sup_{w\in\tilde{W}} \frac{\langle \widetilde{B}\, w, \lambda\rangle^2}{\langle \widetilde{B}\, w, M_{\mathrm{sD}}^{-1}\, \widetilde{B}\, w\rangle}.$$

Since $\widetilde{U}^* = \mathrm{range}(\widetilde{B})$, the above estimate implies

$$\langle F\lambda, \lambda\rangle \;\le\; \omega \sup_{\mu\in\tilde{U}^*} \frac{\langle \mu, \lambda\rangle^2}{\langle \mu, M_{\mathrm{sD}}^{-1}\mu\rangle} \;=\; \omega\,\langle M_{\mathrm{sD}}\lambda, \lambda\rangle,$$

i.e., $\lambda_{\max}(M_{\mathrm{sD}}^{-1} F_{|\tilde{U}^*}) \le \omega$.

We now turn to the denominator in (5.16). Let $\lambda \in \widetilde{U}$ be fixed and $\mu \in \widetilde{U}^*$ arbitrary. Let $v \in \widetilde{W} \cap \mathrm{range}(P_D)$ be the special element from Lemma 5.13 with $\widetilde{B}\, v = \mu$. With similar arguments as above and using that $P_D\, \widetilde{I}\, v = \widetilde{I}\, v$ (cf. Lemma 5.12), we obtain that

$$\langle F\lambda, \lambda\rangle \;=\; \sup_{w\in\tilde{W}} \frac{\langle \widetilde{B}\, w, \lambda\rangle^2}{|\widetilde{I}\, w|_S^2} \;\ge\; \frac{\langle \widetilde{B}\, v, \lambda\rangle^2}{|\widetilde{I}\, v|_S^2} \;=\; \frac{\langle \widetilde{B}\, v, \lambda\rangle^2}{|P_D\, \widetilde{I}\, v|_S^2} \;=\; \frac{\langle \widetilde{B}\, v, \lambda\rangle^2}{\langle \widetilde{B}\, v, M_{\mathrm{sD}}^{-1}\, \widetilde{B}\, v\rangle}.$$

Since $\widetilde{B}\, v = \mu \in \widetilde{U}^*$ was arbitrary, this yields

$$\langle F\lambda, \lambda\rangle \;\ge\; \sup_{\mu\in\tilde{U}^*} \frac{\langle \mu, \lambda\rangle^2}{\langle \mu, M_{\mathrm{sD}}^{-1}\mu\rangle} \;=\; \langle M_{\mathrm{sD}}\lambda, \lambda\rangle,$$

which shows that $\lambda_{\min}(M_{\mathrm{sD}}^{-1} F_{|\tilde{U}^*}) \ge 1$ and thus concludes the proof. □

5.2.2 *Estimates for the* P_D *Operator*

We start with estimates for piecewise constant coefficients, i.e., we generalize the results from Chaps. 2 and 4 to the dual-primal case. We will briefly comment on results for non-resolved multiscale coefficients in Sect. 5.2.4.

The following lemmas elaborate estimates of the form (5.15) for Algorithms A–C. Before, we summarize the needed assumptions.

Assumption 5.19 (bounded case). If $0 \notin \mathscr{I}$, the following assumptions hold.

Assumption 2.53: The triangulations $\mathscr{T}^h(\Omega_i), i \in \mathscr{I}_{\text{FEM}}$ and $\mathscr{T}^h(\partial\Omega_i), i \in \mathscr{I}_{\text{BEM}}$ are quasi-uniform with mesh parameter h_i.

Assumption 2.54: There is a shape regular simplicial coarse triangulation $\mathscr{T}^H(\Omega)$ of Ω, such that each subdomain Ω_i is the union of coarse elements and the number of coarse elements per subdomain is uniformly bounded.

Assumption 2.55: For each subdomain Ω_i, $i \in \mathscr{I}_{\text{BEM}}$, there is a neighborhood $\Omega_i' \supset \overline{\Omega_i}$ and a shape regular simplicial coarse triangulation $\mathscr{T}^H(\Omega_i')$, such that the shape regularity constants and the number of coarse elements in $\mathscr{T}^H(\Omega_i')$ are uniformly bounded.

Assumption 2.56: The Dirichlet boundary Γ_D is the union of facets (faces/edges) of the coarse triangulation $\mathscr{T}^H(\Omega)$ from Assumption 2.54.

Assumption 5.20 (unbounded case). If $0 \in \mathscr{I}$, the following assumptions hold.

Assumption 4.4: The triangulations $\mathscr{T}^h(\Omega_i), i \in \mathscr{I}_{\text{FEM}}$ and $\mathscr{T}^h(\partial\Omega_i), i \in \mathscr{I}_{\text{BEM}}$ (including the case $i = 0$) are quasi-uniform with mesh parameter h_i.

Assumption 4.5: There is a shape regular simplicial coarse triangulation $\mathscr{T}^H(\Omega_{\text{int}})$ of the interior part $\Omega_{\text{int}} = \Omega \setminus \Omega_0$, such that each interior subdomain $\Omega_i \subset \Omega_{\text{int}}$ is the union of coarse elements and the number of coarse elements per subdomain is uniformly bounded.

Assumption 4.6: There exists a bounded Lipschitz domain Ω_{int}' such that (i) $\overline{\Omega_0^c} \subset \Omega_{\text{int}}'$, (ii) The coarse triangulation $\mathscr{T}^H(\Omega_{\text{int}})$ from Assumption 4.5 can be extended to a shape regular simplicial coarse triangulation $\mathscr{T}^H(\Omega_{\text{int}}')$, (iii) The Dirichlet boundary Γ_D is the union of facets (faces/edges) from $\mathscr{T}^H(\Omega_{\text{int}}')$.

The following assumption is the same as Assumptions 2.99 and 2.100 from Chap. 2. Note again, that the whole theory below can be straightforwardly generalized for matrix-valued coefficients with mild anisotropy that are constant in each subdomain, cf. Remark 2.101.

Assumption 5.21. In each subdomain Ω_i, for $i = 0, \dots, s$, there exists a scalar constant $\alpha_i > 0$ such that

$$\mathscr{A}_{|\Omega_i} = \alpha_i \, I,$$

where $\mathscr{A}$ is the coefficient from (5.1) and I denotes the identity matrix. Moreover, the *coefficient scaling* is used, i.e., $\rho_i(x^h) = \alpha_i$, cf. Sect. 2.2.4.2.

Lemma 5.22 (Algorithm B, bounded case). *Let $d \in \{2,3\}$, let $\mathscr{I} = \{1,\dots,s\}$ and let Assumption 5.19 as well as Assumption 5.21 hold. Then for Ψ defined according to Algorithm 5.6,*

$$|P_D \widetilde{I} w|_S^2 \leq C \max_{i=1,\dots,s} (1 + \log(H_i/h_i))^2 \, |\widetilde{I} w|_S^2 \qquad \forall w \in \widetilde{W},$$

where C is independent of h_i, H_i, the values α_i, and the number of subdomains.

Proof. The proof follows the line of [KWD02a] (see also [TW05, Sect. 6.4.3]). Let $w \in \widetilde{W}$ be arbitrary. From the proof of Lemma 2.104, we can conclude that for all $w \in \widetilde{W}$ and for all $i = 1, \ldots, s$,

$$|(P_D w)_i|_{S_i}^2 \lesssim \sum_{\mathscr{G}_i \subset \Gamma} \sum_{j \in \mathscr{N}_{\mathscr{G}_i}} \min(\alpha_i, \alpha_j)\, |\mathscr{H}_i^h(\theta_{\mathscr{G}_i}(w_i - w_j))|_{H^1(\Omega_i)}^2 . \tag{5.17}$$

Recall from the definition of $\widetilde{W}$ from Algorithm 5.6, that for any glob $\mathscr{G}_i \subset \Gamma$,

$$\overline{w_i}^{\mathscr{G}_i} = \overline{w_j}^{\mathscr{G}_i},$$

and so,

$$\begin{aligned} &\min(\alpha_i, \alpha_j)\, |\mathscr{H}_i^h(\theta_{\mathscr{G}_i}(w_i - w_j))|_{H^1(\Omega_i)}^2 \qquad (5.18)\\ &\lesssim \alpha_i\, |\mathscr{H}_i^h(\theta_{\mathscr{G}_i}(w_i - \overline{w_i}^{\mathscr{G}_i}))|_{H^1(\Omega_i)}^2 + \alpha_j\, |\mathscr{H}_i^h(\theta_{\mathscr{G}_i}(w_j - \overline{w_j}^{\mathscr{G}_i}))|_{H^1(\Omega_i)}^2 . \end{aligned}$$

Note that the term above vanishes if $\mathscr{G}_i$ is a subdomain vertex. Using Corollary 2.78 and Lemma 2.87, we get that for $k \in \{i, j\}$,

$$\begin{aligned} &|\mathscr{H}_i^h(\theta_{\mathscr{G}_i}(w_k - \overline{w_k}^{\mathscr{G}_i})|_{H^1(\Omega_i)}^2\\ &\lesssim (1 + \log(\tfrac{H_i}{h_i}))^2\, |\mathscr{H}_k^h w_k|_{H^1(\Omega_k)}^2 + (1 + \log(\tfrac{H_i}{h_i}))\, H_i^{-2}\, \|\mathscr{H}_k^h(w_k - \overline{w_k}^{\mathscr{G}_i})\|_{L^2(\Omega_k)}^2\\ &\lesssim (1 + \log(\tfrac{H_i}{h_i}))^2\, |\mathscr{H}_k^h w_k|_{H^1(\Omega_k)}^2, \qquad (5.19) \end{aligned}$$

provided that $\mathscr{G}_i$ is not a subdomain vertex, where in the last step we have used Lemma 2.70. Combining (5.17)–(5.19), we obtain

$$|(P_D w)_i|_{S_i}^2 \lesssim (1 + \log(\tfrac{H_i}{h_i}))^2 \sum_{j \in \mathscr{N}_i} \alpha_j\, |\mathscr{H}_j^h w_j|_{H^1(\Omega_j)}^2 \lesssim (1 + \log(\tfrac{H_i}{h_i}))^2 \sum_{j \in \mathscr{N}_i} |w_j|_{S_j}^2,$$

where in the last step we have used Corollary 1.94. Since each subdomain has a small number of neighbors, the statement of the lemma follows immediately. □

Lemma 5.23 (Algorithm C, bounded case). *Let $d = 3$, let $\mathscr{I} = \{1, \ldots, s\}$ and let Assumption 5.19 as well as Assumption 5.21 hold. Then for Ψ defined according to Algorithm 5.7,*

$$|P_D \widetilde{I}\, w|_S^2 \le C \max_{i=1,\ldots,s} (1 + \log(H_i/h_i))^2\, |\widetilde{I}\, w|_S^2 \qquad \forall w \in \widetilde{W},$$

where C is independent of h_i, H_i, the values α_i, and the number of subdomains.

Proof. To get the bound for Algorithm C, we only need to modify the proof of Lemma 5.22. For a subdomain face $\mathscr{F}_{ij} \subset \Gamma$, let $\mathscr{E} \subset \partial\mathscr{F}_{ij} \cap (\Gamma \cup \Gamma_D)$ be a subdomain edge such that

$$\overline{w_i}^{\mathscr{E}} = \overline{w_j}^{\mathscr{E}} .$$

Thanks to our assumptions, such an edge always exists. We now replace the corresponding estimates (5.18) and (5.19) in the proof of Lemma 5.23 by

$$\begin{aligned} &\min(\alpha_i, \alpha_j)\, |\mathscr{H}_i^h(\theta_{\mathscr{F}_{ij}}(w_i - w_j))|^2_{H^1(\Omega_i)} \\ &\lesssim \alpha_i\, |\mathscr{H}_i^h(\theta_{\mathscr{F}_{ij}}(w_i - \overline{w_i}^{\mathscr{E}}))|^2_{H^1(\Omega_i)} + \alpha_j\, |\mathscr{H}_i^h(\theta_{\mathscr{F}_{ij}}(w_j - \overline{w_j}^{\mathscr{E}}))|^2_{H^1(\Omega_i)}. \end{aligned} \tag{5.20}$$

and (for $k \in \{i, j\}$)

$$\begin{aligned} &|\mathscr{H}_i^h(\theta_{\mathscr{F}_{ij}}(w_k - \overline{w_k}^{\mathscr{E}})|^2_{H^1(\Omega_i)} \\ &\lesssim (1 + \log(\tfrac{H_i}{h_i}))^2\, |\mathscr{H}_k^h w_k|^2_{H^1(\Omega_k)} + (1 + \log(\tfrac{H_i}{h_i}))\, H_i^{-2}\, \|\mathscr{H}_k^h(w_k - \overline{w_k}^{\mathscr{E}})\|^2_{L^2(\Omega_k)} \\ &\lesssim (1 + \log(\tfrac{H_i}{h_i}))^2\, |\mathscr{H}_k^h w_k|^2_{H^1(\Omega_k)}, \end{aligned} \tag{5.21}$$

respectively. The two estimates above follow again from Corollary 2.78, Lemmas 2.87 and 2.70. The remainder of the proof is then identical. □

Lemma 5.24 (Algorithm A, 2D, bounded case). *Let $d = 2$, let $\mathscr{I} = \{1, \ldots, s\}$ and let Assumption 5.19 as well as Assumption 5.21 hold. Then for Ψ defined according to Algorithm A,*

$$|P_D \widetilde{I}\, w|_S^2 \le C \max_{i=1,\ldots,s} (1 + \log(H_i/h_i))^2\, |\widetilde{I}\, w|_S^2 \qquad \forall w \in \widetilde{W},$$

where the constant $C > 0$ is independent of h_i, H_i, the values of α_i, the number of subdomains.

Proof. As in the proof of Lemma 5.23, we have to adapt the estimates for the subdomain edges. This is done by formally following the proof of Lemma 5.23, but selecting for each subdomain facet (i.e., edge) $\mathscr{F}$ a subdomain vertex $\mathscr{V} \subset \overline{\mathscr{F}} \cup \Gamma_D$ where $w(\mathscr{V}) = w(\mathscr{V})$. □

Lemma 5.25 (Algorithms B, C, unbounded case). *Let $d = 3$, let $\mathscr{I} = \{0, \ldots, s\}$, and let Assumption 5.20 as well as Assumption 5.21 hold. Then for Ψ defined according to Algorithms 5.6 or 5.7,*

$$|P_D \widetilde{I}\, w|_S^2 \le C \max_{i=1,\ldots,s} (1 + \log(H_i/h_i))^2\, |\widetilde{I}\, w|_S^2 \qquad \forall w \in \widetilde{W},$$

where the constant $C > 0$ is independent of h_i, H_i, the values α_i, the number of subdomains, and the number of neighbors of Ω_0. In particular, the estimate is independent of $H_0 = \operatorname{diam}(\Omega_0^c)$.

Proof. Again, the structure of the proof is identical to that of Lemma 5.22. In the unbounded case, however, we have to use Lemma 4.14 in order to get the local estimates (5.17) as well as

$$|(P_D w)_0|_{S_0}^2 \lesssim \sum_{\mathcal{G}_0 \subset \Gamma} \sum_{j \in \mathcal{N}_{\mathcal{G}_0}} \min(\alpha_0, \alpha_j)\, |\mathcal{H}^h_{0,\mathsf{U}_{0,\mathcal{G}_0}}(\theta_{\mathcal{G}_0}(w_0 - w_j))|^2_{H^1(\mathsf{U}_{0,\mathcal{G}_0})}. \tag{5.22}$$

The hidden constants only depend on the shape regularity of the coarse mesh $\mathcal{T}^H(\Omega'_{\text{int}})$ from Assumption 4.6. Recall from Chap. 4 (cf. Lemma 4.17) that for any interface glob $\mathcal{G}_{0j} \subset \Gamma$ and for any auxiliary glob $\mathcal{G}' \subset \overline{\mathcal{G}_{0j}} \cap (\Gamma \cup \Gamma_D)$ with $d_{\mathcal{G}'} \geq d_{\mathcal{G}_{0j}} - 1$ and $\bar{w}_0^{\mathcal{G}'} = \overline{w_j}^{\mathcal{G}'}$,

$$\begin{aligned}
|\mathcal{H}^h_j(\theta_{\mathcal{G}_{0j}}(w_0 - \overline{w_0}^{\mathcal{G}'}))|^2_{H^1(\Omega_j)} &\lesssim (1 + \log(\tfrac{H_j}{h_j}))^2\, |\mathcal{H}^h_{0,\mathsf{U}_{0,\mathcal{G}_{0j}}} w_0|^2_{H^1(\mathsf{U}_{0,\mathcal{G}_{0j}})}, \\
|\mathcal{H}^h_{0,\mathsf{U}_{0,\mathcal{G}_{0j}}}(\theta_{\mathcal{G}_{0j}}(w_j - \overline{w_j}^{\mathcal{G}'}))|^2_{H^1(\mathsf{U}_{0,\mathcal{G}_{0j}})} &\lesssim (1 + \log(\tfrac{H_j}{h_j}))^2\, |\mathcal{H}^h_j w_j|^2_{H^1(\Omega_j)}.
\end{aligned}$$

Using these additional auxiliary estimates, we can follow the proofs of the two previous lemmas and conclude that for both choices Algorithms 5.6 and 5.7,

$$|P_D w|_S^2 \lesssim \max_{j=1,\ldots,s} (1 + \log(\tfrac{H_j}{h_j}))^2 \Big[\sum_{i=1}^{s} |w_j|^2_{S_j} + \underbrace{\alpha_0 \sum_{\mathcal{G}_{0j} \subset \partial\Omega_0 \cap \Gamma} |\mathcal{H}^h_{0,\mathsf{U}_{0,\mathcal{G}_{0j}}} w_0|^2_{H^1(\mathsf{U}_{0,\mathcal{G}_{0j}})}}_{\lesssim |w_0|^2_{S_0}} \Big],$$

where in the last step we have used Lemma 4.15. □

5.2.3 Main Result

The estimates of the P_D operator directly lead to the main theorem of this chapter.

Theorem 5.26. *Let $d = 3$ and let Ψ be defined according to Algorithm 5.6 (Algorithm B) or Algorithm 5.7 (Algorithm C), and let Assumption 5.19 (if Ω is bounded), Assumption 5.20 (if Ω is unbounded), and Assumption 5.21 hold. Then,*

$$\kappa(M_{\text{sD}}^{-1} F) \leq C \max_{i=1,\ldots,s} (1 + \log(H_i/h_i))^2,$$

where the constant $C > 0$ is independent of h_i, H_i, the values α_i, and the number of subdomains. For $0 \in \mathcal{I}$, C does not depend on H_0 or the number of neighbors of Ω_0 either. For $d = 2$ and $0 \notin \mathcal{I}$, the same bound holds as well for Algorithm 5.5 (Algorithm A).

Proof. The statement follows immediately from Lemmas 5.18, 5.22–5.25. □

Remark 5.27. It was shown in [Bre03b] that for bounded domains, the estimate in Theorem 5.26 is sharp in two dimensions. For sharp bounds of wirebasket and Neumann-Neumann preconditioners (related non-overlapping domain decomposition preconditioners) in three dimensions, see [BH03].

Remark 5.28. We see that in contrast to the one-level methods, the analysis for dual-primal methods with unbounded domains is much simpler and based on less restrictive assumptions, because the L^2-terms on the faces of the exterior domain Ω_0 can be eliminated locally using the continuity properties of the space $\widetilde{W}$. Apparently, the hidden coarse problems of the dual-primal methods (see Sect. 5.3) are more powerful than the coarse problems of the one-level methods. Nevertheless, this power is not completely for free and we will discuss the drawbacks in Sect. 5.3.5.2.

5.2.4 Extensions to Multiscale Coefficients

As the attentive reader suspects, Assumption 5.21 can be relaxed. Basically, we need to show the estimate (5.15) for the different definition of S_i, and here the theory from Sect. 3.3.5, p. 171 can be reused. Without loss of generality, assume that we are in the bounded case and that $\mathcal{I}_{\text{FEM}} = \mathcal{I}$. Unfortunately, the analysis of FETI-DP for multiscale coefficients turns out to be even more complicated than that of the one-level methods (cf. Chap. 3). Building on [PS08, PS09, PS11b, PSS12], we can summarize three cases.

Constant Coefficients in the Boundary Layers. The following theorem (cf. [PS09, Theorem 4.1]) is an analogue to Theorem 3.64 (p. 204). Note that a similar statement with slightly sharper estimates has been shown by Dryja and Sarkis, cf. [DS11].

Theorem 5.29. *Let Ω be bounded and let the assumptions of Theorem 3.64, p. 204 be fulfilled. In particular, for each $i \in \mathcal{I}$, the lower bound function $\underline{\alpha}_i$ is constant on the boundary layer $\Omega_{i,\eta}$. Assume further that the space $\widetilde{W}$ is constructed according to Algorithm 5.6 (Algorithm B). Then*

$$\kappa(M_{\text{sD}}^{-1} F) \leq C\, c_{\text{noise}} \max_{i=1,\ldots,s} \left(\frac{H_i}{\eta_i}\right)^2 (1 + \log(\tfrac{\eta_i}{h_i}))^2,$$

where c_{noise} is the constant from Assumption 3.14 (p. 165) and the constant C is independent of α, H_i, η_i, and h_i. In particular, C does not depend on the values of α in the subdomain interiors $\Omega_i \setminus \Omega_{i,\eta}$. If for each $i = 1, \ldots, s$,

$$\alpha(x) \geq \underline{\alpha}_{i|\Omega_{i,\eta}} \qquad \forall x \in \Omega_i,$$

then we have the improved bound

$$\kappa(M_{\text{sD}}^{-1} F) \leq C\, c_{\text{noise}} \max_{i=1,\ldots,s} \frac{H_i}{\eta_i} (1 + \log(\tfrac{H_i}{h_i}))\, (1 + \log(\tfrac{\eta_i}{h_i})),$$

where in two dimensions, the penultimate factor can be replaced by $(1 + \log(\frac{\eta_i}{h_i}))$.

Proof. By adapting the proof of Lemma 3.27, p. 172 with the same technique as in Lemma 5.22, we obtain that for any $w \in \widetilde{W}$,

$$|(P_D w)_i|_{S_i}^2 \leq C \max_{k \in \mathcal{N}_i} \left\| \frac{\overline{\alpha}_k}{\underline{\alpha}_k} \right\|_{L^\infty(\Omega_{k,\eta})} \sum_{j \in \mathcal{N}_i} \left[\left(1 + \log\left(\frac{\eta_i}{h_i}\right)\right)^2 |\mathcal{H}_j^{\alpha,h} w_j|_{H^1(\Omega_{j,\eta}),\underline{\alpha}_j}^2 \right.$$

$$\left. + \sum_{\mathcal{G} \subset \partial\Omega_j \cap \Gamma,\, d_{\mathcal{G}} \geq 1} \frac{1}{\eta_i^2} \left(1 + \log\left(\frac{\eta_i}{h_i}\right)\right) \|\mathcal{H}_j^{\alpha,h} w_j - \overline{w_j}^{\mathcal{G}}\|_{L^2(\Omega_{j,\eta}),\underline{\alpha}_j}^2 \right].$$

Let $\mathcal{G} = \mathcal{F} \subset \partial\Omega_i \cap \Gamma$ be a subdomain facet, then it follows from the proof of Theorem 3.64 and the fact that $\underline{\alpha}_j$ is constant that $C_{P,\underline{\alpha}_j}(\Omega_{i,\eta}, \mathcal{F})^2 \lesssim 1$ and so

$$\frac{1}{\eta_i^2} \|\mathcal{H}_j^{\alpha,h} w_j - \overline{w_j}^{\mathcal{F}}\|_{L^2(\Omega_{j,\eta}),\underline{\alpha}_j}^2 \leq \left(\frac{H_i}{\eta_i}\right)^2 |\mathcal{H}_j^{\alpha,h} w_j|_{H^1(\Omega_j),\underline{\alpha}_j}^2 .$$

If $d = 3$ and $\mathcal{G} = \mathcal{E} \subset \partial\Omega_i \cap \Gamma$ is a subdomain edge, then we need to adapt the technique from Lemma 3.57 (p. 196) and Example 3.61 (p. 198) to show that $C_{P,\underline{\alpha}_j}(\Omega_{i,\eta}, \mathcal{E})^2 \lesssim (1 + \log(\frac{\eta_i}{h_i}))$, and so

$$\frac{1}{\eta_i^2} \|\mathcal{H}_j^{\alpha,h} w_j - \overline{w_j}^{\mathcal{F}}\|_{L^2(\Omega_{j,\eta}),\underline{\alpha}_j}^2 \leq \left(\frac{H_i}{\eta_i}\right)^2 (1 + \log(\tfrac{\eta_i}{h_i}))\, |\mathcal{H}_j^{\alpha,h} w_j|_{H^1(\Omega_j),\underline{\alpha}_j}^2 .$$

Summarizing, this shows that for any $w \in \widetilde{W}$,

$$|(P_D w)_i|_{S_i}^2 \leq C\, c_{\text{noise}} \max_{i=1,\ldots,s} \left[(1 + \log(\tfrac{\eta_i}{h_i}))^2 + \left(\frac{H_i}{\eta_i}\right)^2 (1 + \log(\tfrac{\eta_i}{h_i}))^2 \right] |w|_S^2,$$

from which the first bound for $\kappa(M_{\text{sD}}^{-1} F)$ follows immediately by Lemma 5.18.

For the second case, where $\alpha(x) \geq \underline{\alpha}_{i|\Omega_{i,\eta}}$ for all $x \in \Omega_i$, we adapt the argumentation from the proofs of Theorem 3.64 (p. 204) and Theorem 3.73 (p. 211) and obtain that

$$\begin{aligned}
\frac{1}{\eta_j^2} \|\mathcal{H}_j^{\alpha,h} w_j - \overline{w_j}^{\mathcal{G}}\|_{L^2(\Omega_{j,\eta}),\underline{\alpha}_j}^2 &\leq \underline{\alpha}_j\, |\mathcal{H}_j^{\alpha,h} w_j|_{H^1(\Omega_{j,\eta})}^2 + \underline{\alpha}_j \frac{1}{\eta_j} \|w_j - \overline{w_j}^{\mathcal{G}}\|_{L^2(\partial\Omega_j)}^2 \\
&\lesssim \underline{\alpha}_j \left(1 + \frac{H_j}{\eta_j} \sigma^{d-d_{\mathcal{G}}}(\tfrac{H_j}{h_j})\right) |\mathcal{H}_j^{\alpha,h} w_j|_{H^1(\Omega_j)}^2 \\
&\lesssim \frac{H_j}{\eta_j} \sigma^{d-d_{\mathcal{G}}}(\tfrac{H_j}{h_j})\, |\mathcal{H}_j^{\alpha,h} w_j|_{H^1(\Omega_j),\alpha}^2,
\end{aligned}$$

which implies the second bound. □

The analogous bounds can be obtained for Algorithm 5.7 (Algorithm C), $d = 3$ and for Algorithm 5.5 (Algorithm A), $d = 2$ using the techniques from Lemmas 5.23 and 5.24. For an improved bound for the case that $\alpha(x) \leq \overline{\alpha_i}$ for all $x \in \Omega_i$ see [DS11]. For an extension of the above theory to linear elasticity see [GKR12], numerical examples can be found in [KR07c, KR07b].

Jumps Along Or Across Interfaces. The coefficient varies along the interface, but there are *no* jumps that occur both *along and across* subdomain interfaces. Here, it is proposed to use *weighted* edge and possibly also face averages as primal nodal variables, cf. [Doh03, KR07b]. A definition of the corresponding method together with some robustness theory can be found in [PSS12].

Jumps Along And Across Interfaces. The coefficient is allowed to jump *both along and across* subdomain interfaces. Here, it is yet an open problem how to choose the primal constraints and how to construct a proof of statements comparable to Theorem 3.73, p. 211.

5.2.5 An Estimate for the BDDC Preconditioner

The following lemma shows that the condition number bound of FETI/BETI-DP also provides as a bound for the condition number of BDDC (constructed from the same spaces, weights, etc.). Even more, it has been shown that the spectra of both methods coincide (except for the eigenvalue of one), cf. [MDT05, LW06b, LW07, BS07, MS07b].

Lemma 5.30. *Let* $\Psi = \{\psi_k\}_{k=1}^{n_\Pi}$ *be a linearly independent set of primal nodal variables that controls* $\ker(S)$. *Assume further that there exists a parameter* ω *such that* (5.15) *holds, i.e.,*

$$|P_D \widetilde{I}\, w|_S^2 \;\le\; \omega\, |\widetilde{I}\, w|_S^2 \qquad \forall w \in \widetilde{W}.$$

Then

$$\kappa(M_{\mathrm{BDDC}}^{-1}\, \widehat{S}) \;\le\; \omega.$$

Proof. Proofs can be found e.g. in [MDT05, MS07b]. Here, we make use of Theorem 2.46 (see also [MS07a, Theorem 2] and [Sou10, Theorem 2.5]) with

$$X := \widetilde{W}, \quad \widehat{X} := \widehat{W}, \quad \widehat{Q} := \widetilde{E}_D\,.$$

The theorem implies that

$$\kappa(M_{\mathrm{BDDC}}^{-1}\, \widehat{S}) \;\le\; \sup_{w \in \widetilde{W}} \frac{|\widetilde{E}_D\, w|_S^2}{|\widetilde{I}\, w|_S^2}\,.$$

Since the norm of a non-trivial projection in a Hilbert space depends only on the angle between its kernel and range [IM95], we get that

$$\sup_{w \in \widetilde{W}} \frac{|\widetilde{E}_D\, w|_S^2}{|\widetilde{I}\, w|_S^2} \;=\; \sup_{w \in \widetilde{W}} \frac{|P_D \widetilde{I}\, w|_S^2}{|\widetilde{I}\, w|_S^2}\,,$$

since $\widetilde{E}_D = I - P_D \widetilde{I}$. This concludes the proof. □

The following theorem summarizes the behavior of BDDC under appropriate assumptions.

Theorem 5.31. *Let $d = 3$ and let Ψ be defined according to Algorithm 5.6 (Algorithm B) or Algorithm 5.7 (Algorithm C), and let Assumption 5.19 (if Ω is bounded), Assumption 5.20 (if Ω is unbounded), and Assumption 5.21 hold. Then,*

$$\kappa(M_{\mathrm{BDDC}}^{-1}\,\widehat{S}) \;\leq\; C \max_{i=1,\dots,s} (1 + \log(H_i/h_i))^2,$$

where the constant $C > 0$ is independent of h_i, H_i, the values α_i, the number of subdomains. For $0 \in \mathcal{I}$, C does not depend on H_0 or the number of neighbors of Ω_0 either. For $d = 2$ and $0 \notin \mathcal{I}$, the same bound holds as well for Algorithm 5.5 (Algorithm A).

Proof. The statement follows from Lemmas 5.30, 5.22–5.25. □

Because of Lemma 5.30, the multiscale results from Sect. 5.2.4 carry over immediately to the BDDC preconditioner as well.

5.3 Implementation of FETI/BETI-DP Methods

To roughly summarize the building blocks of a FETI/BETI-DP method, we need

1. The application of S (locally on the subdomain),
2. The application of $\widetilde{B}$, B_D and their transposes,
3. The application of $\widetilde{S}^{-1}$ (which is the most complicated part).

Correspondingly, three different sets of vectors are needed that represent

(a) Functions in W (or their extensions),
(b) The Lagrange multipliers in U,
(c) Functions in $\widetilde{W}$ (or their extensions).

In a BDDC algorithm, we need

1. The application of $\widehat{S}$,
2. The application of $\widetilde{E}_D$ and its transpose,
3. The application of $\widetilde{S}^{-1}$.

Correspondingly, we need vectors representing

(a) Functions in $\widehat{W} \equiv V_D^h(\Gamma_S)$ (or their extensions),
(b) Functions in $\widetilde{W}$.

Depending on the implementation of $\widetilde{E}_D$, one might additionally need vectors representing functions in W. Note that the parallelization of the global space $\widehat{W}$ follows the standard concepts of a parallel finite element code, cf. e.g., [DHL03, Haa99, SBG96].

There are two key issues in the implementation of both FETI/BETI-DP and BDDC: firstly, how to identify the subdomain globs and how to implement the averages, and secondly, how to deal with the space $\widetilde{W}$ and the operator $\widetilde{S}^{-1}$. When using the algebraic averages from Example 5.2 (c), (d), the first issue can be reduced to finding out which *DOFs* belong to a certain subdomain glob. Analogously to Definition 2.48 (ii), we collect all coupling DOFs into equivalence classes. If an equivalence class is associated to a pair of (two) subdomain indices, it represents a subdomain face. If an equivalence class consists only of one DOF, it represents a subdomain vertex. All other classes must be subdomain edges. Note however, that for certain METIS partitions, this classification does not necessarily represent the geometric partition. Also, to control the kernel in the case of elasticity is a non-trivial task. For further reading we refer e.g. to [Les03, Doh03, KR06, ŠČBN12]

The second key issue is the parametrization of the space $\widetilde{W}$. In the following, we will explore direct splittings of the form

$$\widetilde{W} = W_\Pi \oplus \Big(\prod_{i \in \mathscr{I}} W_{\Delta,i} \Big), \tag{5.23}$$

where the *primal subspace* W_Π is a global "coarse" space whose dimension equals the number of global primal variables. Thus, we will parametrize functions $w \in \widetilde{W}$ by a (global) vector representing the component in W_Π as well as local vectors representing the components in $W_{\Delta,i}$. Section 5.3.1 below explores two such splittings together with suitable bases for the corresponding primal subspace W_Π. This leads to a parametrization of $\widetilde{W}$ and consequently of $\widetilde{W}^*$, which is discussed in detail in Sect. 5.3.2. This directly leads to the realization of the operators $\widetilde{I}$ and $\widetilde{I}^\top$, discussed in Sect. 5.3.3 below.

The basic idea to realize $\widetilde{S}^{-1}$ is to split $\widetilde{S}$ according to (5.23). Suppose that we have such a splitting in matrix form, i.e.,

$$\widetilde{\mathbf{S}} = \begin{bmatrix} \mathbf{S}_{\Pi\Pi} & \mathbf{S}_{\Pi\Delta} \\ \mathbf{S}_{\Delta\Pi} & \mathbf{S}_{\Delta\Delta} \end{bmatrix},$$

where $\mathbf{S}_{\Delta\Delta}$ is block-diagonal. Then we can realize the inverse of $\widetilde{\mathbf{S}}$ by a block Cholesky decomposition, i.e.,

$$\widetilde{\mathbf{S}}^{-1} = \begin{bmatrix} \mathbf{I} & 0 \\ -\mathbf{S}_{\Delta\Delta}^{-1}\mathbf{S}_{\Delta\Pi} & \mathbf{I} \end{bmatrix} \begin{bmatrix} \mathbf{T}_\Pi^{-1} & 0 \\ 0 & \mathbf{S}_{\Delta\Delta}^{-1} \end{bmatrix} \begin{bmatrix} \mathbf{I} & -\mathbf{S}_{\Pi\Delta}\mathbf{S}_{\Delta\Delta}^{-1} \\ 0 & \mathbf{I} \end{bmatrix},$$

with the Schur complement $\mathbf{T}_\Pi = \mathbf{S}_{\Pi\Pi} - \mathbf{S}_{\Pi\Delta}\mathbf{S}_{\Delta\Delta}^{-1}\mathbf{S}_{\Delta\Pi}$. This procedure is described in detail in Sect. 5.3.4.

5.3.1 Splitting Into Primal and Dual Subspaces

In this section, we discuss two (different) splittings of the form (5.23). The first one goes along merely with the original FETI-DP method, as proposed in [FLP00, FLL+01] and in the extension in [KWD02a], see also [TW05, Sect. 6.4]. The second splitting is motivated from the BDDC method, proposed in [Doh03], see also [MDT05].

5.3.1.1 A Continuous Primal Subspace

Lemma 5.32. *Let* $\Psi = \{\psi_k\}_{k=1}^{n_\Pi}$ *be a fixed set of linearly independent nodal variables. Then there exists a linearly independent set* $\widehat{\Phi} = \{\widehat{\phi}_k\}_{k=1}^{n_\Pi} \subset \widehat{W}$ *of continuous functions such that*

$$\psi_k(\widehat{\phi}_j) = \delta_{kj}\,. \tag{5.24}$$

Moreover, the subspace

$$\widehat{W}_\Pi := \operatorname{span}\{\widehat{\phi}_k\}_{k=1}^{n_\Pi} \subset \widehat{W} \subset \widetilde{W}, \tag{5.25}$$

is complementary to $\widetilde{W}_\Delta$*, i.e., we have the direct sum*

$$\widetilde{W} = \widehat{W}_\Pi \oplus \widetilde{W}_\Delta\,. \tag{5.26}$$

Due to property (5.24)*, we call* $\widehat{\Phi}$ *a* nodal *basis of* $\widehat{W}_\Pi$*.*

Proof. Recall that the primal nodal variables Ψ are linearly independent and that each of them is associated to a unique glob. This implies that there exist linearly independent functions $p_k \in V^h(\Gamma_S)$, $k = 1,\dots,n_\Pi$ such that the matrix $\mathbf{G} := [\psi_k(p_j)]_{k,j=1}^{n_\Pi}$ is regular (e.g., we can choose $p_k \leftrightarrow \mathbf{p}_k$, where $\mathbf{p}_k$ are the Ritz vectors such that $\psi_k(v) = (\mathbf{p}_k, \mathbf{v})_{\ell^2}$ for $\mathbf{v} \leftrightarrow v \in V_D^h(\Gamma_S)$; in that case $\mathbf{G}$ would even be SPD). Now, we set $\widehat{W}_\Pi := \operatorname{span}\{p_k\}_{k=1}^{n_\Pi}$ and observe that $\dim(\widehat{W}_\Pi) = n_\Pi$. By construction, $\widehat{W}_\Pi \cap \widetilde{W}_\Delta = \{0\}$, and $n_\Pi = \dim(\widehat{W}_\Pi)$ equals the codimension of $\widetilde{W}_\Delta$ with respect to $\widetilde{W}$. Hence, identity (5.26) must hold. Finally, applying $\mathbf{G}^{-1}$ to all the unit vectors in $\mathbb{R}^{n_\Pi}$ gives coefficients with respect to the basis $\{p_k\}_{k=1}^{n_\Pi}$ of $\widehat{W}_\Pi$ which define the nodal basis $\widehat{\Phi}$ with property (5.24). □

Remark 5.33. Note that there are many subspaces $\widehat{W}_\Pi \subset \widehat{W}$ that fulfill (5.26), and one can always find a *nodal* basis of $\widehat{W}_\Pi$, i.e., a basis with property (5.24). As we will see later on, property (5.24) has essential advantages in the implementation, and thus we will require a nodal basis in the sequel. However, the nodal basis as constructed in the proof of Lemma 5.32 does not necessarily have a local support property, which is highly undesirable in practice. In the following we describe how a nodal basis with local support can be constructed. Let the functions $\{p_k\}_{k=1}^{n_\Pi}$ in the proof above be selected such that $p_k(x^h) = 0$ for all $x^h \notin \overline{\mathscr{G}}$ if ψ_k is associated to $\mathscr{G}$. In the case of **G** being block-diagonal, with each block corresponding to a single glob with all its associated nodal variables, the local support of p_k carries over to the nodal basis. The block-diagonal property can be fulfilled for the interior averages from Example 5.2 (a), (c), (e). For example, for Algorithm B with interior averages, the outcome of this procedure (with the specific construction of p_k as in the proof of Lemma 5.32) is $\widehat{W}_\Pi = \operatorname{span}(\{\theta_{\mathscr{V}}\}_{\mathscr{V}\subset\Gamma} \cup \{\theta_{\mathscr{E}}\}_{\mathscr{E}\subset\Gamma} \cup \{\theta_{\mathscr{F}}\}_{\mathscr{F}\subset\Gamma})$, cf. [TW05, Algorithm 6.24] (in that case, **G** is even diagonal). If **G** is *not* block-diagonal, then still in many situations, a local support property can be achieved in the sense that

$$\widehat{\phi}_{k|\partial\Omega_j} = 0 \qquad \text{if } \psi_k \text{ is not associated to } \Omega_j\,.$$

5.3.1.2 An Energy Minimizing Primal Subspace

Lemma 5.34. *Let $\Psi = \{\psi_k\}_{k=1}^{n_\Pi}$ be a fixed set of linearly independent nodal variables that controls $\ker(S)$. Then there exists a distinguished complementary space $\widetilde{W}_\Pi^S \subset \widetilde{W}$ such that we have the S-orthogonal decomposition*

$$\widetilde{W} = \widetilde{W}_\Pi^S \oplus \widetilde{W}_\Delta, \qquad \langle S\, w_\Pi,\, w_\Delta\rangle = 0 \quad \forall w_\Pi \in \widetilde{W}_\Pi^S,\ w_\Delta \in \widetilde{W}_\Delta\,.$$

Proof. Since $\widetilde{W}_\Delta \subset \widetilde{W}$ and since $\widetilde{S}$ is SPD, the orthogonal complement $\widetilde{W}_\Pi^S := \widetilde{W}_\Delta^{\perp_{\widetilde{S}}}$ is unique and well-defined. □

In the following, we describe how to construct a *nodal* basis for $\widetilde{W}_\Pi^S$ with *local* support by solving a series of local subdomain problems, cf. [Doh03]. For each subdomain Ω_i, let

$$n_{\Pi,i} := \#\{k = 1, \ldots, n_\Pi : \psi_k \text{ associated to } \Omega_i\} \tag{5.27}$$

denote the number of primal nodal variables associated to Ω_i and let us agree on a local numbering such that $k(i,j)$ is the global index of the j-th local primal nodal variable on Ω_i. For each $i \in \mathscr{I}$, let $C_i : W_i \to \mathbb{R}^{n_{\Pi,i}}$ be given by

$$(C_i\, v)_\ell = \psi_{k(i,\ell)}(v) \qquad \forall v \in W_i,\ \forall \ell = 1, \ldots, n_{\Pi,i}, \tag{5.28}$$

and let the unit vector $e_{i,j} \in \mathbb{R}^{n_{\Pi_i}}$ be defined by $e_{i,j,\ell} = \delta_{j,\ell}$. For each $i \in \mathscr{I}$ and $j = 1, \ldots, n_{\Pi,i}$ we compute local basis functions $\{\phi_{i,j}\}_{j=1}^{n_{\Pi,i}}$ from the solution of the following saddle point problem,

$$\text{find } (\widetilde{\phi}_{i,j}, \widetilde{\mu}_{i,j}) \in W_i \times \mathbb{R}^{n_{\Pi,i}} : \quad \begin{bmatrix} S_i & C_i^\top \\ C_i & 0 \end{bmatrix} \begin{bmatrix} \widetilde{\phi}_{i,j} \\ \widetilde{\mu}_{i,j} \end{bmatrix} = \begin{bmatrix} 0 \\ e_{i,j} \end{bmatrix}. \tag{5.29}$$

Apparently, $\ker(C_i) = W_{\Delta,i}$. If Ψ controls $\ker(S)$ then $\ker(S_i) \cap \ker(C_i) = \{0\}$. Furthermore, if Ψ is linearly independent (cf. Definition 5.1 (iv)), then $\ker(C_i^\top) = \{0\}$. Therefore, under the assumptions of Lemma 5.34, problem (5.29) is uniquely solvable (cf. Lemma 1.16). Finally, the global basis functions $\widetilde{\phi}_k \in W$, $k = 1, \dots, n_\Pi$ are given by

$$(\widetilde{\phi}_k)_i = \begin{cases} \widetilde{\phi}_{i,j} & \text{if there exists } j \text{ with } k = k(i,j), \\ 0 & \text{else,} \end{cases} \tag{5.30}$$

and they have local support by construction. From Definition 5.3, (5.28), and from the second line in (5.29) we can conclude that indeed $\widetilde{\phi}_k \in \widetilde{W}$. The first line of (5.29) implies that for any $w_i \in W_{\Delta,i}$,

$$\langle S_i \widetilde{\phi}_{i,j}, w_i \rangle = -\langle C_i^\top \widetilde{\phi}_{i,j}, w_i \rangle = -\langle \underbrace{C_i w_i}_{=0}, \widetilde{\phi}_{i,j} \rangle = 0.$$

This shows that $\operatorname{span}\{\widetilde{\phi}_k\}_{k=1}^{n_\Pi}$ and $\widetilde{W}_\Delta$ are S-orthogonal. Since the dimensions of the first and the second space add up the dimension of $\widetilde{W}$, we can conclude that $\widetilde{W}_\Pi^S = \operatorname{span}\{\widetilde{\phi}_k\}_{k=1}^{n_\Pi}$. Finally, from the definitions of $e_{i,j}$, C_i, from the local support property, and from the second line of (5.29), it follows that $\psi_{k(i,j)}(\widetilde{\phi}_{i,\ell}) = \delta_{j\ell}$, and thus implies that the basis is nodal, i.e.,

$$\psi_k(\widetilde{\phi}_\ell) = \delta_{k\ell}\,.$$

We see that the basis $\{\widetilde{\phi}_k\}_{k=1}^{n_\Pi}$ has the flavor of a standard FE basis.

Remark 5.35. In practice, system (5.29) is of course expanded. If $i \in \mathscr{I}_{\text{FEM}}$, we solve

$$\begin{bmatrix} A_i & \overline{C}_i^\top \\ \overline{C}_i & 0 \end{bmatrix} \begin{bmatrix} \overline{\phi}_{i,j} \\ \widetilde{\mu}_{i,j} \end{bmatrix} = \begin{bmatrix} 0 \\ e_{i,j} \end{bmatrix},$$

where $A_i : V_D^h(\Omega_i) \to V_D^h(\Omega_i)^*$ denotes the "stiffness" operator, and $\overline{C}_i$ is the extension from W_i to $V_D^h(\Omega_i)$, see also Remark 2.17, p. 79. If the local dimension $\dim(V_D^h(\Omega_i))$ is not very large, the system matrix can be built and its LU-factorization stored, cf. Sect. 1.2.4.1. For $i \in \mathscr{I}_{\text{BEM}}$, we set

$$A_i := \begin{bmatrix} D_i & \frac{1}{2}I + K_i^\top \\ \frac{1}{2}I + K_i & -V_i \end{bmatrix},$$

see also Remark 2.18. The resulting system for $\overline{\phi}_{i,j}$ is then a two-fold saddle point problem. For small system size, a dense LU-factorization can be computed, and otherwise an $\mathscr{H}$-LU-factorization, cf. Sect. 1.3.7.3.

Remark 5.36. In the case of $i \in \mathscr{I}_{\text{FEM}}$, solving the saddle point system can be reduced to factorizing a small dense matrix and a sparse SPD matrix. This is done by fixing suitable rows/columns in A_i, such that after reordering these special rows/columns last, the upper left block is SPD, cf. [Doh03, Sect. 2.1] and [KW06, Sect. 4.2.2].

5.3.2 Parametrization of $\widetilde{W}$ and $\widetilde{W}^*$

Let W_Π denote either $\widehat{W}_\Pi$ from Sect. 5.3.1.1 or $\widetilde{W}_\Pi^S$ from Sect. 5.3.1.2, and let $\{\phi_k\}_{k=1}^{n_\Pi}$ denote the respective primal *nodal* basis functions ($\widehat{\phi}_k$ or $\widetilde{\phi}_k$ from above). Furthermore, we require that the basis has local support in the sense that

$$\phi_{k|\Omega_j} = 0 \qquad \text{if } \psi_k \text{ is not associated to } \Omega_j\,. \tag{5.31}$$

This property is always fulfilled for the S-orthogonal choice (cf. (5.30)), whereas for the other choice, it may only be fulfilled in special cases (cf. Remark 5.33).

As sketched in the beginning of Sect. 5.3, we will parametrize a function $w \in \widetilde{W}$ by its primal component in W_Π (represented by a vector with respect to the primal basis), and the dual component in $\widetilde{W}_\Delta$. For the dual component, we can

(i) Construct a basis as well (by algebraic means), see e.g. [KR06, KW06, LW06b, Rhe06, Jar10], or
(ii) Use local constraints, see e.g. [KW06], and also [Doh03, MDT05].

Here, we follow the second approach. The first approach is often called *change of basis* or *transformation of basis*.

5.3.2.1 The Primal and Dual Component of a Function

First, we investigate local splittings. Let $W_{\Pi,i}$ denote the restriction of W_Π to subdomain $\partial\Omega_i$. Note that for both choices of W_Π above, we have the direct sum

$$W_i \;=\; W_{\Pi,i} \oplus W_{\Delta,i}\,.$$

Thanks to our assumption (5.31), the space $W_{\Pi,i}$ is spanned by the local basis functions

$$\phi_{i,j} \;:=\; \phi_{k(i,j)|\partial\Omega_i} \in W_i \qquad \text{for } j = 1,\ldots,n_{\Pi,i}\,,$$

where $k(i,j)$ is the global index corresponding to the j-th local primal node (assuming a local numbering). Thus, we can expand any local function $w_{\Pi,i} \in W_{\Pi,i}$ in that basis. Let $\mathbf{w}_{\Pi,i} \in \mathbb{R}^{n_{\Pi,i}}$ denote the corresponding vector of coefficients and let $\Phi_i : \mathbb{R}^{n_{\Pi,i}} \to W_i$ be given by $\Phi_i = [\phi_{i,1}|\cdots|\phi_{i,n_{\Pi,i}}]$, such that

$$w_{\Pi,i} \;=\; \Phi_i\,\mathbf{w}_{\Pi,i}\,, \qquad \mathbf{w}_{\Pi,i} \;=\; C_i\,w_{\Pi,i}\,.$$

The second identity follows from the fact that

$$C_i\Phi_i \;=\; \mathbf{I}, \tag{5.32}$$

which holds because the basis is nodal.

Analogously, we can expand any global function $w_\Pi \in W_\Pi$ in the global basis $\{\phi_k\}_{k=1}^{n_\Pi}$ resulting in the coefficient vector $\mathbf{w}_\Pi = [\psi_k(w_\Pi)]_{k=1}^{n_\Pi}$. The global and local coefficient vectors are of course related: let $\mathbf{A}_i^\top \in \mathbb{R}^{n_{\Pi,i} \times n_\Pi}$ be the Boolean restriction matrix such that

$$\mathbf{w}_{\Pi,i} = \mathbf{A}_i^\top \mathbf{w}_\Pi .$$

Here, we have chosen the letter A, because $\mathbf{A}_i$ is an *assembling* operator (see below). With this notation, we can summarize: *the global primal vector* $\mathbf{w}_\Pi \in \mathbb{R}^{n_\Pi}$ *and its associated function* $w \in W_\Pi$ *are related by*

$$w = \mathbf{\Phi}\,\mathbf{A}^\top \mathbf{w}_\Pi := [\Phi_i\, \mathbf{A}_i^\top \mathbf{w}_\Pi]_{i\in\mathscr{I}} \in W, \tag{5.33}$$

where the block operators $\mathbf{\Phi}$ and $\mathbf{A}$ have to be defined accordingly. Replacing the local basis functions $\phi_{i,j}$ in Φ by their vector representations with respect to a chosen basis of W_i (e.g., the standard nodal basis), we get a transformation rule which implements the natural embedding $E_\Pi : W_\Pi \to W$ in the respective bases.

Having the primal component defined, the dual component is fixed because the sum $W_i = W_{\Pi,i} \oplus W_{\Delta,i}$ is direct. To summarize,

$$w_{\Pi,i} = \Phi_i\, C_i\, w_i\,, \qquad w_{\Delta,i} = w_i - w_{\Pi,i}\,. \tag{5.34}$$

Identity (5.32) implies that indeed $w_{\Delta,i} \in W_{\Delta,i}$. Due to the same identity, it follows that $\Phi_i\, C_i : W_i \to W_{\Pi,i}$ and $I - \Phi_i\, C_i : W_i \to W_{\Delta,i}$ are projections (using Sect. 1.1.3.3 one can show that they are $\Phi_i^\top \Phi_i$-orthogonal).

5.3.2.2 The Primal Component of a Functional

Next, we investigate the *adjoint* $E_\Pi^\top : W^* \to W_\Pi^*$ of the embedding $W_\Pi \subset W$, i.e., for a given functional $f \in W$ we want to extract its primal component as an element of W_Π^*. The restriction $f_\Pi \in W_\Pi^*$ is simply given by the relation

$$\langle f_\Pi, w_\Pi \rangle = \langle f, w_\Pi \rangle \qquad \text{for } w_\Pi \in W_\Pi\,.$$

Thus, the vector $\mathbf{f}_\Pi \in \mathbb{R}^{n_\Pi}$ representing f_Π is given by $(\mathbf{f}_\Pi)_k = \langle f, \phi_k \rangle$. By using formula (5.33), we obtain

$$\mathbf{f}_\Pi = \mathbf{A}\mathbf{\Phi}^\top f,$$

i.e., *the primal component of a functional* $f \in W^*$ *is given by assembling its subdomain contributions,*

$$\mathbf{f}_\Pi = \sum_{i\in\mathscr{I}} \mathbf{A}_i\, \mathbf{f}_{\Pi,i}\,,$$

where $\mathbf{f}_{\Pi,i} = \Phi_i^\top f_i$ or equivalently $\mathbf{f}_{\Pi,i,j} = \langle f_i, \phi_{i,j} \rangle_{\partial\Omega_i}$. Again, by replacing the local basis functions $\phi_{i,j}$ in Φ by their vector representations with respect to a chosen basis of W_i (e.g., the standard nodal basis), we obtain a transformation rule for functionals, implementing the adjoint $E_\Pi^\top : W^* \to W_\Pi^*$ in the respective bases.

5.3.2.3 Splitting Functionals

Let $f_i \in W_i$ be an arbitrary local functional. Motivated from Sect. 5.3.2.2, we define the *local primal component* $f_{\Pi,i} \in W_i^*$ by

$$\langle f_{\Pi,i}, w_i \rangle = \langle f_i, \Phi_i C_i w_i \rangle \qquad \text{for } w_i \in W_i,$$

and the *dual component* by $f_{\Delta,i} := f_i - f_{\Pi,i}$. Equivalently,

$$f_{\Pi,i} = C_i^\top \Phi_i^\top f_i, \tag{5.35}$$

$$f_{\Delta,i} = (I - C_i^\top \Phi_i^\top) f_i = f_i - \sum_{j=1}^{n_{\Pi,i}} \langle f_i, \phi_{i,j} \rangle \psi_{i,j}, \tag{5.36}$$

where $\psi_{i,j}$ denotes the restriction of $\psi_{k(i,j)}$ to W_i^*. Obviously, $C_i^\top \Phi^\top$ and $I - C_i^\top \Phi_i^\top$ are projections and $f_{\Delta,i} \in \operatorname{range}(I - C_i^\top \Phi_i^\top) \subset W_i^*$. In addition to the (unique) additive splitting $f_i = f_{\Pi,i} + f_{\Delta,i}$, one can show that $f_{\Delta,i}$ is orthogonal to $W_{\Pi,i}$ in the duality product, and $f_{\Pi,i}$ to $W_{\Delta,i}$. From Lemma 1.11 we find that $\operatorname{range}(I - C_i^\top \Phi_i^\top)$ is a realization of $W_{\Delta,i}^*$, such that $f_{\Delta,i}$ defined as above is a good representative of the restriction of f_i to $W_{\Delta,i}^*$. Since $\widetilde{W}_\Delta$ is a product space, we can simply set

$$w_\Delta := (I - \mathbf{\Phi}\, C) w \qquad \text{for } w \in \widetilde{W},$$

$$f_\Delta := (I - C^\top \mathbf{\Phi}^\top) f \qquad \text{for } f \in \widetilde{W}^*,$$

where $C = \operatorname{diag}(C_i)_{i \in \mathcal{I}} : W \to \prod_{i \in \mathcal{I}} \mathbb{R}^{n_{\Pi,i}}$.

5.3.3 *Realization of $\widetilde{I}$ and $\widetilde{I}^\top$*

For each $i \in \mathcal{I}$, let $\{\varphi_{i,j}\}_{j=1}^{\dim(W_i)}$ be a given basis for W_i, e.g., the standard nodal basis. To summarize Sect. 5.3.2, a function $w \in \widetilde{W}$ is represented by

- A global vector $\mathbf{w}_\Pi \in \mathbb{R}^{n_\Pi}$ (with respect to the primal basis $\{\phi_k\}_{k=1}^{n_\Pi}$),
- Local functions $w_{\Delta,i} \in W_i$ (represented in the basis $\varphi_{i,j}$) that fulfill the *local constraints* $C_i\, w_{\Delta,i} = 0$,

whereas a function $w \in W$ is simply given by its components $w_i \in W_i$ (represented in the basis $\varphi_{i,j}$). Accordingly, a functional $f \in \widetilde{W}$ is represented by

- A global vector $\mathbf{f}_\Pi \in \mathbb{R}^{n_\Pi}$,
- Local functions $f_{\Delta,i} \in \operatorname{range}(I - C_i^\top \Phi_i^\top)$ (represented in the dual basis of $\varphi_{i,j}$),

whereas a functional $f \in W$ is simply given by its components $f_i \in W_i^*$ (represented in the dual basis of $\varphi_{i,j}$).

5.3.3.1 Realization of $\widetilde{I}$

The discussion in Sect. 5.3.2 reveals that the embedding operator $\widetilde{I} : \widetilde{W} \to W$ is realized by

$$\begin{bmatrix} \mathbf{w}_\Pi \\ [w_{\Delta,i}]_{i\in\mathscr{I}} \end{bmatrix} \mapsto w = \mathbf{\Phi}\,\mathbf{A}^\top \mathbf{w}_\Pi + (I - \mathbf{\Phi}\,C)w_\Delta, \tag{5.37}$$

i.e., firstly the primal DOFs are *distributed* to the subdomains, and secondly the sum of the primal and dual component is formed locally. Note that the operation $-\Phi\, C$ above may be left out, since $w_\Delta \in \ker(C)$. To get the fully vector-oriented transformation rule, we only need to replace the basis functions $\phi_{i,j}$ in Φ_i by their vector representations with respect to the given basis $\varphi_{i,j}$ of W_i, and to replace C by its matrix representation with respect to the same basis.

5.3.3.2 Realization of $\widetilde{I}^\top$

From Sects. 5.3.2.2 and 5.3.2.3 we find that $\widetilde{I}^\top : W^* \to \widetilde{W}^*$ is realized by

$$f \mapsto \begin{bmatrix} \mathbf{f}_\Pi \\ [f_{\Delta_i}]_{i\in\mathscr{I}} \end{bmatrix} = \begin{bmatrix} \mathbf{A}\,\mathbf{\Phi}^\top f \\ (I - C^\top \mathbf{\Phi}^\top) f \end{bmatrix}. \tag{5.38}$$

Comparing with (5.37) highlights the adjointness.

5.3.4 *Realization of $\widetilde{S}^{-1}$*

Let W_Π and $\widetilde{W}_\Delta$ be as above and let $S_{\Pi\Pi}$, $S_{\Delta\Delta}$, $S_{\Pi\Delta}$, and $S_{\Delta\Pi} = S_{\Pi\Delta}^\top$ denote the respective restrictions of S or, equivalently, of $\widetilde{S}$, i.e.,

$$\begin{aligned} \langle S_{\Pi\Pi} v_\Pi, w_\Pi\rangle &= \langle S\, v_\Pi, w_\Pi\rangle && \text{for } v_\Pi, w_\Pi \in W_\Pi, \\ \langle S_{\Pi\Delta} v_\Pi, w_\Delta\rangle &= \langle S\, v_\Pi, w_\Delta\rangle && \text{for } v_\Pi \in W_\Pi,\ w_\Delta \in \widetilde{W}_\Delta, \\ \langle S_{\Delta\Delta} v_\Delta, w_\Delta\rangle &= \langle S\, v_\Delta, w_\Delta\rangle && \text{for } v_\Delta, w_\Delta \in \widetilde{W}_\Delta\,. \end{aligned}$$

A formal block Cholesky decomposition yields

$$\widetilde{S}^{-1} = \begin{bmatrix} S_{\Pi\Pi} & S_{\Pi\Delta} \\ S_{\Delta\Pi} & S_{\Delta\Delta} \end{bmatrix}^{-1} = \begin{bmatrix} I_\Pi & 0 \\ -S_{\Delta\Delta}^{-1} S_{\Delta\Pi} & I_\Delta \end{bmatrix} \begin{bmatrix} T_\Pi^{-1} & 0 \\ 0 & S_{\Delta\Delta}^{-1} \end{bmatrix} \begin{bmatrix} I_\Pi & -S_{\Pi\Delta} S_{\Delta\Delta}^{-1} \\ 0 & I_\Delta \end{bmatrix}, \tag{5.39}$$

where I_Π and I_Δ are the identity operators on W_Π and $\widetilde{W}_\Delta$, respectively, and T_Π is the Schur complement operator given by

$$T_\Pi \;=\; S_{\Pi\Pi} - S_{\Pi\Delta} S_{\Delta\Delta}^{-1} S_{\Delta\Pi}\,.$$

Identity (5.39) will serve as an algorithm to apply $\widetilde{S}^{-1}$, and thus we go through the realization of the operators appearing in the block factorization.

Let us note beforehand that for the particular choice $W_\Pi = \widetilde{W}_\Pi^S$, with the basis that is S-orthogonal to $\widetilde{W}_\Delta$, we have $S_{\Delta\Pi} = 0$ and $S_{\Pi\Delta} = 0$, and so

$$\widetilde{S}^{-1} \;=\; \begin{bmatrix} S_{\Pi\Pi}^{-1} & 0 \\ 0 & S_{\Delta\Delta}^{-1} \end{bmatrix},$$

i.e., the primal and the dual problem can be solved independently.

5.3.4.1 Realization of $S_{\Delta\Delta}^{-1}$

Due to the structure of the space $\widetilde{W}_\Delta$, the operators $S_{\Delta\Delta}$ are localized,

$$S_{\Delta\Delta} \;=\; \operatorname{diag}(S_{\Delta\Delta,i})_{i\in\mathscr{I}},$$

such their inverses can be realized locally (and thus in parallel). As in Sect. 2.2.5.4, the application of $S_{\Delta\Delta,i}$ corresponds to solving a Neumann problem in subdomain Ω_i, however with the local primal constraints. It is not hard to see that for any functional $f_i \in W_{\Delta,i}^*$, the function $v_i = S_{\Delta\Delta,i}^{-1} f_i$ is given as the solution of

$$\begin{bmatrix} S_i & C_i^\top \\ C_i & 0 \end{bmatrix} \begin{bmatrix} v_i \\ \mu_i \end{bmatrix} = \begin{bmatrix} f_i \\ 0 \end{bmatrix}. \tag{5.40}$$

As the discussion after Lemma 5.34 shows, system (5.40) always has a unique solution. The same techniques as discussed in Remarks 5.35 and 5.36 of course apply also to the above system.

5.3.4.2 Realization of $S_{\Pi\Pi}$

Let $\mathbf{S}_{\Pi\Pi,i} \in \mathbb{R}^{n_{\Pi,i}\times n_{\Pi,i}}$ be the primal subdomain matrix with respect to the chosen basis of $W_{\Pi,i}$, given by

$$(\mathbf{S}_{\Pi\Pi,i})_{j\ell} \;=\; \langle S_i\,\phi_{i,j},\, \phi_{i,\ell}\rangle_{\partial\Omega_i},$$

i.e., we basically need to apply S_i to the local basis functions. Thanks to (5.31), the global primal matrix $\mathbf{S}_{\Pi\Pi} \in \mathbb{R}^{n_\Pi}$ defined by $(\mathbf{S}_{\Pi\Pi})_{k\ell} = \langle S\phi_k,\, \phi_\ell\rangle$ is simply given by the subassembly of the subdomain contributions,

$$\mathbf{S}_{\Pi\Pi} = \mathbf{A}\,\mathrm{diag}(\mathbf{S}_{\Pi\Pi,i})_{i\in\mathscr{I}}\,\mathbf{A}^\top .$$

For the basis constructed in (5.29) (but not for that from Lemma 5.32), the computation of the primal subdomain matrices can be simplified because

$$(\mathbf{S}_{\Pi\Pi,i})_{j\ell} = -\widetilde{\mu}_{i,j,\ell},$$

where $\widetilde{\mu}_{i,j}$ are the Lagrange parameters from (5.29).

5.3.4.3 Realization of $S_{\Pi\Delta}$

For a function $v_\Delta \in W_\Delta$, the vector $\mathbf{f}_\Pi$ parametrizing $f_\Pi := S_{\Pi\Delta}v_\Delta$ is simply given by the primal component of Sv_Δ, i.e.,

$$\mathbf{f}_\Pi = \mathbf{A}\,\boldsymbol{\Phi}^\top S\,v_\Delta .$$

Note that the application of $\boldsymbol{\Phi}^\top S$ to v_Δ is performed locally (the local primal component of $S_i v_{\Delta,i}$ is given by $\Phi_i^\top S_i v_{\Delta,i}$).

5.3.4.4 Realization of $S_{\Delta\Pi}$

For a vector $\mathbf{w}_\Pi$ parametrizing $w_\Pi \in W_\Pi$, the functional $g_\Delta := S_{\Delta\Pi}w_\Pi \in W_\Delta^*$ is given by the dual component of $S\,w_\Pi$, i.e.,

$$g_\Delta = (I - C^\top\boldsymbol{\Phi}^\top)S\,\boldsymbol{\Phi}\,\mathbf{A}^\top\mathbf{w}_\Pi .$$

Note that after distributing the entries of $\mathbf{w}_\Pi$ (by applying $\mathbf{A}^\top$), all the operations are local (the dual component of $S_i\,w_{\Pi,i}$ is given by $(I - C_i^\top\Phi_i^\top)S_i\,\Phi_i\,\mathbf{w}_{\Pi,i}$).

5.3.4.5 Realization of T_Π^{-1}

Let the operator $T_\Pi : W_\Pi \to W_\Pi^*$ be represented by a matrix $\mathbf{T}_\Pi$. For the choice $W_\Pi = \widetilde{W}_\Pi^S$ with the S-orthogonal basis, $\mathbf{T}_\Pi = \mathbf{S}_{\Pi\Pi}$. Otherwise, the matrix $\mathbf{T}_\Pi$ is assembled from subdomain contributions,

$$\mathbf{T}_\Pi = \mathbf{A}\,\mathrm{diag}(\mathbf{T}_{\Pi,i})_{i\in\mathscr{I}}\,\mathbf{A}^\top,$$

where for each $i \in \mathscr{I}$, the subdomain matrix $\mathbf{T}_{\Pi,i}$ represents the operator

$$T_{\Pi,i} := S_{\Pi\Pi,i} - S_{\Pi\Delta,i}\,S_{\Delta\Delta,i}^{-1}\,S_{\Delta\Pi,i},$$

whose application should be clear from the above discussion. The columns of $\mathbf{T}_{\Pi,i}$ are given by the vectors $\Phi_i^\top T_{\Pi,i}\,\Phi_i\,\mathbf{e}_j$, where $\mathbf{e}_j \in \mathbb{R}^{n_{\Pi,i}}$ are the unit vectors.

Algorithm 10: For given $(\mathbf{f}_\Pi, f_\Delta) \in \mathbb{R}^{n_\Pi} \times W_\Delta^*$, parametrizing $f \in \widetilde{W}^*$, compute the parametrization $(\mathbf{w}_\Pi, w_\Delta) \in \mathbb{R}^{n_\Pi} \times W_\Delta$ of $w = \widetilde{S}^{-1} f \in \widetilde{W}$

$v_{\Delta,i} = S_{\Delta\Delta,i}^{-1} f_{\Delta,i}$ for all $i \in \mathscr{I}$, via Sect. 5.3.4.1
$\mathbf{g}_{\Pi,i} = \Phi_i^\top S_i v_{\Delta,i}$ for all $i \in \mathscr{I}$
$\mathbf{g}_\Pi = \mathbf{A}\,[\mathbf{g}_{\Pi,i}]_{i\in\mathscr{I}}$
$\mathbf{w}_\Pi = \mathbf{T}_\Pi^{-1}\,(\mathbf{f}_\Pi - \mathbf{g}_\Pi)$
$[\mathbf{u}_{\Pi,i}]_{i\in\mathscr{I}} = \mathbf{A}^\top \mathbf{w}_\Pi$
$\mathbf{h}_{\Delta,i} = (I - C_i^\top \Phi_i^\top) S\, \Phi_i\, \mathbf{u}_{\Pi,i}$ for all $i \in \mathscr{I}$
$w_{\Delta,i} = v_{\Delta,i} - S_{\Delta\Delta,i}^{-1} h_{\Delta,i}$ for all $i \in \mathscr{I}$, via Sect. 5.3.4.1

Algorithm 11: Simplification of Algorithm 10 for the choice of the S-orthogonal basis from Sect. 5.3.1.2

$w_{\Delta,i} = S_{\Delta\Delta,i}^{-1} f_{\Delta,i}$ for all $i \in \mathscr{I}$, via Sect. 5.3.4.1
$\mathbf{w}_\Pi = \mathbf{S}_{\Pi\Pi}^{-1}\mathbf{f}_\Pi$

Once, the global primal matrix $\mathbf{T}_\Pi$ has been assembled, its Cholesky (or LU) factorization can be built and stored. Applying the inverses of the factors in turn realizes T_Π^{-1}. Note that the self-adjointness of S implies the symmetry of $\mathbf{T}_\Pi$.

5.3.4.6 Realization of $\widetilde{S}^{-1}$

For a given functional $f \in \widetilde{W}^*$ the function $w = \widetilde{S}^{-1} f$ is computed according to (5.39). Algorithm 10 displays this procedure in detail. As it can be seen, we need to solve two subdomain problems (involving $S_{\Delta\Delta,i}$) and one "coarse" problem (involving $\mathbf{T}_\Pi^{-1}$). For the special choice of the S-orthogonal basis form Sect. 5.3.1.2, the procedure simplifies to Algorithm 11, where only one subdomain problem needs to be solved.

5.3.5 Comparison of Dual-Primal and One-Level Methods

In this section, we briefly compare dual-primal and one-level methods, first for the case of a bounded computational domain, then for the case of an unbounded one.

5.3.5.1 The Bounded Case

A fine comparison of FETI-DP and the standard one-level FETI methods can be found in [TW05, Sect. 6.4.1]. In the following, we compare the dual primal methods to the one-level methods, including the all-floating methods.

1. Dual-primal algorithms do not require the knowledge of the kernels of the Steklov-Poincaré or subdomain operators in practice. All-floating methods need these kernels, but at least for the operators without the (possibly complicated) Dirichlet boundary conditions incorporated, cf. [DHK06, Of06, OS09].
2. In three dimensions, dual-primal methods require the handling of edge and possibly face constraints, but as Sect. 5.3 shows, this does not lead to complications in the implementation. However, in order to be competitive with the one-level methods in view of the size of the coarse problem, dual-primal methods require a careful choice of the primal DOFs, see Remark 5.8.
3. The coarse matrix in the dual-primal methods is often sparser than that of one-level methods: in the dual-primal methods, the primal degrees of freedom couple to their neighbors, whereas in the one-level methods, the coarse degrees of freedom on a subdomain couple to the neighbors of the neighbors of the subdomain.
4. The same dual-primal algorithms and codes can be applied to problems with and without non-trivial local kernels, e.g., differential operators with zero-order terms. In such cases, one-level methods need a non-trivial adaption, cf. Sect. 2.8.2.1 (p. 150).
5. Dual-primal methods do not require the construction of a scaling operator Q which appears in the coarse problem of one-level methods. The dual-primal coarse problem is determined just from the local operators and the primal nodal variables.
6. If the variational problem to be solved is part of a transient (dynamic) or of a nonlinear (Newton) iteration, then in the dual-primal methods one can set the initial guess for λ in the PCG algorithm to the solution of λ from the previous time or Newton step. Such a strategy is not in general possible for the one-level methods, because the condition $G^\top \lambda_0 = e$ has to be fulfilled, cf. Sect. 2.2.1.4. See also [FCR94] for an acceleration for dynamic problems that is applicable to both one-level FETI and FETI-DP.
7. The known theory of all-floating FETI method for multiscale coefficients is (at the moment) richer than that for dual-primal methods. The later case is still under investigation, cf. [DS11, PSS12, GKR12].

5.3.5.2 The Unbounded Case

From Lemma 5.25, we see that the primal ("coarse") space of dual-primal methods for unbounded domains is more powerful than that of the one-level methods, cf. Remark 5.28. However, this goes along with a possibly high computational cost. If the number of primal DOFs $n_{\Pi,0}$ on $\partial\Omega_0$ stays small, the algorithm will work fine. If $n_{\Pi,0}$ grows, however, we encounter two problems.

(a) The assembling of the local primal matrix $\mathbf{S}_{\Pi\Pi,0}$ involves at least $n_{\Pi,0}$ solves of the single layer potential on $\partial\Omega_0$.

(b) The global primal matrix $\mathbf{S}_{\Pi\Pi}$ (or $\mathbf{T}_{\Pi}$) is not sparse anymore if Ω_0 has a large number of neighbors. At high probability, this sparsity problem can be solved using an $\mathscr{H}$-matrix approximation of $\mathbf{S}_{\Pi\Pi}$ (or $\mathbf{T}_{\Pi}$).

However, these two problem are not present at all in case of one-level methods, since the Steklov-Poincaré operator S_0 does not at all contribute to the coarse matrix $G^\top Q\, G$. When being parallelized, both the one-level and the dual-primal methods can suffer from the lack of load balancing if the number of DOFs on $\partial\Omega_0$ is large. The buffer strategy outlined in Sect. 4.4 can be used to cope with this problem. To summarize, the one-level method can be used well, unless the predicted condition number due to Theorems 4.21 or 4.30 is getting very large (either because of a relatively small coefficient α_0 or a relatively small shape parameter η). In such cases, the dual-primal method might be of greater advantage.

5.4 Inexact Solvers, Multilevel Extensions, and Deflation

There are several ways to incorporate inexact solvers within the FETI-DP or BDDC framework.

- On how to incorporate inexact subdomain solvers within FETI-DP or BDDC see, e.g., [Doh07, LW07, KR07a].
- For inexact solvers for the coarse problem, see [KR07a] as well as the three- and multilevel extensions in, e.g., [Tu07c, Tu07b, Tu11, MSD08, Sou10, ŠMSB12]. For a multilevel BDDC solver library see also http://www.math.cas.cz/~sistek/software/bddcml.html. We note that incorporating a third level in BDDC is more natural than in FETI-DP since the coarse solve of BDDC appears in the *preconditioner*, opposed to FETI DP where it appears in the dual equation.

For a hybrid one-level / dual-primal FETI method see [KR10].

Finally, we note that deflation (cf. [FCM95, FM98] and Sect. 2.8.2.1) can also be used for dual-primal and BDDC methods [Jar10, JKR12, KR12]. This creates a second coarse problem which can speed up the convergence.

References

[AB05] G. Allaire and R. Brizzi. A multiscale finite element method for numerical homogenization. *Multiscale Model. Simul.*, 4(3):790–812, 2005.

[AB06] T. Arbogast and K.J. Boyd. Subgrid upscaling and mixed multiscale finite elements. *SIAM J. Numer. Anal.*, 44:1150–1171, 2006.

[ABDP81] R. E. Alcouffe, Achi Brandt, J. E. Dendy, Jr., and J. W. Painter. The multi-grid method for the diffusion equation with strongly discontinuous coefficients. *SIAM J. Sci. Comput.*, 2(4):430–454, 1981.

[AF03] R. A. Adams and J. J. F. Fournier. *Sobolev Spaces*, volume 140 of *Pure and Applied Mathematics*. Academic Press, Amsterdam, Boston, second edition, 2003.

[AG02] M. Ainsworth and B. Guo. Analysis of iterative sub-structuring techniques for boundary element approximation of the hypersingular operator in three dimensions. *Appl. Anal.*, 81(2):241–280, 2002.

[AGKS08] B. Aksoylu, I. G. Graham, H. Klie, and R. Scheichl. Towards a rigorously justified algebraic preconditioner for high-contrast diffusion problems. *Comput. Visual Sci.*, 11(4-6):319–331, 2008.

[AS97] M. Ainsworth and B. Senior. Aspects of an adaptive hp-finite element method: adaptive strategy, conforming approximation and efficient solvers. *Comput. Methods Appl. Mech. Engrg.*, 150(1-4):65–87, 1997.

[Bas96] P. Bastian. *Parallele Adaptive Mehrgitterverfahren*. Teubner Skripten zur Numerik. B. G. Teubner, Stuttgart, 1996.

[BB08] S. Boerm and J. Bendoraityte. Distributed $\mathscr{H}^2$-matrices for non-local operators. *Comput. Visual Sci.*, 11(4-6):237–249, 2008.

[BBC$^+$06] Y. Bazilevs, L. Beirão da Veiga, J. A. Cottrell, T. J. R. Hughes, and G. Sangalli. Isogeometric analysis: approximation, stability and error estimates for h-refined meshes. *Math. Models Methods Appl. Sci.*, 16(7):1031–1090, 2006.

[BBK$^+$09] D. Balzani, D. Brands, A. Klawonn, O. Rheinbach, and J. Schröder. On the mechanical modeling of anisotropic biological soft tissue and iterative parallel solution strategies. *Archive of Applied Mechanics*, 80(5):479–488, 2009.

[BCLP10] L. Beirão da Veiga, C. Chinosi, C. Lovadina, and L. F. Pavarino. Robust BDDC preconditioners for Reissner - Mindlin plate bending problems and MITC elements. *SIAM J. Numer. Anal.*, 47(6):4214–4238, 2010.

[BCPS12] L. Beirão da Veiga, D. Cho, L. F. Pavarino, and S. Scacchi. BDDC preconditioners for isogeometric analysis. *Math. Models Methods Appl. Sci.*, 2012. accepted for publication.

[BDF$^+$00] M. Bhardwaj, D. Day, C. Farhat, M. Lesoinne, K. Pierson, and D. Rixen. Application of the FETI method to ASCI problems—scalability results on 1000 processors and discussion of highly heterogeneous problems. *Int J. Numer. Meth. Engng.*, 47:513–535, 2000.

C. Pechstein, *Finite and Boundary Element Tearing and Interconnecting Solvers for Multiscale Problems*, Lecture Notes in Computational Science and Engineering 90,
DOI 10.1007/978-3-642-23588-7,

[BDS08] J. Bouchala, Z. Dostál, and M. Sadowská. Theoretically supported scalable BETI method for variational inequalities. *Computing*, 82(1):53–75, 2008.

[BDV97] P. E. Bjørstad, M. Dryja, and E. Vainikko. Additive schwarz methods without subdomain overlap and with new coarse spaces. In R. Glowinski, Z.-C. Shi J. Périaux, and O. Widlund, editors, *Domain Decomposition Methods in Sciences and Engineering (Beijing, 1995)*, pages 141–157. John Wiley & Sons, Chichester, 1997. http://www.ddm.org/DD08/

[Beb00] M. Bebendorf. *Effiziente numerische Lösung von Randintegralgleichungen unter Verwendung von Niedrigrang-Matrizen.* PhD thesis, Universität Saarbrücken, 2000.

[Beb03] M. Bebendorf. A note on the Poincaré inequality for convex domains. *Z. Anal. Anwendungen*, 22(4):751–756, 2003.

[Beb05] M. Bebendorf. Hierarchical LU decomposition based preconditioners for BEM. *Computing*, 74(3):225–247, 2005.

[Beb08] M. Bebendorf. *Hierarchical Matrices – A Means to Efficiently Solve Elliptic Boundary Value Problems*, volume 63 of *Lecture Notes in Computational Science and Engineering*. Springer-Verlag, Berlin, 2008.

[BEL08] S. Beuchler, T. Eibner, and U. Langer. Primal and dual interface concentrated iterative substructuring methods. *SIAM J. Numer. Anal.*, 46(6):2818–2842, 2008.

[BF91] F. Brezzi and M. Fortin. *Mixed and Hybrid Finite Element Methods.* Springer Series in Computational Mathematics. Springer-Verlag, New York, 1991.

[BG86] I. Babuska and B. Q. Guo. The h-p version of the finite element method, part 1: the basic approximation results. *Comp. Mech.*, 1:22–41, 1986.

[BGH06] S. Börm, L. Grasedyck, and W. Hackbusch. Hierarchical matrices. Lecture Note 21/2003, Max-Planck-Institut für Mathematik und Naturwissenschaften, Inselstr. 22, 04103 Leipzig, Germany, 2006. revised version, http://www.mis.mpg.de/preprints/ln/lecturenote-2103.pdf

[BGLV89] J.-F. Bourgat, R. Glowinski, P. Le Tallec, and M. Vidrascu. Variational formulation and algorithm for trace operator in domain decomposition calculations. In T. Chan, R. Glowinski, J. Périaux, and O. Widlund, editors, *Domain Decomposition Methods. Second International Symposium on Domain Decomposition Methods*, pages 3–16. SIAM, Philadelphia, PA, 1989. http://www.ddm.org/DD02/

[BH70] J. H. Bramble and S. R. Hilbert. Estimation of linear functionals on Sobolev spaces with application to Fourier transforms and spline interpolation. *SIAM J. Numer. Anal.*, 7:112–124, 1970.

[BH03] S. C. Brenner and Q. He. Lower bounds for three-dimensional nonoverlapping domain decomposition algorithms. *Numer. Math.*, 93(3):445–470, 2003.

[BHK08] C. Bertoglio, W. Hackbusch, and B. N. Khoromskij. Low rank tensor-product approximation of projected green kernels via sinc-quadratures. Preprint 79/2008, MPI MIS, Leipzig, 2008.

[BJ79] F. Brezzi and C. Johnson. On the coupling of boundary integral and finite element methods. *Calcolo*, 16:189–201, 1979.

[BJN78] F. Brezzi, C. Johnson, and J. C. Nédélec. On the coupling of boundary integral and finite element methods. In *Proceedings of the Fourth Symposium on Basic Problems of Numerical Mathematics (Pilzen, 1978)*, pages 103–114. Charles University, Prague, 1978.

[BK95] S. Buckley and P. Koskela. Sobolev-Poincaré implies John. *Math. Res. Lett.*, 2(5):577–593, 1995.

[BK05] M. Bebendorf and R. Kriemann. Fast parallel solution of boundary integral equations and related problems. *Comput. Visual Sci.*, 8(3-4):121–135, 2005.

[BKRS08] D. Brands, A. Klawonn, O. Rheinbach, and J. Schrder. Modeling and convergence in arterial wall simulations using a parallel FETI solution strategy. *Computer Methods in Biomechanics and Biomedical Engineering*, 11(5):569–583, 2008.

[BM97] C. Bernardi and Y. Maday. Spectral methods. In P. G. Ciarlet and J.-L. Lions, editors, *Handbook of Numerical Analysis, Volume V: Techniques of Scientific Computing (Part 2)*. North-Holland, 1997.

[BMP94] C. Bernardi, Y. Maday, and A. T. Patera. A nonconforming approach to domain decomposition: The mortar element method. In H. Brezis and J.-L. Lions, editors, *College de France Seminar*, pages 13–51. Longman Scientific and Technical Harlow, UK, 1994.

[Boj88] B. Bojarski. Remarks on Sobolev imbedding inequalities. In *Complex analysis, Joensuu 1987*, volume 1351 of *Lecture Notes in Math.*, pages 52–68. Springer-Verlag, Berlin, 1988.

[BP88] J. H. Bramble and J. E. Pasciak. A preconditioning technique for indefinite systems resulting from mixed approximations of elliptic problems. *Mathematics of Computation*, 50(181):1–17, 1988.

[BPS86] J. H. Bramble, J. E. Pasciak, and A. H. Schatz. The construction of preconditioners for elliptic problems by substructuring, I. *Math. Comp.*, 47(175):103–134, 1986.

[BPS87] J. H. Bramble, J. E. Pasciak, and A. H. Schatz. The construction of preconditioners for elliptic problems by substructuring, II. *Math. Comp.*, 49(179):1–16, 1987.

[BPS88] J. H. Bramble, J. E. Pasciak, and A. H. Schatz. The construction of preconditioners for elliptic problems by substructuring, III. *Math. Comp.*, 51(184):415–430, 1988.

[BPS89] J. H. Bramble, J. E. Pasciak, and A. H. Schatz. The construction of preconditioners for elliptic problems by substructuring, IV. *Math. Comp.*, 53(187):1–24, 1989.

[Bra01] D. Braess. *Finite elements. Theory, fast solvers, and applications in solid mechanics.* Cambridge University Press, Cambridge, 2001.

[Bre00] F. Brezzi. Interacting with the subgrid world. In D.F. Griffiths and G.A. Watson, editors, *Numerical Analysis 1999*, pages 69–82. Chapman & Hall/CRC, Boca Raton, 2000.

[Bre02] S. C. Brenner. A new look at FETI. In N. Debit, M. Garbey, R. Hoppe, J. Périaux, D. Keyes, and Y. Kuznetsov, editors, *Domain decomposition methods in science and engineering (Lyon, 2000)*, Theory Eng. Appl. Comput. Methods, pages 41–51. Internat. Center Numer. Methods Eng. (CIMNE), Barcelona, 2002. http://www.ddm.org/DD13/

[Bre03a] S. C. Brenner. An additive Schwarz preconditioner for the FETI method. *Numerische Mathematik*, 94(1):1–31, 2003.

[Bre03b] S. C. Brenner. Analysis of two-dimensional FETI-DP preconditioners by the standard additive Schwarz framework. *Electron. Trans. Numer. Anal.*, 16:165–185, 2003.

[BS87] I. Babuška and M. Suri. The h-p version of the finite element method with quasi-uniform meshes. *RAIRO Modél. Math. Anal. Numér.*, 21(2):199–238, 1987.

[BS00] S. C. Brenner and L.-Y. Sung. Discrete Sobolev and Poincaré inequalities via Fourier series. *East-West J. Numer. Math.*, 8(2):83–92, 2000.

[BS02] S. C. Brenner and L. R. Scott. *The mathematical theory of finite element methods*, volume 15 of *Texts in Applied Mathematics*. Springer-Verlag, New York, second edition, 2002.

[BS07] S. C. Brenner and L. Sung. BDDC and FETI-DP without matrices or vectors. *Comput. Methods Appl. Mech. Engrg.*, 8:1429–1435, 2007.

[BSEH11] M. J. Borden, M. A. Scott, J. A. Evans, and T. J. R. Hughes. Isogeometric finite element data structures based on Bézier extraction of NURBS. *Internat. J. Numer. Methods Engrg.*, 87(1-5):15–47, 2011.

[BX91] J. H. Bramble and J. Xu. Some estimates for a weighted L^2 projection. *Math. Comp.*, 56(194):463–476, 1991.

[BZ00] J. H. Bramble and X. Zhang. The analysis of multigrid methods. In P. G. Ciarlet and J. Lions, editors, *Handbook of Numerical Analysis*, volume VII, pages 173–415. North-Holland, Amsterdam, 2000.

[CCC+03] W. C. Chew, H. Y. Chao, T. J. Cui, C. C. Lu, S. Ohnuki, Y. C. Pan, J. M. Song, S. Velamparambil, and J. S. Zhao. Fast integral equation solvers in computational electromagnetics of complex structures. *Engineering Analysis with Boundary Elements*, 27(8):803–823, 2003.

[CDNQ12] T. Cluzeau, V. Dolean, F. Nataf, and A. Quadrat. Symbolic preconditioning techniques for linear systems of partial differential equations. In R. Bank, R. Kornhuber, and O. Widlund, editors, *Domain Decomposition Methods in Science and Engineering XX*, Lecture Notes in Computational Science and Engineering. Springer-Verlag, 2012. to appear; http://hal.archives-ouvertes.fr/hal-00664092

[CDS03] X.-C. Cai, M. Dryja, and M. Sarkis. Restricted additive Schwarz preconditioners with harmonic overlap for symmetric positive definite linear systems. *SIAM J. Numer. Anal.*, 41(4):1209–1231, 2003.

[CFH+03] T. Chartier, R. D. Falgout, V. E. Henson, J. Jones, T. Manteuffel, S. McCormick, J. Ruge, and P. S. Vassilevski. Spectral amge (ρamge). *SIAM J. Sci. Comput*, 25(1):1–26, 2003.

[CGH10] C.-C. Chu, I. G. Graham, and T.-Y. Hou. A new multiscale finite element method for high-contrast elliptic interface problems. *Math. Comp.*, 79:1915–1955, 2010.

[CGR99] H. Cheng, L. Greengard, and V. Rokhlin. A fast adaptive multipole algorithm in three dimensions. *J. Comput. Phys.*, 155(2):468–498, 1999.

[CGSS00] K. A. Cliffe, I. G. Graham, R. Scheichl, and L. Stals. Parallel computation of flow in heterogeneous media modelled by mixed finite elements. *J. Comput. Phys.*, 164(2):258–282, 2000.

[CH02] Z. Chen and T.Y. Hou. A mixed multiscale finite element method for elliptic problems with oscillating coefficients. *Math. Comp.*, 72:541–576, 2002.

[CHB09] J. A. Cottrell, T. J. R. Hughes, and Y. Bazilevs. *Isogeometric Analysis: Towards Integration of CAD and FEA*. John Wiley & Sons, 2009.

[Chu93] S.-K. Chua. Weighted Sobolev inequalities on domains satisfying the chain condition. *Proc. Amer. Math. Soc.*, 117(2):449–457, 1993.

[Cia87] P. G. Ciarlet. *The finite element method for elliptic problems*, volume 4 of *Studies in Mathematics and its Applications*. North-Holland, Amsterdam, 1987.

[CKL98] C. Carstensen, M. Kuhn, and U. Langer. Fast parallel solvers for symmetric boundary element domain decomposition equations. *Numer. Math.*, 79:321–347, 1998.

[Clé75] P. Clément. Approximation by finite element functions using local regularization. *RAIRO Analyse Numérique*, 9(R-2):77–84, 1975.

[CM94] T. F. Chan and T. P. Mathew. Domain decomposition algorithms. *Acta Numerica*, pages 61–143, 1994.

[Cos87] M. Costabel. Symmetric methods for the coupling of finite elements and boundary elements. In C. A. Brebbia, W. L. Wendland, and G. Kuhn, editors, *Boundary Elements IX*, pages 411–420. Springer-Verlag, Berlin, 1987.

[Cos88] M. Costabel. Boundary integral operators on Lipschitz domains: elementary results. *SIAM J. Math. Anal.*, 19(3):613–626, 1988.

[Cos07] M. Costabel. Some historical remarks on the positivity of boundary integral operators. In M. Schanz and O. Steinbach, editors, *Boundary Element Analysis – Mathematical Aspects and Applications*, volume 29 of *Lecture Notes in Applied and Computational Mechanics*, pages 1–27. Springer-Verlag, Berlin, 2007.

[Cro03] J. M. Cros. A preconditioner for the Schur complement domain decomposition method. In I. Herrera, D. E. Keyes, and O. B. Widlund, editors, *Domain Decomposition Methods in Science and Engineering*. National Autonomous University of Mexico (UNAM), México, 2003. http://www.ddm.org/DD14/

[CS90] M. Costabel and E. P. Stephan. Coupling of finite and boundary element methods for an elastoplastic interface problem. *SIAM J. Numer. Anal.*, 27(5):1212–1226, 1990.

[CW06] S.-K. Chua and R. L. Wheeden. Estimates of best constants for weighted Poincaré inequalities on convex domains. *Proc. London Math. Soc. (3)*, 93(1):197–226, 2006.

[Dah77] B. E. J. Dahlberg. Estimates of harmonic measure. *Arch. Rational Mech. Anal.*, 65(3):275–288, 1977.

[Dau88] M. Dauge. *Elliptic boundary value problems on corner domains*. Springer-Verlag, New York, 1988.

[De 91] Y.-H. De Roeck. *Résolution sur Ordinateurs Multi-Processeurs de Problème d'Elasticité par Décomposition des Domaines*. PhD thesis, Université Paris IX Dauphine, 1991.

[DE12] D. A. Di Pietro and A. Ern. *Mathematical Aspects of Discontinuous Galerkin Methods*, volume 69 of *Mathématiques & Applications (Berlin) [Mathematics & Applications]*. Springer, Heidelberg, 2012.

[Dem07] L. Demkowicz. *Computing with hp-adaptive finite elements. Vol. 1.* Chapman & Hall/CRC Applied Mathematics and Nonlinear Science Series. Chapman & Hall/CRC, Boca Raton, FL, 2007.

[DGS07] M. Dryja, J. Galvis, and M. Sarkis. BDDC methods for discontinuous Galerkin discretization of elliptic problems. *J. Complexity*, 23(4-6):715–739, 2007.

[DGS08] M. Dryja, J. Galvis, and M. Sarkis. Balancing domain decomposition methods for discontinuous Galerkin discretization. In *Domain decomposition methods in science and engineering XVII*, volume 60 of *Lecture Notes in Computational Science and Engineering*, pages 271–278. Springer, Berlin, 2008.

[DGS11] M. Dryja, J. Galvis, and M. Sarkis. N-N solvers for a DG discretization for geometrically nonconforming substructures and discontinuous coefficients. In Y. Huang, R. Kornhuber, O. Widlund, and J. Xu, editors, *Domain Decomposition Methods in Science and Engineering XIX*, volume 78 of *Lecture Notes in Computational Science and Engineering*, pages 27–38. Springer-Verlag, Berlin, 2011.

[DGS12] M. Dryja, J. Galvis, and M. Sarkis. Neumann-Neumann methods for a DG discretization of elliptic problems with discontinuous coefficients on geometrically nonconforming substructures. *Numerical Methods for Partial Differential Equations*, 28(4):1194–1226, 2012.

[DHK+05] Z. Dostál, D. Horák, R. Kučera, V. Vondrák, J. Haslinger, J. Dobiáš, and S. Pták. FETI based algorithms for contact problems: scalability, large displacements and 3D Coulomb friction. *Comput. Methods Appl. Mech. Engrg.*, 194(2–5):395–409, 2005.

[DHK06] Z. Dostál, D. Horák, and R. Kučera. Total FETI – An easier implementable variant of the FETI method for numerical solution of elliptic PDE. *Commun. Numer. Methods Eng.*, 22(12):1155–1162, 2006.

[DHL03] C. C. Douglas, G. Haase, and U. Langer. *A Tutorial on Elliptic PDE Solvers and Their Parallelization.* SIAM, Philadelphia, 2003.

[DKV+10] Z. Dostál, T. Kozubek, V. Vondrák, T. Brzobohatý, and A. Markopoulos. Scalable TFETI algorithm for the solution of multibody contact problems of elasticity. *Internat. J. Numer. Methods Engrg.*, 82(11):1384–1405, 2010.

[DKW08a] C. R. Dohrmann, A. Klawonn, and O. B. Widlund. Domain decomposition for less regular subdomains: Overlapping Schwarz in two dimensions. *SIAM J. Numer. Anal.*, 46(4):2153–2168, 2008.

[DKW08b] C. R. Dohrmann, A. Klawonn, and O. B. Widlund. Extending theory for domain decomposition algorithms to irregular subdomains. In U. Langer, M. Discacciati, O. Widlund, and W. Zulehner, editors, *Domain Decomposition Methods in Science and Engineering XVII*, volume 60 of *Lecture Notes in Computational Science and Engineering*, pages 255–261. Springer-Verlag, Berlin, 2008.

[DL91] Y.-H. De Roeck and P. Le Tallec. Analysis and test of a local domain decomposition preconditioner. In R. Glowinski, Y. A. Kuznetsov, G. A. Meurant, J. Périaux, and O. Widlund, editors, *Fourth International Symposium on Domain Decomposition Methods for Partial Differential Equations*, pages 112–128. SIAM, Philadelphia, PA, 1991. http://www.ddm.org/DD04/

[DNR08] V. Dolean, F. Nataf, and G. Rapin. How to use the Smith factorization for domain decomposition methods applied to the Stokes equation. In *Domain decomposition methods in science and engineering XVII*, volume 60 of *Lecture Notes in Computational Science and Engineering*, pages 477–848. Springer, Berlin, 2008.

[DNR09] V. Dolean, F. Nataf, and G. Rapin. Deriving a new domain decomposition method for the Stokes equation using the Smith factorization. *Math. Comp.*, 78:789–814, 2009.

[DNSS11] V. Dolean, F. Nataf, R. Scheichl, and N. Spillane. Analysis of a two-level schwarz method with coarse spaces based on local Dirichlet-to-Neumann maps. Preprint HAL-00586246, Hyper Articles en Ligne, 2011. submitted.

[Doh03] C. R. Dohrmann. A preconditioner for substructuring based on constrained energy minimization. *SIAM J. Sci. Comput.*, 25(1):246–258, 2003.

[Doh07] C. R. Dohrmann. An approximate BDDC preconditioner. *Numer. Linear Algebra Appl.*, 14(2):149–168, 2007.

[DORH89] L. Demkowicz, J. T. Oden, W. Rachowicz, and O. Hardy. Toward a universal h-p adaptive finite element strategy. I. Constrained approximation and data structure. *Comput. Methods Appl. Mech. Engrg.*, 77(1-2):79–112, 1989.

[Dry87] M. Dryja. A method of domain decomposition for 3-D finite element problems. In R. Glowinski, G. H. Golub, G. A. Meurant, and J. Périaux, editors, *First International Symposium on Domain Decomposition methods for Partial Differential Equations*, pages 43–61. SIAM, Philadelphia, PA, 1987. http://www.ddm.org/DD01/

[DS80] T. Dupont and R. Scott. Polynomial approximation of functions in Sobolev spaces. *Math. Comp.*, 34(150):441–463, 1980.

[DS11] M. Dryja and M. Sarkis. Technical tools for boundary layers and applications to heterogeneous coefficients. In Y. Huang, R. Kornhuber, O. Widlund, and J. Xu, editors, *Domain Decomposition Methods in Science and Engineering XIX*, volume 78 of *Lecture Notes in Computational Science and Engineering*, pages 205–212. Springer-Verlag, Berlin, 2011.

[DSW94] M. Dryja, B. F. Smith, and O. B. Widlund. Schwarz analysis of iterative substructuring algorithms for elliptic problems in three dimensions. *SIAM J. Numer. Anal.*, 31(6):1662–1694, 1994.

[DSW96] M. Dryja, M. V. Sarkis, and O. B. Widlund. Multilevel Schwarz methods for elliptic problems with discontinuous coefficients in three dimensions. *Numer. Math.*, 72:313–348, 1996.

[DW94] M. Dryja and O. B. Widlund. Domain decomposition algorithms with small overlap. *SIAM J. Sci. Comput.*, 15(3):604–620, 1994.

[DW95] M. Dryja and O. B. Widlund. Schwarz methods of Neumann-Neumann type for three-dimensional elliptic finite element problems. *Comm. Pure Appl. Math.*, 48(2):121–155, 1995.

[DW06] L. T. Dechevski and W. L. Wendland. On the Bramble-Hilbert lemma. II. An improved estimate of the embedding constant. *Int. J. Pure Appl. Math.*, 33(4):433–464, 2006.

[DW09] C. R. Dohrmann and O. B. Widlund. An overlapping Schwarz algorithm for almost incompressible elasticity. *SIAM J. Numer. Anal.*, 47(4):2897–2923, 2009.

[DW10] C. R. Dohrmann and O. B. Widlund. Hybrid domain decomposition algorithms for compressible and almost incompressible elasticity. *Internat. J. Numer. Methods Engrg.*, 82:157–183, 2010.

[DW12a] C. R. Dohrmann and O. B. Widlund. An iterative substructuring algorithm for two-dimensional problems in H(curl). *SIAM J. Numer. Anal.*, 50(3):1004–1028, 2012.

[DW12b] C. R. Dohrmann and O. B. Widlund. Some recent tools and a BDDC algorithm for 3D problems in H(curl). In R. Bank, R. Kornhuber, and O. Widlund, editors, *Domain Decomposition Methods in Science and Engineering XX*, Lecture Notes in Computational Science and Engineering. Springer-Verlag, 2012. to appear; http://www.cs.nyu.edu/cs/faculty/widlund/Dohrmann-plenary-final.pdf

[EGLW12] Y. Efendiev, J. Galvis, R. Lazarov, and J. Willems. Robust domain decomposition preconditioners for abstract symmetric positive definite bilinear forms. *ESAIM Math. Model. Numer. Anal.*, 46(5):1175–1199, 2012.

[EH09] Y. Efendiev and T.Y. Hou. *Multiscale Finite Element Methods: Theory and Applications*. Springer-Verlag, New York, 2009.

[EHN96] H. W. Engl, M. Hanke, and A. Neubauer. *Regularization of inverse problems*, volume 375 of *Mathematics and its Applications*. Kluwer Academic Publishers Group, Dordrecht, 1996.

[Eib06] T. Eibner. *Adaptive und randkonzentrierte FEM*. PhD thesis, TU Chemnitz, Chemnitz, Germany, June 2006.

[EM06] T. Eibner and J. M. Melenk. A local error analysis of the boundary concentrated FEM. *IMA J. Numer. Anal.*, 26(4):752–778, 2006.

[EM07a] T. Eibner and J. M. Melenk. An adaptive strategy for hp-FEM based on testing for analyticity. *Comput. Meth.*, 39(5):575–595, 2007.

[EM07b] T. Eibner and J. M. Melenk. Multilevel preconditioning for the boundary concentrated hp-FEM. *Comp. Methods Appl. Mech. Engrg.*, 196(37-40):3713–3725, 2007.

[Eva98] L. C. Evans. *Partial Differential Equations*, volume 19 of *Graduate Studies in Mathematics*. AMS, Providence, RI, 1998.

[FATL05] C. Farhat, P. Avery, R. Tezaur, and J. Li. FETI-DPH: a dual-primal domain decomposition method for acoustic scattering. *J. Comput. Acoust.*, 13(3):499–524, 2005.

[FCM95] C. Farhat, P. Chen, and J. Mandel. A scalable Lagrange multiplier based domain decomposition method for time-dependent problems. *Int. J. Numer. Meth. Engng.*, 38(22):3831–3853, 1995.

[FCMR98] C. Farhat, P. Chen, J. Mandel, and F.-X. Roux. The two-level FETI method part II: Extensions to shell problems, parallel implementation and performance results. *Comput. Methods Appl. Mech. Engrg.*, 155(1–2):153–179, 1998.

[FCR94] C. Farhat, L. Crivelli, and F.-X. Roux. Extending substructure based iterative solvers to multiple load and repeated analyses. *Comput. Meth. Appl. Mech. Engng.*, 117:195–200, 1994.

[FCRR98] C. Farhat, P.-S. Chen, F. Risler, and F.-X. Roux. A unified framework for accelerating the convergence of iterative substructuring methods with Lagrange multipliers. *Int. J. Numer. Meth. Engng.*, 42(2):257–288, 1998.

[FF60] H. Federer and W. H. Fleming. Normal and integral currents. *Ann. of Math.*, 2:482–520, 1960.

[Fie73] M. Fiedler. Algebraic connectivity of graphs. *Czechoslovak Math. J.*, 23(98):298–305, 1973.

[FKS82] E. B. Fabes, C. E. Kenig, and R. P. Serapioni. The local regulartity of solutions of degenerate elliptic equations. *Comm. Partial Differential Equations*, 7(1):77–116, 1982.

[FLL$^+$01] C. Farhat, M. Lesoinne, P. Le Tallec, K. Pierson, and D. Rixen. FETI-DP: A dual-primal unified FETI method I: A faster alternative to the two-level FETI method. *Internat. J. Numer. Methods Engrg.*, 50(7):1523–1544, 2001.

[FLLA05] C. Farhat, J. Li, M. Lesoinne, and P. Avery. A FETI method for a class of indefinite or complex second- or fourth-order problems. In R. Kornhuber, R. H. W. Hoppe, D. E. Keyes, J. Periaux, O. Pironneau, and J. Xu, editors, *Domain Decomposition Methods in Science and Engineering*, volume 40 of *Lecture Notes in Computational Science and Engineering*, pages 19–33. Springer-Verlag, Berlin, 2005. http://www.ddm.org/DD15/

[FLP00] C. Farhat, M. Lesoinne, and K. Pierson. A scalable dual-primal domain decomposition method. *Numer. Linear Algebra Appl.*, 7(7–8):687–714, 2000.

[FM97] C. Farhat and J. Mandel. Scalable substructuring by lagrange multipliers in theory and practice. In P. Bjørstad, M. Espedal, and D. Keyes, editors, *Ninth International Conference: Domain Decomposition Methods in Science and Engineering*, 1997. http://www.ddm.org/DD9/Farhat.pdf

[FM98] C. Farhat and J. Mandel. The two-level FETI method for static and dynamic plate problems part I: An optimal iterative solver for biharmonic systems. *Comput. Methods Appl. Mech. Engrg.*, 155(1–2):129–151, 1998.

[FML00a] C. Farhat, A. Macedo, and M. Lesoinne. A two-level domain decomposition method for the iterative solution of high-frequency exterior Helmholtz problems. *Numer. Math.*, 85(2):283–303, 2000.

[FML$^+$00b] C. Farhat, A. Macedo, M. Lesoinne, F.-X. Roux, F. Magoulès, and A. de La Bourdonnaye. Two-level domain decomposition methods with Lagrange multipliers for the fast iterative solution of acoustic scattering problems. *Comput. Methods Appl. Mech. Engrg.*, 184:213–240, 2000.

[FMR94] C. Farhat, J. Mandel, and F.-X. Roux. Optimal convergence properties of the FETI domain decomposition method. *Comput. Methods Appl. Mech. Engrg.*, 115:365–385, 1994.

[FMT99] C. Farhat, A. Macedo, and R. Tezaur. FETI-H: A scalable domain decomposition method for high frequency exterior Helmholtz problems. In C.-H. Lai, P. E. Bjørstad, M. Cross, and O. B. Widlund, editors, *Domain Decomposition Methods in Science and Engineering: Eleventh International Conference London, UK*, pages 231–241, 1999. http://www.ddm.org/DD11/Farhat.pdf

[FP03] Y. Fragakis and M. Papadrakakis. The mosaic of high performance domain decomposition methods for structural methanics: Formulation, interrelation and numerical efficiency of primal and dual methods. *Comput. Methods Appl. Mech. Engrg.*, 192:3799–3830, 2003.

[FP04] Y. Fragakis and M. Papadrakakis. The mosaic of high performance domain decomposition methods for structural methanics – Part II: Formulation enhancements, multiple right-hand sides and implicit dynamics. *Comput. Methods Appl. Mech. Engrg.*, 193:4611–4662, 2004.

[FR91] C. Farhat and F.-X. Roux. A method of finite element tearing and interconnecting and its parallel solution algorithm. *Int. J. Numer. Meth. Engrg.*, 32(6):1205–1227, 1991.

[FR92] C. Farhat and F.-X. Roux. An unconventional domain decomposition method for an efficient parallel solution of large-scale finite element systems. *SIAM J. Sci. and Stat. Comput.*, 13(1):379–396, 1992.

[FR94] C. Farhat and F.-X. Roux. Implicit parallel processing in structural mechanics. In J. T. Oden, editor, *Computational Mechanics Advances*, volume 2, pages 1–124. North-Holland, Amsterdam, 1994.

[FS97] S. A. Funken and E. P. Stephan. The BPX preconditioner for the single layer potential operator. *Appl. Anal.*, 67(3–4):327–340, 1997.

[GE10a] J. Galvis and Y. Efendiev. Domain decomposition preconditioners for multiscale flows in high contrast media. *Multiscale Model. Simul.*, 8(4):1461–1483, 2010.

[GE10b] J. Galvis and Y. Efendiev. Domain decomposition preconditioners for multiscale flows in high contrast media: reduced dimension coarse spaces. *Multiscale Model. Simul.*, 8(5):1621–1644, 2010.

[GH99] I. G. Graham and M. J. Hagger. Unstructured additive Schwarz-conjugate gradient method for elliptic problems with highly discontinuous coefficients. *SIAM J. Sci. Comput.*, 20(6):2041–2066, 1999.

[GHLS12] I. G. Graham, T. Y. Hou, O. Lakkis, and R. Scheichl (eds.). *Numerical analysis of multiscale problems. Selected papers based on the presentations at the 91st London Mathematical Society symposium, Durham, UK, July 5–15, 2010*, volume 83 of *Lecture Notes in Computational Science and Engineering*. Springer, Berlin, 2012.

[GKM08] I. Georgiev, J. Kraus, and S. Margenov. On the robustness of hierarchical multilevel splittings for discontinuous galerkin rotated bilinear fe problems. RICAM-Report 2008-09, Johann Radon Institut for Computational and Applied Mathematics, Linz, 2008. http://www.ricam.oeaw.ac.at/publications/reports/08/rep08-09.pdf

[GKR12] S. Gippert, A. Klawonn, and O. Rheinbach. Analysis of FETI-DP and BDDC for linear elasticity in 3D with almost incompressible components and varying coefficients inside subdomains. *SIAM J. Numer. Anal.*, 50(5):2208–2236, 2012. accepted for publication; http://www.uni-due.de/numerik/publications.shtml

[GL81] A. George and J. W. H. Liu. *Computer solution of large sparse positive definite systems.* Prentice-Hall, Inc., Englewood Cliffs, N.J., 1981.

[GLS07] I. G. Graham, P. O. Lechner, and R. Scheichl. Domain decomposition for multiscale PDEs. *Numer. Math.*, 106(4):589–626, 2007.

[GR86] V. Girault and P. A. Raviart. *Finite Element Methods for Navier-Stokes Equations.* Springer-Verlag, New York, 1986.

[Gri85] P. Grisvard. *Elliptic Problems on Nonsmooth Domains.* Pitman, Boston, London, Melbourne, 1985.

[Gri92] P. Grisvard. *Singularities in boundary value problems*, volume 22 of *RMA*. Masson, Paris, 1992.

[GS07] I. G. Graham and R. Scheichl. Robust domain decomposition algorithms for multiscale PDEs. *Numerical Methods for Partial Differential Equations*, 23:859–878, 2007.

[GV96] G. H. Golub and C. F. Van Loan. *Matrix Computations.* Johns Hopkins University Press, Baltimore, MD, third edition, 1996.

[GW88] R. Glowinski and M. F. Wheeler. Domain decomposition and mixed finite element methods for elliptic problems. In R. Glowinski, G. H. Golub, G. A. Meurant, and J. Périaux, editors, *First International Symposium on Domain Decomposition Methods for Partial Differential Equations*, pages 144–172. SIAM, Philadelphia, PA, 1988. http://www.ddm.org/DD01/

[Haa99] G. Haase. *Parallelisierung numerischer Algorithmen für partielle Differentialgleichungen*. Teubner, Stuttgart, 1999.

[Hac99] W. Hackbusch. A sparse matrix arithmetic based on $\mathscr{H}$-matrices. *Computing*, 62(2):89–108, 1999.

[Hac03] W. Hackbusch. *Multi-Grid Methods and Applications*. Springer-Verlag, Berlin, second edition, 2003.

[Haj01] P. Hajłasz. Sobolev inequalities, truncation method, and John domains. In *Papers on Analysis*, pages 109–126. Report Univ. Jyväkylä, Dep. Math. Stat. 83, Jyväkylä, 2001.

[HCB05] T. J. R. Hughes, J. A. Cottrell, and Y. Bazilevs. Isogeometric analysis: CAD, finite elements, NURBS, exact geometry and mesh refinement. *Comput. Methods Appl. Mech. Engrg.*, 194(39-41):4135–4195, 2005.

[Heu01] N. Heuer. Additive Schwarz method for the p-version of the boundary element method for the single layer potential operator on a plane screen. *Numer. Math.*, 88(3):485–511, 2001.

[HFMQ98] T. J. R. Hughes, G. R. Feijóo, L. Mazzei, and J.-B. Quincy. The variational multiscale method – a paradigm for computational mechanics. *Comput. Meth. Appl. Mech. Engrg.*, 166:3–24, 1998.

[HHKL97] G. Haase, B. Heise, M. Kuhn, and U. Langer. Adaptive domain decomposition methods for finite elements and boundary elements. In W. L. Wendland, editor, *Boundary Element Topics*, pages 121–147. Springer-Verlag, Berlin, 1997.

[Hil87] E. Hille. *Analytic Function Theory*, volume II. Chelsea Publishing Company, New York, second edition, 1987.

[Hip02] R. Hiptmair. Symmetric coupling for eddy current problems. *SIAM J. Numer. Anal.*, 40:41–65, 2002.

[HK00] W. Hackbusch and B. N. Khoromskij. A sparse $\mathscr{H}$-matrix arithmetic: II. Approximation to multi-dimensional problems. *Computing*, 64(1):21–47, 2000.

[HKD04] J. Haslinger, R. Kučera, and Z. Dostál. An algorithm for the numerical realization of 3D contact problems with Coulomb friction. *J. Comput. Appl. Math.*, pages 164–165, 2004.

[HL92] G. Haase and U. Langer. The non-overlapping domain decomposition multiplicative Schwarz method. *International Journal of Computer Mathematics*, 44:223–242, 1992.

[HLM91a] G. Haase, U. Langer, and A. Meyer. The approximate Dirichlet decomposition method. part I: An algebraic approach. *Computing*, 47:137–151, 1991.

[HLM91b] G. Haase, U. Langer, and A. Meyer. The approximate Dirichlet decomposition method. part II: Application to 2nd-order elliptic B.V.P.s. *Computing*, 47:153–167, 1991.

[HLP10] C. Hofreither, U. Langer, and C. Pechstein. Analysis of a non-standard finite element method based on boundary integral operators. *Electron. Trans. Numer. Anal.*, 37:413–436, 2010.

[HN89] W. Hackbusch and Z. P. Nowak. On the fast matrix multiplication in the boundary element method by panel clustering. *Numer. Math.*, 54(4):463–491, 1989.

[Hof11] C. Hofreither. l_2 error estimates for a nonstandard finite element method on polyhedral meshes. *J. Numer. Math.*, 19(1):27–39, 2011.

[HS98] N. Heuer and E. P. Stephan. Iterative substructuring for hypersingular integral equations in $\mathbf{R}^3$. *SIAM J. Sci. Comput.*, 20(2):739–749, 1998.

[HS01] N. Heuer and E. P. Stephan. An additive Schwarz method for the h-p version of the boundary element method for hypersingular integral equations in $\mathbf{R}^3$. *IMA J. Numer. Anal.*, 21(1):265–283, 2001.

[HS05] Paul Houston and Endre Süli. A note on the design of hp-adaptive finite element methods for elliptic partial differential equations. *Comput. Methods Appl. Mech. Engrg.*, 194(2-5):229–243, 2005.

[HS07] T. J. R Hughes and G. Sangalli. Variational multiscale analysis: the fine-scale Green's function, projection, optimization, localization, stabilized methods. *SIAM J. Num. Anal.*, 45:539–557, 2007.

[HSW00] G. C. Hsiao, O. Steinbach, and W. L. Wendland. Domain decomposition methods via boundary integral equations. *J. Comput. Appl. Math.*, pages 521–537, 2000.

[HW91] G. C. Hsiao and W. L. Wendland. Domain decomposition in boundary element methods. In R. Glowinski, Y. A. Kuznetsov, G. Meurant, J. Périaux, and O. B. Widlund, editors, *Proceedings of the Fourth International Symposium on Domain Decomposition Methods for Partial Differential Equations*, pages 41–49. SIAM, Philadelphia, PA, 1991. http://www.ddm.org/DD04/

[HW08a] J. S. Hesthaven and T. Warburton. *Nodal Discontinuous Galerkin Methods: Algorithms, Analysis, and Applications*, volume 54 of *Texts in Applied Mathematics*. Springer, New York, 2008.

[HW08b] G. C. Hsiao and W. L. Wendland. *Boundary Integral Equations*, volume 164 of *Applied Mathematical Sciences*. Springer-Verlag, Berlin, 2008.

[IM95] I. C. F. Ipsen and C. D. Meyer. The angle between complementary subspaces. *Amer. Math. Monthly*, 102(10):904–911, 1995.

[IN85] T. Iwaniec and C. A. Nolder. Hardy-Littlewood inequality for quasiregular mappings in certain domains in $\mathbb{R}^n$. *Ann. Acad. Sci. Fenn. Ser. A I Math.*, 10:267–282, 1985.

[Jar10] M. Jarošová. *Effective implementation of some algorithms for solving quadratic programming problems*. Doctoral dissertation, Faculty of Electrical Engineering and Computer Science, VŠB-Technical University of Ostrava, 2010. http://am.vsb.cz/theses/jarosova_phd.pdf

[JKR12] M. Jarošová, A. Klawonn, and O. Rheinbach. Projector preconditioning and transformation of basis in FETI-DP algorithms for contact problems. *Math. Comp. Simulation*, 82(10):1894–1907, 2012. to appear, published online Dec. 2010, http://dx.doi.org/10.1016/j.matcom.2010.10.031

[JN80] C. Johnson and J. C. Nédélec. On coupling of boundary integral and finite element methods. *Math. Comp.*, 35(152):1063–1079, 1980.

[Jon81] P. W. Jones. Quasiconformal mappings and extendability of functions in Sobolev spaces. *Acta Math.*, 147:71–88, 1981.

[Kam00] C. Kamath. The FETI level 1 method: theory and implementation. Technical Report UCRLID-138075, Lawrence Livermore National Laboratory, CA, 2000.

[KDW08] H. H. Kim, M. Dryja, and O. B. Widlund. A BDDC method for mortar discretizations using a transformation of basis. *SIAM J. Numer. Anal.*, 47(1):136–157, 2008.

[Kho08a] B. N. Khoromskij. Fast tensor approximation of multi-dimensional convolution with linear scaling. Preprint 36/2008, MPI MIS, Leipzig, 2008.

[Kho08b] B. N. Khoromskij. On tensor approximation of green iterations for kohn-sham equations. *Comput. Visual Sci.*, 11(4–6), 2008.

[Kim07] H. H. Kim. A FETI-DP preconditioner for mortar methods in three dimensions. *Electron. Trans. Numer. Anal.*, 26:103–120, 2007.

[Kim08a] H. H. Kim. A BDDC algorithm for mortar discretization of elasticity problems. *SIAM J. Numer. Anal.*, 46(4):2090–2111, 2008.

[Kim08b] H. H. Kim. Domain decomposition algorithms for mortar discretizations. In *Domain decomposition methods in science and engineering XVII*, volume 60 of *Lecture Notes in Computational Science and Engineering*, pages 81–92. Springer, Berlin, 2008.

[Kim08c] H. H. Kim. A FETI-DP formulation of three dimensional elasticity problems with mortar discretization. *SIAM J. Numer. Anal.*, 46(5):2346–2370, 2008.

[KK98] G. Karypis and V. Kumar. *METIS: A Software Package for Partitioning Unstructured Graphs, Partitioning Meshes, and Computing Fill-Reducing Orderings of Sparse Matrices; Version 4.0*. University of Minnesota, Department of Computer Science / Army HPC Research Center Minneapolis, MN 55455, USA, September 1998. http://people.scs.fsu.edu/~burkardt/pdf/metis.pdf

[KL04] V. G. Korneev and U. Langer. Domain decomposition and preconditioning. In E. Stein, R. de Borst, and T. J. R. Hughes, editors, *Encyclopedia of Computational Mechanics*, volume 1, chapter 22. John Wiley & Sons, 2004.

[KL05] H. H. Kim and C.-O. Lee. A preconditioner for the FETI-DP formulation with mortar methods in two dimensions. *SIAM J. Numer. Anal.*, 42(5):2159–2175, 2005.

[KL10] H. H. Kim and C.-O. Lee. A FETI-DP formulation for the three-dimensional Stokes problem without primal pressure unknowns. *SIAM J. Sci. Comput.*, 32(6):3301–3322, 2010.

[KLP10a] H. H. Kim, C.-O. Lee, and E.-H. Park. A FETI-DP formulation for the Stokes problem without primal pressure components. *SIAM J. Numer. Anal.*, 47(6):4142–4162, 2010.

[KLP10b] H. H. Kim, C.-O. Lee, and E.-H. Park. On the selection of primal unknowns for a FETI-DP formulation of the Stokes problem in two dimensions. *Comput. Math. Appl.*, 60(12):3047–3057, 2010.

[KM03] B. N. Khoromskij and J. M. Melenk. Boundary concentrated finite element methods. *SIAM J. Numer. Anal.*, 41(1):1–36, 2003.

[KM07] J. Kraus and S. Margenov. Multilevel methods for anisotropic elliptic problems. In *Lectures on Advanced Computational Methods in Mechanics*, volume 1 of *Radon Series on Computational and Applied Mathematics*. Walter de Gruyter, Berlin, New York, 2007.

[KNRV11] A. Klawonn, P. Neff, O. Rheinbach, and S. Vanis. FETI-DP domain decomposition methods for elasticity with structural changes: P-elasticity. *ESAIM Math. Model. Numer. Anal.*, 45:563–602, 2011.

[Kny01] A. V. Knyazev. Toward the optimal preconditioned eigensolver: locally optimal block preconditioned conjugate gradient method. *SIAM J. Sci. Comput.*, 23(2):517–541, 2001.

[KPJT12] S. K. Kleiss, C. Pechstein, B. Jüttler, and S. Tomar. IETI – isogeometric tearing and interconnecting. RICAM Report 2012-01, Johann Radon Institute of Computational and Applied Mathematics, Linz, Austria, 2012. accepted for publication in Comput. Methods Appl. Mech. Engrg., 247–248:201–215, 2012. http://www.ricam.oeaw.ac.at/publications/reports/12/rep12-01.pdf

[KPR08] A. Klawonn, L. Pavarino, and O. Rheinbach. Spectral element FETI-DP and BDDC preconditioners with multi-element subdomains. *Comput. Meth. Appl. Mech. Engrg.*, 198(3–4):511–523, 2008.

[KR06] A. Klawonn and O. Rheinbach. A parallel implementation of dual-primal FETI methods for three dimensional linear elasticity using a transformation of basis. *SIAM J. Sci. Comput.*, 28(5):1886–1906, 2006.

[KR07a] A. Klawonn and O. Rheinbach. Inexact FETI-DP methods. *Inter. J. Numer. Methods Engrg.*, 69:284–307, 2007.

[KR07b] A. Klawonn and O. Rheinbach. Robust FETI-DP methods for heterogeneous three dimensional elasticity problems. *Comput. Methods Appl. Mech. Engrg.*, 196:1400–1414, 2007.

[KR07c] A. Klawonn and O. Rheinbach. Some computational results for robust FETI-DP methods applied to heterogeneous elasticity problems in 3D. In D. Keyes and O. Widlund, editors, *Domain Decomposition Methods in Sciences and Engineering XVI*, volume 55 of *Lecture Notes in Computational Science and Engineering*, pages 391–398. Springer-Verlag, Heidelberg, 2007.

[KR10] A. Klawonn and O. Rheinbach. Highly scalable parallel domain decomposition methods with an application to biomechanics. *Z.Angew. Math. Mech.*, 90(1):5–32, 2010.

[KR12] A. Klawonn and O. Rheinbach. Deflation, projector preconditioning, and balancing in iterative substructuring methods: connections and new results. *SIAM J. Sci. Comput.*, 34(1):A459–A484, 2012.

[Kra12] J. Kraus. Additive Schur complement approximation and application to multilevel preconditioning. RICAM Report 2011-22, Johann Radon Institute of Computational and Applied Mathematics, Linz, Austria, 2011. submitted, *SIAM J. Sci. Comput.*, accepted for publication 2012, http://www.ricam.oeaw.ac.at/publications/reports/12/rep11-22.pdf

[KRW07] A. Klawonn, O. Rheinbach, and B. Wohlmuth. Dual-primal iterative substructuring for almost incompressible elasticity. In D. Keyes and O. Widlund, editors, *Domain Decomposition Methods in Sciences and Engineering XVI*, volume 55 of *Lecture Notes in Computational Science and Engineering*, pages 399–406. Springer-Verlag, Heidelberg, 2007.

[KRW08] A. Klawonn, O. Rheinbach, and O. B. Widlund. An analysis of a FETI-DP algorithm on irregular subdomains in the plane. *SIAM J. Numer. Anal.*, 46(5):2484–2504, 2008.

[KS99] G. Karniadakis and S. Sherwin. *Spectral/hp Element Methods for CFD*. Oxford University Press, Oxford, 1999.
[KS02] M. Kuhn and O. Steinbach. Symmetric coupling of finite and boundary element for exterior magnetic field problems. *Math. Methods Appl. Sci.*, 25:357–371, 2002.
[KT09] H. H. Kim and X. Tu. A three-level BDDC algorithm for mortar discretizations. *SIAM J. Numer. Anal.*, 47(2):1576–1600, 2009.
[KW00] A. Klawonn and O. B. Widlund. A domain decomposition method with Lagrange multipliers and inexact solvers for linear elasticity. *SIAM J. Sci. Comput.*, 22(4):1199–1219, 2000.
[KW01] A. Klawonn and O. B. Widlund. FETI and Neumann-Neumann iterative substructuring methods: Connections and new results. *Comm. Pure Appl. Math.*, 54(1):57–90, 2001.
[KW05] A. Klawonn and O. B. Widlund. Selecting constraints in dual-primal FETI methods for elasticity in three dimensions. In R. Kornhuber, R. H. W. Hoppe, D. E. Keyes, J. Periaux, O. Pironneau, and J. Xu, editors, *Domain Decomposition Methods in Science and Engineering*, volume 40 of *Lecture Notes in Computational Science and Engineering*, pages 67–81. Springer-Verlag, Berlin, 2005. http://www.ddm.org/DD15/
[KW06] A. Klawonn and O. B. Widlund. Dual-primal FETI methods for linear elasticity. *Comm. Pure Appl. Math.*, 59(11):1523–1572, 2006.
[KWD02a] A. Klawonn, O. B. Widlund, and M. Dryja. Dual-primal FETI methods for three-dimensional elliptic problems with heterogeneous coefficients. *SIAM J. Numer. Anal.*, 40(1):159–179, 2002.
[KWD02b] A. Klawonn, O. B. Widlund, and M. Dryja. Dual-primal FETI methods with face constraints. In L. Pavarino and A. Toselli, editors, *Recent Developments in Domain Decomposition Methods*, volume 23 of *Lecture Notes in Computational Science and Engineering*, pages 27–40. Springer-Verlag, Berlin, 2002.
[Lan94] U. Langer. Parallel iterative solution of symmetric coupled FE/BE-equations via domain decomposition. *Contemporary Mathematics*, 157:335–344, 1994.
[Le 94] P. Le Tallec. Domain decomposition methods in computational mechanics. In J. T. Oden, editor, *Computational Mechanics Advances*, volume 1, pages 121–220. North-Holland, 1994.
[Les03] M. Lesoinne. A FETI-DP corner selection algorithm for three-dimensional problems. In I. Herrera, D. E. Keyes, O. Widlund, and R. Yates, editors, *Proceedings of the 14th International Conference on Domain Decomposition Methods*, pages 233–240. Mexico, 2003. http://www.ddm.org/DD14/lesoinne.pdf
[Li05] J. Li. A dual-primal FETI method for incompressible stokes equations. *Numer. Math.*, 102(2):257–275, 2005.
[Lie06] M. Liebmann. *A user friendly toolbox for parallel PDE-solvers*. Institut for Mathematics and Scientific Computing, Karl-Franzens-Universität Graz, Austria, 2006. http://paralleltoolbox.sourceforge.net/paralleltoolbox.pdf
[LM72] J. L. Lions and E. Magenes. *Non-Homogeneous Boundary Value Problems and Applications*, volume I. Springer-Verlag, Berlin, 1972.
[LOSZ07a] U. Langer, G. Of, O. Steinbach, and W. Zulehner. Inexact data-sparse boundary element tearing and interconnecting methods. *SIAM J. Sci. Comp.*, 29:290–314, 2007.
[LOSZ07b] U. Langer, G. Of, O. Steinbach, and W. Zulehner. Inexact fast multipole boundary element tearing and interconnecting methods. In D. Keyes and O. Widlund, editors, *Domain Decomposition Methods in Sciences and Engineering XVI*, volume 55 of *Lecture Notes in Computational Science and Engineering*, pages 405–412. Springer-Verlag, Heidelberg, 2007.
[LP98] M. Lesoinne and K. Pierson. An efficient FETI implementation on distributed and shared memory machines with independent number of subdomains and processors. In J. Mandel, C. Farhat, and X. Cai, editors, *Domain Decomposition Methods 10. Proceedings of the 10th International Conference held in Boulder, CO, August 10–14, 1997*, volume 218 of *Contemporary Mathematics*, pages 318–324. AMS, Providence, RI, 1998. http://www.ddm.org/DD10/

[LP06] U. Langer and C. Pechstein. Coupled finite and boundary element tearing and interconnecting solvers for nonlinear potential problems. *ZAMM Z. Angew. Math. Mech.*, 86(12), 2006.

[LP07] U. Langer and C. Pechstein. Coupled FETI/BETI solvers for nonlinear potential problems in (un)bounded domains. In G. Ciuprina and D. Ioan, editors, *Scientific Computing in Electrical Engineering*, volume 11 of *Mathematics in Industry: The European Consortium for Mathematics in Industry*, pages 371–377. Springer-Verlag, Berlin, 2007.

[LP08] U. Langer and C. Pechstein. All-floating coupled data-sparse boundary and interface-concentrated finite element tearing and interconnecting methods. *Comput. Visual Sci.*, 11(4–6):307–317, 2008.

[LPS05] U. Langer, A. Pohoaţă, and O. Steinbach. Dual-primal boundary element tearing and interconnecting methods. Technical report no. 2005/6, Institute for Computational Mathematics, University of Technology, Graz, 2005.

[LS03] U. Langer and O. Steinbach. Boundary element tearing and interconnecting method. *Computing*, 71(3):205–228, 2003.

[LS05] U. Langer and O. Steinbach. Coupled boundary and finite element tearing and interconnecting methods. In R. Kornhuber, R. Hoppe, J. Periaux, O. Pironneau, O. Widlund, and J. Xu, editors, *Domain Decomposition Methods in Science and Engineering*, volume 40 of *Lecture Notes in Computational Science and Engineering*, pages 83–97. Springer-Verlag, Heidelberg, 2005. http://www.ddm.org/DD15/

[Lun06] Y. Lung-An. *Numerical Methods for Exterior Problems*, volume 2 of *Peking University Series in Mathematics*. World Scientific Publishing Co. Pte. Ltd., Hackensack, NJ, 2006.

[LW06a] J. Li and O. Widlund. BDDC algorithms for incompressible Stokes equations. *SIAM J. Numer. Anal.*, 44(6):2432–2455, 2006.

[LW06b] J. Li and O. B. Widlund. FETI-DP, BDDC, and block Cholesky methods. *Internat. J. Numer. Methods Engrg.*, 66(2):250–271, 2006.

[LW07] J. Li and O. B. Widlund. On the use of inexact subdomain solvers for BDDC algorithms. *Comput. Methods Appl. Mech. Engrg.*, 196(8):1415–1428, 2007.

[Man93] J. Mandel. Balancing domain decomposition. *Comm. Numer. Methods Engrg.*, 9(3):233–241, 1993.

[Mat08] T. P. A. Mathew. *Domain decomposition methods for the numerical solution of partial differential equations*, volume 61 of *Lecture Notes in Computational Science and Engineering*. Springer-Verlag, Berlin, 2008.

[Maz60] V. G. Maz'ja. Classes of domains and imbedding theorems for functions spaces. *Soviet Math. Dokl.*, 1:882–885, 1960.

[Maz85] V. G. Maz'ja. *Sobolev spaces*. Springer Series in Soviet Mathematics. Springer-Verlag, Berlin, 1985.

[MB96] J. Mandel and M. Brezina. Balancing domain decomposition for problems with large jumps in coefficients. *Math. Comp.*, 65:1387–1401, 1996.

[McL00] W. McLean. *Strongly Elliptic Systems and Boundary Integral Equations*. Cambridge University Press, Cambridge, UK, 2000.

[MD03] J. Mandel and C. R. Dohrmann. Convergence of a balancing domain decomposition by constraints and energy minimization. *Numer. Lin. Alg. Appl.*, 10(7):639–659, 2003.

[MDT05] J. Mandel, C. R. Dohrmann, and R. Tezaur. An algebraic theory for primal and dual substructuring methods by constraints. *Appl. Numer. Math.*, 54(2):167–193, 2005.

[Mel02] J. M. Melenk. *hp-Finite Element Methods for Singular Pertubations*, volume 1796 of *Lecture Notes in Mathematics*. Springer-Verlag, Berlin, 2002.

[Mon03] P. Monk. *Finite element methods for Maxwell's equations*. Numerical Mathematics and Scientific Computation. Oxford University Press, Oxford, 2003.

[Mor66] C. Morrey. *Multiple Integrals in the Calculus of Variations*. Springer-Verlag, New York, 1966.

[MPI09] MPI: a message-passing interface standard, version 2.2. Technical report, University of Tennessee, Knoxville, Tennessee, 2009. http://www.mpi-forum.org/docs/mpi-2.2/mpi22-report.pdf

[MS07a] J. Mandel and B. Sousedík. Adaptive selection of face coarse degrees of freedom in the BDDC and FETI-DP iterative substructuring methods. *Comput. Methods Appl. Mech. Engrg.*, 196(8):1389–1399, 2007.

[MS07b] J. Mandel and B. Sousedík. BDDC and FETI-DP under minimalist assumptions. *Computing*, 81(4):269–280, 2007.

[MSD08] J. Mandel, B. Sousedík, and C. R. Dohrmann. Multispace and multilevel BDDC. *Computing*, 83(2-3):55–85, 2008.

[MSŠ12] J. Mandel, B. Sousedík, and J. Šístek. Adaptive BDDC in three dimensions. *Math. Comput. Simulation*, 82(10):1812–1831, 2012. to appear, published online Apr. 2011, http://dx.doi.org/10.1016/j.matcom.2011.03.014

[MT96] J. Mandel and R. Tezaur. Convergence of a substructuring method with Lagrange multipliers. *Numer. Math.*, 73(4):473–487, 1996.

[MT01] J. Mandel and R. Tezaur. On the convergence of a dual-primal substructuring method. *Numer. Math.*, 88(3):543–558, 2001.

[MTF99] J. Mandel, R. Tezaur, and C. Farhat. A scalable substructuring method by Lagrange multipliers for plate bending problems. *SIAM J. Numer. Anal.*, 36(5):1370–1391, 1999.

[Muc72] B. Muckenhoupt. Weighted norm inequalities for the Hardy maximal function. *Trans. Amer. Math. Soc.*, 165:207–226, 1972.

[Neč67] J. Nečas. *Les méthodes directes en théorie des équations elliptiques*. Masson et Cie, Éditeurs, Paris, 1967.

[Néd80] J. C. Nédélec. Mixed finite elements in $\mathbb{R}^3$. *Numer. Math.*, 35(3):315–341, 1980.

[Néd86] J. C. Nédélec. A new family of mixed finite elements in $\mathbb{R}^3$. *Numer. Math.*, 50(1):57–81, 1986.

[Nep91a] S. V. Nepomnyaschikh. Mesh theorems on traces, normalizations of function traces and their inversion. *Russion Journal of Numerical Analysis and Mathematical Modelling*, 6(3):233–242, 1991.

[Nep91b] S. V. Nepomnyaschikh. Method of splitting into subspaces for solving elliptic boundary value problems in complex-form domains. *Soviet J. Numer. Anal. Math. Modelling*, 6(2):151–168, 1991.

[Nep92] S. V. Nepomnyaschikh. Domain decomposition and ficticious domains methods for elliptic boundary value problems. In *Fifth International Symposium on Domain Decomposition Methods for Partial Differential Equations*, pages 62–72. SIAM, Philadelphia, PA, 1992.

[Nep07] S. Nepomnyaschikh. Domain decomposition methods. In *Lectures on advanced computational methods in mechanics*, volume 1 of *Radon Series in Computational and Applied Mathematics*, pages 89–159. Walter de Gruyter, Berlin, 2007.

[ODRW89] J. T. Oden, L. Demkowicz, W. Rachowicz, and T. A. Westermann. Toward a universal *h-p* adaptive finite element strategy. II. A posteriori error estimation. *Comput. Methods Appl. Mech. Engrg.*, 77(1-2):113–180, 1989.

[Of06] G. Of. *BETI-Gebietszerlegungsmethoden mit schnellen Randelementverfahren und Anwendungen*. PhD thesis, Universität Stuttgart, Germany, January 2006.

[Of08] G. Of. The all-floating BETI method: Numerical results. In U. Langer, M. Discacciati, D. E. Keyes, O. B. Widlund, and W. Zulehner, editors, *Domain Decomposition Methods in Science and Engineering XVII*, volume 60 of *Lecture Notes in Computational Science and Engineering*, pages 295–302. Springer-Verlag, Berlin, 2008.

[OS09] G. Of and O. Steinbach. The all-floating boundary element tearing and interconnecting method. *J. Num. Math.*, 17(4):277–298, 2009.

[OSU06] G. Of, O. Steinbach, and P. Urthaler. Fast evaluation of Newton potentials in the boundary element method. Technical report no. 2008/3, Department for Numerical Mathematics, Graz University of Technology, Austria, 2006.

[Osw99] P. Oswald. On the robustness of the BPX-preconditioner with respect to jumps in the coefficients. *Math. Comp.*, 68(226):633–650, 1999.

[PAR05] PARDISO. *Parallel Sparse Direct Linear Solver. User Guide Version 3.0.* Computer Science Department, University of Basel, Switzerland, 2005. http://www.computational.unibas.ch/cs/scicomp

[Pav97] L. F. Pavarino. Neumann-Neumann algorithms for spectral elements in three dimensions. *RAIRO Modél. Math. Anal. Numér.*, 31:471–493, 1997.

[Pav07] L. F. Pavarino. BDDC and FETI-DP preconditioners for spectral element discretizations. *Comput. Methods Appl. Mech. Engrg.*, 196(8):1380–1388, 2007.

[Pec08a] C. Pechstein. BETI-DP methods in unbounded domains. In K. Kunisch, G. Of, and O. Steinbach, editors, *Numerical Mathematics and Advanced Applications – ENUMATH 2007*, pages 381–388. Springer-Verlag, Berlin, 2008.

[Pec08b] C. Pechstein. *Finite and boundary element tearing and interconnecting methods for multiscale elliptic partial differential equations.* Doctoral dissertation, Institute of Computational Mathematics, Johannes Kepler University, Linz, Austria, December 2008. http://www.numa.uni-linz.ac.at/Teaching/PhD/Finished/pechstein

[Pec09] C. Pechstein. Boundary element tearing and interconnecting methods in unbounded domains. *Appl. Numer. Math.*, 59(11):2824–2842, 2009.

[Pec12] C. Pechstein. Shape-explicit constants for some boundary integral operators. *Appl. Anal.*, 2012. published online, Dec. 2011, http://dx.doi.org/10.1080/00036811.2011.643781

[Pet89] T. von Petersdorff. *Randwertprobleme der Elastizitätstheorie für Polyeder – Singularitäten und Approximation mit Randelementmethoden.* PhD thesis, Technische Hochschule Darmstadt, Germany, 1989.

[Pie00] K. H. Pierson. *A Family of Domain Decomposition Methods for the Massively Parallel Solution of Computational Mechanics Problems.* PhD thesis, Aerospace Engineering, University of Colorado at Boulder, Boulder, CO, 2000.

[PJF97] K. C. Park, M. R. Justino, Jr., and C. A. Felippa. An algebraically partitioned FETI method for parallel structural analysis: algorithm description. *Int. J. Numer. Meth. Engng.*, 40(15):2717–2737, 1997.

[PS72] C. C. Paige and M. A. Saunders. Computational variants of the Lanczos method for eigenproblems. *J. Numer. Anal.*, 15:801–812, 1972.

[PS08] C. Pechstein and R. Scheichl. Analysis of FETI methods for multiscale PDEs. *Numer. Math.*, 111(2):293–333, 2008.

[PS09] C. Pechstein and R. Scheichl. Scaling up through domain decomposition. *Appl. Anal.*, 88(10–11):1589–1608, 2009.

[PS10] C. Pechstein and R. Scheichl. Robust FETI solvers for multiscale elliptic PDEs. In J. Roos and L. R. J. Costa, editors, *Scientific Computing in Electrical Engineering SCEE 2008*, volume 14 of *Mathematics in Industry*. Springer-Verlag, Berlin, 2010.

[PS11a] A. Pechstein and J. Schöberl. Tangential-displacement and normal-normal-stress continuous mixed finite elements for elasticity. *Math. Models Methods Appl. Sci.*, 21(8):1761–1782, 2011.

[PS11b] C. Pechstein and R. Scheichl. Analysis of FETI methods for multiscale PDEs. Part II: interface variation. *Numer. Math.*, 118(3):485–529, 2011.

[PS11c] C. Pechstein and R. Scheichl. Weighted Poincaré inequalities and applications in domain decomposition. In Y. Huang, R. Kornhuber, O. Widlund, and J. Xu, editors, *Domain Decomposition Methods in Science and Engineering XIX*, volume 78 of *Lecture Notes in Computational Science and Engineering*, pages 197–204. Springer-Verlag, Berlin, 2011.

[PS12a] A. Pechstein and J. Schöberl. Anisotropic mixed finite elements for elasticity. *Int. J. Numer. Meth. Engng.*, 90(2):196–217, 2012.

[PS12b] C. Pechstein and R. Scheichl. Weighted Poincaré inequalities. *IMA J. Numer. Anal.*, 2012. to appear. Published online Oct. 2012, http://dx.doi.org/10.1093/imanum/drs017

[PSS12] C. Pechstein, M. Sarkis, and R. Scheichl. New theoretical robustness results for FETI-DP. In R. Bank, R. Kornhuber, and O. Widlund, editors, *Domain Decomposition Methods in Science and Engineering XX*, Lecture Notes in Computational Science and Engineering. Springer-Verlag, 2012. to appear; http://www.numa.uni-linz.ac.at/~clemens/PechsteinSarkisScheichlDD20.pdf

[PW60] L. E. Payne and H. F. Weinberger. An optimal Poincaré inequality for convex domains. *Arch. Rat. Mech. Anal.*, 5:286–292, 1960.

[PW02] L. F. Pavarino and O. B. Widlund. Balancing Neumann-Neumann methods for incompressible Stokes equations. *Comm. Pure Appl. Math.*, 55(3):302–335, 2002.

[PWZ10] L. F. Pavarino, O. B. Widlund, and S. Zampini. BDDC preconditioners for spectral element discretizations of almost incompressible elasticity in three dimensions. *SIAM J. Sci. Comput.*, 32(6):3604–3626, 2010.

[QV99] A. Quarteroni and A. Valli. *Domain Decomposition Methods for Partial Differential Equations*. Oxford Science Publications, Oxford, 1999.

[RF98a] D. Rixen and C. Farhat. Preconditioning the FETI method for problems with intra- and inter-subdomain coefficient jumps. In P. E. Bjørstad, M. Espedal, and D. E. Keyes, editors, *Proceedings of 9th International Conference on Domain Decomposition*, pages 472–479, 1998. http://www.ddm.org/DD9/Rixen.pdf

[RF98b] F.-X. Roux and C. Farhat. Parallel implementation of direct solution strategies for the coarse grid solvers in 2-level FETI method. In J. Mandel, C. Farhat, and X. Cai, editors, *Domain Decomposition Methods 10. Proceedings of the 10th International Conference held in Boulder, CO, August 10–14, 1997*, volume 218 of *Contemporary Mathematics*, pages 158–173. AMS, Providence, RI, 1998. http://www.ddm.org/DD10/

[RF99] D. Rixen and C. Farhat. A simple and efficient extension of a class of substructure based preconditioners to heterogeneous structural mechanics problems. *Internat. J. Numer. Methods Engrg.*, 44(4):489–516, 1999.

[RFTM99] D. Rixen, C. Farhat, R. Tezaur, and J. Mandel. Theoretical comparison of the FETI and algebraically partitioned FETI methods, and performance comparisons with a direct sparse solver. *Internat. J. Numer. Methods Engrg.*, 46(4):501–533, 1999.

[Rhe02] Oliver Rheinbach. FETI – A dual iterative substructuring method for elliptic partial differential equations. Master's thesis, Mathematisches Institut, Universität zu Köln, Germany, 2002.

[Rhe06] O. Rheinbach. *Parallel Scalable Iterative Substructuring: Robust Exact and Inexact FETI-DP Methods with Applications to Elasticity*. PhD thesis, Universität Essen-Duisburg, 2006.

[Riv08] B. Rivière. *Discontinuous Galerkin Methods for Solving Elliptic and Parabolic Equations: Theory and Implementation*, volume 35 of *Frontiers in Applied Mathematics*. SIAM, Philadelphia, PA, 2008.

[Rix97] D. Rixen. Substructuring and dual methods in structural analysis. PhD thesis, Collection des Publications de la Faculté des Sciences Appliquées no. 175, Université de Liège, Belgium, 1997.

[Rja90] S. Rjasanow. *Vorkonditionierte iterative Auflösung von Randelementgleichungen für die Dirichlet-Aufgabe*. Wissenschaftliche Schriftenreihe der Technischen Universität Chemnitz, Chemnitz, 1990.

[ROD89] W. Rachowicz, J. T. Oden, and L. Demkowicz. Toward a universal h-p adaptive finite element strategy. III. Design of h-p meshes. *Comput. Methods Appl. Mech. Engrg.*, 77(1-2):181–212, 1989.

[Rog06] L. G. Rogers. Degree-independent Sobolev extension on locally uniform domains. *J. Funct. Anal.*, 235(2):619–665, 2006.

[Rok85] V. Rokhlin. Rapid solution of integral equations of classical potential theory. *J. Comput. Phys.*, 60(2):187–207, 1985.

[Rou09] F.-X. Roux. A FETI-2LM method for non-matching grids. In *Domain Decomposition Methods in Science and Engineering XVIII*, volume 70 of *Lecture Notes in Computational Science and Engineering*, pages 121–128. Springer-Verlag, Heidelberg, 2009.

[RS85] J. Ruge and K. Stüben. Efficient solution of finite difference and finite element equations by algebraic multigrid (AMG). In D. J. Paddon and H. Holstein, editors, *Multigrid Methods for Integral and Differential Equations*, IMA Conference Series, pages 169–212. Clarendon Press, Oxford, 1985.

[RS07] S. Rjasanow and O. Steinbach. *The Fast Solution of Boundary Integral Equations*. Mathematical and Analytical Techniques with Applications to Engineering. Springer, New York, 2007.

[RT77] P. A. Raviart and J. M. Thomas. Primal hybrid finite element methods for 2nd order elliptic equations. *Math. Comp.*, 31(138):391–413, 1977.

[RT01] F. Rapetti and A. Toselli. A FETI preconditioner for two-dimensional edge element approximations of Maxwell's equations on nonmatching grids. *SIAM J. Sci. Comput.*, 23(1):92–108, 2001.

[Saa03] Y. Saad. *Iterative Methods for Sparse Linear Systems*. SIAM, Philadelphia, 2003.

[Sar93] M. V. Sarkis. Two-level Schwarz methods for nonconforming finite elements and discontinuous coefficients. In N. D. Melson, T. A. Manteuffel, and S. F. McCormick, editors, *Proceedings of the Sixth Copper Mountain Conference on Multigrid Methods*, volume 2, pages 543–566. NASA, Hampton, VA, 1993.

[Sar94] M. V. Sarkis Martins. *Schwarz Preconditioners for Elliptic Problems with Discontinuous Coefficients Using Conforming and Non-Conforming Elements*. Technical report, Courant Institute of Mathematical Sciences, 1994. TR-671, Department of Computer Science.

[Sar97] M. Sarkis. Nonstandard coarse spaces and Schwarz methods for elliptic problems with discontinuous coefficients using non-conforming elements. *Numer. Math.*, 77(3):383–406, 1997.

[SBG96] B. F. Smith, P. E. Bjørstad, and W. Gropp. *Domain Decomposition: Parallel Multilevel Methods for Elliptic Partial Differential Equations*. Cambridge University Press, 1996.

[ŠČBN12] J. Šístek, M. Čertíková, P. Burda, and J. Novotný. Face-based selection of corners in 3D substructuring. *Math. Comput. Simulation*, 82(10):1799–1811, 2012 to appear, published online Sept. 2011, http://dx.doi.org/10.1016/j.matcom.2011.06.007

[Sch98a] R. Schneider. *Multiskalen- und Wavelet-Matrixkompression*. Advances in Numerical Mathematics. B. G. Teubner, Stuttgart, 1998.

[Sch98b] Ch. Schwab. *p- and hp-Finite Element Methods: Theory and Applications in Solid and Fluid Mechanics*. Oxford University Press, New York, 1998.

[SDH04] E. Stein, R. De Borst, and T. J. R. Hughes. *Encyclopedia of Computational Mechanics, Volume 2: Solids and Structures*. John Wiley & Sons, 2004.

[SDH$^+$11] N. Spillane, V. Dolean, P. Hauret, F. Nataf, C. Pechstein, and R. Scheichl. Abstract robust coarse spaces for systems of PDEs via gereralized eigenproblems in the overlap. NuMa-Report 2011-07, Institute of Computational Mathematics, Johannes Kepler University Linz, 2011.

[SG04] O. Schenk and K. Gärtner. Solving unsymmetric sparse systems of linear equations with PARDISO. *Journal of Future Generation Computer Systems*, 20(3):475–487, 2004.

[SG06] O. Schenk and K. Gärtner. On fast factorization pivoting methods for sparse symmetric indefinite systems. *Electron. Trans. Numer. Anal.*, 23:158–179, 2006.

[Sin08] A. Sinwel. *A new family of mixed finite elements for elasticity*. Doctoral dissertation, Institute of Computational Mathematics, Johannes Kepler University, Linz, Austria, December 2008. http://www.numa.uni-linz.ac.at/Teaching/PhD/Finished/sinwel

[SK98] D. Stefanica and A. Klawonn. The FETI method for mortar finite elements. In *Proceedings of the Eleventh International Conference on Domain Decomposition Methods*, pages 121–129, 1998. http://www.ddm.org/DD11/Stefanica.pdf

[SK04] H.-R. Schwarz and N. Köckler. *Numerische Mathematik*. Teubner, Stuttgart, 2004.

[SM08] B. Sousedík and J. Mandel. On the equivalence of primal and dual substructuring preconditioners. *Electron. Trans. Numer. Anal.*, 31:384–402, 2008.

[ŠMSB12] J. Šístek, J. Mandel, B. Sousedík, and P. Burda. Parallel implementation of multilevel BDDC. In *Proceedings of ENUMATH 2011*, 2012. submitted, http://arxiv.org/abs/1201.5991

[ŠNM$^+$10] J. Šístek, J. Novotný, J. Mandel, M. Čertíková, and P. Burda. BDDC by a frontal solver and the stress computation in a hip joint replacement. *Math. Comp. Simulation*, 80(6):1310–1323, 2010.

[Sob91] S. L. Sobolev. *Some applications of functional analysis in mathematical physics*, volume 91 of *Translations of Mathematical Monographs*. AMS, Providence, RI, 1991. Original Russian version published by Leningrad State University, 1950.

[Sou10] B. Sousedík. *Adaptive-Multilevel BDDC*. PhD thesis, Department of Mathematical and Statistical Sciences, University of Colorado Denver, 2010. http://www-bcf.usc.edu/~sousedik/papers/BSthesisUS.pdf

[Sou12] B. Sousedík. Multiscale BDDC for a saddle-point problem. UCD/CCM Report 303, Center for Computational Mathematics, University of Colorado at Denver, 2012. submitted, http://arxiv.org/abs/1109.0580v1

[SS11] S. Sauter and C. Schwab. *Boundary Element Methods*, volume 39 of *Springer series in Compuational Mathematics*. Springer-Verlag, Berlin, 2011.

[ŠSB+11] J. Šístek, B. Sousedík, P. Burda, J. Mandel, and J. Novotný. Application of the parallel BDDC preconditioner to the Stokes flow. *Comput. & Fluids*, 46(1):429–435, 2011.

[ST98] E. P. Stephan and T. Tran. Domain decomposition algorithms for indefinite hypersingular integral equations: the h and p versions. *SIAM J. Sci. Comput.*, 19(4):1139–1153 (electronic), 1998.

[Ste70] E. M. Stein. *Singular Integrals and Differentiability Properties of Functions*, volume 30 of *Princeton Math Series*. Princeton University Press, Princeton, New Jersey, 1970.

[Ste00] O. Steinbach. OSTBEM – A boundary element software package. Technical report, University of Stuttgart, 2000.

[Ste01] D. Stefanica. A numerical study of FETI algorithms for mortar finite element methods. *SIAM J. Sci. Comput.*, 23(4):1135–1160, 2001.

[Ste03a] O. Steinbach. Artificial multilevel boundary element preconditioners. *Proc. Appl. Math. Mech.*, 3:539–542, 2003.

[Ste03b] O. Steinbach. *Stability estimates for hybrid coupled domain decomposition methods*, volume 1809 of *Lecture Notes in Mathematics*. Springer-Verlag, Heidelberg, 2003.

[Ste08] O. Steinbach. *Numerical Approximation Methods for Elliptic Boundary Value Problems. Finite and Boundary Elements*. Springer-Verlag, New York, 2008.

[Ste11] O. Steinbach. A note on the stable one-equation coupling of finite and boundary elements. *SIAM J. Numer. Anal.*, 49(4):1521–1531, 2011.

[SV07] R. Scheichl and E. Vainikko. Additive Schwarz with aggregation-based coarsening for elliptic problems with highly variable coefficients. *Computing*, 80(4):319–343, 2007.

[SVZ11] R. Scheichl, P. S. Vassilevski, and L. T. Zikatanov. Weak approximation properties of elliptic projections with functional constraints. *Multiscale Modeling and Simulation and Analysis*, 9(4):1677–1699, 2011.

[SVZ12] R. Scheichl, P. S. Vassilevski, and L. T. Zikatanov. Mutilevel methods for elliptic problems with highly varying coefficients on non-aligned coarse grids. *SIAM J. Numer. Anal.*, 50(3):1675–1694, 2012. to appear, Preprint LLNL-JRNL-404462, Lawrence Livermore National Laboratory, CA.

[SW01] O. Steinbach and W. L. Wendland. On C. Neumann's method for second-order elliptic systems in domains with non-smooth boundaries. *J. Math. Anal. Appl.*, 262(2):733–748, 2001.

[SW11a] O. Steinbach and M. Windisch. Robust boundary element domain decomposition solvers in acoustics. In Y. Huang, R. Kornhuber, O. Widlund, and J. Xu, editors, *Domain Decomposition Methods in Science and Engineering XIX*, volume 78 of *Lecture Notes in Computational Science and Engineering*, pages 277–284. Springer-Verlag, Heidelberg, 2011.

[SW11b] O. Steinbach and M. Windisch. Stable boundary element domain decomposition methods for the Helmholtz equation. *Numer. Math.*, 118(1):171–195, 2011.

[SW12] O. Steinbach and M. Windisch. Stable BETI methods in electromagnetics. In R. Bank, R. Kornhuber, and O. Widlund, editors, *Domain Decomposition Methods in Science and Engineering XX*, Lecture Notes in Computational Science and Engineering. Springer-Verlag, 2012. to appear.

[SZ90] L. R. Scott and S. Zhang. Finite element interpolation of non-smooth functions satisfying boundary conditions. *Math. Comp.*, 54:483–493, 1990.

[Tez98] R. Tezaur. *Analysis of Lagrange multiplier based domain decomposition*. PhD thesis, University of Colorado at Denver, 1998. http://www-math.cudenver.edu/graduate/thesis/rtezaur.pdf

[TK01] A. Toselli and A. Klawonn. A FETI domain decomposition method for edge element approximations in two dimensions with discontinous coefficients. *SIAM J. Numer. Anal.*, 39(3):932–956, 2001.

[TL08] X. Tu and J. Li. A balancing domain decomposition method by constraints for advection-diffusion problems. *Comm. App. Math. and Comp. Sci.*, 3(1), 2008.

[Tos99] A. Toselli. *Domain Decomposition for Vector Field Problems*. PhD thesis, Courant Institute of Mathematical Sciences, New York University, New York, USA, May 1999.

[Tos00] A. Toselli. Neumann-Neumann methods for vector field problems. *Electron. Trans. Numer. Anal.*, 11:1–24, 2000.

[Tos01] A. Toselli. FETI domain decomposition methods for scalar advection-diffusion problems. *Comput. Methods Appl. Mech. Engrg.*, 190(43–44):5759–5776, 2001.

[Tos06] A. Toselli. Dual-primal FETI algorithms for egde finite-element approximations in 3D. *IMA J. Numer. Anal.*, 26:96–130, 2006.

[Tu05] X. Tu. A BDDC algorithm for a mixed formulation of flow in porous media. *Electron. Trans. Numer. Anal.*, 20:164–179, 2005.

[Tu07a] X. Tu. A BDDC algorithm for flow in porous media with a hybrid finite element discretization. *Electron. Trans. Numer. Anal.*, 26:146–160, 2007.

[Tu07b] X. Tu. Three-level BDDC in three dimensions. *SIAM J. Sci. Comp.*, 29(4):1759–1780, 2007.

[Tu07c] X. Tu. Three-level BDDC in two dimensions. *Int. J. Numer. Meth. Engng.*, 69(1):33–59, 2007.

[Tu11] X. Tu. A three-level BDDC algorithm for saddle point problems. *Numer. Math.*, 119:189–217, 2011.

[TV03] A. Toselli and X. Vasseur. A numerical study on Neumann-Neumann and FETI methods for *hp*-approximations on geometrically refined boundary layer meshes in two dimensions. *Comput. Methods Appl. Mech. Engrg.*, 192:4551–4579, 2003.

[TV04] A. Toselli and X. Vasseur. A numerical study on Neumann-Neumann and FETI methods for *hp*-approximations boundary layer meshes in three dimensions. *IMA J. Numer. Anal.*, 24(1):123–156, 2004.

[TV06] A. Toselli and X. Vasseur. A numerical study on Neumann-Neumann methods for *hp* approximations on geometrically refined boundary layer meshes ii. Three-dimensional problems. *ESAIM Math. Model. Numer. Anal.*, 40(1):99–122, 2006.

[TW05] A. Toselli and O. B. Widlund. *Domain Decomposition Methods – Algorithms and Theory*, volume 34 of *Springer Series in Computational Mathematics*. Springer-Verlag, Berlin, 2005.

[Urt08] P. Urthaler. Schnelle Auswertung von Volumenpotentialen in der Randelementmethode. Master's thesis, Graz University of Technology, Graz, 2008. http://www.numerik.math.tu-graz.ac.at/berichte/Bericht0108.pdf

[Van82] H. Van der Vorst. A generalized Lanczos scheme. *Math. Comp.*, 39:559–561, 1982.

[Vas08] P. S. Vassilevski. *Multilevel block factorization preconditioners*. Springer-Verlag, New York, 2008.

[VMB96] P. Vaněk, J. Mandel, and M. Brezina. Algebraic multigrid by smoothed aggregation for 2nd and 4th order elliptic problems. *Computing*, 56(3):179–196, 1996.

[VSM99] C. Vuik, A. Segal, and J. A. Meijerink. An efficient preconditioned CG method for the solution of a class of layered problems with extreme contrasts in the coefficients. *J. Comput. Phys.*, 152(1):385–403, 1999.

[VSMW01] C. Vuik., A. Segal., J. A. Meijerink, and G. T. Wijma. The construction of projection vectors for a deflated ICCG method applied to problems with extreme contrasts in the coefficients. *J. Comput. Phys.*, 172(2):426–450, 2001.

[VV09] A. Veeser and R. Verfürth. Explicit upper bounds for dual norms of residuals. *SIAM J. Numer. Anal.*, 47(3):2387–2405, 2009.

[VV11] A. Veeser and R. Verfürth. Poincaré constants of finite element stars. *IMA J. Numer. Anal.*, 32(1):40–47, 2011.

[Wei98] J. Weickert. *Anisotropic diffusion in image processing*. Teubner, Stuttgart, Germarny, 1998. out of print, free download http://www.mia.uni-saarland.de/weickert/Papers/book.pdf

[Win10] M. Windisch. *Boundary Element Tearing and Interconnecting for Acoustig and Electromagnetic Scattering*. Doctoral dissertation, TU Graz, Graz, Austria, December 2010. appeared in the Monographic Series TU Graz: Computation in Engineering and Science, vol. 11, 2011, http://fmysql.tu-graz.ac.at/~karl/verlagspdf/buch_windisch_10062011.pdf

[WX94] J. Wang and R. Xie. Domain decomposition for elliptic problems with large jumps in coefficients. In *Proceedings of Conference on Scientific and Engineering Computing*, pages 74–86. National Defense Industry Press, Beijing, China, 1994.

[XZ08] J. Xu and Y. Zhu. Uniform convergent multigrid methods for elliptic problems with strongly discontinuous coefficients. *Math. Mod. Meth. Appl. Sci.*, 18(1):77–105, 2008.

[Zag06] S. Zaglmayr. *High order finite elements for electromagnetic field computation*. PhD thesis, Johannes Kepler University Linz, Austria, 2006. http://www.numa.uni-linz.ac.at/Teaching/PhD/Finished/zaglmayr-diss.pdf

[Zei90] E. Zeidler. *Nonlinear Functional Analysis and Its Applications – Part 2B: Nonlinear Monotone Operators*. Springer-Verlag, New York, 1990.

[ZKB77] O. C. Zienkiewicz, D. M. Kelly, and P. Bettess. The coupling of the finite element method and boundary solution procedures. *Int. J. Numer. Meth. Eng.*, 11(2):355–375, 1977.

[ZKB79] O. C. Zienkiewicz, D. M. Kelly, and P. Bettess. Marriage a la mode – the best of both worlds (Finite elements and boundary integrals). In R. Glowinski, E. Y. Rodin, and O. C. Zienkiewicz, editors, *Energy Methods in Finite Element Analysis*, chapter 5, pages 81–106. Wiley and Son, London, 1979.

[ZP08] V. V. Zhikov and S. E. Pastukhova. Homogenization of degenerate elliptic equations. *Sib. Math. J.*, 49(1):80–101, 2008.

[Zul02] W. Zulehner. Analysis of iterative methods for saddle point problems: A unified approach. *Mathematics of Computation*, 71:479–505, 2002.

[Zul11] W. Zulehner. Nonstandard norms and robust estimates for saddle point problems. *SIAM J. Matrix Anal. Appl.*, 32(2):536–560, 2011.

[Zum03] G. Zumbusch. *Parallel Multilevel Methods – Adaptive Mesh Refinement and Loadbalancing*. B. G. Teubner, Stuttgart, Leipzig, Wiesbaden, 2003.

List of Symbols

Operators, Accents, Norms, etc.

$\lesssim$	less or equal up to constant factor, p. 116
$\gtrsim$	greater or equal up to constant factor, p. 116
$\eqsim$	equivalent up to constant factors, p. 116
$\oplus$	sum of two spaces
$\prod_{i=1}^n V_i$	product space $\{(v_1, \ldots, v_n) : v_i \in V_i\}$
$\#(Z)$	cardinality of a set Z
$\lvert p \rvert$	Euclidean norm of a vector or absolute value of a real number
$\lvert v \rvert_{H^k(D)}$	Sobolev norm of v on domain D, Definition 1.17, p. 10
$\lvert v \rvert_{H^1(D),\alpha}$	weighted H^1-norm of v on domain D, Definition 3.3, p. 160
$\lVert v \rVert_{L^p(D)}$	Lebesgue norm of v on domain/manifold D, (1.10), p. 8
$\lVert v \rVert_{L^2(D),\alpha}$	weighted L^2-norm of v on domain D, Definition 3.3, p. 160
$\lVert\!\lvert \cdot \rvert\!\rVert_{H^1(\Omega)^*}$	special norm, (1.14), p. 17
1_Ω	constant function on domain/manifold Ω with value 1
$p \cdot q$	Euclidean inner product of vectors p, q
$(p, q)_{\ell^2}$	Euclidean inner product of vectors p, q
$A^\dagger$	pseudo inverse, Definition 1.13, p. 6
$A^\top$	transpose of a matrix / adjoint of a linear operator
V^*	dual of a Banach space V
U°	polar of a subspace U, Definition 1.7, p. 4
$U^{\perp_Q}$	orthogonal complement of subspace U w.r.t. $(\cdot,\cdot)_Q$
∂^α	partial derivative w.r.t. multi-index α
$\partial f / \partial x_i$	(1st order) partial derivative w.r.t. x_i

C. Pechstein, *Finite and Boundary Element Tearing and Interconnecting Solvers for Multiscale Problems*, Lecture Notes in Computational Science and Engineering 90,
DOI 10.1007/978-3-642-23588-7,

Greek Letters

α	scalar (isotropic) coefficient or weight function,
α_{coarse}	coefficient that varies on a coarse scale η, Sect. 3.1.2, p. 159
$\alpha_D^{\max}$	maximum of α in domain D, Definition 3.4, p. 160
$\alpha_D^{\min}$	minimum of α in domain D, Definition 3.4, p. 160
α_i	constant coefficient in subdomain Ω_i, Assumption 2.99, p. 134
$\alpha_{\min}$	global lower bound of the coefficient $\mathscr{A}$, (1.22), p. 21
α_{noise}	coefficient varying on fine scale, small noise level, p. 159
$\delta_j^\dagger$	weighted counting function, (2.53), p. 91
Δ	Laplace operator
∇	gradient operator
γ_0	trace operator, Theorem 1.23, p. 13
$\gamma_1^{\mathscr{A}}$	Neumann trace operator, Theorem 1.37, p. 20
$\gamma_1^{\mathscr{A}}(u, f)$	generalized Neumann trace, Theorem 1.40, p. 21
Γ	interface, Table 2.1, p. 65
Γ^h	set of nodes on the interface
Γ_D	Dirichlet boundary ($\Gamma_D = \overline{\Gamma}_D$)
Γ_D^h	set of nodes on the Dirichlet boundary
Γ_N	Neumann boundary
Γ_N^h	set of non-coupling nodes on the Neumann boundary, (2.76), p. 115
Γ_S	skeleton, Table 2.1, p. 65
Γ_S^h	set of nodes on the skeleton
η	(a) Chapter 3: mesh parameter, see η_i, $\eta_{\min}$, $\eta_{\max}$
	(b) Chapter 4: geometrical parameter, Assumption 4.23, p. 232
η_i	local mesh parameter, $h_i \le \eta_i \le H_i$, Assumption 3.13, p. 165
$\eta_{\max}$	mesh parameter, (3.37), p. 191
$\eta_{\min}$	mesh parameter, (3.37), p. 191
$\lambda_{\min}(A)$	minimal eigenvalue of a self-adjoint matrix A
$\lambda_{\max}(A)$	maximal eigenvalue of a self-adjoint matrix A
Λ	skeleton cover, Definition 3.10, p. 164
Λ_i	local cover, $\Lambda_i = \Lambda \cap \Omega_i$, Definition 3.10, p. 164
φ_{x^h}	nodal basis function associated to node x^h
$\rho_i(x^h)$	scalings, p. 91
$\sigma^j(x)$	indicator function, Definition 2.68, p. 120
$\theta_{\mathscr{G}_i}$	cut-off function in $V^h(\partial\Omega_i)$ associated to glob $\mathscr{G}_i$, Definition 2.51, p. 114
$\vartheta_{\mathscr{G}_i}$	cut-off function in $V^h(\Omega_i)$ associated to glob $\mathscr{G}_i$ Definition 2.72, p. 122
Ω	computational domain
Ω_0	exterior part, p. 216

continued on next page

Greek Letters – continued from previous page

Ω_0^c	complement of exterior part, p. 216
Ω^{ext}	exterior of computation domain Ω^{int}
Ω_i	subdomain, (2.3), p. 65
Ω^{int}	computational domain, interior to its bounday
Ω_{int}	interior part, p. 216
Ω'_{int}	extension of interior part, Assumption 4.6, p. 223
Π_0	projection used in Chap. 4, (4.14), p. 228
Π^h	Scott-Zhang quasi-interpolant, Lemma 1.45, p. 27
Π_{int}	projection used in Chap. 4, (4.14), p. 228

Other Symbols in Alphabetical Order

$a(\cdot,\cdot)$	bilinear form, pp. 2, 64, 217
$a_i(\cdot,\cdot)$	bilinear form on subdomain Ω_i, (2.4), p. 65
$\mathscr{A}$	diffusion coefficient, p. 18
$\underset{x\in S}{\operatorname{argmin}} f(x)$	element $x^* \in S$ where $f(x^*) = \min_{x\in S} f(x)$; uniqueness assumed
B	jump operator in classical formulation, (2.23), p. 73
B	jump operator in all-floating formulation, (2.48), p. 85
$\widetilde{B}$	jump operator for the dual-primal formulation, (5.7), p. 254
B_D	weighted jump operator, (2.55)–(2.56), p. 92
c_0	constant related to V and D, Lemma 1.77, p. 47
$\bar{c}_a$	boundedness constant of $a(\cdot,\cdot)$, Theorem 1.1, p. 2
$\underline{c}_a$	coercivity constant of $a(\cdot,\cdot)$, Theorem 1.1, p. 2
c_K	contraction constant of K, Lemma 1.77, p. 47
c_{noise}	noise level, Assumption 3.14, p. 165
C	(a) generic constant (b) $\operatorname{diag}(C_i)_{i\in\mathscr{I}}$, p. 275
$C_F(\Omega, \Gamma_D)$	Friedrichs constant, Lemma 1.31, p. 16
C_i	constraint operator, (5.28), p. 271
$C_P(\Omega)$	Poincaré constant, Lemma 1.27, p. 15
$C_{P,\alpha}(\Omega)$	WPI constant, weight α, Definition 3.23, p. 170
$C_{P,\alpha}(\Omega; h)$	discrete WPI constant, weight α, Definition 3.23, p. 170
$C_P(Y, X)$	Poincaré constant, average over X, Definition 2.67(i), p. 120
$C_P(Y, X; h)$	discrete Poincaré constant, Definition 2.67(i), p. 120
$C_{P,\alpha}(Y, X)$	WPI constant, average over X, Definition 3.33, p. 183
$C_P(Y, X, W)$	Poincaré constant, L^2-norm over W, Definition 2.67(ii), p. 120
$C_P(Y, X, W; h)$	discrete Poincaré constant, Definition 2.67(ii), p. 120

continued on next page

Other Symbols in Alphabetical Order – continued from previous page

C_{SZ}	Scott-Zhang stability constant, Lemma 1.45, p. 27
$\mathscr{C}^\infty(\Omega)$	space of infinitely differentiable functions on Ω
$\mathscr{C}_0^\infty(\Omega)$	space of $\mathscr{C}^\infty(\Omega)$ functions with compact support in Ω
d	spatial dimension (2 or 3)
d_X	dimension of manifold X
$\mathrm{diam}(\Omega)$	diameter of a domain/manifold Ω
$\dim(V)$	dimension of a finite-dimensional space V
div	divergence operator
$\mathrm{dist}(x, \Omega)$	smallest distance from a point x from a domain/manifold Ω
D	hypersingular integral operator, Table 1.2, p. 45
D	generic domain (Chap. 3)
$\mathscr{D}(\Omega)$	$\mathscr{C}_0^\infty(\Omega)$ with sequential convergence, p. 9
$\mathscr{D}^*(\Omega)$	set of Schwartz distributions on Ω, p. 9f
e	η-edge, Definition 3.17, p. 168
E_D	projection operator, (2.69), p. 107
$\widehat{E}_D$	projection operator, (2.60), p. 102
$\widetilde{E}_D$	projection operator, (5.13), p. 258
$\mathscr{E}$	(a) extension operator, Theorem 1.23, p. 13 (b) subdomain edge (see below)
$\mathscr{E}_i$	subdomain edge of subdomain Ω_i, p. 114
$\mathscr{E}^h$	set of nodes that lie on $\mathscr{E}$, Definition 2.50, p. 114
f	functional, (2.1), p. 64
f_i	functional, (2.4), p. 65
f_Ω	source function, (2.2), p. 64
f	η-facet, Definition 3.17, p. 168
F	(a) $F = B\, S^\dagger B^\top$, (2.35), p. 78 (one-level methods) (b) $F = \widetilde{B}\, \widetilde{S}^{-1}\, \widetilde{B}^\top$, (5.10), p. 255 (dual-primal methods)
$\mathscr{F}_i$	subdomain facet of subdomain Ω_i, p. 114
$\mathscr{F}^h$	set of nodes that lie on $\mathscr{F}$, Definition 2.50, p. 114
g	η-glob, Definition 3.17, p. 168
g	functional, (2.26), p. 74
g_D	Dirichlet data, (2.2), p. 64
g_i	functional, (2.26), p. 74
g_N	Neumann data, (2.2), p. 64
G	$G = B\, R$, (2.35), p. 78
$\mathscr{G}_i$	glob of subdomain Ω_i, Definition 2.48 p. 113
$\mathscr{G}^h$	set of nodes that lie within $\mathscr{G}$, Definition 2.50, p. 114
h	mesh parameter, Sect. 1.2.3.1, p. 23
h_i	mesh parameter on subdomain Ω_i, Assumption 2.53, p. 115

continued on next page

Other Symbols in Alphabetical Order – continued from previous page

H_0	$= \mathrm{diam}(\Omega_0^c)$, (4.4), p. 218
H_i	$= \mathrm{diam}(\Omega_i)$, Table 2.1, p. 65
$\mathscr{H}$	harmonic extension, Definition 1.41, p. 21
$\mathscr{H}^a$	PDE-harmonic extension, Definition 1.41, p. 21
$\mathscr{H}_i$	harmonic extension on Ω_i, p. 69
$\mathscr{H}_i^{\alpha,h}$	discrete PDE-harmonic extension on Ω_i, p. 161
$\mathscr{H}_i^h$	discrete harmonic extension on Ω_i, (3.2), p. 122
$H^k(\Omega)$	Sobolev space of integer order k, Definition 1.17, p. 10
$H_0^k(\Omega)$	closure of $\mathscr{C}_0^\infty(\Omega)$ under $\\|\cdot\\|_{H^k(\Omega)}$, Definition 1.17, p. 10
$H^s(\Omega)$	Sobolev space of real order s, Definition 1.19, p. 11
$H^s(\widetilde{\Gamma})$	Sobolev space on hypersurface $\widetilde{\Gamma}$, Definition 1.20, p. 11
$H_0^s(\Omega)$	closure of $\mathscr{C}_0^\infty(\Omega)$ under $\\|\cdot\\|_{H^s(\Omega)}$, Definition 1.19, p. 11
$H_{\mathrm{pw}}^s(\Gamma)$	$L^2(\Gamma)$-functions that are picewise H^s on the (flat) facets of the hypersurface Γ, Definition 1.88, p. 52
$H^{1/2}(\Gamma_S)$	trace space on skeleton, (2.5), p. 65
$H_{00}^{1/2}(\widetilde{\Gamma})$	Sobolev space on hypersurface $\widetilde{\Gamma}$, Definition 1.22, p. 12
$H^{-1/2}(\widetilde{\Gamma})$	dual of $H_{00}^{1/2}(\widetilde{\Gamma})$, p. 12
$\mathscr{I}$	set of subdomain indices, p. 248
$\mathscr{I}_{\mathrm{BEM}}$	set of BEM subdomain indices, p. 67, p. 219
$\mathscr{I}_{\mathrm{FEM}}$	set of FEM subdomain indices, p. 67, p. 219
I^h	nodal interpolator, Definition 2.52, p. 115
$J(v)$	Ritz energy functional, Lemma 1.3, p. 2
$\ker(A)$	kernel (nullspace) of linear operator A
K	double layer potential operator, Table 1.2, p. 45
K	(a) FEM stiffness matrix, (1.28), p. 25 (b) BEM matrix corresponding to K, (1.63), p. 55
ℓ^*	index of subregion with largest coefficient, Definition 3.34, p. 184
$L^p(\Omega)$	Lebesgue space of p-th power integrable functions on Ω, p. 8
$L^\infty(\Omega)$	Lebesgue space (bounded essential supremum), p. 8
$L_+^\infty(\Omega)$	subset of positive $L^\infty(\Omega)$ functions, Definition 3.1, p. 159
$\mathrm{meas}_m(X)$	measure of the m-dimensional domain/manifold X
M^{-1}	general FETI/BETI preconditioner, Sect. 2.2.4, p. 89
M_{sD}^{-1}	scaled Dirichlet preconditioner, (2.57), p. 92
M	BEM mass matrix, (1.63), p. 55
n	(a) unit normal on a boundary (b) number of subregions $Y^{(\ell)}$ within Y, Sect. 3.4.1, p. 183
n_i	number of subregions in Λ_i, (3.4), p. 164
n_Π	global number of primal constraints, Lemma 5.32, p. 270

continued on next page

Other Symbols in Alphabetical Order – continued from previous page

$n_{\Pi,i}$	number of primal constraints associated to Ω_i, (5.27), p. 271
N	Newton potential, Definition 1.41, p. 21
N_i	Newton potential on subdomain Ω_i, p. 65
$\mathcal{N}_{\mathsf{g}}$	set of subdomain indices sharing g, Definition 3.17, p. 168
$\mathcal{N}_{\mathcal{G}}$	set of subdomain indices sharing $\mathcal{G}$, Definition 2.48 (ii), p. 113
$\mathcal{N}_i$	set of subdomain indices neighboring to Ω_i, Definition 2.12, p. 76
$\mathcal{N}_{x^h}$	set of subdomain indices sharing x^h, Definition 2.12, p. 76
$\mathcal{P}_k$	space of polynomials of total degree $\leq k$
P	projection operator, (2.38), p. 80
P_D	projection operator, (2.66), p. 107
P_Z	projection operator, (2.50), p. 87
$\widehat{P}_0$	projection operator, (2.61), p. 102
Q	scaling operator, p. 80, Sect. 2.2.4.4, and Sect. 2.6.3
Q_{diag}	special diagonal choice of Q, p. 137ff, p. 175ff
$Q_{\text{diag}}^{(2)}$	special diagonal choice of Q, (3.20), p. 176
$Q_{\text{diag}}^{(3)}$	special diagonal choice of Q, (3.21), p. 176
$q_i(x^h)$	weight occurring in Q_{diag}, Definition 2.107, p. 137; (3.21)–(3.19), p. 176; (4.8)–(4.10), p. 221
$q_i^{(2)}(x^h)$	weight from $Q_{\text{diag}}^{(2)}$, (3.20), p. 176
$q_i^{(3)}(x^h)$	weight from $Q_{\text{diag}}^{(3)}$, (3.21), p. 176
range(A)	range (image) of a linear operator A
$\mathbb{R}_{\text{sym}}^{d\times d}$	set of symmetric and real-valued $d \times d$ matrices
s	number of (interior) subdomains, (2.3), p. 65; Sect. 4.1.2, p. 217
S	(a) Steklov Poincaré operator on single domain, Definition 1.41, p. 21 (b) block operator diag(S_i), (2.25), p. 74
$\widehat{S}$	restriction of S to $\widehat{W}$, (2.59), p. 102; Sect. 5.1.4, p. 258
S^{corr}	correction operator, (1.53), p. 49
S_i	approximate Steklov-Poincaré operator, p. 65
S^{int}	interior Steklov-Poincaré operator, p. 49
S^{ext}	exterior Steklov-Poincaré operator, p. 49
$\mathcal{T}^h(\Omega)$	triangulation of domain Ω with mesh parameter h
$\mathcal{T}^h(\Gamma)$	triangulation of surface Γ with mesh parameter h
U	space of Lagrange multipliers, p. 74; Table 2.2, p. 81
$\widetilde{U}$	factor space for the dual-primal methods, (5.11), p. 255

continued on next page

Other Symbols in Alphabetical Order – continued from previous page

U_{ad}	space of admissible increments, (2.37), p. 79
$\widetilde{U}_{\mathrm{ad}}$	$= U_{\mathrm{ad}/\ker(B^\top)}$, (2.42), p. 82
$\widetilde{U}^*_{\mathrm{ad}}$	range$(P^\top) \cap$ range(B), (2.43), p. 82
$U^*(x, y)$	fundamental solution of the Laplacian, (1.46), p. 43
$\mathscr{U}_{\mathscr{G}}$	glob patch of the glob $\mathscr{G}$, Definition 2.86, p. 129
V	η-vertex, Definition 3.17, p. 168
V	single layer potential operator, Table 1.2, p. 45
$\widetilde{V}$	single layer potential, (1.47), p. 44
$V^h(\Omega)$	FE space on domain Ω, (1.25), p. 24
$V^h_D(\Omega)$	$V^h(\Omega)$ with homogeneous Dirichlet conditions on Γ_D, p. 25
$V^h(\Gamma_S)$	FE space on skeleton Γ_S, (2.10), p. 66
$V^h_D(\Gamma_S)$	$V^h(\Gamma_S)$ with homogeneous Dirichlet conditions on Γ_S, (2.12), p. 68
$\mathscr{V}_i$	subdomain vertex of subdomain Ω_i, p. 114
$\mathscr{V}^h$	$= \{\mathscr{V}\}$, Definition 2.50, p. 114
W	product space $\prod_{i=1}^s W_i$, p. 73, p. 85, p. 249
W_i	subdomain trace space of finite element functions; (a) (2.20), p. 73 (classical formulation) (b) (2.45), p. 85 (all-floating formulation) (c) (5.3), p. 249 (dual-primal formulation)
$W_i^\perp$	subspace of W_i of codimension one or zero, (2.86), p. 135
$\widehat{W}$	$= \ker(B)$, space of continuous functions, (2.24), p. 74
$\widehat{W}_\Pi$	special primal subspace, (5.25), p. 270
$\widetilde{W}^S_\Pi$	special primal subspace, Lemma 5.34, p. 271
$\widetilde{W}$	(a) double layer potential, (1.48), p. 44 (b) subspace of W with primal constraints, (5.4), p. 251
$\widetilde{W}_\Delta$	dual subspace of $\widetilde{W}$, Definition 5.5, p. 251
$Y_i^{(\ell)}$	subregions of the local cover Λ_i, (3.4), p. 164
z_w	special element from Lemma 2.44, p. 109
$Z^h(\Gamma)$	space of piecewise constants, (1.57), p. 52

Index

C. Pechstein, *Finite and Boundary Element Tearing and Interconnecting Solvers for Multiscale Problems*, Lecture Notes in Computational Science and Engineering 90,
DOI 10.1007/978-3-642-23588-7, © Springer-Verlag Berlin Heidelberg 2013

Editorial Policy

1. Volumes in the following three categories will be published in LNCSE:

i) Research monographs
ii) Tutorials
iii) Conference proceedings

Those considering a book which might be suitable for the series are strongly advised to contact the publisher or the series editors at an early stage.

2. Categories i) and ii). Tutorials are lecture notes typically arising via summer schools or similar events, which are used to teach graduate students. These categories will be emphasized by Lecture Notes in Computational Science and Engineering. **Submissions by interdisciplinary teams of authors are encouraged.** The goal is to report new developments – quickly, informally, and in a way that will make them accessible to non-specialists. In the evaluation of submissions timeliness of the work is an important criterion. Texts should be well-rounded, well-written and reasonably self-contained. In most cases the work will contain results of others as well as those of the author(s). In each case the author(s) should provide sufficient motivation, examples, and applications. In this respect, Ph.D. theses will usually be deemed unsuitable for the Lecture Notes series. Proposals for volumes in these categories should be submitted either to one of the series editors or to Springer-Verlag, Heidelberg, and will be refereed. A provisional judgement on the acceptability of a project can be based on partial information about the work: a detailed outline describing the contents of each chapter, the estimated length, a bibliography, and one or two sample chapters – or a first draft. A final decision whether to accept will rest on an evaluation of the completed work which should include

- at least 100 pages of text;
- a table of contents;
- an informative introduction perhaps with some historical remarks which should be accessible to readers unfamiliar with the topic treated;
- a subject index.

3. Category iii). Conference proceedings will be considered for publication provided that they are both of exceptional interest and devoted to a single topic. One (or more) expert participants will act as the scientific editor(s) of the volume. They select the papers which are suitable for inclusion and have them individually refereed as for a journal. Papers not closely related to the central topic are to be excluded. Organizers should contact the Editor for CSE at Springer at the planning stage, see *Addresses* below.

In exceptional cases some other multi-author-volumes may be considered in this category.

4. Only works in English will be considered. For evaluation purposes, manuscripts may be submitted in print or electronic form, in the latter case, preferably as pdf- or zipped ps-files. Authors are requested to use the LaTeX style files available from Springer at http://www.springer.com/authors/book+authors/helpdesk?SGWID=0-1723113-12-971304-0 (Click on Templates → LaTeX → monographs or contributed books).

For categories ii) and iii) we strongly recommend that all contributions in a volume be written in the same LaTeX version, preferably LaTeX2e. Electronic material can be included if appropriate. Please contact the publisher.

Careful preparation of the manuscripts will help keep production time short besides ensuring satisfactory appearance of the finished book in print and online.

5. The following terms and conditions hold. Categories i), ii) and iii):

Authors receive 50 free copies of their book. No royalty is paid.
Volume editors receive a total of 50 free copies of their volume to be shared with authors, but no royalties.

Authors and volume editors are entitled to a discount of 33.3 % on the price of Springer books purchased for their personal use, if ordering directly from Springer.

6. Commitment to publish is made by letter of intent rather than by signing a formal contract. Springer-Verlag secures the copyright for each volume.

Addresses:

Timothy J. Barth
NASA Ames Research Center
NAS Division
Moffett Field, CA 94035, USA
barth@nas.nasa.gov

Michael Griebel
Institut für Numerische Simulation
der Universität Bonn
Wegelerstr. 6
53115 Bonn, Germany
griebel@ins.uni-bonn.de

David E. Keyes
Mathematical and Computer Sciences
and Engineering
King Abdullah University of Science
and Technology
P.O. Box 55455
Jeddah 21534, Saudi Arabia
david.keyes@kaust.edu.sa

and

Department of Applied Physics
and Applied Mathematics
Columbia University
500 W. 120 th Street
New York, NY 10027, USA
kd2112@columbia.edu

Risto M. Nieminen
Department of Applied Physics
Aalto University School of Science
and Technology
00076 Aalto, Finland
risto.nieminen@aalto.fi

Dirk Roose
Department of Computer Science
Katholieke Universiteit Leuven
Celestijnenlaan 200A
3001 Leuven-Heverlee, Belgium
dirk.roose@cs.kuleuven.be

Tamar Schlick
Department of Chemistry
and Courant Institute
of Mathematical Sciences
New York University
251 Mercer Street
New York, NY 10012, USA
schlick@nyu.edu

Editor for Computational Science
and Engineering at Springer:
Martin Peters
Springer-Verlag
Mathematics Editorial IV
Tiergartenstrasse 17
69121 Heidelberg, Germany
martin.peters@springer.com

Lecture Notes in Computational Science and Engineering

1. D. Funaro, *Spectral Elements for Transport-Dominated Equations.*
2. H.P. Langtangen, *Computational Partial Differential Equations.* Numerical Methods and Diffpack Programming.
3. W. Hackbusch, G. Wittum (eds.), *Multigrid Methods V.*
4. P. Deuflhard, J. Hermans, B. Leimkuhler, A.E. Mark, S. Reich, R.D. Skeel (eds.), *Computational Molecular Dynamics: Challenges, Methods, Ideas.*
5. D. Kröner, M. Ohlberger, C. Rohde (eds.), *An Introduction to Recent Developments in Theory and Numerics for Conservation Laws.*
6. S. Turek, *Efficient Solvers for Incompressible Flow Problems.* An Algorithmic and Computational Approach.
7. R. von Schwerin, ***M**ulti **B**ody **S**ystem **SIM**ulation.* Numerical Methods, Algorithms, and Software.
8. H.-J. Bungartz, F. Durst, C. Zenger (eds.), *High Performance Scientific and Engineering Computing.*
9. T.J. Barth, H. Deconinck (eds.), *High-Order Methods for Computational Physics.*
10. H.P. Langtangen, A.M. Bruaset, E. Quak (eds.), *Advances in Software Tools for Scientific Computing.*
11. B. Cockburn, G.E. Karniadakis, C.-W. Shu (eds.), *Discontinuous Galerkin Methods.* Theory, Computation and Applications.
12. U. van Rienen, *Numerical Methods in Computational Electrodynamics.* Linear Systems in Practical Applications.
13. B. Engquist, L. Johnsson, M. Hammill, F. Short (eds.), *Simulation and Visualization on the Grid.*
14. E. Dick, K. Riemslagh, J. Vierendeels (eds.), *Multigrid Methods VI.*
15. A. Frommer, T. Lippert, B. Medeke, K. Schilling (eds.), *Numerical Challenges in Lattice Quantum Chromodynamics.*
16. J. Lang, *Adaptive Multilevel Solution of Nonlinear Parabolic PDE Systems.* Theory, Algorithm, and Applications.
17. B.I. Wohlmuth, *Discretization Methods and Iterative Solvers Based on Domain Decomposition.*
18. U. van Rienen, M. Günther, D. Hecht (eds.), *Scientific Computing in Electrical Engineering.*
19. I. Babuška, P.G. Ciarlet, T. Miyoshi (eds.), *Mathematical Modeling and Numerical Simulation in Continuum Mechanics.*
20. T.J. Barth, T. Chan, R. Haimes (eds.), *Multiscale and Multiresolution Methods.* Theory and Applications.
21. M. Breuer, F. Durst, C. Zenger (eds.), *High Performance Scientific and Engineering Computing.*
22. K. Urban, *Wavelets in Numerical Simulation.* Problem Adapted Construction and Applications.

23. L.F. Pavarino, A. Toselli (eds.), *Recent Developments in Domain Decomposition Methods.*

24. T. Schlick, H.H. Gan (eds.), *Computational Methods for Macromolecules: Challenges and Applications.*

25. T.J. Barth, H. Deconinck (eds.), *Error Estimation and Adaptive Discretization Methods in Computational Fluid Dynamics.*

26. M. Griebel, M.A. Schweitzer (eds.), *Meshfree Methods for Partial Differential Equations.*

27. S. Müller, *Adaptive Multiscale Schemes for Conservation Laws.*

28. C. Carstensen, S. Funken, W. Hackbusch, R.H.W. Hoppe, P. Monk (eds.), *Computational Electromagnetics.*

29. M.A. Schweitzer, *A Parallel Multilevel Partition of Unity Method for Elliptic Partial Differential Equations.*

30. T. Biegler, O. Ghattas, M. Heinkenschloss, B. van Bloemen Waanders (eds.), *Large-Scale PDE-Constrained Optimization.*

31. M. Ainsworth, P. Davies, D. Duncan, P. Martin, B. Rynne (eds.), *Topics in Computational Wave Propagation.* Direct and Inverse Problems.

32. H. Emmerich, B. Nestler, M. Schreckenberg (eds.), *Interface and Transport Dynamics.* Computational Modelling.

33. H.P. Langtangen, A. Tveito (eds.), *Advanced Topics in Computational Partial Differential Equations.* Numerical Methods and Diffpack Programming.

34. V. John, *Large Eddy Simulation of Turbulent Incompressible Flows.* Analytical and Numerical Results for a Class of LES Models.

35. E. Bänsch (ed.), *Challenges in Scientific Computing - CISC 2002.*

36. B.N. Khoromskij, G. Wittum, *Numerical Solution of Elliptic Differential Equations by Reduction to the Interface.*

37. A. Iske, *Multiresolution Methods in Scattered Data Modelling.*

38. S.-I. Niculescu, K. Gu (eds.), *Advances in Time-Delay Systems.*

39. S. Attinger, P. Koumoutsakos (eds.), *Multiscale Modelling and Simulation.*

40. R. Kornhuber, R. Hoppe, J. Périaux, O. Pironneau, O. Wildlund, J. Xu (eds.), *Domain Decomposition Methods in Science and Engineering.*

41. T. Plewa, T. Linde, V.G. Weirs (eds.), *Adaptive Mesh Refinement – Theory and Applications.*

42. A. Schmidt, K.G. Siebert, *Design of Adaptive Finite Element Software.* The Finite Element Toolbox ALBERTA.

43. M. Griebel, M.A. Schweitzer (eds.), *Meshfree Methods for Partial Differential Equations II.*

44. B. Engquist, P. Lötstedt, O. Runborg (eds.), *Multiscale Methods in Science and Engineering.*

45. P. Benner, V. Mehrmann, D.C. Sorensen (eds.), *Dimension Reduction of Large-Scale Systems.*

46. D. Kressner, *Numerical Methods for General and Structured Eigenvalue Problems.*

47. A. Boriçi, A. Frommer, B. Joó, A. Kennedy, B. Pendleton (eds.), *QCD and Numerical Analysis III.*

48. F. Graziani (ed.), *Computational Methods in Transport.*

49. B. Leimkuhler, C. Chipot, R. Elber, A. Laaksonen, A. Mark, T. Schlick, C. Schütte, R. Skeel (eds.), *New Algorithms for Macromolecular Simulation.*

50. M. Bücker, G. Corliss, P. Hovland, U. Naumann, B. Norris (eds.), *Automatic Differentiation: Applications, Theory, and Implementations.*

51. A.M. Bruaset, A. Tveito (eds.), *Numerical Solution of Partial Differential Equations on Parallel Computers.*

52. K.H. Hoffmann, A. Meyer (eds.), *Parallel Algorithms and Cluster Computing.*

53. H.-J. Bungartz, M. Schäfer (eds.), *Fluid-Structure Interaction.*

54. J. Behrens, *Adaptive Atmospheric Modeling.*

55. O. Widlund, D. Keyes (eds.), *Domain Decomposition Methods in Science and Engineering XVI.*

56. S. Kassinos, C. Langer, G. Iaccarino, P. Moin (eds.), *Complex Effects in Large Eddy Simulations.*

57. M. Griebel, M.A Schweitzer (eds.), *Meshfree Methods for Partial Differential Equations III.*

58. A.N. Gorban, B. Kégl, D.C. Wunsch, A. Zinovyev (eds.), *Principal Manifolds for Data Visualization and Dimension Reduction.*

59. H. Ammari (ed.), *Modeling and Computations in Electromagnetics: A Volume Dedicated to Jean-Claude Nédélec.*

60. U. Langer, M. Discacciati, D. Keyes, O. Widlund, W. Zulehner (eds.), *Domain Decomposition Methods in Science and Engineering XVII.*

61. T. Mathew, *Domain Decomposition Methods for the Numerical Solution of Partial Differential Equations.*

62. F. Graziani (ed.), *Computational Methods in Transport: Verification and Validation.*

63. M. Bebendorf, *Hierarchical Matrices.* A Means to Efficiently Solve Elliptic Boundary Value Problems.

64. C.H. Bischof, H.M. Bücker, P. Hovland, U. Naumann, J. Utke (eds.), *Advances in Automatic Differentiation.*

65. M. Griebel, M.A. Schweitzer (eds.), *Meshfree Methods for Partial Differential Equations IV.*

66. B. Engquist, P. Lötstedt, O. Runborg (eds.), *Multiscale Modeling and Simulation in Science.*

67. I.H. Tuncer, Ü. Gülcat, D.R. Emerson, K. Matsuno (eds.), *Parallel Computational Fluid Dynamics 2007.*

68. S. Yip, T. Diaz de la Rubia (eds.), *Scientific Modeling and Simulations.*

69. A. Hegarty, N. Kopteva, E. O'Riordan, M. Stynes (eds.), *BAIL* 2008 – *Boundary and Interior Layers.*

70. M. Bercovier, M.J. Gander, R. Kornhuber, O. Widlund (eds.), *Domain Decomposition Methods in Science and Engineering XVIII.*

71. B. Koren, C. Vuik (eds.), *Advanced Computational Methods in Science and Engineering.*

72. M. Peters (ed.), *Computational Fluid Dynamics for Sport Simulation.*

73. H.-J. Bungartz, M. Mehl, M. Schäfer (eds.), *Fluid Structure Interaction II - Modelling, Simulation, Optimization.*

74. D. Tromeur-Dervout, G. Brenner, D.R. Emerson, J. Erhel (eds.), *Parallel Computational Fluid Dynamics 2008.*

75. A.N. Gorban, D. Roose (eds.), *Coping with Complexity: Model Reduction and Data Analysis.*

76. J.S. Hesthaven, E.M. Rønquist (eds.), *Spectral and High Order Methods for Partial Differential Equations.*

77. M. Holtz, *Sparse Grid Quadrature in High Dimensions with Applications in Finance and Insurance.*

78. Y. Huang, R. Kornhuber, O.Widlund, J. Xu (eds.), *Domain Decomposition Methods in Science and Engineering XIX.*

79. M. Griebel, M.A. Schweitzer (eds.), *Meshfree Methods for Partial Differential Equations V.*

80. P.H. Lauritzen, C. Jablonowski, M.A. Taylor, R.D. Nair (eds.), *Numerical Techniques for Global Atmospheric Models.*

81. C. Clavero, J.L. Gracia, F.J. Lisbona (eds.), *BAIL 2010 – Boundary and Interior Layers, Computational and Asymptotic Methods.*

82. B. Engquist, O. Runborg, Y.R. Tsai (eds.), *Numerical Analysis and Multiscale Computations.*

83. I.G. Graham, T.Y. Hou, O. Lakkis, R. Scheichl (eds.), *Numerical Analysis of Multiscale Problems.*

84. A. Logg, K.-A. Mardal, G. Wells (eds.), *Automated Solution of Differential Equations by the Finite Element Method.*

85. J. Blowey, M. Jensen (eds.), *Frontiers in Numerical Analysis - Durham 2010.*

86. O. Kolditz, U.-J. Gorke, H. Shao, W. Wang (eds.), *Thermo-Hydro-Mechanical-Chemical Processes in Fractured Porous Media - Benchmarks and Examples.*

87. S. Forth, P. Hovland, E. Phipps, J. Utke, A. Walther (eds.), *Recent Advances in Algorithmic Differentiation.*

88. J. Garcke, M. Griebel (eds.), *Sparse Grids and Applications.*

89. M. Griebel, M.A. Schweitzer (eds.), *Meshfree Methods for Partial Differential Equations VI.*

90. C. Pechstein, *Finite and Boundary Element Tearing and Interconnecting Solvers for Multiscale Problems.*

For further information on these books please have a look at our mathematics catalogue at the following URL: `www.springer.com/series/3527`

Monographs in Computational Science and Engineering

1. J. Sundnes, G.T. Lines, X. Cai, B.F. Nielsen, K.-A. Mardal, A. Tveito, *Computing the Electrical Activity in the Heart.*

For further information on this book, please have a look at our mathematics catalogue at the following URL: www.springer.com/series/7417

Texts in Computational Science and Engineering

1. H. P. Langtangen, *Computational Partial Differential Equations.* Numerical Methods and Diffpack Programming. 2nd Edition
2. A. Quarteroni, F. Saleri, P. Gervasio, *Scientific Computing with MATLAB and Octave.* 3rd Edition
3. H. P. Langtangen, *Python Scripting for Computational Science.* 3rd Edition
4. H. Gardner, G. Manduchi, *Design Patterns for e-Science.*
5. M. Griebel, S. Knapek, G. Zumbusch, *Numerical Simulation in Molecular Dynamics.*
6. H. P. Langtangen, *A Primer on Scientific Programming with Python.* 3rd Edition
7. A. Tveito, H. P. Langtangen, B. F. Nielsen, X. Cai, *Elements of Scientific Computing.*
8. B. Gustafsson, *Fundamentals of Scientific Computing.*
9. M. Bader, *Space-Filling Curves.*
10. M. Larson, F. Bengzon, *The Finite Element Method: Theory, Implementation and Applications.*

For further information on these books please have a look at our mathematics catalogue at the following URL: www.springer.com/series/5151

MIX
Papier aus verantwortungsvollen Quellen
Paper from responsible sources
FSC® C105338

FSC
www.fsc.org

If you have any concerns about our products,
you can contact us on
ProductSafety@springernature.com

In case Publisher is established outside the EU,
the EU authorized representative is:
Springer Nature Customer Service Center GmbH
Europaplatz 3, 69115 Heidelberg, Germany

Printed by Libri Plureos GmbH
in Hamburg, Germany